Lecture Notes in Mathematics 1637

Editors:
A. Dold, Heidelberg
F. Takens, Groningen

T0255595

Springer
Berlin
Heidelberg
New York
Barcelona
Budapest
Hong Kong
London
Milan
Paris
Santa Clara
Singapore
Tokyo

Bruce Hunt

The Geometry of some special Arithmetic Quotients

Springer

Author

Bruce Hunt
MPI für Mathematik in den Naturwissenschaften
Inselstr. 22–26
D-04103 Leipzig, Germany

Library of Congress Cataloging-in-Publication Data

Hunt, Bruce, 1958-
   The geometry of some special arithmetic quotients / Bruce Hunt.
      p.   cm. -- (Lecture notes in mathematics ; 1637)
   Includes bibliographical references and index.
   ISBN 3-540-61795-7 (softcover : alk. paper)
   1. Threefolds (Algebraic geometry) 2. Moduli theory.
3. Surfaces, Algebraic.   I. Title.   II. Series: Lecture notes in
mathematics (Springer-Verlag) ; 1637.
QA3.L28 no. 1637
[QA573]
510 s--dc20
[516.3'52]
                                                        96-41832
                                                           CIP

Mathematics Subject Classification (1991):
14J30, 14K10, 14K25, 11F55, 22E40, 32M15, 12E12

ISSN 0075-8434
ISBN 3-540-61795-7 Springer-Verlag Berlin Heidelberg New York

© Springer-Verlag Berlin Heidelberg 1996
Printed in Germany

Typesetting: Camera-ready TEX output by the author
SPIN: 10479879   46/3144-543210 - Printed on acid-free paper

A space section of the invariant quintic $\mathcal{I}_5$

More pictures in living color are available at the WWW site:

http://www.mathematik.uni-kl.de/~wwwagag/Galerie.html

Contents

Appendices   255

List of Tables

List of Figures

Chapter 0

Introduction

... daß man eine ... Gruppe ... an die Spitze stellen und den Stoff nach den Untergruppen ordnen soll, die in der Gesamtgruppe erhalten sind, ... die Gruppentheorie als ordnendes Prinzip im Wirrsal der Erscheinungen zu benutzen.

— Felix Klein

Geometry is one of the oldest and most basic branches of mathematics, as is algebra. Nowhere is the interplay between the two more pronounced than in group theory, and that interplay, with group theory acting as a mediator between geometry and algebra, is the theme of this book. Group theory had its genesis in a decidedly algebraic context, solving algebraic equations (Galois theory). It was Felix Klein in his "Erlanger Programm"[1] who put group theory at the basis of geometry. At that time (1872) he had been pursuing studies with Sophius Lie on one-parameter families of algebraic curves[2], whose invariants, as they had noticed, were correlated. This was the advent of continuous groups, here essentially one-parameter subgroups of the automorphism group of the projective plane, $PGL(3, \mathbb{C})$. Klein advocated considering geometry as "invariance properties under a group of automorphisms", and using groups as a basis of classifications of objects, which otherwise do not seem easily related with one another (like formulas for elliptic functions). We shall take this standpoint and consider the geometry of a very special kind of object, namely that of arithmetic quotients of bounded symmetric domains. We shall require groups at several different levels.

Level 1: *Real Lie groups and symmetric spaces.* A Riemannian manifold X is said to be symmetric, if at any point $x \in X$ there is a symmetry σ_x. A symmetry is an automorphism of X (i.e., a diffeomorphism preserving the Riemannian

---

[1] "Vergleichende Betrachtungen über neuere geometrische Forschungen" was the title of the talk

[2] Über diejenigen ebenen Kurven, welche durch ein geschloßenes System von einfach unendlich vielen vertauschbaren Transformationen in sich übergehen

structure) which is involutive and has x as an isolated fixed point. The auto-morphism group of a symmetric space X, $\mathrm{Aut}(X)$, is a real Lie group, and the symmetry at each point of X defines an involution of $\mathrm{Aut}(X)$ by $g \mapsto \sigma_x^{-1} \circ g \circ \sigma_x$. X decomposes into a product $X = X_1 \times \cdots \times X_s$, where each X_i is irreducible; on each irreducible component the curvature is negative, zero or positive. This occurs when $\mathrm{Aut}(X)$ is non-compact simple, abelian or compact simple, respec-tively. The abelian case yields the Euclidean geometry, and the other cases yield a correspondence:

$$\{\text{real simple Lie groups}\} \longleftrightarrow$$
$$\{\text{irreducible symmetric spaces of negative or positive curvature}\}\,.$$

Classifying the symmetric spaces amounts to a classification of involutary auto-morphisms of compact, simple Lie groups, and was first accomplished by É. Car-tan in 1926.

Level 2: *Discrete subgroups of Lie groups and locally symmetric spaces.* Let $G = \mathrm{Aut}(X)$ be as in Level 1, and let $\Gamma \subset G$ be a discrete subgroup, which acts properly discontinuously on X. This assures that the quotient $\Gamma \backslash X$ is a Hausdorf space, and we assume, henceforth, it is of finite volume; Γ is then called a lattice in G. The notion of symmetric spaces of Level 1 can be expressed locally by the condition: the curvature tensor is parallel with respect to the Levi-Cevita connection, and a Riemannian manifold Y is called locally symmetric if this condition is satisfied. The universal cover $\tilde{Y}$ of Y is a symmetric space as in Level 1, and of course $Y = \pi_1(Y) \backslash \tilde{Y}$, with the fundamental group $\pi_1(Y)$ acting properly discontinuously. If $\tilde{Y}$ is compact, then $\pi_1(Y)$ is finite. Interesting things occur if $\tilde{Y}$ is non-compact, so we get a correspondence, specializing that of Level 1,

$$\left\{\begin{array}{l}\text{pairs } (G,\Gamma), \ \Gamma \text{ a lattice in } G, \ G \text{ a non-compact}\\ \text{semisimple real Lie group}\end{array}\right\} \longleftrightarrow$$
$$\left\{\begin{array}{l}\text{locally symmetric spaces of non-positive cur-}\\ \text{vature and finite volume}\end{array}\right\}.$$

Remark: We must formulate this in terms of semisimple groups, since the quotient $\Gamma \backslash X$ may be irreducible, even if the domain X is reducible.

Level 3: *Normal subgroups of finite index in discrete groups and locally sym-metric spaces with automorphism groups.* Discrete subgroups $\Gamma \subset G$ as in Level 2 tend to have lots of normal subgroups of finite index, $\Gamma' \subset \Gamma$. Consider the locally symmetric spaces $\Gamma \backslash X$ and $\Gamma' \backslash X$. Since Γ' is normal in Γ, the fi-nite group Γ/Γ' acts on $\Gamma' \backslash X$ with quotient $\Gamma \backslash X$, i.e., the natural morphism $\pi_{\Gamma'|\Gamma} : \Gamma' \backslash X \longrightarrow \Gamma \backslash X$ is a Galois cover. Assuming Γ is torsion free it is étale, while if Γ has torsion this cover will be branched. In this way we get locally symmetric spaces with automorphism groups (here Γ/Γ') and a correspondence

$$\left\{\begin{array}{l}\text{triples } (G,\Gamma,\Gamma'), \text{ with } (G,\Gamma) \text{ as in}\\ \text{Level 2}, \ \Gamma' \vartriangleleft \Gamma, \ [\Gamma : \Gamma'] < \infty\end{array}\right\} \longleftrightarrow$$
$$\left\{\begin{array}{l}\text{locally symmetric spaces of non-positive curvature and of}\\ \text{finite volume with automorphism group}\end{array}\right\}.$$

The more interesting the group Γ/Γ' is, the more interesting the automorphism group of $\Gamma'\backslash X$ is.

Our concern in this book is basically with Level 3; we are interested in the geometry of particular locally symmetric spaces with interesting automorphism group. We will be placing two conditions on these data, namely we assume the discrete group Γ is *arithmetic* and the symmetric space X is *hermitian*; these two conditions are logically independent. Consider the first. Assuming the discrete group Γ is arithmetic necessitates introducing a new object into the picture, an algebraic group, which we may assume is defined over $\mathbb{Q}$. This algebraic group, call it $G_{\mathbb{Q}}$, is, compared with the real Lie group, a rather mysterious object: it is an algebraic scheme defined over $\mathbb{Q}$. In fact, the real Lie group is just "a tiny part" of $G_{\mathbb{Q}}$, namely the group of $\mathbb{R}$-valued points of the algebraic group:

$$G_{\mathbb{Q}}(\mathbb{R}) \cong G.$$

This very statement shows that, as far as notation is concerned, things can get very confusing in this business, and we must be very careful in choosing notation and making statements.

A discrete group $\Gamma \subset G$ is arithmetic, if there is a $\mathbb{Q}$-group $G_{\mathbb{Q}}$ with $\Gamma \subset G_{\mathbb{Q}}(\mathbb{Q}) \subset G_{\mathbb{Q}}(\mathbb{R}) = G$, and a rational representation $\varrho : G_{\mathbb{Q}} \longrightarrow GL(V_{\mathbb{Q}})$ such that $\varrho^{-1}(GL(V_{\mathbb{Z}}))$ and Γ are commensurable. This presupposes a choice of lattice $\mathcal{L} = V_{\mathbb{Z}}$ such that $V_{\mathbb{Q}} = \mathcal{L} \otimes \mathbb{Q}$, and this is, or course, not entirely canonical. That is why the notion "preserves a lattice" is only well-defined on a commensurability class of groups. Now, note that since $G_{\mathbb{Q}}$ defines G, it also defines X, and this leads us to refine our levels of groups as follows:

Level 0: start with a semisimple algebraic group $G_{\mathbb{Q}}$ defined over $\mathbb{Q}$;

Level 1: the group of $\mathbb{R}$-points of $G_{\mathbb{Q}}$ is a real semisimple Lie group, $G_{\mathbb{Q}}(\mathbb{R}) = G$, and defines a symmetric space X;

Level 2: choose a lattice $\mathcal{L} \subset V_{\mathbb{Q}}$ and a representation $\varrho : G_{\mathbb{Q}} \longrightarrow GL(V_{\mathbb{Q}})$; this defines an arithmetic group

$$G_{\mathcal{L}} := \{g \in GL(V_{\mathbb{Q}}) \,\big|\, g(\mathcal{L}) \subset \mathcal{L}\};$$

Level 3: as above, normal subgroups $\Gamma' \subset G_{\mathcal{L}}$ determine arithmetic quotients with automorphism group.

There is a subtle point about algebraic groups which we mention here. Even if the group $G_{\mathbb{Q}}$ is simple over $\mathbb{Q}$ (that is, has no normal $\mathbb{Q}$-subgroups), there will, in general, be a finite field extension $k|\mathbb{Q}$ for which the lifted group G_k is no longer simple, but rather only semisimple, a product of simple groups. Consequently, the real group $G_{\mathbb{R}}$, and hence the symmetric space X, is also a product. In particular, it can and does happen that some factors of $G_{\mathbb{R}}$ may be *compact*, so this goes beyond the description we originally started with. It was proven quite early by Borel and Harish-Chandra that the existence of a compact factor of $G_{\mathbb{R}}$ implies that $G_{\mathbb{Q}}$ is anisotropic, and this in turn implies that any quotient $\Gamma\backslash X$ for an arithmetic group $\Gamma \subset G_{\mathbb{Q}}(\mathbb{Q})$ is *compact*. A deep theorem of Margulis states that if the $\mathbb{R}$-rank of $G_{\mathbb{R}}$ is ≥ 2, then *any* discrete subgroup

$\Gamma \subset G_\mathbb{R}$ is *automatically* arithmetic. So the condition "Γ is arithmetic" is only a condition for the rank 1 real Lie groups.

The second assumption we will be making, which is much more serious, is that X is *hermitian* symmetric. By definition, this means X has a complex structure compatible with the symmetric Riemannian structure, in other words that X is Kähler, with the Riemannian part of the hermitian metric being symmetric. In particular, X is Kähler homogenous, and this implies that the compact dual $\check{X}$ of X is an *algebraic* variety, and in fact it is a rational variety. There is a natural, group theoretic embedding $X \subset \check{X}$, displaying X as an (homogenous) open (in the Euclidean topology) subset of an (homogenous) algebraic variety. But it turns out that this assumption implies much more. First, taking the topological closure (in the Euclidean topology) of X in $\check{X}$ defines the *boundary* of X; this decomposes into irreducible pieces (holomorphic arc components), defining the *boundary components* of X. Then the following facts hold:

  i) each maximal parabolic of $G_\mathbb{Q}$ is the normalizer of a boundary component;

 ii) for $\Gamma \subset G_\mathbb{Q}$ arithmetic, the quotient $\Gamma \backslash X$ can be compactified to a normal, analytic space $(\Gamma \backslash X)^*$, which is projective algebraic $(\Gamma \backslash X)^* \subset \mathbb{P}^N$;

iii) the embedding $(\Gamma \backslash X)^* \subset \mathbb{P}^N$ is given by modular forms (with the usual exception of dimension 1 factors);

 iv) there is a smooth compactification $\overline{\Gamma \backslash X}$ which resolves the singularities of $\Gamma \backslash X^*$, and for which $\Delta = \overline{\Gamma \backslash X} - \Gamma \backslash X$ is a normal crossings divisor.

Items ii) and iii) are the Baily-Borel embedding, iv) the toroidal compactification. These results display the fact that in this case the locally symmetric space $\Gamma \backslash X$ is an *algebraic* object, or more precisely, an object of algebraic geometry. And so we arrive at one of the main themes of this book: the geometry of arithmetic quotients of bounded symmetric domains is geometry in the sense of *algebraic geometry*. Let us pause for a moment to explain this statement. Generally speaking, a locally symmetric space is an object in the category of Riemannian manifolds, so geometry of them is clearly geometry in the sense of Riemannian geometry. Everything is expressed in terms of curvature, and the geometry *is* the geometry of that curvature tensor. On the other hand, in algebraic geometry, there is no curvature tensor to consider. Rather, one considers embeddings in projective space (like $\Gamma \backslash X^* \subset \mathbb{P}^N$) and their properties: singular locus, hyperplane sections (subvarieties), invariants (of the isomorphism class under $PGL(N+1, \mathbb{C})$), inflection points and the like. If a variety V has an interesting automorphism group, this usually induces a projective representation of the group, and there will be some "invariant configuration" in the ambient $\mathbb{P}^N$, of which V is only one aspect. This is what we understand by geometry when we speak of algebraic geometry of arithmetic quotients of bounded symmetric domains.

We first explain the geometry of the boundary. The Satake compactification X_Γ^* (we will from now on use the notation of the text: $\mathcal{D}$ denotes the non-compact hermitian symmetric space, $\check{\mathcal{D}}$ its compact dual, $X_\Gamma = \Gamma \backslash \mathcal{D}$ an arithmetic quotient, X_Γ^* (respectively $\overline{X}_\Gamma$) the Satake compactification (respectively a toroidal embedding)) is a disjoint union

$$X_\Gamma^* = X_\Gamma \cup V_1 \cup \ldots \cup V_s,$$

with each V_i an arithmetic quotient of lower rank than that of X_Γ, say $V_i = \Gamma_i \backslash \mathcal{D}_i$, $\Gamma_i \subset \Gamma$. On the other hand, for the Baily-Borel embedding $X_\Gamma^* \subset \mathbb{P}^N$, the *singular locus* (under mild assumptions on Γ) coincides with $X_\Gamma^* - X_\Gamma = V_1 \cup \cdots \cup V_s$, the union of lower-dimensional varieties. If one wishes, these lower-dimensional subvarieties can be turned into divisors, by means of $\pi_\Gamma : \overline{X}_\Gamma \longrightarrow X_\Gamma^*$, under which $\pi_\Gamma^{-1}(V_i) =: B_i$ is a divisor and a fibre space over V_i. The neighborhood of B_i in $\overline{X}_\Gamma$ is described by the normal bundle $N_{\overline{X}_\Gamma} \overline{B}_i$ of the closure $\overline{B}_i$ of B_i in $\overline{X}_\Gamma$. Now the compactification $\overline{X}_\Gamma$ is not unique, but it will be birationally unique, and if V_i has some group of automorphisms H_i then B_i is unique up to an equivariant birational equivalence.

There is a simplicial complex associated with this situation, the Tits building of Γ, $\mathcal{T}(\Gamma)$, whose vertices correspond to the components V_i and whose j-simplices correspond to j-flags of components in the closures (Satake compactifications) of the others, $V_{i_1} \subset V_{i_2}^* \subset \cdots \subset V_{i_j}^*$. This complex, together with the descriptions of the individual components $V_i = \Gamma_i \backslash \mathcal{D}_i$, completely determines the boundary. But by its very definition (which is in terms of parabolic subgroups), $\mathcal{T}(\Gamma)$ thus relates the (geometric) boundary to a group theoretic problem, that of parabolic subgroups. Indeed, each Γ_i is the intersection of Γ and a factor L_i of the Levi component of a unique parabolic $P_i \subset G_\mathbb{Q}$, $\Gamma_i = L_i \cap \Gamma$. But if we consider the smooth model $\overline{X}_\Gamma$, then in fact, we can get the entire parabolic lattice $\Gamma_{P_i} = P_i \cap \Gamma$ by considering an ε-neighborhood of B_i in $\overline{X}_\Gamma$ (with respect to any smooth Riemannian metric on the Riemannian manifold $\overline{X}_\Gamma$). In a nutshell, we have

$$\{\text{singular locus of } X_\Gamma^*\} \longleftrightarrow \{\text{parabolic subgroups of } \Gamma\}.$$

For an algebraic group (semisimple, say) G, a *symmetric* subgroup is an algebraic subgroup defined by a closed symmetric set of roots. If G is of hermitian type, i.e., if the symmetric space associated with $G(\mathbb{R})$ is hermitian symmetric, then a symmetric subgroup M is *hermitian symmetric*, if $\mathcal{D}_M \subset \mathcal{D}$ is a hermitian symmetric subspace. Finally, if M is hermitian symmetric and defined over $\mathbb{Q}$, we call it $\mathbb{Q}$-*hermitian symmetric*. This is the notion one requires on a subgroup M to be able to conclude that for a lattice $\Gamma \subset G$, the arithmetic subgroup $\Gamma_M = M \cap \Gamma$ determines an algebraic subvariety X_{Γ_M} of X_Γ.

Having the notion of $\mathbb{Q}$-hermitian symmetric subgroup $M \subset G$, it is canonical to define the modular subgroups. It is a general property that Γ_M is an arithmetic subgroup of M, and given an explicit description of M and Γ, one gets an explicit description of Γ_M. A description of M is well known. A description of possible arithmetic groups is sketched in Appendix A. Generally speaking this is given by a pair $(V, \mathcal{L})$, where V is a D-vector space and $\mathcal{L}$ is a Δ-lattice, where $\Delta \subset D$ is a fixed maximal order in D, where D is a division algebra over an algebraic number field. Hence the description depends only on the classification of maximal orders in divison algebras and is generally not too much more complicated than the classification of the Q-groups themselves (see Theorems A.5.2 and A.5.3).

We then turn to our original object of study, Level 3 above, arithmetic quotients with "nice" automorphism group. More precisely, we wish to apply the general theory of arithmetic quotients to give a conceptual understanding of an incredible set of examples, which are the primary object of interest. To do this the most natural way of viewing things is in terms of moduli spaces. Deligne has shown that the bounded symmetric domains are parameter spaces of a certain representation theoretic problem, which, in particular cases, is a known geometric moduli problem. Indeed, in his intensive studies of moduli spaces associated with the moduli problem of isomorphism classes of abelian varieties with polarization, given endomorphism ring and level structure, Shimura (much earlier, in the 1960's) gave a complete list of domains and groups which occur in this manner. In this list, all domains of types $I_{p,q}$, II_n, III_n occur. The domains of type IV_n are period domains of pure Hodge structures of weight 2 and type $(1, n, 1)$. So in fact, all except the exceptional domains occur in this way. We give a real quick review of Shimura's theory in Chapter 1, and in Chapter 2 we study the split over $\mathbb{R}$ case, in which a maximal Q-split torus is also a maximal $\mathbb{R}$-split one. This is the easiest case, in which the well-known geometry of the domain $\mathcal{D}$ is reflected in the geometry of the quotient. Most of the material of Chapters 1 and 2 is more or less well-known, but we give them a unified treatment, and, for example, Shimura's theory is very easy to formulate. The result in Chapter 2 on Janus-like varieties is recent; proofs have appeared in [J].

In Chapter 3 we come to "real geometry", and study particular algebraic varieties, say $X \subset \mathbb{P}^N$, which turn out to be Baily-Borel embeddings of arithmetic quotients, say $X = X_\Gamma^*$. There are, generally speaking, two general approaches to this kind of problem. The first (and standard) method utilizes *automorphic forms* (usually theta functions) to display explicit embeddings $X_\Gamma^* \subset \mathbb{P}^N$. The second (and non-standard) approach is the question of *uniformization*. That is, we take as given a singular algebraic variety $X \subset \mathbb{P}^N$ with smooth locus $X_{sm} \subset X$ and inquire as to the universal cover $\widetilde{X}_{sm}$ of X_{sm}. Note that this inverts Baily-Borel: $X_\Gamma \subset X_\Gamma^*$ is the smooth locus, and $\widetilde{X}_\Gamma = \mathcal{D}$. Although we review the work of Igusa and Coble as well as recent results of v. Geemen giving explicit embeddings by means of theta functions, we adhere to the second approach and try to get uniformization results "without automorphic forms". Now, it turns out this works with present technology only for ball quotients, which consequently give our most important examples. We now give a brief description of the examples studied in Chapter 3. First of all, we note that all the examples are related with one another and in all examples the automorphism group of the arithmetic quotient is Σ_6, the symmetric group on 6 letters. In all cases we consider the following questions:

   i) the geometry of $X \subset \mathbb{P}^N$;

   ii) description of X as an arithmetic quotient, $X = X_\Gamma$;

iii) the ensuing moduli interpretation of X.

First, at the beginning of Chapter 3 we introduce some very useful geometric objects: arrangements of hyperplanes in projective space, especially those defined by Weyl groups of simple Lie groups. We describe the notion of Fermat cover, an abelian branched cover of $\mathbb{P}^N$ with branch locus the given arrangement. The particular arrangement $\mathcal{A}(W(\mathbf{A_n}))$, the arrangement in $\mathbb{P}^{n-1}$ of $\binom{n}{2}$ hyperplanes defined by the Weyl group of $\mathbf{A_n}$, is related to the hypergeometric differential equation on $\mathbb{P}^{n-1}$. These notions are used again and again throughout the rest of the book.

We start with the tetrahedron in $\mathbb{P}^3$, in other words with the arrangement $\mathcal{A}(W(\mathbf{A_4}))$ of ten planes: the faces and symmetry planes of the tetrahedron in three-space. Blowing up the four corners and center of the tetrahedron and blowing down ten copies of $\mathbb{P}^1 \times \mathbb{P}^1$ to ordinary double points, the result is the Segre cubic $\mathcal{S}_3 \subset \mathbb{P}^4$, a cubic hypersurface. The *dual* variety is the Igusa quartic $\mathcal{I}_4 \subset \mathbb{P}^4$. The *Hessian* variety of $\mathcal{S}_3$ is the Nieto quintic $\mathcal{N}_5$, a beautiful variety considered in the paper [BN]. The Hessian variety of the Igusa quartic is a hypersurface of degree 10 which we denote by $\mathcal{W}_{10}$. Finally, we consider the Coble variety $\mathcal{Y}^3$, the double cover of $\mathbb{P}^4$ branched along the Igusa quartic $\mathcal{I}_4$, and the double cover $\mathcal{W}$ of $\mathbb{P}^4$ branched along $\mathcal{W}_{10}$. All of these examples fit nicely into our framework. We list them, together with ii) and iii) above.

1 Segre cubic $\mathcal{S}_3$: $\mathcal{S}_3$ is the Satake compactification of a ball quotient, $\mathcal{S}_3 \cong (\Gamma_{\sqrt{-3}}(\sqrt{-3}) \backslash \mathbb{B}_3)^*$, where $\Gamma_{\sqrt{-3}}(\sqrt{-3}) \subset U(3,1; \mathcal{O}_{\mathbb{Q}(\sqrt{-3})})$ is the principal congruence subgroup of level $\sqrt{-3}$. This is the moduli space of principally polarized abelian fourfolds with complex multiplication by $\mathbb{Q}(\sqrt{-3})$ of signature $(3,1)$ and a level $\sqrt{-3}$ structure. From the isomorphism $\mathbb{Q}(\sqrt{-3}) \cong \mathbb{Q}(\varrho)$, where ϱ is a cube root of unity, these abelian fourfolds can be identified as the Jacobians of genus 4 Picard curves, trigonal curves $C_{x_\lambda} = \{\prod_{\lambda=1}^6 (x - x_\lambda) = y^3\}$, and here the level $\sqrt{-3}$ structure is given by an ordering of the six branch points x_λ.

2 Igusa quartic $\mathcal{I}_4$: $\mathcal{I}_4$ is the Satake compactification of a Siegel modular threefold, $\mathcal{I}_4 \cong (\Gamma(2) \backslash \mathbb{S}_2)^*$, where $\mathbb{S}_2$ is the Siegel space of degree 2 and $\Gamma(2) \subset Sp(4, \mathbb{Z})$ is the principal congruence subgroup of level 2. This is the moduli space of principally polarized abelian surfaces with a level 2 structure. Since, on a Zariski open subset, principally polarized abelian surfaces are Jacobians of genus 2 curves, and since any genus 2 curve is hyperelliptic, $D_{x_\lambda} = \{\prod_{\lambda=1}^6 (x - x_\lambda) = y^2\}$, this is just the space of genus 2 hyperelliptic curves with an ordering of the six branch points. The isomorphism on a Zariski open subset between $\mathcal{S}_3$ and $\mathcal{I}_4$ (which ensues due to projective duality of the two varieties) corresponds to identifying the branch points of C_{x_λ} and D_{x_λ}.

---

[3]denoted $\mathcal{Y}$ for Yoshida variety; in [MSY] this variety was rediscovered, but historically it is Coble's variety.

3 Nieto quintic $\mathcal{N}_5$: $\mathcal{N}_5$ has some Zariski open subset, which is the moduli space of Kummer surfaces of abelian surfaces with a (2,6) polarization. The precise relation between the Satake compactification and $\mathcal{N}_5$ is not quite understood.

4 $\mathcal{W}_{10}$: ?

5 Coble's variety $\mathcal{Y}$: $\mathcal{Y}$ is the Satake compactification of $\Gamma(2)\backslash\mathcal{D}$, where $\mathcal{D}$ is of type $\mathbf{IV_4}$ and $\Gamma(2) \subset SO(4,2)$ is the principal congruence subgroup. This is the moduli space of the set of K3-surfaces which are double covers of $\mathbb{P}^2$ branched over six ordered lines. The branch locus of $\mathcal{Y} \longrightarrow \mathbb{P}^4$, a copy of $\mathcal{I}_4$, corresponds to the six lines being tangent to a conic, so the double covers are (projections of) Kummer surfaces. But dually $\mathcal{Y}$ is also the moduli space of ordered sets of six points in $\mathbb{P}^2$, and hence (birationally at least) the moduli space of cubic surfaces with a marking of the 27 lines.

6 $\mathcal{W}$: ?

I have the feeling that $\mathcal{W}_{10}$ and $\mathcal{W}$ are also arithmetic quotients, although in these cases there are other (birationally equivalent) models, which it seems more natural to study. These are the matters discussed in Chapter 3. Although we give several new proofs of some of these results, most of the material of Chapter 3 may be considered known. All in all, the "nice" automorphism group in this chapter is Σ_6.

In fact, there is only one other "nice" automorphism group which we study: the Weyl group of $\mathbf{E_6}$. The remainder of the book is devoted to some of the beautiful geometry which arises from $W(E_6)$. Let us remark that one of the reasons why these two groups, Σ_6 and $W(E_6)$, are particularly interesting is the descriptions they possess in terms of *different* Chevally groups:

$$\Sigma_6 \cong PU(3,1;\mathbb{F}_3) \cong PSp(4,\mathbb{Z}/2\mathbb{Z})$$

$$G_{25,920} \cong PU(3,1;\mathbb{F}_4) \cong PSp(4,\mathbb{Z}/3\mathbb{Z}),$$

where $G_{25,920} \subset W(E_6)$ is the simple subgroup of index two consisting of all *even* elements. Another important peculiarity of this situation is the fact that the imaginary quadratic field $\mathbb{Q}(\sqrt{-3})$ is cyclotomic. This accounts for the fact that the four-dimensional abelian varieties are Jacobians.

We begin Chapter 4 with a leisurely review of cubic surfaces and the 27 lines. It seems helpful to understand the combinatorics of the 27 lines in order to understand the Weyl group $W(E_6)$; in particular, the notation coming from the 27 lines is invaluable. Much of the interesting geometry arises from the difference between the root lattice and the weight lattice, which occur in this setting as 36 roots $\longleftrightarrow$ 36 double sixes of lines, and 27 fundamental weights $\longleftrightarrow$ 27 lines. Most of this material is known, in fact much of it is very classical. However, it was difficult to give references, so we present enough of the theory to introduce notations.

Another objective in the chapter on cubic surfaces is to discuss in great detail a classical problem, which introduces all kinds of lovely geometry, but which is purely algebraic: solving the equation for the 27 lines. This problem, although rather irrelevant from the point of view of modern mathematics, was of great importance in the last century to the development of the theory of modular forms, and so, bears the embryo of the modern theory of arithmetic quotients. The story of the general equation of fifth degree is well-known, ultimately culminating in F. Klein's book *Vorlesungen über das Ikosaeder*. Here geometry, analysis and algebra were brought to bear, showing not only *how* one can solve an equation of fifth degree, but also the *relationships* between the different methods which had been used previously. Similar methods turned out to apply to the said degree 27 equation, as noticed first by Klein, and this lead to, among other things, the development of the invariant theory of several finite collineation groups (unitary reflection groups in projective space). All of this laid the grounds for a theory of arithmetic groups, so its discussion also entails a historical journey, which certainly many readers will enjoy.

The equation of degree 27 is derived in the following manner. Thinking of each of the 27 lines as defining a point on the Grassmanian $G(2, 4) \subset \mathbb{P}^5$, one can *project* to 27 points onto a line – giving the roots of a degree 27 equation. It is not difficult to find the equation; the tricky thing is to find the roots. Burkhardt in 1890 demonstrated how, following Klein's suggestions, the equation could be solved, involving the calculation of the periods of an abelian surface. Later on Coble reconsidered the problem, essentially by considering the Galois group $W(E_6)$ not as a reflection group, but as a Cremona group, i.e., as a (finite) group of birational automorphisms of the plane $\mathbb{P}^2$.

In the last two chapters we consider two very beautiful varieties, the Burkhardt quartic threefold $\mathcal{B}_4 \subset \mathbb{P}^4$, whose automorphism group is $G_{25,920}$, and the invariant quintic fourfold $\mathcal{I}_5 \subset \mathbb{P}^5$, whose automorphism group is $W(E_6)$. Now both of these varieties are rational, and so from the point of view of classification of algebraic varieties, not particularly interesting. But it turns out they are rich in structure of very interesting moduli interpretations, so from our point of view they are *very* interesting.

Our main goal in Chapter 5 is to give a new and complete proof of the following theorem, for which several proofs are already known, but only one having been written down to date (given in [Ge]):

Theorem 0.0.1 (5.6.1) *Let $\mathcal{B}_4^0 = \mathcal{B}_4 - \{45 \text{ nodes}\}$ (=smooth locus of $\mathcal{B}_4$). $\mathcal{B}_4^0$ is the moduli space of principally polarized abelian fourfolds with complex multiplication by $\mathbb{Q}(\sqrt{-3})$, of signature (3,1) and a level 2 structure. $\mathcal{B}_4$ is the Baily-Borel embedding of this arithmetic quotient.*

The proof given by v. Geemen uses directly the theta functions for the Picard modular group $\Gamma_{\sqrt{-3}}(2) \subset PU(3, 1; \mathcal{O}_{\mathbb{Q}(\sqrt{-3})})$; in particular, the fact is used that, because the ring $\mathcal{O}$ has non-trivial units ($\mathcal{O}^*_{\mathbb{Q}(\sqrt{-3})} \cong \mathbb{Z}/6\mathbb{Z}$), the abelian four-folds which occur in the mentioned family have automorphisms. Then, applying representation theory of $G_{25,920}$ to this, he gets a description of the quotient

$\Gamma_{\sqrt{-3}}(2)\backslash\mathbb{B}_3$ as a hypersurface in $\mathbb{P}^4$, invariant under $G_{25,920}$, and then checks the degree is ≤ 5, since $\mathcal{B}_4$ is the *only* invariant in these degrees. Our approach is, as we have mentioned above, to prove the result "just looking at $\mathcal{B}_4$". Our procedure requires two steps:

Step I: prove that $\mathcal{B}_4^0$ is a ball quotient;

Step II: identify the group.

For Step I we utilize the argument with the Kähler-Einstein metric and Chern numbers to reduce the question to a computation of Chern numbers. This is the only point in the book where we make (implicit) use of any curvature tensor at all. It would be highly desirable to find an alternate proof of some kind, for this Step I. On the other hand, Step II fits really nicely into our program.

To make this proof self-contained, we give a complete presentation of the required Chern number arguments in Chapter 5. Although known to experts, I could not find a reference for this material. We apply this to $\mathcal{B}_4$, and show that the existence of 40 j-planes on $\mathcal{B}_4$, which are known to be surface ball quotients, together with a Chern number proportionality concerning their normal bundles, imply the desired result that $\mathcal{B}_4^0$ is a ball quotient (for a group with torsion, which is what makes the discussion somewhat difficult). In other words, we have a sublocus on $\mathcal{B}_4$ which "looks like" a union of modular subvarieties on an arithmetic quotient, and it *turns out* that they are.

We should mention that the projective dual of $\mathcal{B}_4$, like the dual of $\mathcal{S}_3$, has a different moduli interpretation, which was (more or less) proved long ago by Coble. He applied this in [C1] to give an alternative solution to the problem of solving the equation of 27th degree discussed above. In the process he showed that $(\Gamma(3)\backslash\mathbb{S}_2)^*$ is naturally embedded in $\mathbb{P}^9$, and projecting into an invariant (under the action of $G_{25,920}$) $\mathbb{P}^4$, the image is the dual of $\mathcal{B}_4$. This proof once again is the approach using the theta functions to get the Baily-Borel embedding.

A very interesting fact, which we discover in our study of $\mathcal{B}_4$, is that the Kummer variety of $E_\varrho \times E_\varrho$ (where ϱ is a 3rd root of unity and $E_\varrho = \mathbb{C}/\mathbb{Z} \oplus \varrho\mathbb{Z}$) can be embedded (birationally) as a 12-nodal quartic surface containing 16 lines lying on $\mathcal{B}_4$. There are 45 of these 12-nodal quartics on $\mathcal{B}_4$, and they, too, are modular subvarieties such that the 12 nodes are the cusps. Hence, this gives a new kind of Janus-like variety, and shows that anything that can happen, will.

Finally, in the last chapter we unveil an object of unparalleled beauty — the invariant quintic. Rather than classical, this object was "discovered" in recent times and has been studied with high technology, mostly with Macaulay, but also with Reduce, for example. Especially in this chapter, but also throughout, when we refer to some "calculation" without further specification, we generally are referring to some such computer calculation. As far as possible we have given other proofs, but for some results there seems no alternative to "brute computation". Of course, Macaulay computes in finite characteristic, but as long as the characteristic is generic, most results (dimensions of linear systems, degrees of generators of ideals, etc.) will be the same as in characteristic 0, so we make use of them also. The story of the "discovery" of $\mathcal{I}_5$ is amusing. Partially motivated by the first proof that $\mathcal{B}_4$ is the Satake compactification of a Picard modular threefold, v. Geemen studied theta functions for higher-dimensional Picard varieties, and proved there is a four-dimensional $G_{25,920}$-invariant hypersurface of degree 10 in $\mathbb{P}^5$, which is the Satake compactification of an arithmetic quotient of a domain of type $\mathbf{I}_{2,2}$. Although not familiar with $W(E_6)$ at the time[4], I had run across the $W(E_6)$-invariant polynomials in $\mathbb{P}^5$, and so we (v. Geemen, v. Straten and I) asked Macaulay about the singular locus of different degree 10 invariants, looking for 45 lines (which the $\mathbf{I}_{2,2}$-quotient should have as compactification locus). To start with we looked at $\mathcal{I}_5$, and *its* singular locus was a curve of degree 120. This got Duco (v. Straten) and me interested in this variety, and together with Macaulay we got many of the results on the $W(E_6)$ geometry in $\mathbb{P}^5$ presented in Chapter 6. Originally a joint work on this topic was planned, but as the story of $\mathcal{I}_5$ got more and more confusing, we never got around to this. In particular, it seemed natural to believe that $\mathcal{I}_5$ *is* the moduli space of semistable marked cubic surfaces (which turns out to be not quite true), since it is a unique invariant of $W(E_6)$ of lowest degree in $\mathbb{P}^5$. Then Duco realized that the family of tangent (to $\mathcal{I}_5$) hyperplane sections have a natural cubic surface associated with them, coming from the infinitesimal variation of Hodge structures in $H^{2,1}$. The precise relation is still unknown today. But in the mean time we do have a much better understanding of the situation.

The main results of Chapter 6 show that $\mathcal{I}_5$ is (closely) related to virtually all the examples in Chapter 3. The Segre cubic is the resolving divisor of 36 triple points of $\mathcal{I}_5$, while there are 36 hyperplane sections (dual to the 36 triple points) isomorphic to $\mathcal{N}_5$. Similarly, $\mathcal{I}_4$ and $\mathcal{W}_{10}$ are related to the dual variety. But by far the most intriguing is the relation to the Coble variety $\mathcal{Y}$, which is the main result of Chapter 6. As we just mentioned, $\mathcal{I}_5$ comes equipped with a natural family of cubic surfaces over it, while $\mathcal{Y}$ *is* the moduli space of (marked) cubic surfaces, so it is natural to expect a close relationship. $\mathcal{I}_5$ has a biregular action of $W(E_6)$, $\mathcal{Y}$ has a *birational* action of $W(E_6)$, which becomes regular on some modification of $\mathcal{Y}$. We may project $\mathcal{I}_5$ from one of the 36 triple points; this breaks the biregular symmetry group from $W(E_6)$ to Σ_6. A birational model of $\mathcal{I}_5$, $\mathcal{I}_{5,p}^{\%}$ (where the triple point p is blown up and certain linear subspaces are blown down) is a 2:1 cover of $\mathbb{P}^4$, branched along the union $\mathcal{S}_3 \cup \mathcal{N}_5$. On the other hand, the rational map given by the Jacobian ideal of $\mathcal{S}_3$, $\varphi : \mathbb{P}^4 \dashrightarrow \mathbb{P}^4$, is of degree 6. If we consider $\varphi^{-1}(\mathcal{I}_4)$, then it is the union of $\mathcal{S}_3$ and another Σ_6-invariant quintic $\mathcal{P}_5$, which has 15 singular lines and some additional singular points. We consider the diagram

$$\begin{array}{ccc} \mathcal{Z} & \dashrightarrow & \mathcal{Y} \\ \downarrow & & \downarrow \\ \mathbb{P}^4 & \overset{\varphi}{\dashrightarrow} & \mathbb{P}^4, \end{array}$$

and observe that the branch locus of $\mathcal{Z} \longrightarrow \mathbb{P}^4$ is just $\mathcal{S}_3 \cup \mathcal{P}_5$. Since $\mathcal{I}_{5,p}^{\%} \longrightarrow \mathbb{P}^4$ is a double cover branched along $\mathcal{S}_3 \cup \mathcal{N}_5$, we obtain the following.

---

[4]Although $G_{25,920}$ is of index 2 in $W(E_6)$, the geometry differs significantly: the 36 reflections are not in $G_{25,920}$.

Theorem 0.0.2 (Theorem 6.4.13) $\mathcal{I}_5$ *is* Σ_6-*equivariantly birational to a* Σ_6-*equivariant deformation of the 6:1 cover* $\mathcal{Z}$ *of* $\mathcal{Y}$. *The deformation takes place in the space of double covers of* $\mathbb{P}^4$ *whose branch locus consists of the union of* S_3 *and a* Σ_6-*symmetric quintic hypersurface.*

It is known that there is a desingularization $\mathcal{C}$ of $\mathcal{Y}$ (the *Naruki cross ratio variety*) on which $W(E_6)$ acts biregularly. Then the 6:1 map $\mathcal{Z} - - \to \mathcal{Y}$ lifts, as in the following diagram

$$
\begin{array}{ccc}
\widetilde{\mathcal{Z}} & - - \to & \mathcal{C} \\
\downarrow & & \downarrow \\
\mathcal{Z} & \longrightarrow & \mathcal{Y},
\end{array}
$$

and on $\widetilde{\mathcal{Z}}$ we have 36 copies of a threefold birational to S_3 and 36 copies of a variety birational to $\mathcal{P}_5$. Since on a desingularization of $\mathcal{I}_5$ we have also 36 copies of the same birational model of S_3 and 36 copies of a variety birational to $\mathcal{N}_5$, it is natural to suspect that $\mathcal{N}_5$ and $\mathcal{P}_5$ are Σ_6-equivariantly birational and moreover that the desingularizations $\widetilde{\mathcal{I}}_5$ and $\widetilde{\mathcal{Z}}$ are birational, or even isomorphic. If this were the case, then we would know that $\mathcal{I}_5$ is (birational to a smooth compactification of) an arithmetic quotient defined by a subgroup Γ of index 6 in the group $\Gamma(2)$ which is not normal.

As a by-product we get the following result concerning moduli spaces of Calabi-Yau threefolds:

Theorem 0.0.3 (Corollary 6.4.9) *The family of hyperplane sections of* $\mathcal{I}_5$ *passing through one of the 36 triple points p is, via projection, a family of Calabi-Yau threefolds which are degenerations of double octics.*

In other words, the family of quintic hypersurfaces is connected with the moduli space of double octics, along a four-dimensional sublocus.

Let us now make a few remarks about the entire set of examples. First of all, they are (almost) all hypersurfaces. For a Baily-Borel embedding to be a hypersurface, the singularities on X_Γ^* must be hypersurface singularities — which they usually are not. Note that all of the examples where X_Γ^* is a normal hypersurface (i.e., excluding the dual of $\mathcal{B}_4$), the arithmetic groups contain torsion. Roughly speaking, this is experimental evidence of a statement like: Γ has torsion $\longleftrightarrow$ the singularities of X_Γ^* are very mild (hypersurface, complete intersection); Γ torsion free $\longleftrightarrow$ the singularities are not so mild. Whether this has some general validity would seem to be quite a difficult problem.

Secondly, all examples have interesting decomposing hyperplane sections, and the components of these reducible sections are modular subvarieties. Let $\mathcal{X}$ be one of our varieties, and let:

$$
\begin{array}{rcl}
\mu & = & \text{\# of hyperplanes } H, \text{ such that } H \cap \mathcal{X} \text{ decomposes} \\
\nu & = & \text{\# of linear subspaces on } \mathcal{X} \text{ cut out by them} \\
\tau & = & \text{\# spaces in each } H \cap \mathcal{X}
\end{array}
$$

Then the results are summed up in Table 0.   So this behavior at least looks

Table 0. Special hyperplane sections in the examples.

$\mathcal{X}$	μ	ν	τ	rest of $H \cap \mathcal{X}$
$\mathcal{S}_3$	15	15	3	−
$\mathcal{I}_4$	10	10 $\mathbb{P}^1 \times \mathbb{P}^{1\prime}s$	1	−
$\mathcal{N}_5$	15	15	3	quadric surface
	15	15	5	−
$\mathcal{W}_{10}$	10	10 $\mathbb{P}^1 \times \mathbb{P}^{1\prime}s$	1	−
	?	?	?	?
$\mathcal{B}_4$	40	40	4	−
$\mathcal{B}_4^{\vee}$	45	45 $\mathbb{P}^1 \times \mathbb{P}^{1\prime}s$	1	?
$\mathcal{I}_5$	27	45	5	−
$\mathcal{I}_5^{\vee}$	36	120	10	$\mathcal{I}_4$

like it can be expected in general for arithmetic quotients whose Baily-Borel embedding is as a hypersurface.

Some other curiosities of the examples are the following. In the appendix we gather a few definitions from classical algebraic geometry and invariant theory, which are not so common nowadays. The Hessian variety $\mathbf{Hess}(V)$ of a hypersurface V meets V along the parabolic divisor, which is the locus of points where the tangent hyperplane section has a singularity worse than an ordinary double point at the point of tangency; it gets blown down to a singular locus of the dual variety under the duality map. The following strange behavior occurs in our examples:

i) $\mathcal{S}_3 \cap \mathbf{Hess}(\mathcal{S}_3) = 15$ planes;

ii) $\mathcal{I}_4 \cap \mathbf{Hess}(\mathcal{I}_4) = 10$ quadric surfaces;

iii) $\mathcal{B}_4 \cap \mathbf{Hess}(\mathcal{B}_4) = 40$ planes;

iv) $\mathcal{I}_5 \cap \mathbf{Hess}(\mathcal{I}_5) = 45 \; \mathbb{P}^3$'s.

Furthermore, each of these intersections consists of modular subvarieties, viewing the varieties as arithmetic quotients. Also in the appendix we define the Steinerian of a hypersurface: it is the locus of singular points of the quadric polars with respect to V, which are singular. (The Hessian is the locus of points for which the quadric polar is singular, and the vertices of these cones cut out the Steinerian.) The following is even stranger:

i) $\mathcal{B}_4$ is self-Steinerian;

ii) $\mathcal{I}_5$ is self-Steinerian.

It is really not understood what self-Steinerian means geometrically, but has something to do with the parabolic divisor. For i) we have a conceptual proof, due to Coble; in the latter case this has just been checked computationally.

Many of the results described in this book were known classically to the geometers of the last century; it is legitimate to ask about the merit of our presentation. In many respects our goal is not in presenting new results at all, it is to give an understanding of these results through a modern, coherent treatment. For example in the third chapter, where the various varieties which occur, each beautiful in its own right, are derived from a common geometric configuration: the tetrahedron in $\mathbb{P}^3$. The fourth chapter presents no new results at all; still a modern presentation of these contents seems lacking. The heart of the book consists of Chapters 5 and 6, which, in addition to presenting results already known classically from a modern perspective, also give some new results. In particular, our proof of the fact that the Burkhardt quartic is a ball quotient is new, and is of more general interest than to just this example. Indeed, the method combines the characterization of ball quotients as given by Yau's theorem with algebro-geometric arguments to deduce the fact that B is a ball quoitent, and moreover, deduce the group. Finally, Chapter 6 contains completely new material, which has only one small flaw – we do not yet know whether the quintic is actually an arithmetic quotient itself.

We also define in section 6.4.1 another birational model of $\mathcal{I}_5$, which we call the cuspidal model $\widehat{\mathcal{I}}_5$. We present quite a bit of evidence which indicates that $\widehat{\mathcal{I}}_5$ is the Satake compactification of a ball quotient, but we have not yet proved this. Taken together, these two interpretations lead us to the conjecture that $\mathcal{I}_5$ (which is a conjecturally birational to the known type **IV** quotient $\mathcal{Z}$ and known to be (Σ_6-equivariantly) birational to the conjectured ball quotient $\widehat{\mathcal{I}}_5$) is a Janus-like algebraic variety.

This work was completed during my stay at the University of Kaiserslautern. I am indebted to the Fachbereich Mathematik there and in particular to Günter Trautmann for generous support.

Chapter 1

Moduli spaces of PEL structures

In this chapter we introduce the basis for the moduli interpretation of most arithmetic quotients, in terms of abelian varieties with given polarization, endomorphism ring and level structure. This is Shimura's theory of PEL structures, developed in [Sh2] and [Sh3] and summarized in [Sh4]. Although quite well-known, we present very briefly this relation, in order to fix notations for the rest of the book. In all later examples (from the next chapter on), all groups will be what is called "split over $\mathbb{R}$", in which a maximal k-split torus is also a maximal $\mathbb{R}$-split one. However this is perhaps not so representative, so we also take some time in this chapter to consider a case which is, roughly speaking, the *opposite* of the split over $\mathbb{R}$ type. This is the case of *hyperbolic planes*, which is considered in detail in [H2]. This case is included to give the reader the flavor of a more "typical" example of arithmetic quotients. Our later examples may in fact be very untypical, but nonetheless very beautiful, something which is also true of the hyperbolic planes.

1.1 Shimura's construction

1.1.1 Endomorphism rings

Let V be an abelian variety over $\mathbb{C}$, $End(V)$ the endomorphism ring and

$$End_{\mathbb{Q}}(V) = End(V) \otimes_{\mathbb{Z}} \mathbb{Q}$$

the endomorphism algebra. A polarization, i.e., a linear equivalence class of ample divisors giving a projective embedding of V, gives rise to a positive involution on $End_{\mathbb{Q}}(V)$, the so-called Rosati involution:

$$\begin{aligned} \varrho : End_{\mathbb{Q}}(V) &\longrightarrow End_{\mathbb{Q}}(V) \\ \phi &\mapsto \phi^{\varrho}. \end{aligned} \tag{1.1}$$

This can be defined, for example, by viewing the polarization as an *isogeny* $\mathcal{C}$: $V \longrightarrow V^\vee$, where $V^\vee$ is the dual abelian variety. The degree of this isogeny is the degree of the polarization. Then for any $\alpha \in End_\mathbb{Q}(V)$, one sets $\alpha^\varrho := \mathcal{C}^{-1} \circ \alpha^\vee \circ \mathcal{C}$ (see [Mi], §17 for a presentation along these lines). The Rosati involution gives rise to a positive definite bilinear form $(\alpha, \beta) \mapsto Tr(\alpha \circ \beta^\varrho)$ (*loc. cit.* 17.3).

If A is a central simple algebra over $\mathbb{Q}$, an involution $*$ on A is called *positive*, if $tr_{A|\mathbb{Q}}(x \cdot x^*) > 0$ for all $x \in A$, $x \neq 0$, where $tr_{A|\mathbb{Q}}$ denotes the reduced trace (see Definition A.1.9). Assuming $(A, *)$ to be simple with positive involution, the $\mathbb{R}$-algebra $A(\mathbb{R})$ is isomorphic to one of the following (see [Sh2], Lemma 1)

(i) $M_r(\mathbb{R})$ with involution $X^* = {}^t X$;

(ii) $M_r(\mathbb{C})$ with involution $X^* = {}^t \overline{X}$, where $^-$ is complex conjugation;

(iii) $M_r(\mathbb{H})$ with involution $X^* = {}^t \overline{X}$, where $^-$ is quaternionic conjugation.

The algebras A occuring in (i) and (iii) are central simple over $\mathbb{R}$, while those of (ii) are central simple over $\mathbb{C}$. The $\mathbb{Q}$-algebra A itself is a $\mathbb{Q}$-form of one of these. The central simple algebras A over $\mathbb{Q}$ are known to be the $M_n(D)$, where D is a division algebra over $\mathbb{Q}$ (see section A.1.2). If the algebra A has a positive involution, the same holds for D. The division algebras D which can occur are also known.

Proposition 1.1.1 *Let D be a division algebra over $\mathbb{Q}$ with a positive involution. Then D occurs in one of the following cases:*

I. A totally real algebraic number field k;

II. D a totally indefinite quaternion algebra over k;

III. D a totally definite quaternion algebra over k;

IV. D is central simple with a $K|k$ involution of the second kind (see Definition A.1.3), where K is an imaginary quadratic extension of k.

In case III the canonical involution on D is the unique positive involution, while in case II the positive involutions correspond to $x \in D$ such that x^2 is totally negative in k. If the algebra D has an involution of the second kind (see Theorem A.1.14) it is easy to see that it admits a positive one. It follows from the fact that $End_\mathbb{Q}(V)$ is a semisimple algebra over $\mathbb{Q}$ with a positive involution that each simple factor is a total matrix algebra $M_n(D)$, with D as in the proposition.

1.1.2   Abelian varieties with given endomorphisms

Let $(A, *)$ be a semisimple algebra over $\mathbb{Q}$ with positive involution, and let

$$\Phi : A \longrightarrow GL(n, \mathbb{C}) \tag{1.2}$$

be a faithful representation. Shimura considers data $\mathcal{P} = (V, \mathcal{C}, \theta)$ and $\{A, \Phi, *\}$ and defines the notion of *polarized abelian variety of type* $\{A, \Phi, *\}$ by the conditions:

(i) V is an abelian variety over $\mathbb{C}$, $\mathcal{C}$ is a polarization;

(ii) $\theta : A \xrightarrow{\sim} End_{\mathbb{Q}}(V)$ is an algebra isomorphism, and for
$\theta(x) : \tilde{V} \longrightarrow \tilde{V}$ (the~denoting the universal cover, i.e., $\tilde{V}$ is
a complex vector space) one has $\theta(x) = \Phi(x)$;   (1.3)

(iii) the involution ϱ determined by $\mathcal{C}$ as in (1.1) coincides on
$\theta(A)$ with the involution coming from $(A, *)$, i.e. $\theta(x)^\varrho = \theta(x^*)$.

The condition (ii) is to be understood as follows. Fixing an isomorphism

$$\psi : V \cong \mathbb{C}^n / \Lambda, \qquad (1.4)$$

each $a \in End_{\mathbb{Q}}(V)$ is represented by a linear transformation of $\mathbb{C}^n$ preserving Λ; that is each a can be represented by a matrix, and $\theta(x) = a$ is the matrix corresponding to $x \in A$ via θ. Recall also that a complex torus $\mathbb{C}^n / \Lambda$ is an abelian variety if and only if there exists a *Riemann form* : each positive $(1,1)$ form ω gives rise to a skew symmetric matrix (q_{ij}):

$$\omega = \sum q_{ij} dx_i \wedge dx_j,$$

where the x_i are canonical coordinates on $\mathbb{C}^n$. Hence if we fix a positive divisor $C \subset \mathcal{C}$, it determines an involution as in (1.1) *and* a Riemann form $E_C(x, y)$ on $\mathbb{C}^n / \Lambda$, and these are related by

$$E_C(\psi(a)x, y) = E_C(x, \psi(a^\varrho)y), \qquad (1.5)$$

where for $a \in End_{\mathbb{Q}}(V)$, $\psi(a)$ denotes the matrix representation for a arising from the identification $\psi : V \cong \mathbb{C}^n / \Lambda$ in (1.4).

Let $(V, \mathcal{C}, \theta)$ be an abelian variety of type $(D, \Phi, *)$ with D a division algebra, so that $(D, *)$ is one of the algebras of Proposition 1.1.1. In the notations used there, put

$$[k : \mathbb{Q}] = g, \quad [D : K] = d^2, \text{ if } D \text{ is of type IV} \qquad (1.6)$$

defining the numbers g and d. Let $n = \dim(V)$; then, assuming D to be a division algebra, $2n$ is a multiple of $[D : \mathbb{Q}]$, i.e., $2n = [D : \mathbb{Q}]m$. Note that $[D : \mathbb{Q}] = g$ for type I, $[D : \mathbb{Q}] = 4g$ for types II and III, while $[D : \mathbb{Q}] = 2d^2 g$ if D is of type IV. For the existence of $(V, \mathcal{C}, \theta)$ of type $(D, \Phi, *)$, certain restrictions are placed on Φ; we assume these are fulfilled. So under the isomorphism θ, each $x \in D$ is represented by the matrix $\Phi(x)$. This makes the lattice Λ with $V \cong \mathbb{C}^n / \Lambda$, tensored with $\mathbb{Q}$, a (left) D-module, i.e.,

$$Q := \mathbb{Q} \cdot \Lambda = \sum_1^m \Phi(D) \cdot x_i \qquad (1.7)$$

for a suitable set of vectors x_i. But this is the same as saying there exists a $\mathbb{Z}$-lattice $\mathcal{M} \subset D$, such that

$$\Lambda = \{\sum_1^m \Phi(a_i)x_i \big| (a_1,\ldots,a_m) \in \mathcal{M}\}. \tag{1.8}$$

If D is central over K, then $\mathcal{M}$ is clearly also an $\mathcal{O}_K$-lattice in D. Now the integrality of the Riemann form can be expressed in terms of $tr_{D|K}$:

$$E_C(\sum_1^m \Phi(a_i)x_i, \sum_1^m \Phi(b_j)x_j) = tr_{D|K}(\sum_{i,j} a_i t_{ij} b_j^*), \tag{1.9}$$

and $T = (t_{ij}) \in M_m(D)$ is a skew-hermitian matrix:

$$T^* = -T, \tag{1.10}$$

where T^* denotes the matrix (t_{ji}^*), where $*$ is the involution on D. For the lattice $\mathcal{M}$ one has

$$tr_{D|K}(\mathcal{M}T\mathcal{M}^*) \subset \mathbb{Z}. \tag{1.11}$$

1.1.3   Arithmetic groups

Hence to each $(V, \mathcal{C}, \theta)$ of type $(D, \Phi, *)$ one gets a $*$-skew hermitian $T \in M_m(D)$ and a lattice $\mathcal{M} \subset D$. To this situation there is a naturally associated $\mathbb{Q}$-group. On the vector space D^m we consider

$$G(D,T) := \{g \in GL(D^m) \big| gTg^* = T\}, \tag{1.12}$$

the symmetry group of the $*$-skew hermitian form determined by T. It is now easy to determine the $\mathbb{R}$-group:

$$G(D,T)(\mathbb{R}) = \begin{cases} \text{Type I:} & Sp(m,\mathbb{R}) \times \cdots \times Sp(m,\mathbb{R}) \ (m \text{ is even}) \\ \text{Type II:} & Sp(2m,\mathbb{R}) \times \cdots \times Sp(2m,\mathbb{R}) \\ \text{Type III:} & SO^*(2m) \times \cdots \times SO^*(2m) \\ \text{Type IV:} & U(p_1,q_1) \times \cdots \times U(p_g,q_g), \end{cases} \tag{1.13}$$

where the number of factors is in each case g, and $p_\nu + q_\nu = md$, and (p_ν, q_ν) is the signature corresponding to the νth real prime. These groups are described in more detail in sections A.3.3, A.3.4, A.3.2 and A.3.5-A.3.6, respectively. For each ν, there is a matrix W_ν which transforms T_ν into the standard form, i.e.,

$$W_\nu T_\nu^{-1} {}^t W_\nu = \begin{pmatrix} 0 & 1_l \\ -1_l & 0 \end{pmatrix}, l = \frac{m}{2} \text{ for Type I}, l = m \text{ for Type II} \tag{1.14}$$

$$W_\nu T_\nu^{-1} W_\nu^* = -i\begin{pmatrix} -1_m & 0 \\ 0 & 1_m \end{pmatrix}, \text{ Type III}; \tag{1.15}$$

$$W_\nu(iT_\nu^{-1})W_\nu^* = \begin{pmatrix} 1_{p_\nu} & 0 \\ 0 & -1_{q_\nu} \end{pmatrix}, \text{ Type IV}. \tag{1.16}$$

Let $\mathcal{D} = \mathcal{D}_{(D,T)}$ denote the domain determined by $G(D,T)(\mathbb{R})$ (actually a particular unbounded realization of this domain, see [Sh2], 2.6). Then $\mathcal{D} = \prod \mathcal{D}_\nu$, and $z_\nu \in \mathcal{D}_\nu$ gives rise to a normalized period (i.e., one of the form $(1,\Omega)$) for an abelian variety, by setting $X_\nu = Y_\nu \overline{W}_\nu$, where

$$Y_\nu = \begin{pmatrix} z_\nu & 1_l \\ \overline{z}_\nu & 1_l \end{pmatrix}, l = \frac{m}{2}, \text{ Type I}, l = m, \text{ Type II;} \qquad (1.17)$$

$$Y_\nu = \begin{pmatrix} -z_\nu & 1_m \\ 1_m & \overline{z}_\nu \end{pmatrix}, \text{ Type III;} \qquad (1.18)$$

$$Y_\nu = \begin{pmatrix} 1_{p_\nu} & z_\nu \\ {}^t\overline{z}_\nu & 1_{q_\nu} \end{pmatrix}, \text{ Type IV.} \qquad (1.19)$$

The matrix X_ν determines m vectors $x_1, \dots x_m$ of $\mathbb{C}^n$ (in a rather complicated fashion, see formulas (17)-(20) in [Sh2]), which determine a lattice $\Lambda = \Lambda(z, T, \mathcal{M})$ by the formula in equations (1.7)-(1.8) above.

Note that the representation Φ contains the representations $\chi_\nu =$ projection on the νth real factor with multiplicities. For Type IV, $p_\nu + q_\nu = md$, and $p_\nu =$multiplicity of χ_ν while $q_\nu =$multiplicity of $\overline{\chi}_\nu$. For things to work out one must therefore assume, in case of Type IV, that iT_ν^{-1} has the *same* signature (p_ν, q_ν) as occurs in Φ. With this restriction, the following holds:

Theorem 1.1.2 ([Sh2], Thm. 1) *For every $z \in \mathcal{D} = \mathcal{D}_{(D,T)}$, and every lattice $\mathcal{M} \subset D$, we get a polarized abelian variety $V_z = \mathbb{C}^n/\Lambda(z, T, \mathcal{M})$ of type $(D, \Phi, *)$, and conversely, every such V is of the form $V = \mathbb{C}^n/\Lambda(z, T, \mathcal{M})$ for some $z \in \mathcal{D}_{(D,T)}$, $\mathcal{M} \subset D$ a lattice.*

The lattice $\mathcal{M} \subset D$ gives rise to an arithmetic subgroup

$$\Gamma = \Gamma_{(D,T,\mathcal{M})} = \{g \in G(D,T) \big| g\mathcal{M} \subset \mathcal{M}\} \qquad (1.20)$$

as discussed in section A.5. If one defines an isomorphism $\phi : V_z \longrightarrow V_{z'}$ of two abelian varieties of type $(D, \Phi, *)$ as an isomorphism of the underlying varieties, such that $\phi^{-1}(C') = C$ and $\phi\theta(a) = \theta'(a)\phi$, for all $a \in D$, then one has

Theorem 1.1.3 ([Sh2], Thm. 2) *The arithmetic quotient $X_\Gamma = \Gamma\backslash\mathcal{D}_{(D,T)}$ is in one-to-one correspondence with the set of isomorphism classes of abelian varieties $V_z = \mathbb{C}^n/\Lambda(z, T, \mathcal{M})$ of type $\{(D, \Phi, *), (T, \mathcal{M})\}$.*

Moreover, one calls two such pairs $(T_1, \mathcal{M}_1)$, $(T_2, \mathcal{M}_2)$ *equivalent*, if $\exists_{U \in M_m(D)}$, such that $UT_2U^* = \delta T_1$ for some positive $\delta \in \mathbb{Q}$ and $\mathcal{M}_1 U = \mathcal{M}_2$. Equivalent pairs give rise to isomorphic families of abelian varieties ([Sh2], Prop. 4). Summing up, *-skew hermitian matrices $T \in M_m(D)$ determine certain $\mathbb{Q}$-groups, lattices $\mathcal{M} \subset D$ determine certain arithmetic groups, and the corresponding arithmetic quotients are moduli spaces for certain families of abelian varieties.

Remark 1.1.4 The complex multiplication by $\mathcal{M}$ describes the endomorphism ring. The *automorphisms* determined by $\mathcal{M}$ are the invertible elements, i.e., $\text{Aut}(V) = \mathcal{M}^*$, the group of units.

1.1.4   PEL structures

One can also accomodate level structures in this settup, introduced in [Sh3], cf. also [Sh4]. This is done by fixing s points $y_1, \ldots, y_s$ in the D-module Q, as in (1.7), and s points $t_1, \ldots, t_s$ of the abelian variety V. One requires that the map ψ of (1.4) maps the y_i onto the t_i. More precisely,

Definition 1.1.5 Let Q be a D-vector space of dimension m, and $\mathcal{M} \subset Q$ a $\mathbb{Z}$-lattice. Consider a conglomeration:

$$\mathcal{T} := \{(D, \Phi, *), (Q, T, \mathcal{M}); y_1, \ldots, y_s\},$$

where $(D, \Phi, *)$ is as above, Q is a D-vector space with lattice $\mathcal{M} \subset Q$, T is a $*$-skew hermitian (D-valued) form on Q, and y_i are points in Q. This is called a PEL-*type*. Consider a conglomeration:

$$\mathfrak{Q} := \{(V, \mathcal{C}, \theta); t_1, \ldots, t_s\},$$

where $(V, \mathcal{C}, \theta)$ is a polarized abelian variety with analytic coordinate θ as above and t_i are points of *finite order* on V. This is called a PEL-*structure*. Then $\mathfrak{Q}$ is *of type* $\mathcal{T}$, if there exists a commutative diagram

$$
\begin{array}{ccccccccc}
0 & \longrightarrow & \mathcal{M} & \longrightarrow & Q(\mathbb{R}) & \longrightarrow & Q(\mathbb{R})/\mathcal{M} & \longrightarrow & 0 \\
 & & \downarrow & & f\downarrow & & \downarrow & & \\
0 & \longrightarrow & \Lambda & \longrightarrow & \mathbb{C}^n & \overset{\psi}{\longrightarrow} & V & \longrightarrow & 0
\end{array}
\qquad (1.21)
$$

satisfying the conditions

   (i) ψ gives a holomorphic isomorphism (strictly speaking, this is the ψ^{-1} of above);

  (ii) f is an $\mathbb{R}$-linear isomorphism, and $f(\mathcal{M}) = \Lambda$;

 (iii) $C \in \mathcal{C}$ determines a Riemann form E_C as in (1.5);

 (iv) $f(ax) = \Phi(a)f(x)$, and $\Phi(a)$ defines $\theta(a)$ for every $a \in D$ as (1.3), (ii);

  (v) $\psi \circ f(y_i) = t_i$, $i = 1, \ldots, s$.

Note that the finite set of points y_i and t_i come both equipped with a form; on the former the form T, and the Riemann form E_C on the latter. These forms are preserved under the isomorphism. The conditions (iii), (iv) and (v) correspond respectively to polarization, endomorphism ring and level structure.

There is a natural notion of isomorphism of abelian varieties with PEL structures. Given two PEL-structures $\mathfrak{Q}$ and $\mathfrak{Q}'$, an isomorphism $\phi : V \longrightarrow V'$ is an *isomorphism* from $\mathfrak{Q}$ to $\mathfrak{Q}'$, if $\phi\theta(a) = \theta'(a)\phi$ for all $a \in D$, and $\phi(t_i) = t_i'$ for all i.

Definition 1.1.6 A PEL-type $\mathcal{T}$ is *equivalent* to a PEL-type $\mathcal{T}'$, if $D = D'$, $* = *'$, $s = s'$, Φ and Φ' are equivalent as representations of D, and there is a D-linear automorphism μ of Q such that $T'(x\mu, y\mu) = T(x, y)$, $\mathcal{M}\mu = \mathcal{M}'$, $y_i\mu \equiv$

Table 1.1: Rational groups for PEL-structures.

	Type II	Type III	Type IV
D	A totally indefinite quaternion algebra over $\mathbb{Q}$	a totally definite quaternion algebra over $\mathbb{Q}$	simple division algebra, central over K, an imaginary quadratic extension of $\mathbb{Q}$, with an involution of the second kind. One may assume D to be a cyclic algebra
d	2	2	d
$\dim(V)$	$2m$	$2m$	d^2m
Tits index	$C_{m,s}^{(2)}$	$^iD_{m,s}^{(2)}, i = 1,2$	$^2A_{dm-1,s}^{(d)}$

Here the $\mathbb{Q}$-rank is s, and this is the Witt index of the $\pm$hermitian form. The Tits index is the index listed in [T1].

$y_i' \mathrm{mod} \mathcal{M}'$ for all i. If Ω is of type $\mathcal{T}$, then Ω is also of type $\mathcal{T}'$ if and only if $\mathcal{T}$ and $\mathcal{T}'$ are equivalent. A PEL-type $\mathcal{T}$ is called *admissible*, if there exists at least one PEL-structure of that type.

One has an anolgue of Theorems 1.1.2 and 1.1.3 in this situation also.

Theorem 1.1.7 ([Sh4], Thm. 3) *For every admissible PEL-type $\mathcal{T}$ there exists a bounded symmetric domain $\mathcal{D}$ (this is the same domain as in Theorem 1.1.2) such that the statement of Theorem 1.1.2 holds in this situation, and every PEL-structure Ω of type $\mathcal{T}$ occurs in this family.*

Now define a corresponding arithmetic group as follows:

$$\Gamma = \{g \in G(D,T) \mid \mathcal{M}g = \mathcal{M}, \ y_ig \equiv y_i \mathrm{mod} \mathcal{M}, \ i = 1,\ldots,s\} \qquad (1.22)$$

Then the analogue of Theorem 1.1.3 is

Theorem 1.1.8 ([Sh4], Thm. 4) *Two members of the family of Theorem 1.1.7 corresponding to points $z_1, z_2 \in \mathcal{D}$ are isomorphic if and only if $z_1 = \gamma(z_2)$ for some $\gamma \in \Gamma$.*

In Table 1.1 we list the data for each of the cases II, III and IV of 1.14.

From Table 1.1 it is easy to determine the arithmetic groups which occur. For a lattice $\mathcal{M} \subset V$, let $\mathcal{O}_\mathcal{M}$ be the maximal order it defines in $M_m(D)$. Then this $\mathcal{O}_\mathcal{M}$ determines an arithmetic subgroup $G_\mathcal{M}$ as in section A.5.2.

1.2   Moduli spaces

1.2.1   Moduli functors

The right way to formulate problems about isomorhism classes of polarized varieties with some additional structure is in terms of functors and their representability. So instead of considering just the set of isomorphism classes of some polarized variety $(X, \mathcal{C})$, one considers the functor associating to a scheme S the set

$$\mathcal{F}(S) = \left\{ \begin{array}{l} \mathcal{X} \longrightarrow S \text{ a family of varieties such that for all } s \in S, \\ (\mathcal{X}_s, \mathcal{C}_{|\mathcal{X}_s}) \text{ is of a given type} \end{array} \right\} /isom.$$

and poses the question of representability of this functor.

Definition 1.2.1 A *moduli problem* is given by

- a class $\mathfrak{F}(k)$ of (isomorphism classes of) *objects* of polarized varieties $(X, \mathcal{C})$;

- *families of objects*, $(f : \mathcal{X} \longrightarrow S, \mathcal{C})$, f flat, proper, $\mathcal{C}$ invertible on $\mathcal{X}$, such that for all $s \in S$, the pair $(\mathcal{X}_s, \mathcal{C}_{|\mathcal{X}_s}) \in \mathfrak{F}(k)$;

- a notion of *isomorphism* of families;

- a functor

$$\mathcal{F}(S) := \{(f : \mathcal{X} \longrightarrow S, \mathcal{C}) \text{ families of objects in } \mathfrak{F}(k)\} /isom.$$

The functor $\mathcal{F}$ is called the moduli functor of the problem.

1.2.2   Examples

1. We work in the category of k-schemes for an arbitrary algebraically closed field k, for which the notion of polarized abelian scheme makes sense. One defines the following functors:

   - $\mathbf{A}_g$: **Schemes** $\longrightarrow$ **Sets**, $\mathbf{A}_g(S) = \{$isomorphism classes of principally polarized abelian schemes over S of relative dimension $g\}$. Here the class $\mathfrak{A}_g(k)$ consists of pairs $(X, \mathcal{C})$, where X is an abelian scheme and $\mathcal{C}$ is a principal polarization.

   - $\mathbf{A}_{g,d,n}$: **Schemes** $\longrightarrow$ **Sets**, $\mathbf{A}_{g,d,n}(S) = \{$isomorphism classes of abelian schemes of relative dimension g with a polarization of degree d^2 and a level n structure$\}$. A *level n structure* is given by choosing a basis $(\sigma_1, \ldots, \sigma_{2g})$ of the points of order n, i.e., of the kernel of multiplication by n, which induce a symplectic isomorphism $X_{[n]} \cong (\mathbb{Z}/n\mathbb{Z})^{2g}/S$. Here the class $\mathfrak{A}_{g,d,n}(k)$ consists of triples (X, λ, σ), where X is an abelian scheme, λ is a polarization of degree d^2 and $\sigma = (\sigma_1, \ldots, \sigma_{2g})$ is a basis of the kernel of multiplication by n on X.

   – $\mathbf{A}^*_{g,n}$: **Schemes** $\longrightarrow$ **Sets**, $\mathbf{A}^*_{g,n}(S) = \{$isomorphism classes of principally polarized abelian schemes of relative dimension g with a level n structure which lifts$\}$. A *level n structure which lifts* is given by an isomorphism $\alpha : X_{[n]} \cong (\mathbb{Z}/n\mathbb{Z})^g \times \mu_n^g /S$ such that α^*(standard symplectic pairing) = Weil pairing. The class $\mathfrak{A}^*_{g,n}(k)$ consists of triples (X, λ, α), where X is an abelian scheme of dimension g, λ is a principal polarization and α is an isomorphism of $X_{[n]}$ with $(\mathbb{Z}/n\mathbb{Z})^g \times \mu_n^g$ as above.

In the following we will only describe the class $\mathfrak{F}(k)$, and it will be clear what the corresponding functor should be. Also the notion of isomorphism should be clear in each case, one preserving all the structure.

2. We now shift our attention to the category of $\mathbb{C}$-schemes and complex abelian varieties. Given an admissible PEL type

$$\mathcal{T} = \{(D, \Phi, *), (Q, T, \mathcal{M}), y_1, \ldots, y_s\},$$

we consider the class of abelian varieties $\{(V, \mathcal{C}, \theta), t_1, \ldots, t_s\}$ which have a PEL structure of type $\mathcal{T}$, as defined in the previous section. Let $\mathbf{A}_{\mathcal{T}}$ denote the corresponding moduli functor.

3. As a special case of the preceeding consider the functor $\mathbf{A}^*_{2,2}$ corresponding to a full level 2 structure which lifts on principally polarized complex abelian surfaces (see section 3.3) A slight variant of this is given by considering the class of Kummer surfaces $K_s = A_s/i$ with a polarization as a quartic surface, if A_s does not split. We let $\mathbf{K}_2$ denote this functor. We also consider $\mathbf{A}^*_{2,3}$, for principally polarized abelian surfaces with a level 3 structure which lifts (see section 5.3 and 5.7.1), and a variant of $\mathbf{K}_2$ describing a "level 3 structure on Kummer surfaces". Consider the class of objects (see section 5.3.3) which are triples $\{(X, \mathcal{C}), (X', \mathcal{C}'), (X'', \mathcal{C}'')\}$, where X is a Kummer surface as above (polarization as a quartic surface), X' is a Weddle surface (determining the polarization $\mathcal{C}'$) which is a birational image of X, and X'' is a sextic K3 in $\mathbb{P}^4$ with 10 nodes, a birational image of X (again "sextic K3" determines the polarization $\mathcal{C}''$). We let $\mathbf{K}_{2,3}$ denote the corresponding moduli functor.

4. In the category of $\mathbb{C}$-schemes consider the class of triples $(X, \mathcal{L}, \phi)$, where X is a K3 surface, $\mathcal{L}$ is a polarization of degree e and ϕ is an isomorphism $\phi : Pic(X) \longrightarrow \Lambda$ with a fixed sublattice Λ of the K3 lattice (see 2.20), Λ orthogonal to $\mathcal{L}$. Let $\mathbf{K3}_{(e,\Lambda)}$ denote the corresponding functor.

To introduce a level structure, we require an ordered isomorphism: if $\phi : Pic(X) \longrightarrow \Lambda$ is the given isomorphism, with basis $e_1, \ldots, e_m$ mapping in any order to a fixed basis $f_1, \ldots, f_m$ of Λ, an ordered isomorphism requires $\phi(e_i) = f_i$, $i = 1, \ldots, m$ (see Proposition 2.3.3). We let $\mathbf{K3}_{(e,\Lambda,e_1,\ldots,e_m)}$ denote the corresponding functor.

Let $PH_{2,2}$ be the projective Heisenberg group acting on $\mathbb{P}^4$ (see (3.70)), and consider the class of $PH_{2,2}$-invariant quartic surfaces containing 16 lines (see Theorem 3.4.16). Let $\mathbf{K3}_{PH_{2,2}}$ denote the corresponding functor.

5. Let V be a $\mathbb{Q}$-vector space, $\mathcal{L} \subset V$ a lattice, and consider the class of objects consisting of Hodge structures of weight 2 on $V_\mathbb{C}$ (see Proposition 2.3.1):
$$V_\mathbb{C} = V^{2,0} \oplus V^{1,1} \oplus V^{0,2},$$
which are deformations of a fixed Hodge structure. Let $\mathbf{H}_\mathcal{L}$ denote the corresponding functor. Here one can also introduce level structures by placing additional assumptions on the Hodge structure, for example preserving a fixed ordering on $H^2(X, \mathbb{Z})$. A special case of this is the functor $\mathbf{K3}_{(e,\Lambda,e_1,\dots,e_m)}$ defined above.

Remark 1.2.2 This does not precisely coincide with the moduli functor defining one of the Shimura varieties in the sense of Deligne for the domain $\mathcal{D}$ of type **IV** which arises here, the latter in general not being connected. Compare also Remark 1.2.9 below.

6.   a) Consider the class of complex curves given as follows. Let

$$\nu, \mu_0, \dots, \mu_{n+1} \in \mathbb{Q} \text{ such that } \sum \mu_i = 2, \ (1 - \mu_i - \mu_j)^{-1} \in \mathbb{Z} \cup \infty$$

(see Theorem 3.1.2) and consider the class of curves given by equations

$$C_{t_1,\dots,t_n} = \{y^\nu = x^{\mu_0}(x-1)^{\mu_{n+1}}(x-t_1)^{\mu_1} \cdots (x-t_n)^{\mu_n}\}.$$

   b) Introduce a level structure by requiring an isomorphism to preserve the order of the n branch points $t_1, \dots, t_n$.

   c) As a special case of the above take $(\nu, \mu_0, \dots, \mu_{n+1}) = (3, 1, 1, 1, 1, 1, 1)$; In this case the curves $C_{t_1,\dots,t_4}$ are so-called Picard curves (see Remark 3.2.8).

7.   a) Consider the class of *cubic surfaces* with at most ordinary double points. Let $\mathbf{S}$ denote the corresponding functor.

   b) Consider the class of pairs (S, ϕ), where S is a cubic surface as above and ϕ is a *marking* of the cubic surface, i.e., an isomorphism $\phi : Pic^0(S) \longrightarrow \Lambda$ of the orthogonal complement of the hyperplane section in the Picard group with the $\mathbf{E_6}$-lattice (see Theorem 4.1.14). Let $\mathbf{S}_\Lambda$ denote the moduli functor.

   c) A slight variant of this was suggested by E. Looijenga. Consider pairs (S, C) consisting of a semistable Del Pezzo surface S of degree 3 with a cuspidal cubic curve C; let $\mathbf{S'}$ denote this functor. One has a level structure as in b) with functor $\mathbf{S'}_\Lambda$. Looijenga has shown that the last functor has a corresponding moduli space (defined in the next section) which is the invariant quintic of Chapter 6.

8. The objects consist of sets of k points in $\mathbb{P}^n$. A level structure is introduced by considering *ordered* sets of k points in $\mathbb{P}^n$ (see (3.83)).

9. Consider the class of objects (see section 6.4.3) consisting of double covers of $\mathbb{P}^3$ which are branched over an octic which splits into a cubic and a quintic surface. This determines the polarization of course. Consider also the subclass consisting of those double covers for which the cubic and the quintic intersect in the union of 15 lines. Finally, consider the level structure given by requiring a marking of the 15 lines of intersection.

1.2.3 Coarse and fine moduli spaces

Recall that a functor $\mathcal{F} : \mathbf{Schemes}/k \longrightarrow \mathbf{Sets}$ is called *representable* by a scheme M if for all schemes S, there is an isomorphism

$$\Theta(S) : \mathcal{F}(S) \longrightarrow \mathrm{Hom}(S, M).$$

Definition 1.2.3 A *fine moduli scheme* (space) M for a moduli functor $\mathcal{F}$ is a scheme M which represents the functor $\mathcal{F}$. Equivalently, there is a *universal family* $(g : \mathcal{X} \longrightarrow M, \mathcal{L}) \in \mathcal{F}(M)$, i.e., such that for all families of objects $(f : \mathcal{Y} \longrightarrow S, \mathcal{H}) \in \mathcal{F}(S)$ there is a unique morphism $\tau : S \longrightarrow M$ such that f is the pullback family:

$$
\begin{array}{ccc}
\mathcal{Y} & \xrightarrow{\tilde{\tau}} & \mathcal{X} \\
f \downarrow & & \downarrow \\
S & \xrightarrow{\tau} & M,
\end{array}
$$

and $\tilde{\tau}^*(\mathcal{L}) = \mathcal{H}$.

A fine moduli space is unique if it exists, but in general such do not exist. A useful more general notion is

Definition 1.2.4 A *coarse moduli space* M for a moduli functor $\mathcal{F}$ is a scheme M together with a natural transformation of functors

$$\Theta : \mathcal{F} \longrightarrow \mathrm{Hom}(-, M)$$

satisfying:

a) $\Theta(Spec(k)) : \mathcal{F}(k) \longrightarrow M(k)$ is a bijection.

b) Given a scheme B and a natural transformation $\chi : \mathcal{F} \longrightarrow \mathrm{Hom}(-, B)$, there is a unique natural transformation $\psi : \mathrm{Hom}(-, M) \longrightarrow \mathrm{Hom}(-, B)$ with $\chi = \psi \circ \Theta$.

The first condition determines M as a topological space (point set); condition b) determines the *scheme* structure on M uniquely. Again, coarse moduli spaces are unique if they exist, but they in general do not exist. The main interest is in such moduli functors for which (at least) a coarse moduli space does exist.

1.2.4   Moduli spaces of abelian varieties

For the moduli functors of abelian varieties mentioned above coarse moduli spaces do exist. The existence is proven in [M1] by means of geometric invariant theory.

Theorem 1.2.5   *(a) For any $g, d, n \in \mathbf{N}_+$, the coarse moduli scheme $\mathcal{A}_{g,d,n}$ of the functor $\mathbf{A}_{g,d,n}$ exists. It is faithfully flat over $Spec\mathbb{Z}[1/n]$ and quasi-projective over $Spec\mathbb{Z}[1/np]$ for any prime number p. Furthermore, if $n \geq 3$, then $\mathcal{A}_{g,d,n}$ actually represents $\mathbf{A}_{g,d,n}$, i.e., is a fine moduli space, and is smooth over $Spec\mathbb{Z}[1/nd]$.*

   (b) The coarse moduli scheme $\mathcal{A}_{g,n}^*$ of the functor $\mathbf{A}_{g,n}^*$ always exists. It is faithfully flat over $Spec\mathbb{Z}$ and quasi-projectuve over $Spec\mathbb{Z}[1/p]$ for any prime number p. If $n \geq 3$, $\mathcal{A}_{g,n}^*$ is a fine moduli scheme, smooth over $Spec\mathbb{Z}[1/n]$.

Note that $\mathcal{A}_{g,1}^* = \mathcal{A}_{g,1,1}$ which is usually denoted just $\mathcal{A}_g$. A peculiar fact about the spaces $\mathcal{A}_{g,d,n}$ $(n > 1)$ is that they are not connected, while $\mathcal{A}_{g,n}^*$ is; we will get back to this in a moment.

We now return to the context of interest for us in this book: $k = \mathbb{C}$, and the abelian schemes are complex abelian varieties. In the cases where the fine moduli space exists, we can actually construct the universal family. For this we need the Siegel space of degree g,

$$\mathbb{S}_g = \{ Z \in M_g(\mathbb{C}) | Z = {}^t Z, \ \mathrm{Im}(Z) > 0 \},$$

as well as the principal congruence subgroup of level n, defined as the kernel of the following exact sequence

$$1 \longrightarrow \Gamma(n) \longrightarrow Sp(2g, \mathbb{Z}) \longrightarrow Sp(2g, \mathbb{Z}/n\mathbb{Z}) \longrightarrow 1. \qquad (1.23)$$

It is not trivial, but well-known, that the morphism on the right is surjective. One considers the semi-direct product $\mathcal{G} = Sp(2g, \mathbb{Z}) \ltimes \mathbb{Z}^{2g}$, where $Sp(2g, \mathbb{Z})$ acts on $\mathbb{Z}^{2g}$ by matrix multiplication. Then $\mathcal{G}$ acts on the product $\mathbb{S}_g \times \mathbb{C}^g$ as follows:

$$\mathcal{G} \ni \gamma = (M, \alpha) : (Z, z) \mapsto ((AZ + B)(CZ + D)^{-1}, (CZ + D)^{-1}(z + Z\alpha_1 + \alpha_2)),$$

where $M = \begin{pmatrix} A & B \\ C & D \end{pmatrix}$, $\alpha = {}^t(\alpha_1, \alpha_2)$, $\alpha_i \in \mathbb{Z}^g$. For a subgroup $\Gamma \subset Sp(2g, \mathbb{Z})$ of finite index, the group $\mathcal{G}_\Gamma := \Gamma \ltimes \mathbb{Z}^{2g}$ acts similarly on $\mathbb{S}_g \times \mathbb{C}^g$, and if Γ is *torsion free*, then the quotient

$$\mathcal{X}_\Gamma := \mathcal{G}_\Gamma \backslash \mathbb{S}_g \times \mathbb{C}^g$$

is actually *smooth*, and comes equipped with a projection onto $X_\Gamma := \Gamma \backslash \mathbb{S}_g$,

$$\mathcal{X}_\Gamma \longrightarrow X_\Gamma,$$

and this is a universal family. More precisely, since for $n \geq 3$, $\Gamma(n)$ is torsion free, we have an isomorphism

$$\mathcal{A}_{g,n}^*(\mathbb{C}) \cong X_{\Gamma(n)}$$

of complex manifolds, and $\mathcal{X}_{\Gamma(n)} \longrightarrow X_{\Gamma(n)}$ is the universal family. Furthermore, by Baily-Borel [BB] it is known that $X_{\Gamma(n)}$ is actually a quasi-projective variety, hence

Proposition 1.2.6 For $n \geq 3$, the space $X_{\Gamma(n)}$ is a fine moduli space for the moduli functor $\mathbf{A}_{g,n}^*$.

Similarly, the space $\mathcal{A}_{g,d,n}(\mathbb{C})$ is a finite disjoint union of copies of $X_{\Gamma(n)}$. As in our work we are concerned only with arithmetic quotients of connected symmetric domains, it is convenient to formulate the moduli problem in such a way that the corresponding moduli space is connected. One way to do this is to use the functor $\mathbf{A}_{g,n}^*$ as described above. However, if we are concerned only with complex abelian varieties, we may also pose the moduli problem as follows.

Definition 1.2.7 A *full level n stucture* on a complex abelian variety A is a symplectic isomorphism

$$\mathrm{Ker}(m_n) \cong (\mathbb{Z}/n\mathbb{Z})^{2g},$$

where $m_n : A \longrightarrow A$, $m_n(z) = nz$ is the multiplication by n map, which *lifts* to a symplectic isomorphism

$$H_1(A, \mathbb{Z}) \cong \mathbb{Z}^{2g} \bmod(n).$$

It is easy to see that for complex abelian varieties this is equivalent to the moduli problem described by the functor $\mathbf{A}_{g,n}^*$: the Weil pairing is the one that lifts (see [M2], §24, p. 237). Next we note that this notion is a special case of Shimura's PEL-structures. Here the triples $\{(D, \Phi, *), (Q, T, \mathcal{M})\}$ are as follows. $D = \mathbb{Q}$, Φ is the inclusion of $\mathbb{Q}$ in $\mathrm{End}^0(A)$, $*$ is trivial, $Q = \mathbb{Q}^{2g}$, T is the standard symplectic form and $\mathcal{M} = \mathbb{Z}^{2g}$ is the standard lattice. Furthermore, $y_1, \ldots, y_s \in \mathbb{Q}^{2g}$ are the n-division points, i.e., given by the $\left(\frac{k_1}{n}, \ldots, \frac{k_{2g}}{n}\right)$, $1 \leq k_i \leq n$ with respect to a $\mathbb{Z}$-basis of $\mathcal{M}$. Let (V, C) be a principally polarized abelian g-fold, with diagram

$$
\begin{array}{ccccccccc}
0 & \longrightarrow & \mathcal{M} & \longrightarrow & Q(\mathbb{R}) & \longrightarrow & Q(\mathbb{R})/\mathcal{M} & \longrightarrow & 0 \\
 & & \downarrow & & f \downarrow & & \downarrow & & \\
0 & \longrightarrow & \Lambda & \longrightarrow & \mathbb{C}^g & \overset{\psi}{\longrightarrow} & V & \longrightarrow & 0
\end{array}
$$

and $V \cong \mathbb{C}^g/\Lambda$, $f(\mathcal{M}) = \Lambda$. Then $H_1(V, \mathbb{Z}) = \Lambda = f(\mathcal{M})$ gives a symplectic isomorphism such that, letting $(t_1, \ldots, t_s)$, $t_i \in V$ denote the n-division points, we have

$$\psi \circ f(y_i) = t_i.$$

This in turn implies the sympectic isomorphism $\mathrm{Ker}(m_n) \cong (\mathbb{Z}/n\mathbb{Z})^{2g}$ lifts to a symplectic isomorphism

$$H_1(V, \mathbb{Z}) \cong \mathbb{Z}^{2g} \bmod(n).$$

Corollary 1.2.8 *The moduli spaces for* PEL*-structures (assuming they exist) in the sense of Shimura are connected.*

Remark 1.2.9 Shimura varieties in the sense of Deligne (see [D]) are not necessarily connected, but a connected component of that space is the moduli space of this corollary. Shimura varieties in the sense of Deligne have *canonical models* which are projective (in general not irreducible) schemes defined over certain number fields. For example, the space $\mathcal{A}_{g,1,n}(\mathbb{C})$ is not connected if $n > 1$, and its canonical model is not defined over $\mathbb{Q}$. For the connected moduli space $\mathcal{A}^*_{g,n}(\mathbb{C})$ we will see it has a Baily-Borel embedding which is defined over $\mathbb{Q}$ for $g = 2$, $n = 2$ (section 3.3) and $n = 3$ (Chapter 5). This shows that our models are not necessarily connected components of the corresponding canonical model. □

We now show that the arithmetic quotients $\Gamma\backslash\mathcal{D}$ occuring in the formulation of Theorems 1.1.3 and 1.1.8 are moduli spaces (in the sense of Definitions 1.2.3 and 1.2.4) of the moduli functors $\mathbf{A}_{\mathcal{T}}$ descibed above. The difference between coarse moduli space and fine moduli space in this situation is the question whether Γ has torsion or not. So let us first consider Theorem 1.1.8.

Theorem 1.2.10 *Let a* PEL*-type* $\mathcal{T} = \{(D, \Phi, *), (Q, T, \mathcal{M}), y_1, \ldots, y_s\}$ *be given,* $\mathcal{D}$ *the corresponding domain,* Γ *the group (1.22), and let* X_Γ *denote the quotient* $X_\Gamma = \Gamma\backslash\mathcal{D}$. *Then if* Γ *is torsion-free,* X_Γ *is a fine moduli space of the moduli functor* $\mathbf{A}_{\mathcal{T}}$ *described in section 1.2.2, 2.*

Proof: We have seen that the family of abelian varieties $\mathcal{X}_\Gamma \longrightarrow X_\Gamma$ can be constructed if Γ is torsion free, which together with the statement of Theorem 1.1.8, is a universal family. Hence it suffices to show that X_Γ is really a scheme. This follows from [BB], Theorem 10.11, which implies that X_Γ is a quasi-projective variety, which in fact is smooth since Γ is torsion free. □

Corollary 1.2.11 *Let* $\mathcal{D}$, Γ *be as in the theorem,* $\Gamma' \subset \Gamma$ *any torsion free subgroup of finite index. Then* $X_{\Gamma'} = \Gamma'\backslash\mathcal{D}$ *is a fine moduli space for a moduli functor with the same polarization and endomorphism structure as above and some level structure.*

Corollary 1.2.12 *Let* $\mathcal{D} = \mathcal{D}_{(D,T)}$ *be as in Theorems 1.1.2 and 1.1.3,* $\Gamma = \Gamma_{(D,T,\mathcal{M})}$ *the discrete subgroup (1.20) (resp* $\mathcal{D}$, Γ *as in the Theorem 1.2.10, but* Γ *not torsion free). Then* $X_\Gamma = \Gamma\backslash\mathcal{D}$ *is a coarse moduli space of the moduli functor of isomorphism classes of complex abelian varieties of type* $\{(D, \Phi, *), (Q, T, \mathcal{M})\}$ *(resp. of the moduli functor* $\mathbf{A}_{\mathcal{T}}$).

Proof: By imposing an additional level structure we get a PEL moduli problem and a torsion free subgroup $\Gamma' \subset \Gamma$ of finite index; the space $X_{\Gamma'}$ is by Theorem 1.2.10 a fine moduli space. We have a (branched) cover $X_{\Gamma'} \longrightarrow X_\Gamma$, which, together with the statement of Theorem 1.1.3 implies the result. □

1.3 Hyperbolic planes

The example we present here is roughly speaking the opposite of the "split over $\mathbb{R}$" case, in which a maximal k-split torus is also maximal $\mathbb{R}$-split. The "opposite" of this is where we have a small group, $SU(1,1)$, with arbitrarily large coefficients, namely elements of a simple division algebra. This is the case of hyperbolic planes. The k-split torus is always one-dimensional, while the $\mathbb{R}$-split torus gets arbitrarily large. We now define these groups.

1.3.1 The rational groups

First we fix a division algebra D over k with involution which will be denoted by $x \mapsto \bar{x}$; more precisely, consider D as in one of the following cases:

1) $d = 1$: k is a totally real number field of degree f over $\mathbb{Q}$, $K|k$ is an imaginary quadratic extension, $D := K$, the involution is the Galois automorphism of K over k, and $V := K^2$. The hyperbolic plane is denoted (K^2, h).

2) $d = 2$: k as in 1), D a totally indefinite quaternion division algebra, central simple over k with the canonical involution, $V := D^2$. The hyperbolic plane is denoted (D^2, h).

3) $d \geq 2$: k as above, $K|k$ an imaginary quadratic extension, D a central simple division algebra over K with a $K|k$-involution of the second kind, $V := D^2$; this is again denoted (D^2, h).

Then let V be a two-dimensional right D-vector space, $H : V \times V \longrightarrow D$ a hermitian form, which we assume is *isotropic* . Then the maximal isotropic subspaces are lines (over D), and one can find a basis of V with respect to which h is given by the standard matrix $H = \begin{pmatrix} 0 & 1 \\ 1 & 0 \end{pmatrix}$. The algebraic groups defined by this setup are the

$$U(D^2, h) = \left\{ g \in GL(2, D) \big| gHg^* = H \right\}, \quad SU(D^2, h) = U(D^2, h) \cap SL(2, D).$$
(1.24)

The equations defining $U(D^2, h)$ are then

$$U(D^2, h) = \left\{ g = \begin{pmatrix} a & b \\ c & d \end{pmatrix} \big| a\bar{d} + b\bar{c} = 1, a\bar{b} + b\bar{a} = c\bar{d} + d\bar{c} = 0 \right\}.$$
(1.25)

The additional equation defining $SU(D^2, h)$ can be written in terms of determinants, using Dieudonné's theory of determinants over skew fields, (see [Art], p. 157)

$$\det(g) = n_{D|k}(ad - aca^{-1}b) = 1,$$
(1.26)

where $n_{D|k}$ denotes the reduced norm (see Definition A.1.9). These groups have indices ${}^2A_{1,1}^{(1)}$, $C_{2,1}^{(2)}$ and ${}^2A_{2d-1,1}^{(d)}$. It was discovered in [H2] that certain subalgebras of D give rise to interesting, non-trivial subgroups of the algebraic groups $G_D := U(V, h)$. If D has an involution of the second kind, we have the splitting field L ($= K$ for $d = 1$), which is an imaginary quadratic extension of the totally real field ℓ ($= k$ for $d = 1$). Suppose then $d = 2$, D quaternionic, say $D = (\ell/k, \sigma, b)$, where $\ell = k(\sqrt{a})$, $e^2 = b \in k^*$; in this case ℓ is a splitting field. Recall the conjugation σ is given by

$$(z_1 + \sqrt{a}z_2)^\sigma = z_1 - \sqrt{a}z_2,$$

so the element $c = diag(\sqrt{a}, -\sqrt{a})$ representing $\sqrt{a} \in \ell$ satisfies $c^\sigma = -c$, while $e^\sigma = e$. Consequently, the relation (A.1) for c is $ec = c^\sigma e = -ce$, and

$$(ec)^2 = (ec)(-ce) = -ec^2e = -ab, \qquad (1.27)$$

so $k(ec) \cong k(\sqrt{-ab})$. If we assume, as we may, that $a > 0, b > 0$ (otherwise replace $-ab$ in what follows by the negative one of a, b), then $L := \sqrt{-ab}$ is an imaginary quadratic extension of k which is a subfield of D. So in all cases (1), 2) and 3)) we have an imaginary quadratic extension of ℓ (cses 1) and 3)) or k (case 2)), L, which is a subfield of D. Then one has

Proposition 1.3.1 *Let L be the following field: case 1), for $\mathbb{Q} \subset k' \subset k$ and $K = k(\sqrt{-\eta})$ set $L = k'(\sqrt{-\eta})$; case 2), $L = k(\sqrt{-ab})$; case 3), $L = \ell(\sqrt{-\eta})$. Then the following are $\mathbb{Q}$-subgroups of G_D, which we denote by G_L:*

$$G_L := U(L^2, h) \subset U(D^2, h) = G_D.$$

Also, by means of the isomorphism $SU(L^2, h) \cong SL_2(k')$, $SL_2(k)$ and $SL_2(\ell)$ in the respective cases 1), 2) and 3) (see [H2], 2.3), we have

1)

$$SL_2(k') \cong \left\{ \begin{pmatrix} \alpha & 2\beta/\sqrt{-\eta} \\ \gamma\sqrt{-\eta}/2 & \delta \end{pmatrix} \Big| \alpha, \beta, \gamma, \delta \in k', \alpha\delta - \gamma\beta = 1 \right\}$$
$$= SU(L^2, h) \subset SU(K^2, h).$$

2)

$$SL_2(k) \cong \left\{ \begin{pmatrix} \alpha & 2\beta/\sqrt{-ab} \\ \gamma\sqrt{-ab}/2 & \delta \end{pmatrix} \Big| \alpha, \beta, \gamma, \delta \in k, \alpha\delta - \gamma\beta = 1 \right\}$$
$$= SU(L^2, h) \subset SU(D^2, h).$$

3)

$$SL_2(\ell) \cong \left\{ \begin{pmatrix} \alpha & 2\beta/\sqrt{-\eta} \\ \gamma\sqrt{-\eta}/2 & \delta \end{pmatrix} \Big| \alpha, \beta, \gamma, \delta \in \ell, \alpha\delta - \gamma\beta = 1 \right\}$$
$$= SU(L^2, h) \subset SU(D^2, h).$$

The corresponding groups and domains are given by

Proposition 1.3.2 *The real groups $G_D(\mathbb{R})$ are the following.*

1) $d = 1$, $G_D(\mathbb{R}) = (U(1,1))^f$, $SG_D(\mathbb{R}) \cong (SL(2,\mathbb{R}))^f$.

2) $d = 2$, $G_D(\mathbb{R}) = SG_D(\mathbb{R}) = (Sp(4,\mathbb{R}))^f$.

3) $d \geq 2$, $G_D(\mathbb{R}) = (U(d,d))^f$, $SG_D(\mathbb{R}) = (SU(d,d))^f$.

Theorem 1.3.3 *The hermitian symmetric domains defined by the $G_D(\mathbb{R})$ as in Proposition 1.3.2*

- *are tube domains, products of irreducible components of the types $\mathrm{I}_{1,1}$, III_2, $\mathrm{I}_{d,d}$ in the cases 1), 2) and 3), respectively.*

- *For any rational parabolic $P \subset G_D$, the corresponding boundary component F such that $P(\mathbb{R}) = N(F)$ is a point.*

Let G_L denote the subgroup $SL_2(k')$, $SL_2(k)$ or $SL_2(\ell)$ in the respective cases of Proposition 1.3.1. The subdomain determined by the subgroups G_L is given by

Proposition 1.3.4 *We have a commutative diagram*

$$\begin{array}{ccc} G_L(\mathbb{R}) & \hookrightarrow & G_D(\mathbb{R}) \\ \downarrow & & \downarrow \\ \mathcal{D}_L & \hookrightarrow & \mathcal{D}_D, \end{array}$$

where $\mathcal{D}_L$ and $\mathcal{D}_D$ are hermitian symmetric domains, and the maps $G(\mathbb{R}) \longrightarrow \mathcal{D}$ are the natural projections. Moreover, the subdomains $\mathcal{D}_L$ are as follows:

1) $d = 1$; We now assume that the extension k/k' (cf. 1.3.1 1)) is Galois. Then, if $\deg_\mathbb{Q} k' = f'$, $f/f' = m$, we have $\mathcal{D}_L \cong (\mathfrak{H})^{f'}$ and $\mathcal{D}_D \cong ((\mathfrak{H})^m)^{f'}$ and the embedding $\mathcal{D}_L \subset \mathcal{D}_D$ is given by $\mathfrak{H} \hookrightarrow (\mathfrak{H})^m$ diagonally, and the product of this f' times.

2) $d = 2$; $\mathcal{D}_L \cong \begin{pmatrix} \tau_1 & 0 \\ 0 & b^{\zeta_1}\tau_1 \end{pmatrix} \times \cdots \times \begin{pmatrix} \tau_1 & 0 \\ 0 & b^{\zeta_f}\tau_1 \end{pmatrix}$, where $\zeta_i : k \longrightarrow \mathbb{R}$ denote the distinct real embeddings of k.

3) $d \geq 2$; $\mathcal{D}_L \cong \begin{pmatrix} \tau_1 & & 0 \\ & \ddots & \\ 0 & & \tau_d \end{pmatrix}^f$.

1.3.2 Arithmetic groups

As in section A.5.2, we introduce arithmetic groups in terms of orders of D. Let $\Delta \subset D$ be a fixed maximal order, and set

$$\Gamma_\Delta := G_D(K) \cap M_2(\Delta). \tag{1.28}$$

If we view $\Delta^2 \subset D^2$ as a lattice, then this is also described as

$$\Gamma_\Delta = \left\{ g \in G_D(K) \big| g(\Delta^2) \subset \Delta^2 \right\}.$$

From the second description it is clear that Γ_Δ is an *arithmetic* subgroup. We will denote the quotient of the domain $\mathcal{D}_D$ by this arithmetic subgroup by

$$X_{\Gamma_\Delta} = X_\Delta = \Gamma_\Delta \backslash \mathcal{D}_D. \tag{1.29}$$

Now assume $d \geq 2$ and recall the subfield $L \subset D$ and subgroups $G_L \subset G_D$ of Proposition 1.3.1. By Proposition 1.3.4 these give rise to subdomains of the domain $\mathcal{D}_D$. Consider, in G_L, the discrete group $\Gamma_{\mathcal{O}_L} \subset G_L(K)$.

Lemma 1.3.5 *We have for any maximal order* $\Delta \subset D$,

$$\Gamma_\Delta \cap G_L = \Gamma_{\mathcal{O}_L}.$$

Proof: This follows from the definitions and the fact that $\Delta \cap L = \mathcal{O}_L$.   □

Let M_L denote the arithmetic quotient

$$M_L = \Gamma_{\mathcal{O}_L} \backslash \mathcal{D}_L. \tag{1.30}$$

It follows from Proposition 1.3.4 that we have a commutative diagram

$$\begin{array}{ccc} \mathcal{D}_L & \hookrightarrow & \mathcal{D}_D \\ \downarrow & & \downarrow \\ M_L & \longrightarrow & X_{\Gamma_\Delta}. \end{array} \tag{1.31}$$

We now describe the moduli interpretation of the arithmetic quotient X_{Γ_Δ}. The data $(D, \Phi, *)$ will be given in our cases as follows. D is our central simple division algebra over K with a $K|k$-involution, and the representation $\Phi : D \longrightarrow M_N(\mathbb{C})$ is obtained by base change from the natural operation of D on D^2 by right multiplication. Explicitly,

$$\Phi : D \longrightarrow End_D(D^2, D^2) \otimes_\mathbb{Q} \mathbb{R} \;\cong\; M_2(D) \otimes_\mathbb{Q} \mathbb{R} \tag{1.32}$$
$$\cong\; M_2(D \otimes_\mathbb{Q} \mathbb{R}) \cong M_2(\mathbb{R}^N) \cong M_N(\mathbb{C}),$$

where $N = \dim_\mathbb{Q} D = 2f,\ 4f,\ 2d^2 f$ in the cases 1), 2) and 3), respectively. The involution $*$ on D will be our involution, which we still denote by $x \mapsto \bar{x}$. Then a $^-$-skew hermitian matrix $T \in M_2(D)$ will be one such that $T = -T^*$, where $(t_{ij})^* = (\bar{t}_{ji})$, the canonical involution on $M_2(D)$ induced by the involution on D. Note that for any $c \in D^*$ such that $c = -\bar{c}$, the matrix $T = cH$ (H our hyperbolic matrix $\left(\begin{smallmatrix} 0 & 1 \\ 1 & 0 \end{smallmatrix} \right)$) has this property. To be more specific, then, we set

1) $d = 1$: $T = \sqrt{-\eta} H = \left(\begin{array}{cc} 0 & \sqrt{-\eta} \\ \sqrt{-\eta} & 0 \end{array} \right).$

2) $d = 2$: $T_a = \sqrt{a} H = \left(\begin{array}{cc} 0 & \sqrt{a} \\ \sqrt{a} & 0 \end{array} \right)$ or $T_b = eH = \left(\begin{smallmatrix} 0 & e \\ e & 0 \end{smallmatrix} \right)$, where $e = \left(\begin{smallmatrix} 0 & 1 \\ b & 0 \end{smallmatrix} \right).$

3) $d \geq 2$: $T = \sqrt{-\eta} H = \left(\begin{array}{cc} 0 & \sqrt{-\eta} \\ \sqrt{-\eta} & 0 \end{array} \right).$

$$\tag{1.33}$$

Since two such forms T are equivalent when they are scalar multiples of one another, assuming T of the form in (1.33) is no real restriction. Finally the lattice $\mathcal{M}$ will be $\Delta^2 \subset D^2$. Then $D^2 \otimes_{\mathbb{Q}} \mathbb{R} \cong \mathbb{C}^N$, N as above, and for "suitable" vectors $x_1, x_2 \in D^2$, the lattice

$$\Lambda_x = \{\Phi(a_1)x_1 + \Phi(a_2)x_2 \,|\, (a_1, a_2) \in \Delta^2\} \tag{1.34}$$

gives rise to an abelian variety $A_x = \mathbb{C}^N / \Lambda_x$.

Theorem 1.3.6 (Corollary 1.2.12) *The arithmetic quotient X_{Γ_Δ} is the moduli space of isomorphism classes of abelian varieties determined by the PEL-data:*

$$\{(D, \Phi, {}^-), \ (D^2, T, \Delta^2)\},$$

where Φ is given in (1.32), T in (1.33).

The corresponding classes of abelian varieties can be described as follows:

1) $d = 1$. Here we have *two* families, relating from the isomorphism of Proposition 2.3 of [H2]. The first, for $D = k$, $D^2 = k^2$ yields $D_{\mathbb{R}}^2 \cong \mathbb{C}^f$, and we have abelian varieties of dimension f with real multiplication by k. Secondly, for $D = K$, $D^2 = K^2$, we have abelian varieties of dimension $2f$ with complex mulitplication by K, with signature $(1,1)$, that is, for each eigenvalue χ of the differential of the action, also $\overline{\chi}$ occurs. If $K = k(\sqrt{-\eta})$, then setting $K' = \mathbb{Q}(\sqrt{-\eta})$ we have $k \otimes_{\mathbb{Q}} K' \cong K$, hence $k^2 \otimes_{\mathbb{Q}} K' \cong K^2$ and $k_{\mathbb{R}}^2 \otimes K'_{\mathbb{R}} \cong K_{\mathbb{R}}^2$, giving the relation between the $\mathbb{Q}$-vector spaces and their real points. Moreover, $\mathcal{O}_k \otimes_{\mathbb{Z}} \mathcal{O}_{K'} \cong \mathcal{O}_K$, and if

$$\Lambda_{x,k} = \{\sum_1^2 \Phi(a_i)x_i \,|\, (a_1, a_2) \in \mathcal{O}_k^2\} \tag{1.35}$$

is a lattice giving an abelian variety with mulitiplication by k,

$$A_x := \mathbb{C}^f / \Lambda_{x,k},$$

then $\Lambda_{x,k} \otimes_{\mathbb{Z}} \mathcal{O}_{K'} = \Lambda_{x,K}$ is a lattice in $\mathbb{C}^{2f}$, and determines an abelian variety

$$A'_x := \mathbb{C}^{2f} / \Lambda_{x,K}. \tag{1.36}$$

From the fact that $\Gamma_{\mathcal{O}_K}$ is a subgroup of finite index in $\Gamma_{\mathcal{O}_k}$, we have a finite cover

$$\Gamma_{\mathcal{O}_K} \backslash \mathfrak{H}^f \longrightarrow \Gamma_{\mathcal{O}_k} \backslash \mathfrak{H}^f. \tag{1.37}$$

Remark 1.3.7 It turns out that this case is one of the exceptions of Theorem 5 in [Sh2], denoted case d) there. The actual endomorphism ring of the generic member of the family is larger than K:

Theorem 1.3.8 ([Sh2], Prop. 18) *The endomorphism ring E of the generic element of the family (1.37) is a totally indefinite quaternion algebra over k, having K as a quadratic subfield.*

In our situation, the totally indefinite quaternion algebra E over k is constructed as the cyclic algebra $E = (K/k, \sigma, \lambda)$, where $\lambda = -u^{-1}v$, if the matrix T of (1.33) is diagonalized $T = \begin{pmatrix} u & 0 \\ 0 & v \end{pmatrix}$. So in our case we have $\lambda = 1$ and hence the algebra E is split; the corresponding abelian variety is isogenous to a product of two copies of a simple abelian variety B with real multiplication by k, as has been described already above. The conclusion follows from our choice of T, i.e., of hyperbolic form. It would seem one gets more interesting quaternion algebras by choosing different hermitian forms (which, by the way, will also lead to other polarizations).

2) $d = 2$. $D = (\ell/k, \sigma, b) = (a, b)$ is a totally indefinite quaternion algebra, cental simple over k, with canonical involution. We have $D_\nu \cong M_2(\mathbb{R})$, and $D \otimes_{\mathbb{Q}} \mathbb{R} \cong \mathbb{R}^4$, while $M_2(D) \otimes_{\mathbb{Q}} \mathbb{R} \cong M_2(\mathbb{R}^4) \cong M_4(\mathbb{C})$. Let $\Delta \subset D$ be a maximal order, $\Gamma_\Delta \subset G_D$ the corresponding arithmetic group. Two vectors $x_1, x_2 \in D^2$ arising from a point in the domain $\mathbb{S}_2$ (Siegel space of degree 2) determine a lattice Λ_x as in (1.35), with $(a_1, a_2) \in \Delta^2$, and $A_x = \mathbb{C}^{4f}/\Lambda_x$ is the corresponding abelian variety.

3) $d \geq 2$. In this case D is the cyclic algebra of degree d over K, and the abelian varieties are of dimension $2d^2 f$.

1.3.3 Modular subvarieties

Now considering also the modular subvarieties, one finds their moduli interpretation to be the following:

Proposition 1.3.9 *In case 2), the abelian varieties parameterized by the modular subvariety M_L are isogenous to products of two abelian varieties of dimensions $2f$ with complex multiplication by L.*

Proposition 1.3.10 *In case 3), the abelian varieties parameterized by the modular subvariety M_L are isogenous to the product of d abelian varieties of dimension $2df$ with complex multiplication by the field L.*

Of course, there are also other modular subvarieties, not considered in the discussion above, for example, corresponding to splittings $A_x \cong A_{x_1} \times A_{x_1}$, i.e., for which both factors coincide, and for which A_{x_1} has multiplication by D. In fact, consider the group:

$$N = \left\{ g \in G \,\middle|\, g = \begin{pmatrix} a & 0 \\ 0 & d \end{pmatrix} \right\}; \tag{1.38}$$

we have $a\bar{d} = 1$ or $a = (\bar{d})^{-1}$, hence $N(\mathbb{R}) \cong SU(p, q)$, $p + q = d$, and N is an anisotropic form. For $d = 2$, we have $N(\mathbb{R}) \cong SU(1, 1)$, and this is just the group, anisotropic over $\mathbb{Q}$, which describes abelian surfaces A with multiplication by the quaternion algebra D. For $\Gamma_N \subset N(\mathbb{Q})$ arithmetic, the compact curve $\Gamma_N \backslash \mathcal{D}_N$ is known in that case as a *Shimura curve*.

Consider also, in each of the G_L or SG_L, the subgroup of Galois-invariant elements,

$$SG_{min} := (SG_L)^{Gal(\overline{\mathbb{Q}}|\mathbb{Q})}$$

By means of the isomorphisms $SG_L \cong SL_2(k')$, $SL_2(k)$ and $SL_2(\ell)$ in each of the respective cases, we see that $SG_{min} \subset SG_L$ is diagonally embedded, $G_{min} \cong SL_2(\mathbb{Q})$. Given an arithmetic group $\Gamma_\Delta \subset G_D(\mathbb{Q})$, the intersection $S\Gamma_{min} := SG_{min} \cap \Gamma_\Delta$ is an arithmetic subgroup, hence commensurable with $SL_2(\mathbb{Z})$. In particular, $S\Gamma_{min} \backslash \mathbb{S}_1$ is a moduli space of elliptic curves with some level structure. Let $\sigma_1 = id, \ldots, \sigma_s$ be the s embeddings of k', k or ℓ into $\mathbb{R}$. Then $G_L(\mathbb{R}) \cong \prod G_{\sigma_i}$, and each $G_{\sigma_i} \cong SL_2(\mathbb{R})$. Hence we get a subdomain $\mathbb{S}_1 \hookrightarrow \mathcal{D}_L \hookrightarrow \mathcal{D}_D$, and the image of the first inclusion is invariant under the natural Galois action. Let M_{min} denote the corresponding modular curve. From the above one easily deduces the following moduli interpretation of M_{min}.

Proposition 1.3.11 *The moduli interpretation of M_{min} is as follows. For $x \in M_L$, let A_x isogenous to $A_{x_1} \times A_{x_1}$ (respectively A_x isogenous to $A_{x_1} \times \cdots \times A_{x_1}$ (d times)) be the splitting of Proposition 1.3.9 (respectively 1.3.10). Then if $x \in M_{min}$, A_{x_1} is isogenous to $E_\tau \times E_\tau$, 2f times, for some elliptic curve E_τ.*

Chapter 2

Arithmetic quotients

In this chapter we will study several types of arithmetic quotients from a more general point of view. In each case we are interested in some of the following information on the quotients:

- The moduli interpretation.

- Degenerations of the moduli and singularities of the quotients.

- Modular subvarieties and their moduli interpretation.

- Level structures and the coverings they define.

It is the existence of certain rational subgroups of the original algebraic group which give rise to modular subvarieties. For simplicity we consider here only the cases we refer to as "split over $\mathbb{R}$", in which a maximal k-split torus is also a maximal $\mathbb{R}$-split one. The domains which will be occuring are $\mathbf{III_n}$ (Siegel space), $\mathbf{I_{p,1}}$ (complex ball) and $\mathbf{IV_n}$. The natural moduli interpretation occuring in the first two cases is that of abelian varieties with real (respectively complex) multiplication, whereas in the third case, a natural moduli interpretation is in terms of K3-surfaces (for $n \leq 19$).

An interesting question which occurs in this situation is the following. Given an arithmetic quotient X_Γ, and letting X_Γ^* and $\overline{X}_\Gamma$ denote the Satake and a toroidal compactification, respectively, is the structure of locally symmetric space on $\overline{X}_\Gamma$ unique? This is the question which has been studied in a concrete example in [J], and it turns out that the following occurs. There are domains $\mathcal{D}$, $\mathcal{D}'$, not isomorphic, and arithmetic groups Γ, Γ' such that

$$\overline{X} := \overline{X}_\Gamma = \overline{X}_{\Gamma'}.$$

The Zariski open subsets $X_\Gamma \subset \overline{X}_\Gamma$ and $X_{\Gamma'} \subset \overline{X}$ are different, and as these complements $\Delta = \overline{X} - X_\Gamma$, $\Delta' = \overline{X} - X_{\Gamma'}$ are normal crossings divisors, this gives rise to interesting geometric configurations. This occurs in particular for normal subgroups $\Gamma' \subset \Gamma$ of finite index, which give rise to coverings $X_{\Gamma'} \longrightarrow X_\Gamma$ of the arithmetic quotients. This is what makes its presence in the examples of [J], and we sketch this in the last section of this chapter.

2.1 Siegel modular varieties

2.1.1 The groups

This is the best known and easiest case. We consider only the absolutely simple split over $\mathbb{R}$ case here. An arithmetic group $\Gamma \subset Sp(2n, \mathbb{R})$ is in this case commensurable with $Sp(2n, \mathbb{Z})$, the Siegel modular group. The arithmetic quotient X_Γ is the moduli space of n-dimensional polarized abelian varieties (this is Type I in Shimura's notation of Proposition 1.1.1, with $k = \mathbb{Q}$), with some level structure. The principally polarized case without level structure is given by, as is well-known, the Siegel modular group itself. The moduli interpretation of these quotients is the simplest possible, corresponding to Shimura's Theorem 1.1.3 with principal polarization and generic endomorphism ring isomorphic to $\mathbb{Z}$.

Theorem 2.1.1 *The quotient $Sp(2n, \mathbb{Z})\backslash \mathcal{D}$, with $\mathcal{D}$ the domain of type III_n, is a coarse moduli space of isomorphism classes of principally polarized abelian varieties of dimension n.*

The level structures are precisely as defined in Definition 1.2.7, which we repeat here for the convenience of the reader in the notions to be used from now on.

Definition 2.1.2 A *full level N stucture* on a complex abelian variety A is a symplectic isomorphism

$$\mathrm{Ker}(m_N) \cong (\mathbb{Z}/N\mathbb{Z})^{2n},$$

where $m_N : A \longrightarrow A$, $m_N(z) = Nz$ is the multiplication by N map, which *lifts* to a symplectic isomorphism

$$H_1(A, \mathbb{Z}) \cong \mathbb{Z}^{2n} \bmod(N).$$

Again, the equivalence of two such full level N structures is given in terms of the *principal congruence subgroup*, $\Gamma(N)$, defined as the kernel in the exact sequence (see (1.23)):

$$1 \longrightarrow \Gamma(N) \longrightarrow Sp(2n, \mathbb{Z}) \longrightarrow Sp(2n, \mathbb{Z}/N\mathbb{Z}) \longrightarrow 1. \tag{2.1}$$

We already discussed in section 1.2.4 that this is indeed a special case of level structure as defined by Shimura (Definition 1.1.5). From the Definition 2.1.2 it is clear that two such choices differ by an element $\gamma \in Sp(2n, \mathbb{Z}/N\mathbb{Z})$, so the following statement about moduli spaces follows immediately from Theorem 1.2.10 (if $N \geq 3$) and Corollary 1.2.12 otherwise.

Theorem 2.1.3 *The quotient $X_{\Gamma(N)} := \Gamma(N)\backslash \mathcal{D}$, with $\mathcal{D}$ the Siegel space of degree n as in 2.1.1, is a coarse moduli space of isomorphism classes of principally polarized complex abelian n-folds, together with a full level N structure, i.e., there is an isomorphism $X_{\Gamma(N)} \cong \mathcal{A}^*_{n,N}(\mathbb{C})$. For $N \geq 3$, this is even a fine moduli space, i.e., there exists a universal object $\mathcal{X}_N \longrightarrow X_{\Gamma(N)}$ whose fibres are these abelian n-folds.*

Proof: The first statement is a special case of Corollary 1.2.12, while the second follows from Theorem 1.2.10 and the fact that for $N \geq 3$, the group $\Gamma(N)$ is torsion free.   □

2.1.2   Compactifications

The Satake compactification X_Γ^* is also easy to describe, at least if Γ is torsion free. It is well-known that the number of boundary components for $Sp(2n, \mathbb{Z})$ is $\nu_r(Sp(2n, \mathbb{Z})) = 1$ for $r = 0, \ldots, n-1$, which is easily seen, since there is, for each r, up to $Sp(2n, \mathbb{Z})$-equivalence, only a single isotropic subspace in each dimension. These form a single flag, so the cusp number is $c(Sp(2n, \mathbb{Z})) = 1$. From this it now follows easily that

Lemma 2.1.4 *The number of boundary components $\nu_r(\Gamma(N))$ is given by the following formula:*

$$\nu_r(\Gamma(N)) = \#\{(n-r)\text{-dimensional isotropic subspaces } W \subset (\mathbb{Z}/N\mathbb{Z})^{2n}\},$$

where the term isotropic means with respect to the natural symplectic form on $(\mathbb{Z}/N\mathbb{Z})^{2n}$.

The degenerations of abelian varieties corresponding to the boundary components are also relatively well-understood. Let $z \in F_r$, where F_r is a rank r boundary component. The z determines a principally polarized abelian r-fold A^r, and the degeneration is given by compactifying a quasi-abelian variety

$$1 \longrightarrow (\mathbb{C}^*)^{n-r} \longrightarrow B \longrightarrow A^r \longrightarrow 1. \tag{2.2}$$

To describe *exactly* degenerate fibres, one must pass to a toroidal compactification $\overline{X}_\Gamma$, and the degeneration will then *depend on the compactification used*.

Throughout this book, the notation $\overline{X}_\Gamma$ for a Siegel modular variety X_Γ will denote the Igusa desingularization [Ig]. For $r = n - 1$, there are divisors on $\overline{X}_\Gamma$ which are families of abelian r-folds along the rank r boundary component, the number of which is given by $\mu_{n-1}(\Gamma(N))$ as above.

2.1.3   Modular subvarieties

Here we consider, for $1 \leq r < n$, the subgroup (with respect to a basis

$$(e_1, f_1, \ldots, e_n, f_n)$$

instead of the usual $(e_1, \ldots, e_n, f_1, \ldots, f_n)$)

$$
\begin{aligned}
N = N_{r,n-r} &= \left\{ \begin{pmatrix} A & 0 \\ 0 & B \end{pmatrix} \in Sp(2n, \mathbb{R}) \middle| A \in M_{2r}(\mathbb{R}),\ B \in M_{2n-2r}(\mathbb{R}) \right\} \\
&\cong Sp(2r, \mathbb{R}) \times Sp(2n - 2r, \mathbb{R});
\end{aligned} \tag{2.3}
$$

the corresponding modular subgroup is $\Gamma_N = \Gamma \cap N_{r,n-r}$ of Γ. Let $\mathcal{D}_N$ be the symmetric subdomain defined by N; then

$$\mathcal{D}_N \cong \mathbb{S}_r \times \mathbb{S}_{n-r}, \qquad (2.4)$$

where we are denoting by $\mathbb{S}_k$ the Siegel space of degree k, i.e., the hermitian symmetric space of type III_k (in the usual unbounded realization). There are finitely many Γ-equivalence classes of subgroups of $Sp(2n, \mathbb{Z})$ conjugate to Γ_N. Let $X_{\Gamma_N} \subset X_\Gamma$ denote the subvariety coming from one of these subgroups.

Lemma 2.1.5 *The modular subvariety $X_{\Gamma_N} \subset X_\Gamma$ is a moduli space of principally polarized abelian n-folds A (with level structure determined by Γ), which split $A \cong A_1 \times A_2$, where $dim(A_1)=r$ and $dim(A_2)=n-r$.*

Proof: This is well-known, and is clear, as $z \in \mathcal{D}_N$ means that z has the form
$z = \begin{pmatrix} z_1 & 0 \\ 0 & z_2 \end{pmatrix}$ with $z_1 \in \mathbb{S}_r$, $z_2 \in \mathbb{S}_{n-r}$. $\qquad \square$

Now suppose that for a boundary variety $Y \subset X_\Gamma^*$, $Y \cong \Gamma_r \backslash \mathbb{S}_r$ and $Y \subset X_{\Gamma_N}^*$. Then Y corresponds to degenerations which have an r-dimensional abelian variety component and an $(n-r)$-dimensional $\mathbb{C}^*$-factor, in the notation of equation (2.2).

Similarly as in Lemma 2.1.4 we can describe the class numbers of modular subvarieties for $\Gamma = \Gamma(N)$ in terms of finite geometry.

Lemma 2.1.6 $\mu_\nu(\Gamma(N)) = \# \left\{ \begin{array}{l} \text{non-degenerate pairs } \{\delta, \delta^\perp\} \subset \\ (\mathbb{Z}/N\mathbb{Z})^{2n}, \text{ rank of } \delta = 2n - 2\nu, \\ \text{rank } \delta^\perp = 2\nu. \end{array} \right\}.$

Proof: This follows from the fact that $\mu_\nu(Sp(2n, \mathbb{Z})) = 1$ and the exact sequence defining $\Gamma(N)$. $\qquad \square$

2.1.4 Commensurable subgroups

Let $(A, \mathcal{C})$ be a polarized abelian variety, where we do not assume $\mathcal{C}$ to be principal. The Riemann form associated to $\mathcal{C}$ as in (1.5) can be represented by a non-degenerate, integral, skew symmetric matrix T as in (1.9) and (1.11). (In the case at hand, the involution * is just transpose, as there is no conjugation). It is well-known that T is then conjugate over $\mathbb{Z}$ to a unique

$$\Delta = \begin{pmatrix} 0 & \delta \\ -\delta & 0 \end{pmatrix}, \quad \delta = \begin{pmatrix} \delta_1 & & \\ & \ddots & \\ & & \delta_n \end{pmatrix}. \qquad (2.5)$$

Definition 2.1.7 Let $(A, \mathcal{C})$ be an abelian variety with polarization $\mathcal{C}$, and T the skew symmetric matrix of (1.11), conjugate to Δ as in (2.5). Then one says that A has a $(\delta_1, \ldots, \delta_n)$-polarization.

Then two such abelian varieties A_{τ_1}, A_{τ_2} will be isomorphic, if and only if there is some $\gamma \in Sp_\Delta(2n, \mathbb{Z})$ with $\gamma(\tau_1) = \tau_2$. Here Sp_Δ is the *paramodular group*

$$
\begin{aligned}
Sp_\Delta(2n, \mathbb{Q}) &= \{g \in GL(2n, \mathbb{Q}) \big| g\Delta^t g = \Delta\}, \\
Sp_\Delta(2n, \mathbb{Z}) &= Sp_\Delta(2n, \mathbb{Q}) \cap GL(2n, \mathbb{Z}).
\end{aligned}
\tag{2.6}
$$

This group acts on Siegel space $\mathbb{S}_n$ by

$$
M : \tau \mapsto (A\tau + B\delta)(C\tau + D\delta)^{-1}\delta, \quad M = \begin{pmatrix} A & B \\ C & D \end{pmatrix}, \tag{2.7}
$$

and by Theorem 1.1.3 we have

Proposition 2.1.8 $X_{(\delta_1,\dots,\delta_n)} := PSp_\Delta(2n, \mathbb{Z})\backslash\mathbb{S}_n$ *is a moduli space of pairs* (A, C) *of abelian n-folds with a* $(\delta_1, \dots, \delta_n)$*-polarization.*

Note that the arithmetic group (2.6) is a subgroup of $Sp_\Delta(2n, \mathbb{Q})$, not of the usual symplectic group $Sp(2n, \mathbb{Q})$. It is, however, *conjugate (over $\mathbb{Q}$) to one,* just as $Sp_\Delta(2n, \mathbb{Q})$ is conjugate to $Sp(2n, \mathbb{Q})$:

$$
Sp_\Delta(2n, \mathbb{Q}) = \begin{pmatrix} 1 & 0 \\ 0 & \delta \end{pmatrix} (Sp(2n, \mathbb{Q})) \begin{pmatrix} 1 & 0 \\ 0 & \delta^{-1} \end{pmatrix}. \tag{2.8}
$$

Note that the condition (2.6) on a matrix $M = \begin{pmatrix} A & B \\ C & D \end{pmatrix}$ to be in Sp_Δ can also be expressed as:

$$
A\delta^t B = B\delta^t A, \quad C\delta^t D = D\delta^t C, \quad A\delta^t - B\delta^t C = \delta. \tag{2.9}
$$

This clearly imposes certain divisibility properties on M, for example, if $n = 2$, $\delta = \begin{pmatrix} 1 & 0 \\ 0 & p \end{pmatrix}$, then ([HKW], 1.10):

$$
M \equiv \begin{pmatrix} * & * & * & * \\ 0 & a & 0 & b \\ * & * & * & * \\ 0 & c & 0 & d \end{pmatrix} \mod(p), \quad ad - bc \equiv 1 \mod(p).
$$

On the other hand, conjugating with $\mathbf{D} = \begin{pmatrix} 1 & 0 \\ 0 & \delta \end{pmatrix}$ in (2.8) clearly introduces denominators in certain of the matrix entries, for example in the above considered case ([HKW], 1.14):

$$
\Gamma_\Delta^1 := \mathbf{D} Sp_\Delta(4, \mathbb{Z})\mathbf{D}^{-1} = \left\{ g \in Sp(4, \mathbb{Q}), g \in \begin{pmatrix} \mathbb{Z} & \mathbb{Z} & \mathbb{Z} & p\mathbb{Z} \\ p\mathbb{Z} & \mathbb{Z} & p\mathbb{Z} & p\mathbb{Z} \\ \mathbb{Z} & \mathbb{Z} & \mathbb{Z} & p\mathbb{Z} \\ \mathbb{Z} & \frac{1}{p}\mathbb{Z} & \mathbb{Z} & \mathbb{Z} \end{pmatrix} \right\}.
\tag{2.10}
$$

This explicitly displays Γ^1_Δ as a commensurable subgroup of $Sp(4,\mathbb{Q})$, and the same can be done in the general case. One has in this case a natural level structure, given as follows. Consider the lattice Λ of (1.8) on which the Riemann form is integer valued by (1.11). The *dual lattice* is defined as

$$\Lambda^\vee = \{\lambda \in \Lambda \otimes \mathbb{R} \,\big|\, E_C(x,\lambda) \in \mathbb{Z}, \,\forall_{x \in \Lambda}\}. \qquad (2.11)$$

Then $\Lambda^\vee/\Lambda$ is finite and one sees easily that in fact

Lemma 2.1.9 $\Lambda^\vee/\Lambda \cong (\mathbb{Z}/(\delta_1))^2 \oplus \cdots \oplus (\mathbb{Z}/(\delta_n))^2$.

Proof: From the expression (1.9) for E_C we see that for a symplectic basis $e_1,\ldots,e_n, f_1 \ldots, f_n$ of Λ with $E_C(e_i,f_j) = \delta_{ij}\delta_i$, it holds that $E_C(\frac{1}{\delta_i}e_i, f_i) = \frac{1}{\delta_i}E_C(e_i,f_i) = 1 \in \mathbb{Z}$, and $E_C(e_i, \frac{1}{\delta_i}f_i) = \frac{1}{\delta_i}E_C(e_i,f_i) = 1 \in \mathbb{Z}$. $\qquad\square$

Definition 2.1.10 ([HKW] for $n = 2$) A *canonical level structure* on a $(\delta_1,\ldots,\delta_n)$-polarized abelian n-fold is a symplectic isomorphism

$$\Lambda^\vee/\Lambda \cong (\mathbb{Z}/(\delta_1))^2 \oplus \cdots \oplus (\mathbb{Z}/(\delta_n))^2,$$

where the right hand side has the symplectic form given by $J = \begin{pmatrix} 0 & 1 \\ -1 & 0 \end{pmatrix}$.

It follows that the subgroup in Γ^1_Δ preserving the level structure of 2.1.10 is given by:

$$\begin{aligned} \Gamma_\Delta &= \{g \in \Gamma^1_\Delta \,\big|\, g \text{ induces the identity on } \Lambda^\vee/\Lambda\} \qquad (2.12) \\ &= \{g \in \Gamma^1_\Delta \,\big|\, gv \equiv v \bmod(\Lambda), \,\forall_{v \in \Lambda^\vee}\}. \end{aligned}$$

Of course the second expression only makes sense once we have fixed a basis, allowing $g \in \Gamma_\Delta$ to act on the vector space $\Lambda \otimes \mathbb{Q}$. It is clear that (2.12) introduces certain divisibilities in the matrix entries of Γ^1_Δ, and one could hope that $\Gamma_\Delta \subset Sp(2n,\mathbb{Z})$. In the $n = 2$, $(1,p)$ case considered above this is in fact true:

Lemma 2.1.11 ([HKW], 1.14) *If* $n = 2$, $\delta = \begin{pmatrix} 1 & 0 \\ 0 & p \end{pmatrix}$, *then* $\Lambda \cong \mathbb{Z}^4$, $\Lambda^\vee \cong \mathbb{Z} \oplus \frac{1}{p}\mathbb{Z} \oplus \mathbb{Z} \oplus \frac{1}{p}\mathbb{Z}$, *and*

$$\Gamma_\Delta = \left\{ g \in Sp(4,\mathbb{Q}) \,\Big|\, g - 1 \in \begin{pmatrix} \mathbb{Z} & \mathbb{Z} & \mathbb{Z} & p\mathbb{Z} \\ p\mathbb{Z} & p\mathbb{Z} & p\mathbb{Z} & p^2\mathbb{Z} \\ \mathbb{Z} & \mathbb{Z} & \mathbb{Z} & p\mathbb{Z} \\ \mathbb{Z} & \mathbb{Z} & \mathbb{Z} & p\mathbb{Z} \end{pmatrix} \right\} \subset Sp(4,\mathbb{Z}).$$

Corollary 2.1.12 *There is a ramified cover* $X_{\Gamma_\Delta} \longrightarrow X_{Sp(4,\mathbb{Z})}$ *of degree* $\frac{1}{2}p(p^4 - 1) = |[Sp(4,\mathbb{Z}) : \Gamma_\Delta]|$.

Now as for level structures, one has, by the canonical level structure, something like "half a full level p structure" (see Definition 2.1.2). Let q be another prime, relatively prime to p. Let us denote by Γ_J the Siegel modular group $Sp(4, \mathbb{Z})$. Then we have

Proposition 2.1.13 *The covering of Corollary 2.1.12 lifts to level q-structures, i.e., the following diagram commutes:*

$$
\begin{array}{ccc}
X_{\Gamma_\Delta(q)} & \longrightarrow & X_{\Gamma_J(q)} \\
\downarrow & & \downarrow \\
X_{\Gamma_\Delta} & \longrightarrow & X_{\Gamma_J}
\end{array} \quad ,
$$

where $\Gamma_\Delta(q) = \{ g \in \Gamma_\Delta \big| g \equiv 1 \, mod(q) \}$.

Proof: The vertical covers are defined by assigning different level q-structures to the set of q-torsion points, while the horizontal arrows are maps labeling only a part of the p-torsion points. By assumption p and q are relatively prime, so these two labelings commute. Note that the horizontal maps are *not* Galois covers, while the vertical arrows are, with group $PSp(4, \mathbb{Z}/q\mathbb{Z})$. However, Γ_Δ and Γ_J have a common normal subgroup of finite index in both, call it $\Gamma_{\Delta,J}$, and in the diagram

$$
\begin{array}{ccc}
 & X_{\Gamma_{\Delta,J}} & \\
 \alpha \swarrow & & \searrow \beta \\
X_{\Gamma_\Delta} & \longrightarrow & X_{\Gamma_J} ,
\end{array}
$$

(2.13)

both α and β are Galois. Pulling back to level q:

(2.14)

the argument above shows that the top face:

$$
\begin{array}{ccc}
X_{\Gamma_{\Delta,J}(q)} & \longrightarrow & X_{\Gamma_J(q)} \\
\downarrow & & \downarrow \\
X_{\Gamma_{\Delta,J}} & \longrightarrow & X_{\Gamma_J}
\end{array}
$$

(2.15)

is a commutative square of group operations. The commutativity of 2.1.13 is hence a consequence of (2.14).   □

Now let $X_{(1,p)}$ be as in Proposition 2.1.8, and note that $\Gamma_\Delta(q) \subset \Gamma_J(q)$, yielding the following diagram:

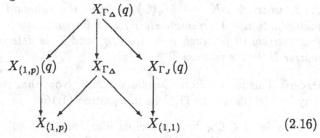

$$\text{(2.16)}$$

Corollary 2.1.14 *There is a correspondence* $X_{(1,p)}(q) \rightrightarrows X_{\Gamma_J}(q)$. $\qquad\square$

2.2 Picard modular varieties

2.2.1 The groups

Let $K|k$ be a totally imaginary quadratic extension of a totally real algebraic number field k, $[k : \mathbb{Q}] = g$. Let V be a K-vector space, of dimension n, and let h be a hermitian form on V. Let $\mathcal{L} \subset V$ be an $\mathcal{O}_K$-lattice and $G_\mathcal{L}$ the corresponding arithmetic subgroup. The hermitian form h is given by a hermitian matrix $H \in M_n(K)$ with $H = H^*$ ($^* = $ conjugate transpose).

Definition 2.2.1 A *Picard modular group* is an arithmetic group $G_\mathcal{L}$ as above, $G_\mathcal{L} \subset G(V, H)$, where the signature of H_ν is $(n-1, 1)$ or $(n, 0)$ for all real primes $\nu \in \Sigma_\infty(k)$ and $\mathcal{L} \subset V$ is an $\mathcal{O}_K$-lattice.

So the group depends on $(K, H, \mathcal{L})$. In the particular case that $k = \mathbb{Q}$ there is only one real embedding, so the condition on the signature can be checked by considering H as a real matrix and diagonalizing. We remark that the classification of hermitian forms on V is roughly given as follows:

Lemma 2.2.2 *Hermitian forms on an n-dimensional K-vector space V are classified by the determinant (see below), and the signatures of H_ν, ν such that, writing $K = k(\sqrt{\delta})$ with $\delta \in k$, then $\delta_\nu < 0$.*

Proof: The determinant $det(h)$ is defined as follows. For a matrix H representing h in some basis, $det(H) \in k$ (since $h(x, x) \in k \; \forall_{x \in V}$). If S is a base change matrix, $H' = SHS^*$, then $det(H') = det(S)det(H)det(S^*)$, and it follows that $det(h)$ is well-defined as an element in $k^\times/N_{K|k}(K^\times)$. Then, considering the trace form q_h, i.e., $q_h : V \longrightarrow k$, $q_h(x) = h(x, x)$, the classification up to isometry is reduced to that of the symmetric bilinear form q_h, and its Witt invariant is determined by K, $det(h)$ and the signature at the mentioned primes (see [Sch], p. 350, 1.6 iv)). $\qquad\square$

Let $\Gamma \subset G_\mathcal{L}$ be a subgroup of finite index and let X_Γ denote the corresponding modular variety.

Proposition 2.2.3 X_Γ *is a moduli space of complex abelian n-folds* $(A, \mathcal{C}, \theta)$ *with a* PEL-*structure of Type* $\{(K, \Phi, -), (T, \mathcal{L})\}$ *in the notations of Theorem 1.1.3, where* $\Phi : K \longrightarrow M_n(\mathbb{C})$ *contains the standard representation* χ_ν *with multiplicity* $n - 1$ *(respectively n) and its conjugate* $\overline{\chi}_\nu$ *with multiplicity one (respectively 0) for each* $\nu = 1, \dots, g$, *and* T *is determined by the hermitian matrix* H *by the relation* $T^{-1} = -iH$.

Proof: This follows from Corollary 1.2.12. Note that the hermitian matrix H_ν may be identified with iT_ν^{-1} as in equation (1.16). $\qquad\qquad\square$

Now let $\mathfrak{a} \in \mathcal{O}_K$ be a principal ideal, say $\mathfrak{a} = (a)$. One has the multiplication map $m_a : V \longrightarrow V, x \mapsto ax$ which maps the lattice $\mathcal{L}$ into itself. This then descends to a map $m_a : A \longrightarrow A$ for any abelian variety in the family of Proposition 2.2.3.

Definition 2.2.4 The *points of order a* are the elements in the kernel of m_a. A *level a-structure* on A is the choice of isomorphism

$$\mathcal{L}/\mathfrak{a}\mathcal{L} \xrightarrow{\sim} Ker(m_a)$$

preserving the natural hermitian forms on both sides, which lifts as in Definition 1.2.7.

Once again, this is a special case of Shimura's PEL-structure. We can also define the *principal congruence subgroup* $\Gamma(\mathfrak{a}) \subset G_\mathcal{L}$ by the following sequence

$$1 \longrightarrow \Gamma(\mathfrak{a}) \longrightarrow G_\mathcal{L} \longrightarrow G(\mathcal{L}/\mathfrak{a}\mathcal{L}, \overline{h}) \longrightarrow 1, \qquad\qquad (2.17)$$

where the group $G(\mathcal{L}/\mathfrak{a}\mathcal{L}, \overline{h})$ is the set of matrices in $GL(\mathcal{L}/\mathfrak{a}\mathcal{L})$ preserving the hermitian form $\overline{h}$ induced on $(\mathcal{L}/\mathfrak{a}\mathcal{L})$ by h on $\mathcal{L}$.

Remark 2.2.5 Suppose $\mathcal{L}$ is free, i.e., $\mathcal{L} \cong (\mathcal{O}_K)^n$. Then $(\mathcal{L}/\mathfrak{a}\mathcal{L}) \cong (\mathcal{O}_K/\mathfrak{a}\mathcal{O}_K)^n$. Let $N(\mathfrak{a})$ denote the $K|k$ norm. Then $\mathcal{O}_K/\mathfrak{a}\mathcal{O}_K$ is a finite ring of order $N(\mathfrak{a})$. If $\mathfrak{a}$ is a maximal ideal, then $\mathcal{O}_K/\mathfrak{a}\mathcal{O}_K$ is actually a field, of order $N(\mathfrak{a})$. In this case the group $G(\mathcal{L}/\mathfrak{a}\mathcal{L}, \overline{h})$ may be viewed as a Chevally group, i.e., of Lie type over a finite field, and will then be a *simple* finite group.

In general, any lattice can be written $\mathcal{L} \cong \mathfrak{b} \oplus \underbrace{\mathcal{O}_K \oplus \cdots \oplus \mathcal{O}_K}_{n-1 \text{ factors}}$. In this case we have $\mathcal{L}/\mathfrak{a}\mathcal{L} \cong \mathfrak{b}/\mathfrak{a}\mathfrak{b} \oplus \mathcal{O}_K/\mathfrak{a}\mathcal{O}_K \oplus \cdots \oplus \mathcal{O}_K/\mathfrak{a}\mathcal{O}_K$. So, for example, if it happens that $\mathfrak{b} = \mathfrak{a}$, then $\mathcal{L}/\mathfrak{a}\mathcal{L} \cong \mathfrak{b}/\mathfrak{b}^2 \oplus (\mathcal{O}_K/\mathfrak{a}\mathcal{O}_K)^{n-1}$.

From the description above we get immediately

Proposition 2.2.6 *Let* $\Gamma(\mathfrak{a})$ *be the principal congruence subgroup of level* $\mathfrak{a}$ *of (2.17). Then the arithmetic quotient* $X_{\Gamma(\mathfrak{a})}$ *is a moduli space of abelian varieties as in Proposition 2.2.3 with a (full) level* $\mathfrak{a}$-*structure.*

Moreover, one sees easily that if $\sqrt{N(\mathfrak{a})} \geq 3$ then $\Gamma(\mathfrak{a})$ is torsion free and $X_{\Gamma(\mathfrak{a})}$ is actually smooth and a fine moduli space.

2.2.2 Compactification

The Satake compactification of X_Γ is affected by adding points to the open variety X_Γ, $X_\Gamma^* = X_\Gamma \dot\cup \text{cusps}$. To determine the number of cusps, let us assume that $\mathcal{L}$ is free and consider a subgroup $\Gamma \subset G_{\mathcal{L}}$ of finite index, where $G_{\mathcal{L}} \cong U(n-1, 1; \mathcal{O}_K)$. The number of cusps is then clearly the number of Γ-equivalence classes of isotropic vectors in V with respect to h. If $k = \mathbb{Q}$, this number has been calculated in [Ze].

Definition 2.2.7 The class number $c_m(K)$ is the number of isometry classes of *definite* unimodular lattices in (K^m, E_m), where E_m is the standard positive definite hermitian form. One sets $c_0(K) := 1$.

Fixing an isotropic vector $v \in V$, one can split off a hyperbolic plane E containing v such that $V = E \perp V_0$, and $(V_0, h_{|V_0})$ is anisotropic (see A.1.2). Then $(\mathcal{L} \cap E) \perp (\mathcal{L} \cap V_0)$ is a decomposition of the lattice $\mathcal{L}$, and the correspondence $v \mapsto E \mapsto \mathcal{L} \cap V_0$ determines a *definite* unimodular lattice for each isotropic $v \in V$. Furthermore, the number of Γ-equivalence classes of v mapping to a given $\mathcal{L} \cap V_0$ is 2^{t-1}, where $t = \#$ prime divisors of the discriminant of K. For $\Gamma = U(n-1, 1; \mathcal{O}_K)$, this results in

Theorem 2.2.8 ([Ze], Theorem 3.9) $c(\Gamma)$, *the number of Γ-cusps for $\Gamma = U(n-1, 1; \mathcal{O}_K)$, is given by*

$$c(\Gamma) = 2^{t-1} c_{n-2}(K),$$

where t is the number of prime divisors of the discriminant of K, and $c_{n-2}(K)$ is the class number of Definition 2.2.7.

Of course, the same argument as in the proof of Lemma 2.1.4 leads to the following.

Corollary 2.2.9 *Let $\mathfrak{a} = (a) \subset \mathcal{O}_K$ be a principal ideal, and $\Gamma(\mathfrak{a}) \subset U(n-1, 1; \mathcal{O}_K)$ the principal congruence subgroup of level $\mathfrak{a}$. Then the number of cusps of $\Gamma(\mathfrak{a})$ is*

$$c(\Gamma(\mathfrak{a})) = c(\Gamma) \cdot \# \text{ isotropic vectors in } (\mathcal{O}_K / \mathfrak{a} \mathcal{O}_K)^n.$$

We now consider toroidal compactifications of X_Γ. It is well-known that in the case of $U(p, 1)$, the five-factor decomposition of a maximal parabolic (see, for example, [SC], III §4.1) takes the form $\mathcal{R} \cong GL(1, \mathbb{R})$, $\mathcal{Z} \cong \mathbb{R}$, $V_P \cong \mathbb{C}^{n-2}$, and $\Gamma \cap P$ defines a lattice in V_P, say $\mathcal{L}_P \subset V_P$. Then $V_P / \mathcal{L}_P$ is an abelian variety, and it, or a quotient of it, if there is torsion in $\Gamma \cap P$, is glued into the cusp by the normal bundle, which in turn is determined by a generator of $\Gamma \cap \mathcal{Z}$, and this yields a smooth compactification (assuming Γ is torsion free). At any rate the compactification divisors are disjoint, and are abelian varieties if Γ is a torsion free subgroup. So summing up we have

Proposition 2.2.10 *Let $\Gamma \subset U(n-1, 1 : \mathcal{O}_K)$ be a torsion free subgroup. Then $\overline{X}_\Gamma - X_\Gamma$ consists of a disjoint union of abelian varieties, all of which have complex multiplication by K.*

2.2.3   Modular subvarieties

Let (V, h) be as above, and consider an *anisotropic* vector $v \in V$; the orthocomplement with respect to h, $v^{\perp} \subset V$, is of dimension $n - 1$, and suppose that $(h_{|v^{\perp}})_{\nu}$ has signature $(n - 2, 1)$ for all real primes ν for which h_{ν} has signature $(n - 1, 1)$. Let N_v be the normalizer of the pair $(v, v^{\perp})$, or, equivalently, of the direct sum decomposition $v \oplus v^{\perp} = V$, in $G(V, h)$. Then, since $(h_{|v^{\perp}})_{\nu}$ has signature $(n - 2, 1)$ for the real primes ν just mentioned, we have

$$N_v(\mathbb{R}) \cong (U(n - 2, 1) \times U(1))^{g_1} \times N_v^c, \tag{2.18}$$

where g_1 is the number of ν for which h_{ν} has signature $(n - 1, 1)$, and N_v^c is compact. Given an $\mathcal{O}_K$ lattice $\mathcal{L} \subset V$, there is a basis $(x_1, \ldots, x_n)$ of V such that $\mathcal{L} = \mathfrak{a}x_1 \oplus \mathcal{O}_K x_2 \oplus \cdots \oplus \mathcal{O}_K x_n$. Assume $\mathfrak{a} = \mathcal{O}_K$, and take one of the x_i which is anisotropic as the v above. Then $\mathcal{L} \cap N_v := N_{\mathcal{L}}$ is a lattice in N_v, and determines a modular subvariety $X_{N_v} \subset X_{\Gamma}$.

Proposition 2.2.11 *The modular subvariety* $X_{N_v} \subset X_{\Gamma}$ *is a moduli space of abelian n-folds with complex multiplication by $\mathcal{O}_K$, which split:*

$$A^n \cong A^{n-1} \times A^1,$$

where A^1 is an elliptic curve with complex multiplication (which is rigid) and A^{n-1} is an abelian $(n-1)$-fold whose modulus determines a unique point of X_{N_v}.

Proof: This is easily seen by writing down the period for a $z \in \mathbb{B}_{n-2} \subset \mathbb{B}_{n-1}$ (where $\mathbb{B}_r$ denotes the r-dimensional complex ball). In equation (1.19), $z_{\nu} = (\tau, z^{(2)}, \ldots, z^{(n-1)})$ with $\tau \in \mathcal{O}_K \cap \mathbb{S}_1$ ($\cong \mathbb{B}_1$) determining the period of A^1, and the $(n - 2)$-tuple $(z^{(2)}, \ldots, z^{(n-1)})$ determines the period of A^{n-1}. $\square$

2.3   Arithmetic quotients of domains of type IV$_n$

2.3.1   The groups

Let V be a k-vector space of dimension n, k a totally real field, and b a symmetric bilinear form on V. Let $G(V, b)$ be the symmetry group, and $G_{\mathbb{Q}} = Res_{k|\mathbb{Q}} G(V, b)$ the $\mathbb{Q}$ group it defines. We assume that $G_{\mathbb{Q}}$ is of hermitian type, so that the signature of b_{ν} is $(n - 2, 2)$ or b_{ν} is definite for all real primes ν. It is easy to see that $G_{\mathbb{Q}}$ is (simple) split over $\mathbb{R}$ if and only if $k = \mathbb{Q}$ and b has Witt index 2. This is the case we consider here. The classification of such forms is well-known; since we require the Witt index to be 2, two such forms are equivalent over $\mathbb{Q}$ $\iff det(b) = det(b')$, where $det(b)$ is to be viewed as an element of $\mathbb{Q}^{\times}/(\mathbb{Q}^{\times})^2$.

Now let $\mathcal{L} \subset V$ be a (maximal) lattice, and let $G_{\mathcal{L}}$ be the arithmetic group it defines, $\Gamma \subset G_{\mathcal{L}}$ a subgroup of finite index. We first remark on the moduli interpretation of the arithmetic quotient X_{Γ}.

Proposition 2.3.1 X_{Γ} *is a moduli space of (pure) Hodge structures of weight 2 on V with $h^{2,0} = 1$ (and $h^{1,1} = dim(X_{\Gamma})$) with respect to the lattice $\mathcal{L} \subset V$ and some level structure, defined by the moduli functor $\mathbf{H}_{\mathcal{L}}$ of section 1.2.2, 5.*

Proof: The symmetry group of such a Hodge structure is of real type $SO(n - 2, 2)$, and $G(V, b)$ is a $\mathbb{Q}$-form in which $G_{\mathcal{L}}$ is an arithmetic subgroup. Since the corresponding "period" (i.e., position of the varying complex subspace $H^{1,1}$ in $H^2_{\mathbb{C}}$) is clearly the same exactly when the two periods differ by an element of $G_{\mathcal{L}}$, while Γ defines a level structure of some kind, the result follows. □

This proposition is often used in the study of polarized K3-surfaces, which have a pure Hodge structure of type (1,19,1). In fact, for each polarization degree (i.e., the number C^2 for the ample divisor C on the K3-surface which gives the projective embedding) $2e$, $e \geq 1$, one has an arithmetic group Γ_e such that the arithmetic quotient X_{Γ_e} is the moduli space of K3-surfaces with the given polarization. Recall the *Picard number* ϱ is the rank of the group of algebraic cycles, i.e., $\varrho = rk_{\mathbb{Z}} H^2(S, \mathbb{Z}) \cap H^{1,1}$. Recall also the functor $\mathbf{K3}_{(e,\Lambda)}$ of section 1.2.2, 4. Then one has the following.

Proposition 2.3.2 *The dimension of the moduli space of K3 surfaces for the moduli functor* $\mathbf{K3}_{(e,\Lambda)}$, *where* Λ *has rank* ϱ, *is 20-ϱ.*

Proof: Recall that for a K3-surface $H^1(S, \Theta) \cong H^1(S, \Omega^1)$, so $H^{1,1}(S)$ may be viewed as the tangent space of the local deformation space, which should be thought of as a varying complex subspace of $H^2(S, \mathbb{C})_{prim}$, while $H^2(S, \mathbb{Z})$ is fixed. Let $\mathcal{A} = H^2(S, \mathbb{Z}) \cap H^{1,1}$ be the lattice of algebraic cycles, $\mathcal{T} = H^2(S, \mathbb{Z}) \cap (H^{2,0}(S) \oplus H^{0,2}(S))$ the lattice of transcendental cycles. We have $rk_{\mathbb{Z}} \mathcal{A} = rk_{\mathbb{Z}}(H^2(S, \mathbb{Z})) - rk_{\mathbb{Z}} \mathcal{T}$ in general and $rk_{\mathbb{Z}} \mathcal{A} = \varrho$ by assumption, so $\varrho = 22 - rk_{\mathbb{Z}} \mathcal{T}$, while the moduli space is defined by the group $G(V', b')$, where $V' = \mathcal{A}^{\perp} \otimes \mathbb{Q}$, since we are requiring $\mathcal{A}$ to be preserved. (Recall that for an algebraic cycle C the integral $\int_C \omega$ over the holomorphic two-form ω vanishes, hence the algebraic cycles contribute nothing to the periods). Thus $G(\mathbb{R}) = SO(20 - \varrho, 2)$, giving rise to a domain of type $\mathbf{IV}_{20-\varrho}$. □

Of course in this particular case, the lattice $\mathcal{L} \subset V$ is very special; it is even and unimodular, and as is well-known, decomposes as

$$\mathcal{L} \cong <-2e> \oplus \mathbf{H}^2 \oplus \mathbf{E}_8{}^2, \tag{2.19}$$

where $\mathbf{H}$ is the two-dimensional hyperbolic lattice and $\mathbf{E}_8$ is the negative of the root lattice of type $\mathbf{E}_8$. Let us remark that the compactifications of these arithmetic quotients has been carried out in the thesis [Sc]. In the rest of this section we will study a particularly interesting family of K3's and discuss different aspects of the theory for this special case.

2.3.2   A four-dimensional family of K3-surfaces

The family of K3-surfaces to be described here is the set of surfaces which are double covers of $\mathbb{P}^2$ branched along the union of six lines. Recall that there is a 19-dimensional family of K3-surfaces which are double covers of the plane branched along a sextic curve; they are smooth as long as the sextic is smooth, and generically have Picard number 1. An arrangement of six lines in $\mathbb{P}^2$ is

a maximally singular sextic; there are 15 intersection points of the six lines
(if they are in general position), and each such gives rise to an A_1-singularity
on the double cover. Resolving the 15 double points introduces 15 exceptional
curves with self intersection number -2, so together with the pullback of the
generic line, this gives 16 independent cycles on the surface: $\varrho = 16$. Hence the
transcendental lattice $\mathcal{T}$ has rank four, so by Proposition 2.3.2, the moduli space
is four-dimensional. Let

$$\Gamma = \{g \in G(\mathcal{T}_{\mathbb{R}}, b)(\; \cong \; SO(4,2)) \big| g(\mathcal{T}) \subset \mathcal{T} \text{ and } g \text{ lifts to an isometry of } \mathcal{L}\},$$
$$(2.20)$$

where b is the intersection form on $H^2(S, \mathbb{Z})$, extended to $\mathbb{R}$, then restricted to
$\mathcal{T}_{\mathbb{R}}$, and $\mathcal{L}$ is the K3 lattice (2.19). This is clearly an arithmetic subgroup, and
by Proposition 2.3.2, the arithmetic quotient $X_\Gamma = \Gamma \backslash \mathcal{D}$ is the four-dimensional
moduli space. We list some of the interesting loci for this family. Let L be the
given arrangement, $L = l_1 \cup \ldots \cup l_6$, and let $t_p :=$ the number of p-fold points
of the arrangement, i.e., the number of points at which p of the lines meet, and
let $\pi : S \longrightarrow \mathbb{P}^2$ denote the (singular) double cover.

2.3.2.1   Three-dimensional loci

1) Suppose there is a conic which is *tangent* to all six lines. Then the inverse image of this quadric is a $\mathbb{P}^1$, which, as is easily checked, has self-
   intersection number $4 - 6 = -2$, so the double cover has 16 exceptional
   cycles, hence $\varrho = 17$. It is in fact easy to see that the surface S is in
   this case a classical Kummer surface , i.e., a quartic surface in $\mathbb{P}^3$ with 16
   nodes which is the Kummer of a principally polarized abelian surface A_S.
   The projection from the node gives the double cover $\pi : S \longrightarrow \mathbb{P}^2$, and the
   tangent conic is the image of the (blown up) node used to project. The
   abelian surface is the Jacobian of a genus two curve (see Lemma B.5.1),
   and this curve is the double cover of the conic, *branched at the six points
   of tangency.*

2) If $t_3 = 1$, $t_2 = 12$, then the three-fold point induces an A_2-singularity
   on the double cover which is resolved by two $\mathbb{P}^1$'s, so there are now $2+12$
   exceptional $\mathbb{P}^1$'s and the hyperplane section. We have the following picture.

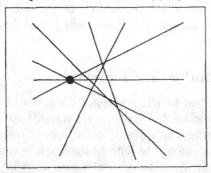

There are in fact three more exceptional $\mathbb{P}^1$'s, which are the inverse image on the double cover
of the three lines which pass
through the triple point and one
of the three double points not lying on a line through the triple
point. It is easy to see that these
three double points are independent parameters of such arrangements, so this defines a three-

dimensional family, so by Proposition 2.3.2, we have $\varrho = 17$ for the generic member of this family.

2.3.2.2 Two-dimensional loci

3) If $t_3 = 2, t_2 = 9$, there are two possibilities. Suppose first that the two three-fold points do *not* lie on a line. Then we have the following picture.

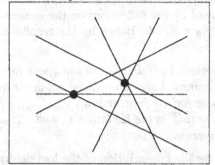

This gives rise to two isolated A_2-singularities. The inverse image of the line joining the two three-fold points is also an exceptional $\mathbb{P}^1$. In this case the generic double cover has $\varrho = 18$.

4) It may also occur that both three-fold points lie on a line, but in this case we also have $\varrho = 18$, i.e., a two-dimensional family.

2.3.2.3 One-dimensional loci

5) If $t_4 = 1$, then the double cover has an *elliptic* singularity over the point, so is not K3. Hence this is a genuine *degeneration* of the K3, i.e., belongs to the boundary of the compactification. It turns out that then a line must also be double, so that the double cover has two components.

6) As a further specialization of 4) it may happen that there are three triple points. Since four of the lines may be choosen to be fixed (for example $x_0 = x_1 = x_2 = 0$, $x_0 - x_1 = 0$), there is only one modulus, given for example by the intersection point of the two variable lines. Here we have $\varrho = 19$.

2.3.2.4 Zero-dimensional loci

7) If three lines are taken, each *double*, then the double cover splits into two copies of $\mathbb{P}^2$. This is the closure of the set of degenerations of type 5).

8) The arrangement is the complete quadrilateral.

It is known that the *Fermat* cover (not the double cover) of this arrangement is Shioda's elliptic modular surface of level 4, $S(4)$, so it follows that the double cover is isogenous to $S(4)$, i.e., a quotient of $S(4)$ by a group isomorphic to $(\mathbb{Z}/2\mathbb{Z})^4$. This is the most special K3-surface in the family and has $\varrho = 20$.

2.3.2.5 Level 2 structure

Now consider, in addition to the above data, a level 2 structure. Geometrically this amounts to fixing an *order* of the six lines. It is the moduli space of K3-surfaces with a fixed isometry of a certain lattice (isomorphic to the Picard lattice of a general member of the family) into the Picard lattice. In [MSY] it is shown by explicit computation that the subgroup $\Gamma(2)$ is the group generated by reflections on the "roots" of $\mathcal{T}$, that is the integral elements of norm -2. Furthermore they show that Γ is generated by the reflections on the elements of norm -2 or -4, and that $\Gamma/\Gamma(2) \cong \Sigma_6 \times \mathbb{Z}/2\mathbb{Z}$. Hence by the results 2.7.1, 2.7.7, 2.8.2 of [MSY] we have

Proposition 2.3.3 *The arithmetic quotient $\Gamma(2)\backslash\mathcal{D}$ is a moduli space for K3-surfaces which are double covers of $\mathbb{P}^2$, branched over an ordered set of six lines. Equivalently, this is a coarse moduli space for the functor $\mathbf{K3}_{(e,\Lambda,e_1,\ldots,e_6)}$, where $e = 1$, $\Lambda = \mathcal{A}$, the orthogonal complement to $\mathcal{T}$ in the K3 lattice $\mathcal{L}$, and $e_1,\ldots,e_6$ is an ordered set of six cycles which represent the six lines.*

We refer the reader to [MSY] for a detailed description of the loci described above, of the periods and of the corresponding Picard-Fuchs equations (and much more). In Table 2.1 we give a description of the loci, exhibiting the dual graph of the six lines (i.e., the graph containing a vertex for each line, two vertices lying on a line $\Longleftrightarrow$ the corresponding lines meet), as well as the number of loci, and the names given to them in [MSY].

2.4 Janus-like algebraic varieties

Consider the question of *uniqueness* of structure of arithmetic quotient, by which we mean the following. Suppose we are given an algebraic variety X, and we know that X is an arithmetic quotient, say $X = X_\Gamma = \Gamma\backslash\mathcal{D}$. Is then X_Γ the *only* such structure on X? Or can there be a different domain $\mathcal{D}'$ and group Γ' such that $X = X_{\Gamma'} = \Gamma'\backslash\mathcal{D}'$? Assuming X_Γ is a compact quotient and Γ is torsion free, this cannot be the case. Indeed, the ratios of the Chern numbers are characteristic for the domain $\mathcal{D}$, by Hirzebruch proportionality, so X can be uniformized only by the domain $\mathcal{D}$. However, if we allow $X = \overline{X}_\Gamma$ for a non-compact quotient X_Γ, the question becomes very interesting.

Definition 2.4.1 A projective algebraic variety X is called *Janus-like*[1], if there exist domains $\mathcal{D}$, $\mathcal{D}'$ and groups Γ, Γ' such that

$$X = \overline{X}_\Gamma = \overline{X}_{\Gamma'}.$$

We will introduce an interesting set of such varieties in this section, which are considered in detail in [J]. Later on we will find more examples of Janus-like varieties.

---

[1]The reader will recall that the Roman god Janus was depicted as one having two faces. This name is also the origin of the word January, for the month looking back into the last as well as forward towards the next year.

Table 2.1: Loci of a four dimensional family of K3 surfaces

The notations $X^{\{ijk;klm;mni;jln\}}$, etc, are taken from [MSY]; the arrows indicate inclusions among the various loci. The symbol ⊚ means a double line.

2.4.1   The arithmetic quotients

The two $\mathbb{Q}$-groups involved are $G_{\mathbb{Q}} = Sp(4,\mathbb{Q})$ and $G_{\mathbb{Q}} = U(3,1;K)$, $K = \mathbb{Q}(\sqrt{-3})$; the corresponding domains (or more precisely, the abstract hermitian symmetric spaces of non-compact type, i.e., without specifying a particular bounded or unbounded realization), which we denote by $\mathbb{S}_2$ and $\mathbb{B}_3$, respectively, are of type $\mathbf{III_2}$ and $\mathbf{I_{3,1}}$, respectively. For arithmetic subgroups $\Gamma \subset Sp(4,\mathbb{Q})$ (respectively $\Gamma_{\sqrt{-3}} \subset U(3,1;K)$), the moduli interpretation of the arithmetic quotient X_Γ (respectively $X_{\Gamma_{\sqrt{-3}}}$) is given by Theorem 2.1.1, and by 2.1.3 if Γ is a principal congruence subgroup (respectively is given by Proposition 2.2.3, and by 2.2.6 for a principal congruence subgroup). There are two very special circumstances about the arithmetic quotients involved, which make this story work:

  i) In both cases, the abelian varieties are *Jacobians* of certain curves, and

  ii) The modular subvarieties on both arithmetic quotients are *divisors*, i.e., of codimension one.

We consider the Igusa desingularization indexIgusa desingularization $\overline{X}_\Gamma$ in the Siegel case and a compactification as in Proposition 2.2.10 in the Picard case.

2.4.1.1   Combinatorics

We have given a description of the boundary and modular subvarieties of $\overline{X}_\Gamma$ in terms of finite geometry in 2.1.4 and 2.1.6 (respectively of $\overline{X}_{\Gamma_{\sqrt{-3}}}$ in Corollary 2.2.9). As a matter of notation, we introduce the following: for a lattice Γ_K in $U(3,1;\mathcal{O}_K)$ we let $Y_{\Gamma_K} = \Gamma_K\backslash\mathbb{B}_3$, $Y^*_{\Gamma_K} = (\Gamma_K\backslash\mathbb{B}_3)^*$, $\overline{Y_{\Gamma_K}} = \overline{(\Gamma_K\backslash\mathbb{B}_3)}$. We write $X(N)$ for $X_{\Gamma(N)}$, and similarly for $X^*(N)$, $\overline{X}(N)$, $Y(n)$, $Y^*(n)$, and $\overline{Y}(n)$. We consider the following ideals (where the rest class rings are fields or products of fields):

In $\mathbb{Z}$:	$N = 2$	3	6		$\mathbb{F}_2$	$\mathbb{F}_3$	$\mathbb{F}_2 \times \mathbb{F}_3$
In $\mathcal{O}_K$:	$n = \sqrt{-3}$	2	$2\sqrt{-3}$		$\mathbb{F}_3$	$\mathbb{F}_4$	$\mathbb{F}_3 \times \mathbb{F}_4$

Then we have the following isomorphisms. For $N \in \mathbb{Z}$, we have $P\Gamma/P\Gamma(N) \cong PSp(4,\mathbb{Z}/N\mathbb{Z})$, and for $n \in \mathcal{O}_K$, we have

$$P\Gamma_K/P\Gamma_K(n) \cong PU(3,1;\mathcal{O}_K/n\mathcal{O}_K).$$

We observe that $\mathcal{O}_K/n\mathcal{O}_K$ is a finite ring of cardinality the norm of n, and if n is a prime it is a field (as K is a principal ideal domain). This is in particular the case if $n = \sqrt{-3}$, of norm 3, and if $n = 2$, of norm 4. In the first case the conjugation on $\mathcal{O}_K$ descends to the trivial automorphism of $\mathbb{F}_3$ (of course), in the second to the unique non-trivial automorphism of $\mathbb{F}_4$ (as may easily be checked).

Proposition 2.4.2 *i)* $PSp(4, \mathbb{F}_2) \cong PU(3, 1; \mathbb{F}_3) \cong \Sigma_6$, *the symmetric group on six elements.*

ii) $PSp(4, \mathbb{F}_3) \cong PU(3, 1; \mathbb{F}_4) \cong G_{25,920}$, *the simple group of order 25,920.*

iii) $PSp(4, \mathbb{F}_2 \times \mathbb{F}_3) \cong PU(3, 1; \mathbb{F}_3 \times \mathbb{F}_4) \cong \Sigma_6 \times G_{25,920}$.

Proof: For i) and ii) see [C] pages 4 and 26; these immediately imply iii). □

The groups $P\Gamma/P\Gamma(N)$ (respectively $P\Gamma_K/P\Gamma_K(n)$) are the Galois groups of the ramified covers $X(N) \longrightarrow X(1)$ (respectively $Y(n) \longrightarrow Y(1)$), which in fact extend to $\overline{X(N)} \longrightarrow \overline{X(1)}$ (respectively $\overline{Y(n)} \longrightarrow \overline{Y(1)}$).

These isomorphisms give, accordingly, certain coincidences of the number of isotropic vectors, non-degenerate subspaces, etc. Let $\nu_r(\Gamma)$ denote the number of cusps corresponding to the standard parabolic P_r, and $\mu_s(\Gamma)$ the class numbers of symmetric subgroups conjugate to the standard symmetric N_s. These numbers can be calculated, as just mentioned, in terms of finite geometry. We have

Lemma 2.4.3 *The class numbers* $\nu_1(\Gamma(N))$ *(respectively* $\nu_1(\Gamma_{\sqrt{-3}}(n))$*) and* $\mu_1(\Gamma(N))$ *(respectively* $\mu_1(\Gamma_{\sqrt{-3}})$*) are given in the following table:*

N	n	$\nu_1(\Gamma(N)) = \mu_1(\Gamma_{\sqrt{-3}})$	$\mu_1(\Gamma(N)) = \nu_1(\Gamma_{\sqrt{-3}}(n))$
1	1	1	1
2	$\sqrt{-3}$	15	10
3	2	40	45
6	$2\sqrt{-3}$	600	450

Proof: We first show that the values of $\nu_1(\Gamma(N))$ and $\nu_1(\Gamma_{\sqrt{-3}}(n))$ are correct.

$\nu_1(\Gamma(1))=1$: Note that every line in $\mathbb{Q}^4$ is isotropic, as we have a symplectic form here. Each such is generated by a primitive vector in $\mathbb{Z}^4$, well-defined up to multiplication by ± 1 (the group of units in $\mathbb{Z}$). Thus this claim is given by the well-known fact that $Sp(4, \mathbb{Z})$ operates transitively on those vectors.

$\nu_1(\Gamma(2))=15$: There are 15 non-zero elements in $(\mathbb{F}_2)^4$, and ± 1 acts trivially.

$\nu_1(\Gamma(3))=40$: There are 80 non-zero elements in $(\mathbb{F}_3)^4$, and ± 1 acts effectively.

$\nu_1(\Gamma_{\sqrt{-3}}(1)) = 1$: This is proved in [Ze], p. 31.

$\nu_1(\Gamma_{\sqrt{-3}}(\sqrt{-3}))=10$: We may take $\mathbb{F}_3 = \{0, 1, -1\}$; note that the units of $\mathcal{O}_K$ act transitively on the non-zero elements of $\mathbb{F}_3$, and conjugation is trivial. We are looking for (x_1, x_2, x_3, x_4) with $x_1^2 + x_2^2 + x_3^2 - x_4^2 = 0$. If $x_4 = 0$, x_1, x_2, x_3 may be choosen arbitrarily, non-zero, giving eight possibilities, and if $x_4 = \pm 1$, exactly one of x_1, x_2 and x_3 must be non-zero, giving six choices, and $(8 + 2 \cdot 6)/2 = 10$. Then $\nu_1(\Gamma_{\sqrt{-3}}(\sqrt{-3})) = 10 \cdot \nu_1(\Gamma_{\sqrt{-3}}(1))=10$.

$\nu_1(\Gamma_{\sqrt{-3}}(2)) = 45$: We may take $\mathbb{F}_4 = \{0, 1, \varrho, \overline{\varrho}\}$; note that the units of $\mathcal{O}_K$ act transitively on the non-zero elements and conjugation is non-trivial. We are looking for $(x_1, ..., x_4)$ with $x_1 \overline{x}_1 + x_2 \overline{x}_2 + x_3 \overline{x}_3 - x_4 \overline{x}_4 = 0$. If $x_4 = 0$, exactly one of x_1, x_2, x_3 must be zero and the others arbitrary non-zero elements, giving 27 possibilities, and if $x_4 = 1$, ϱ, or $\overline{\varrho}$, either exactly one of x_1, x_2 and x_3

is non-zero, and may be arbitrary, giving nine choices, or all of them are non-zero and arbitrary, giving 27 choices; $(27+3(9+27))/3=45$, and $\nu_1(\Gamma_{\sqrt{-3}}(2)) = 45 \cdot \nu_1(\Gamma_{\sqrt{-3}}(1))=45$.

Clearly

$$\nu_1(\Gamma(6)) = \nu_1(\Gamma(2))\nu_1(\Gamma(3))$$

and

$$\nu_1(\Gamma_{\sqrt{-3}}(2\sqrt{-3})) = \nu_1(\Gamma_{\sqrt{-3}}(2))\nu_1(\Gamma_{\sqrt{-3}}(\sqrt{-3})).$$

Note that the values $\nu_1(\Gamma(N))$ are known and can be found, for example in Yamazaki's paper [Y1]; the general formula is

$$\nu_1(\Gamma(N)) = \frac{1}{2}N^4 \prod_{p|N}\left(1 - \frac{1}{p^4}\right),$$

which for $N = 2,3,6$ gives the numbers calculated above. Also the numbers $\mu_1(\Gamma(N))$ are given in [Y1], in Lemma 3:

$$\mu_1(\Gamma(N)) = \frac{1}{2}N^4 \prod_{p|N}\left(1 + \frac{1}{p^2}\right),$$

and for $N = 2,3,6$ this gives the values listed in the second column above.

To treat the $\mu_1(\Gamma_{\sqrt{-3}}(n))$ we remark that if $v \in V$ (where V is the four-dimensional vector space over the residue class ring) is an anisotropic vector, then $v^\perp$ will have the property that the hermitian form restricted to $v^\perp$ will be isotropic, and $v^\perp$ is of codimension one. Furthermore, here again we have $\Gamma_{\sqrt{-3}}$ acting transitively on the set of anisotropic vectors, so to calculate μ_1 is the same as to calculate the number of anisotropic one-dimensional subspaces divided by the order of the center.

- $n = \sqrt{-3}$: $\mathbb{F}_3 = \{0,1,-1\}$, each one-dimensional subspace contains two non-zero vectors, so the number of anisotropic one-dimensional subspaces is equal to the number of (anisotropic vectors − 1)÷2, and since ± 1 acts transitively, $\mu_1 = (3^4 - 1 - 20) \div 4 = 15$.

- $n = 2$: $\mathbb{F}_4 = \{0,1,\varrho,\bar{\varrho}\}$, each one-dimensional subspace contains three non-zero elements, hence the number of anisotropic subspaces is one third the number of anisotropic non-zero vectors, and the center is trivial, i.e., $\mu_1 = (4^4 - 1 - 3 \cdot 45) \div 3 = 40$.

$\square$

2.4.1.2 Isomorphisms

In addition to the above combinatorial identifications of the two situations we also have geometric identifications. These are given by Janus-like *surfaces*, namely, the compactifying divisors and the modular subvarieties, which, as we already mentioned, are also divisors in these cases. Let us denote the compactification divisors by $\Delta(N)$ (respectively $\Delta_{\sqrt{-3}}(n)$) and the modular subvarieties by $X_{\Gamma_M(N)}$ (respectively by $Y_{\Gamma_M\sqrt{-3}(n)}$). Then we have the following isomorphisms:

Theorem 2.4.4 *We have the following isomorphisms of algebraic surfaces:*

i) $\Delta(N) \cong \overline{Y}_{\Gamma_{M\sqrt{-3}}(n)}$ *for the level pairs* $(N,n) = (1,1)$, $(2,\sqrt{-3})$, $(3,2)$, *and* $(6,2\sqrt{-3})$;

ii) $\overline{X}_{\Gamma_M(N)} \cong \Delta_{\sqrt{-3}}(n)$ *for the same level pairs as above.*

In case i) the surfaces are the well-known Shioda elliptic modular surfaces of level N, for $N = 3, 6$, and the Kummer modular surface $S(2)$ for $N = 2$. In case ii) the surfaces are abelian varieties $E_\varrho \times E_\varrho$ for $n = 2\sqrt{-3}$, and quotients thereof for the other levels.

The proof of this theorem is quite a bit of work; it is carried out in full in [J]. For the first three level pairs the surfaces are certain known rational surfaces, and one must identify certain configurations of curves on them. The level pair $(6, 2\sqrt{-3})$ is particularly tricky, and one must somehow show that the open elliptic modular surface $S(6)^0 = \{S(6)$ without the 36 sections$\}$, is a ball quotient. In [J] we do this with Chern number arguments (see section 5.5 below), as we know of no other method. This same kind of argument is used to prove the Theorem 2.4.11 below.

2.4.1.3 Families of curves

2.4.1.3.1 Picard curves   Consider the following family of trigonal curves:

$$y^3 = \prod_{i=1}^{6}(x - \xi_i). \tag{2.21}$$

These curves have genus 4 and, being (Galois-) trigonal, have automorphisms of order 3 given by $(y, x) \mapsto (\varrho y, x)$. Note that $\mathbb{Q}(\varrho) = \mathbb{Q}(\sqrt{-3}) = K$; since this automorphism of the curve passes on to the Jacobian $\mathcal{J}$, $\mathrm{Aut}(\mathcal{J}) \supset \mathbb{Z}_3$. One easily sees that there is $\Phi : K \longrightarrow GL(4, \mathbb{C})$ with $\Phi(\mathcal{O}_{\mathbb{Q}(\varrho)}) \subset \mathrm{End}(\mathcal{J})$, hence the Jacobian variety of this trigonal curve has complex multiplication by K. Furthermore, $\Phi : K \longrightarrow GL(4, \mathbb{C})$ is equivalent to the one given by Shimura's theory Proposition 2.2.3, and it follows from 2.2.3 and Shimura's Theorem 1.1.8, after working out what the level structure means that we have

Corollary 2.4.5 *The quotient $Y(n)$ is a moduli space of (Galois-) trigonal curves (Picard curves) of genus 4 and with a level n structure. The modular subvarieties $(\Gamma_K(n))_2 \backslash \mathbb{B}_2$ parameterize those genus 4 curves which split into a component of genus 3 (Picard curve of genus 3, i.e., whose Jacobian has complex multiplication) and a component of genus 1 (an elliptic curve with complex multiplication) meeting at a point.*

For the family of Picard curves, one of the level structures involved has an easy geometric interpretation:

Observation 2.4.6 *A level $\sqrt{-3}$ structure corresponds to an ordering of the six points ξ_i over which the trigonal curve in (2.21) is branched. Hence the moduli space of ordered sets of six distinct points in $\mathbb{P}^1$ is isomorphic to a Zariski open subset of $Y(\sqrt{-3})$.*

To legitimate this we make the following remarks. Let p be a branch point; clearly $3p=0$ (in the Jacobian). Hence $\sqrt{-3}p$ is a point of order $\sqrt{-3}$, i.e., $\sqrt{-3}p \in Ker(\phi_{\sqrt{-3}})$. Recall that the fibre of $\tau \in Y(1)$ under $Y(\sqrt{-3}) \longrightarrow Y(1)$ is the set of level $\sqrt{-3}$-structures; since the symmetric group Σ_6 is the Galois group, it is clear that an ordering of the six branch points determines an inverse image of τ in $Y(\sqrt{-3})$ and hence a level structure.

2.4.1.3.2   Hyperelliptic curves   The generic principally polarized abelian surface is the Jacobian of a non-singular genus 2 curve. Every such curve is hyperelliptic and so is given by an equation

$$y^2 = \prod_{i=1}^{6}(x - \xi_i), \tag{2.22}$$

and since the branch points are points of order 2, we observe

Observation 2.4.7 *A level 2 structure of the hyperelliptic curve (2.22) corresponds to an ordering of the six branch points.*

We thus see

Corollary 2.4.8 *$X(2)$ and $Y(\sqrt{-3})$ have Zariski open sets which are isomorphic.*

Remark: The 15 copies of $\mathbb{S}_1 \times \mathbb{S}_1$-quotients on $X(2)$ correspond to curves (2.22) for which three of the $\{\xi_i\}$ coincide; these are curves which split into two elliptic curves; the corresponding Jacobians are products $E_1 \times E_2$. On the Zariski open set in $X(2)$ which is the complement of the union of these divisors all $\{\xi_i\}$ are distinct.

2.4.1.4   Configurations of divisors

Finally we consider the union of the boundary components and modular divisors on each of the arithmetic quotients.

Definition 2.4.9 On the compactified arithmetic quotient $\overline{X}(N)$ we define the divisor:

$$\mathcal{D}(N) = \bigcup_{i=1}^{\nu_1(N)} (\Delta(N))_i \bigcup \bigcup_{j=1}^{\mu_1(N)} (\overline{X}_{\Gamma_M(N)})_j.$$

On the compactified arithmetic quotient $\overline{Y}(n)$ we define the divisor:

$$\mathcal{E}(n) = \bigcup_{i=1}^{\nu_1(n)} (\Delta_{\sqrt{-3}}(n))_i \bigcup \bigcup_{j=1}^{\mu_1(n)} (\overline{Y}_{\Gamma_{M\sqrt{-3}}(n)})_j.$$

Then we have

Theorem 2.4.10 *We have the following isomorphisms between normal crossings divisors:*

i) $\mathcal{D}(2) \cong \mathcal{E}(\sqrt{-3})$;

ii) $\mathcal{D}(3) \cong \mathcal{E}(2)$;

iii) $\mathcal{D}(6) \cong \mathcal{E}(2\sqrt{-3})$.

Proof: By Theorem 2.4.4 we have isomorphisms amongst the components in each case. By Lemma 2.4.3 we have the same number of components of each type. As to the intersections, one sees by inspection that they are also the same, at least combinatorially (i.e., the dual graphs are isomorphic). It remains to show that the intersection curves are really the same on each copy. But this is clear from our discussions of the surfaces above: on the modular subvarieties $X_{\Gamma_M(N)}$, which are elliptic modular surfaces, we have intersections which are curves which are either singular fibres or sections; on the compactification divisors $\Delta(N)$ the curves are of the form {cusp}×curve or curve×{cusp}, where "cusp" refers to one of the $\Gamma(N)$ cusps on the corresponding curve. □

2.4.2 The theorem

The purpose of this section is to sketch the proof, given in full in [J], of the following theorem.

Theorem 2.4.11 *We have the following diagram of modular varieties containing the $\overline{X(N)}$, and $\overline{Y(n)}$:*

with isomorphisms as indicated, plus the fibre product isomorphism $\overline{X(6)} \xrightarrow{\cong} \overline{Y(2\sqrt{-3})}$.

Proof: $\overline{X(2)} \longrightarrow \overline{Y(\sqrt{-3})}$:

From Corollary 2.4.8 we know that $X(2)$ and $Y(\sqrt{-3})$ have isomorphic open subsets. This Zariski open set corresponds to hyperelliptic (respectively Picard) curves (2.22) (respectively (2.21)), such that *all ξ_i are distinct*. From the remark following 2.4.8 in the hyperelliptic case this is the complement of the union of modular subvarieties. Letting $\mathcal{V} \subset \overline{X(2)}$ denote this Zariski open subset we thus have $\overline{X(2)} - \mathcal{D}(2) = \mathcal{V}$. From Corollary 2.4.5 we see that the Zariski open subset on $Y(\sqrt{-3})$ is the complement of the modular subvarieties; hence, denoting this set again by $\mathcal{V}$, we have $\overline{Y(\sqrt{-3})} - \mathcal{E}(\sqrt{-3}) = \mathcal{V}$. On the other hand, by Theorem 2.4.10, we have $\mathcal{D}(2) = \mathcal{E}(\sqrt{-3})$. Therefore, to show that the isomorphism of $\mathcal{V}$ (onto itself) extends to each compactification, it is enough to show it extends to a homeomorphism of the compact varieties. The isomorphisms are given explicitly by the "coordinates" ξ_i. Therefore the question is how this correspondence looks at $\mathcal{D}(2)$ and $\mathcal{E}(\sqrt{-3})$. These maps are easily explicitly described (see the discussion in section 2.4.3.3 below), from which continuity follows.

$\overline{X(1)} \longrightarrow \overline{Y(1)}$:

These are the quotients of $\overline{X(2)}$ and $\overline{Y(\sqrt{-3})}$ by $PSp(4, \mathbb{Z}/2\mathbb{Z})$ and $PU(3, 1; \mathbb{F}_3)$ respectively, and both of these groups are, by Proposition 2.4.2, isomorphic to Σ_6. These groups preserve the configurations $\mathcal{D}(2)$ and $\mathcal{E}(\sqrt{-3})$ used above. There is an indexing under which the 15 boundary components $\Delta(N)$ (respectively the 15 modular components $Y_{\Gamma_{\sqrt{-3}}(n)}$) correspond to unordered pairs of elements of $\{1,...,6\}$, and under which the 10 modular divisors $X_{\Gamma_M(N)}$ (respectively the 10 boundary components $\Delta_{\sqrt{-3}}$) correspond to unordered pairs of unordered triples of elements of $\{1,...,6\}$, and Σ_6 operates as the permutations of $\{1,...,6\}$. We need to show that Σ_6 has only one possible action preserving the configurations $\mathcal{D}(2)$ and $\mathcal{E}(\sqrt{-3})$, and it suffices to check the action of a generator on the tangent space to a fixed point. This can be done (see [J] for details).

$\overline{X(6)} \longrightarrow \overline{Y(2\sqrt{-3})}$:

This can be proved using the same method as in the proof of Theorem 3.19 in [J], where a similar result is proved for surfaces. First let $X^0(6)$ be $X(6)$ with the surfaces $X_{\Gamma_M(6)}$ removed (the translates of the diagonal).

Lemma 2.4.12 $X^0(6)$ *is a ball quotient.*

Proof: Here we must prove the equality (see section 5.5)

$$3\overline{c}_1^3(\overline{X(6)}, \Delta) = 8\overline{c}_1\overline{c}_2(\overline{X(6)}, \Delta)$$

for the logarithmic Chern numbers, where $\Delta = \cup X_{\Gamma_M(6)}$. Since the abelian surfaces $X_{\Gamma_M(6)}$ are disjoint, one sees easily (using adjunction):

$$\overline{c}_1(\overline{X(6)}, \Delta) = c_1(\overline{X(6)}) - \Delta,$$

$$\overline{c}_2(\overline{X(6)}, \Delta) = c_2(\overline{X(6)}).$$

Furthermore, applying adjunction to each component $X_{\Gamma_M(6)}$ of Δ yields

$$c_1(\overline{X(6)}) \cdot X_{\Gamma_M(6)} = c_1(\overline{X_{\Gamma_M(6)}}) + (\overline{X_{\Gamma_M(6)}})^2 = (\overline{X_{\Gamma_M(6)}})^2,$$

so $c_1(\overline{X(6)}) \cdot \Delta = \Delta^2$. Hence for the numbers,

$$
\begin{aligned}
\bar{c}_1^3 &= c_1^3(\overline{X(6)}) - 3c_1^2(\overline{X(6)}) \cdot \Delta + 3c_1(\overline{X(6)}) \cdot \Delta^2 - \Delta^3 \\
&= c_1^3(\overline{X(6)}) - \Delta^3 \\
\bar{c}_1 \bar{c}_2(\overline{X(6)}, \Delta) &= c_1 c_2(\overline{X(6)}),
\end{aligned}
$$

so the equality to be shown is

$$3c_1^3(\overline{X(6)}) - 8c_1 c_2(\overline{X(6)}) = 3(\Delta)^3.$$

All intersection numbers involved can be taken from Yamazaki's paper [Ya]. Let $\overline{X(N)}$ be the Igusa desingularization of the Siegel 3-fold of level N. Then

$$
\begin{aligned}
c_1^3(\overline{X(N)}) &= \frac{N^7 \cdot \prod_{p|N}(1 - p^{-2})(1 - p^{-4})}{2^6 \cdot 3 \cdot 5}[-9N^3 + 360N - 880] \\
c_1 c_2(\overline{X(N)}) &= \frac{N^7 \cdot \prod_{p|N}(1 - p^{-2})(1 - p^{-4})}{2^4 \cdot 3 \cdot 5}[-N^3 + 30N - 60] \\
(\Delta)^3 &= \frac{N^{10} \cdot \prod_{p|N}(1 - p^{-2})(1 - p^{-4})}{2^6 \cdot 3^2}
\end{aligned}
$$

Inserting $N = 6$ into the above we get

$$3(-119520) - 8(-69120) = 3 \cdot 64800,$$

which checks. This proves the lemma. $\qquad\square$

Now that we know $X^0(6)$ is a ball quotient (compactified by adding the abelian surfaces $X_{\Gamma_M(6)}$), let $\Gamma_?$ be the lattice in $PU(3,1)$ such that $X^0(6) = \Gamma_? \backslash \mathbb{B}_3$; from the cover $X^0(6) \longrightarrow X^0(2)$ we see that $\Gamma_? \subset PU(3,1; \sqrt{-3}\mathcal{O}_K)$. It is a normal subgroup with quotient $G_{25,920}$, and we strongly suspect $\Gamma_? = PU(3,1; 2\sqrt{-3}\mathcal{O}_K)$. To prove this, we start with a parabolic. Let $P_?$ denote the parabolic in $\Gamma_?$ for a cusp, and $P \subset PU(3,1; 2\sqrt{-3}\mathcal{O}_K)$ the parabolic of a cusp (all parabolics are isomorphic since $\Gamma_?$, $\Gamma_{\sqrt{-3}}(2\sqrt{-3})$ are normal subgroups and the class number of $\Gamma_{\sqrt{-3}}$ is one). We have the following exact sequences:

$$1 \longrightarrow \Gamma_? \xrightarrow{\phi_?} \Gamma_{\sqrt{-3}} \longrightarrow G_{25,920} \longrightarrow 1$$

$$1 \longrightarrow \Gamma_{\sqrt{-3}}(2\sqrt{-3}) \xrightarrow{\phi} \Gamma_{\sqrt{-3}} \longrightarrow G_{25,920} \longrightarrow 1$$

Lemma 2.4.13 $P_? \cong P$, $\phi_?(P_?) = \phi(P)$.

Proof: A parabolic in $PU(3,1)$ looks as follows (here we are working in a certain unbounded realization of the ball, cf. [J]):

$$P_\infty = \left\{ \begin{pmatrix} 1 & 0 & 0 & x_1 \\ 0 & 1 & 0 & x_2 \\ \sqrt{-3}\bar{x}_1 & \sqrt{-3}\bar{x}_2 & 1 & y + \frac{\sqrt{-3}}{2}\Sigma|x_i|^2 \\ 0 & 0 & 0 & 1 \end{pmatrix} \middle| x_i \in \mathbb{C},\ y \in \mathbb{R} \right\}.$$

As in the surface case, $P_\infty \cap \Gamma$ acts on $\mathbb{C}^* \times \mathbb{C}^2$ by dividing $\mathbb{C}^2$ by the lattice generated by the x_1, x_2 and acts on $\mathbb{C}^*$ (normal direction) by a lattice with generator some $q \in \mathbb{R}$. One sees clearly that the quotient of $\mathbb{C}^2$ is a product, $\mathbb{C}^2/\Lambda = \mathbb{C}/\mathcal{O}_K \times \mathbb{C}/\mathcal{O}_K$, and that its normal bundle $N \longrightarrow \mathbb{C}/\mathcal{O}_K \times \mathbb{C}/\mathcal{O}_K$ is $N \cong p_1^* N_1 \otimes p_2^* N_2$, where p_i are the projections onto the factors and N_i are the bundles over $\mathbb{C}/\mathcal{O}_K$ which are the normal bundles in the surface case. Furthermore, these two data suffice to determine the parabolic, and since both are checked for the surface case in the proof of 2.4.4 (here being just the product situation), the lemma follows.   □

Next, consider a fixed cusp ∞ with compactifying surface

$$S_\infty \cong \Delta_{\sqrt{-3}}(2\sqrt{-3}) \cong E_\varrho \times E_\varrho.$$

There are 24 subballs $B_i \cong Y_{\Gamma_{\sqrt{-3}}(2\sqrt{-3})}$ which intersect S_∞, each along a curve of the form $\{p_i\} \times E_\varrho$ or $E_\varrho \times \{q_i\}$, $i = 1, ..., 12$, where the p_i and q_i are the 12 points of order two and 3 used in the proof of [J], 3.15 on the first and second copies of E_ϱ in S_∞. Each B_i is uniformized by a subball $\mathbb{B}_2 \subset \mathbb{B}_3$ which passes through ∞. Such a $\mathbb{B}_2$ is a linear section of $\mathbb{B}_3$, hence is determined by its tangent directions at ∞, or by its intersection with S_∞. Let $N(B_i)$, $i = 1, ..., 24$ be the stabilizers of the B_i, $N(B_i) \subset PU(3,1;2\sqrt{-3}\mathcal{O}_K)$, $(N_?)_i \subset \Gamma_?$ the corresponding stabilizers in $\Gamma_?$.

Lemma 2.4.14 $\phi(N(B_i)) \cong \phi_?((N_?)_i)$, $i = 1, ..., 24$.

Proof: As in the surface case this will follow if we can verify that the 24 subballs $\mathbb{B}_2 \subset \mathbb{B}_3$ corresponding to the $N(B_i)$ (resp. the $(N_?)_i$) are the same: each subball is a linear section of the three-dimensional ball, so is determined by its tangent direction at the cusp. One requires also an argument that any subballs through the cusp are $\Gamma_?$ equivalent if and only if they are equivalent under the stabilizer of the cusp, i.e., under the parabolic. See [J], 3.14 and 3.15 for details. To show that the 24 subballs are the same, it is sufficient to note that the divisors on S_∞ (given by the intersections) are the same in both cases.   □

Finally, we have here as in the surface case (cf. [J], 3.16)

Lemma 2.4.15 $\phi_?(\Gamma_?)$ *is generated by* $\phi_?(P_?)$ *and* $\phi_?((N_?)_i)$.   □

From the last three lemmas we get $\phi_?(\Gamma_?) = \phi(\Gamma_{\sqrt{-3}}(2\sqrt{-3}))$, and the claim $\overline{X(6)} \xrightarrow{\sim} \overline{Y(2\sqrt{-3})}$ follows from this.

$$\overline{X(3)} \xrightarrow{\sim} \overline{Y(2)}:$$

This follows from $\overline{X(6)} \xrightarrow{\sim} \overline{Y(2\sqrt{-3})}$, Σ_6 acting on both spaces in a natural way, as soon as we have identified both group actions. This is done as above by computing eigenvalues at fixed points, hence we omit the explicit verification. $\square$

Corollary 2.4.16 *Consider the isomorphisms* $\overline{X(N)} \xrightarrow{\sim} \overline{Y(n)}$. *Under each such the boundary components of $\overline{X(N)}$ are mapped to the modular subvarieties of $\overline{Y(n)}$, and the modular subvarieties of $\overline{X(N)}$ are mapped to the boundary components of $\overline{Y(n)}$.*

Remark: The really interesting isomorphism of the theorem is

$$\overline{X(6)} \xrightarrow{\sim} \overline{Y(2\sqrt{-3})}.$$

This is because in both cases the lattices act freely, hence the open quotients $X(6)$ and $Y(2\sqrt{-3})$ fulfill logarithmic proportionality with $\overline{c}_1^3/\overline{c}_{12} = 9/4$ for $X(6)$ and $=8/3$ for $Y(2\sqrt{-3})$.

2.4.3 A remarkable duality

Corollary 2.4.16 shows that the isomorphisms of Theorem 2.4.11 actually have more structure than just isomorphisms: there is obviously a duality of some sort involved, between boundary components and modular subvarieties, or as will be discussed in later chapters, projective duality.

2.4.3.1 Modular forms

For a lattice Γ and domain $\mathcal{D}$ as above, let $\mathcal{R}_k(\Gamma)$ be the space of Γ-modular forms of weight k, by which we mean the space of Γ-invariant sections of the equivariant line bundle $\mathcal{L}^{\otimes k}$ on $\mathcal{D}$ defined by the kth power of the isotropy representation $(\mathrm{ad}_{\mathfrak{k}})^{\otimes k}$ on $\mathfrak{g}$. It is well-known that the Baily-Borel embedding of X_Γ^* is given by $Proj(\bigoplus_k \mathcal{R}_k(\Gamma))$, hence the coordinate ring of X_Γ^* in the sense of algebraic geometry is $\bigoplus_k \mathcal{R}_k(\Gamma)$. Consider the dual variety of X_Γ^*, let $\bigoplus_k D_k$ be its coordinate ring. We say $\bigoplus_k \mathcal{R}_k(\Gamma)$ and $\bigoplus_k D_k$ are *projectively dual* rings. Then Theorem 2.4.11 and the results on the varieties $\mathcal{S}_3$ and $\mathcal{B}_4$, the Segre cubic and Burkhardt quartic threefolds in the following chapters (see Theorem 3.2.5, Corollary 3.3.7, Theorem 5.6.1 and Theorem 5.7.2 below) imply

Proposition 2.4.17 *For the following levels, the rings of modular forms for $\Gamma(N)$ are projectively dual to the ring of modular forms for $\Gamma_{\sqrt{-3}}(n)$: N=2 and $n = \sqrt{-3}$; N=3 and n=2.*

Question 2.4.18 *Is the same true for N=6 and $n=2\sqrt{-3}$?*

We have the bundle $\mathcal{L}^{\otimes k}$ on $\mathcal{D}$, and the space of Γ-invariant sections,

$$H^0(\mathcal{D}, \mathcal{L}^{\otimes k})^\Gamma \cong H^0(\Gamma \backslash \mathcal{D}, \mathcal{L}_\Gamma^{\otimes k}),$$

where $\mathcal{L}_\Gamma \longrightarrow X_\Gamma$ is the quotient of $\mathcal{L}$. This can be translated into sections of bundles on the toroidal compactification $\overline{X}_\Gamma$. Namely, there is a unique extension of $\mathcal{L}_\Gamma$ to a bundle $\overline{\mathcal{L}}_\Gamma \longrightarrow \overline{X}_\Gamma$ which is utilized in [M3] to derive the propoertionality property for $\overline{X}_\Gamma$. The space of sections are then bijectively related

$$H^0(X_\Gamma, \mathcal{L}_\Gamma^{\otimes k}) \longleftrightarrow H^0(\overline{X}_\Gamma, \overline{\mathcal{L}}_\Gamma^{\otimes k}).$$

In fact, $\mathcal{L}_\Gamma \cong \Omega^1_{X_\Gamma}$, while $\overline{\mathcal{L}}_\Gamma \cong \Omega^1_{\overline{X}_\Gamma}(\Delta_\Gamma)$, where $\Delta_\Gamma = \overline{X}_\Gamma - X_\Gamma$ is the compactification divisor, and the bundle is that of the differential forms with logarithmic poles along Δ_Γ.

Setting $R_k(\Gamma) = H^0(X_\Gamma, \mathcal{L}_\Gamma^{\otimes k})$, the following is a consequence of Proposition 2.4.17.

Proposition 2.4.19 *The following rings are projectively dual:*

$$\bigoplus_k H^0(\overline{X(N)}, \Omega_{\overline{X(N)}}(log)^{\otimes k}), \quad \bigoplus_k H^0(\overline{Y(n)}, \Omega_{\overline{Y(n)}}(log)^{\otimes k}),$$

with N, n as in 2.4.17.

Let us consider now cusp forms, i.e., modular forms of weight k vanishing along the compactification divisors. Hence these are sections of the line bundle $\Omega_{\overline{X}_\Gamma}(log)^{\otimes k-1} \otimes \Omega_{\overline{X}_\Gamma}$. In particular, cusp forms of weight 1 are just sections of the canonical bundle. For $N = 2, 3$; $n = \sqrt{-3}$, 2 there are none such, but for $N = 6$ the number has been calculated by Weissauer, the result being 2906. Note that in this case there is no distinction between the Picard and Siegel modular cases. Therefore the calculation being done for $\Gamma(6)$, we get for free the number of cusp forms of weight 1 for $\Gamma_{\sqrt{-3}}(2\sqrt{-3})$: it is 2906. In fact, as a consequence of Proposition 2.4.19 we can state

Proposition 2.4.20 *The following $\mathbb{C}$-vector spaces are isomorphic:*

$$\left\{ \begin{array}{c} \text{cusp forms of } \Gamma(6) \\ \text{of weight 1} \end{array} \right\} \cong \left\{ \begin{array}{c} \text{cusp forms of } \Gamma_{\sqrt{-3}}(2\sqrt{-3}) \\ \text{of weight 1} \end{array} \right\}.$$

2.4.3.2   Metrics

Consider a bounded symmetric domain $\mathcal{D}$ with its Bergmann metric. For a lattice Γ acting freely on $\mathcal{D}$, this metric, being $G_\mathbb{R}$-invariant and in particular Γ-invariant, descends to the quotient $\Gamma\backslash\mathcal{D}$ to give a metric there. By Mumford's extension of the proportionality principle to the non-compact case, this metric extends to a metric on $\overline{X}_\Gamma$ with logarithmic singularities along the compactification divisor. To apply this we stick with $\overline{X(6)} \cong \overline{Y(2\sqrt{-3})}$, since as mentioned above, both groups act freely in this case. On this variety, let us call it $\overline{X}$, we have two natural Bergmann metrics with logarithmic singularities, along the two different compactification divisors. Now note that also the modular subvarieties have a geometric interpretation in terms of the Bergmann metrics. In fact, the subdomains $\mathbb{S}_1 \times \mathbb{S}_1 \subset \mathbb{S}_2$ and $\mathbb{B}_2 \subset \mathbb{B}_3$ are *totally geodesic* submanifolds with respect to the Bergmann metrics on $\mathbb{S}_2$ and $\mathbb{B}_3$. We call the modular subvarieties dealt with above as "the integral totally geodesic locus" of the corresponding Bergmann metric. Note that in both cases studied, this locus is of codimension one[2], in fact, a normal crossing divisor. Hence we have on the quotients two normal crossings divisors. The first is given by the modular subvarieties $\overline{X}_{\Gamma_M(N)}$ (resp. $\overline{Y}_{\Gamma_{M,\sqrt{-3}}(n)}$), the second by the $\Delta(N)$ (resp. $\Delta_{\sqrt{-3}}(n)$), all notations as in Theorem 2.4.4. The conclusion is

Proposition 2.4.21 *Let g_1 and g_2 be the Bergmann metrics, extended to $\overline{g_1}$, $\overline{g_2}$ on $\overline{X}$. Then these two metrics are* dual, *that is*

$$\left\{ \begin{array}{c} \text{locus of logarithmic} \\ \text{singularities of } \overline{g_1} \end{array} \right\} \cong \left\{ \begin{array}{c} \text{the integral totally geodesic} \\ \text{locus of } \overline{g_2} \end{array} \right\},$$

where the isomorphism of divisors is induced by the isomorphism $\overline{X}(6) \longrightarrow \overline{Y}(2\sqrt{-3})$. The same is true for $\overline{g_2}$, $\overline{g_1}$.

2.4.3.3 Moduli

We describe here degenerations occuring in our families of curves of genus 4 and genus 2 respectively (here the degeneration does not see the level structure – the degenerations are independent of the level). Hence we assume $N = 2$, $n = \sqrt{-3}$.). We employ the following notations, the equations (2.21) and (2.22) defining the curves:

D_i=divisors $\Delta(N) \cong Y_{\Gamma_{M\sqrt{-3}}(n)}$:	two of the ξ_k coincide, i=1,...,15	
$D_{ij} = D_i \cap D_j$:	two pairs of the ξ_k coincide	
$D_{ijk} = D_i \cap D_j \cap D_k$:	three pairs of the ξ_k coincide	
E_t=divisors $X_{\Gamma_M(N)} \cong \Delta_{\sqrt{-3}}(n)$:	three of the ξ_k coincide, t=1,...,10	
$E_{ti} = D_i \cap E_t$:	one pair and one triple of the ξ_k coincide	
$E_{tij} = D_{ij} \cap E_t$:	two triples of the ξ_k coincide	

The reader should have no difficulty verifying these degenerations. To complete the proof of Theorem 2.4.11, we used the fact that the dependency on the parameters ξ_k is continuous. As an example we describe this on the generic singular locus, i.e., along D_i and E_t. All degenerations are listed in the Table 2.1.

2.4.3.3.1 Picard case

- D_i: If one pair of the ξ_k coincide, we have a three-fold cover, branched at five points, one of them double. Without restricting generality, we can assume one of the points (say the double zero) is at ∞. The resulting cover splits into two components, one of them a smooth genus 3 curve ($\mathbb{Z}/3\mathbb{Z}$-trigonal) and the other an elliptic curve. Note that the Jacobian of this curve still has complex multiplication, and, in addition, both the

---

[2]this is a special feature of the case at hand; in general the modular subvarieties of Siegel space quotients (other than degree 2) are not divisors.

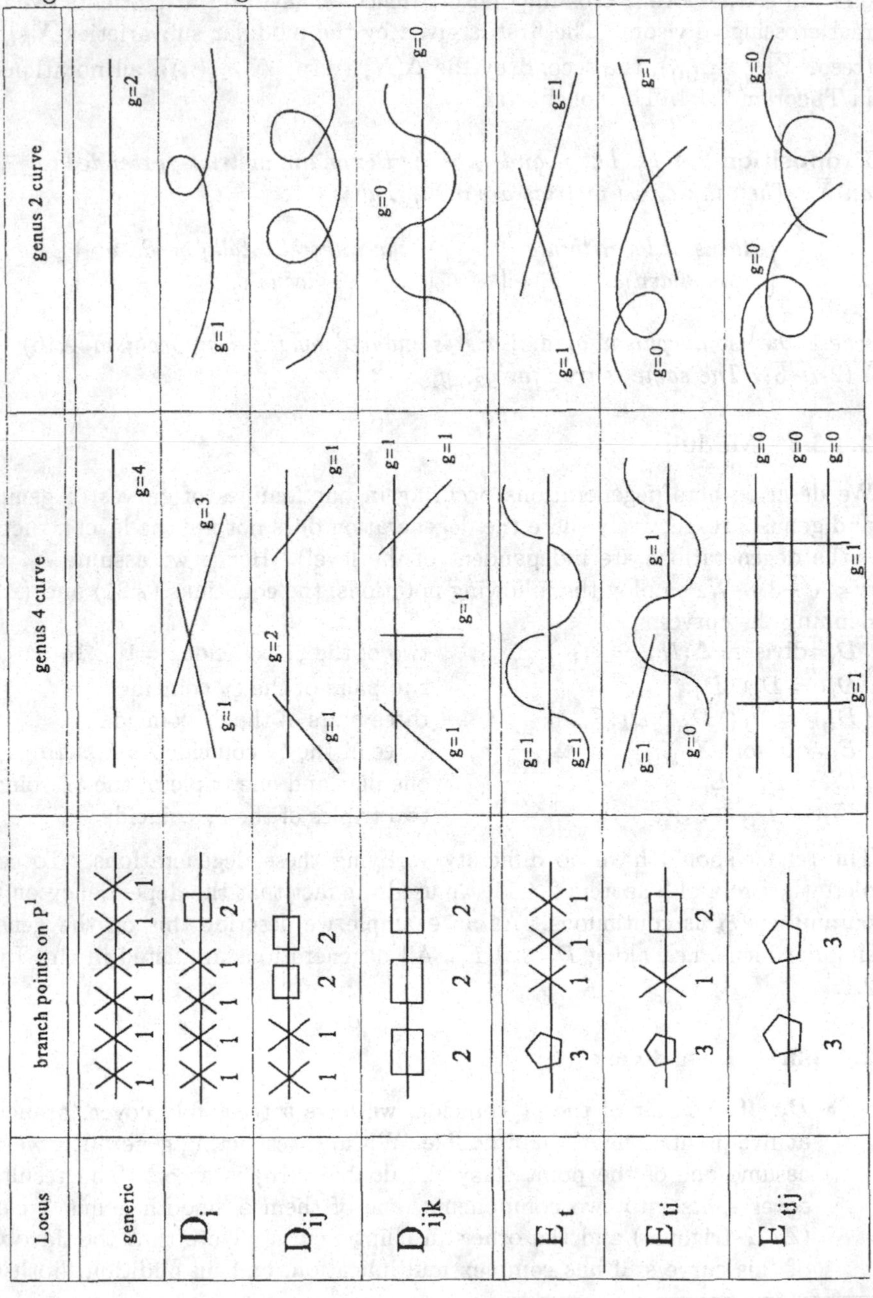

Figure 2.1: The degenerations of the Picard and hyperelliptic curves

genus 3 and the elliptic curve have an extra automorphism. This means, in particular, that the elliptic component is the curve E_ϱ, with no variation in the family. More generally, one easily sees that *all* the elliptic curves in the degenerations corresponding to the divisors D_i are this particular elliptic curve.

- E_t: Here the cover splits into two elliptic curves, but the automorphism is lost, i.e., one just has an endomorphism. The parameter (moduli) of the elliptic curve is given by the double ratio of the four branch points.

2.4.3.3.2 Siegel case

- D_i: If two of the ξ_k coincide, then we have a double cover with five branch points; this is a genus 1 curve with one double point. Since D_i is the Kummer modular surface, its points can be given coordinates (z, w) with $z \in \Gamma(2)\backslash\mathbb{S}_1$, $w \in E_\varrho/(\mathbb{Z}/3\mathbb{Z}) \cong \mathbb{P}^1$. Then z is the moduli point of the elliptic curve, and w gives the double point.

- E_t: When three of the ξ_k coincide, the curve is $y^2 = \prod_1^3(x - \xi_k)$, an elliptic curve. More precisely, let

$$y^2 = \prod_{k=1}^{6}(x - \xi_k)$$

be the original equation, and write it as

$$y^2 = (x - \xi_1)(x - \xi_2)(x - \xi_3)(x - \lambda\xi_4)(x - \lambda\xi_5)(x - \lambda\xi_6);$$

the degeneration is then given by letting $\lambda \longrightarrow \infty$. The limit curve is $y^2 = \prod_1^3(x - \xi_k)$, with a fourth branch point at infinity, and changing variable to $\widetilde{x} := \lambda x$ we get

$$y^2 = (\widetilde{x} - \lambda\xi_1)(\widetilde{x} - \lambda\xi_2)(\widetilde{x} - \lambda\xi_3)(\widetilde{x} - \xi_4)(\widetilde{x} - \xi_5)(\widetilde{x} - \xi_6),$$

which for $\lambda \longrightarrow \infty$ becomes $y^2 = \prod_4^6(\widetilde{x} - \xi_k)$, with fourth branch point at infinity. Therefore, over $\xi \in E_t$ the corresponding degeneration consists of the two elliptic curves above, meeting at their common branch point.

Chapter 3

Projective embeddings of modular varieties

In this chapter the examples we study are no longer "abstract" quotients, but will be given completely concrete embeddings as projective varieties. From this point onward the "geometry" of the situation will be increasingly the "projective algebraic geometry" of the situation. In fact, the algebraic varieties themselves (mostly hypersurfaces) are worthy of study in their own right, without any reference to arithmetic quotients. This arises from the "level 3" mentioned in the introduction: introducing level structures to the moduli interpretation adds projective symmetry to the quotients. In the cases studied in this chapter, the symmetry group will be the symmetric group Σ_6.

A rather striking and unexpected result of the study of the individual cases is that the geometry of the embedding is strongly controlled by the Hessian variety of the variety itself; in many cases it turns out that the Hessian variety of an arithmetic quotient is (birational to) an arithmetic quotient. This is true for the varieties $\mathcal{S}_3$ (studied in this chapter), $\mathcal{B}_4$ (studied in the next chapter), and it seems also possible for $\mathcal{I}_5$, studied in the final chapter. We conjecture it to be true also for the variety $\mathcal{I}_4$, also dealt with in this chapter. It is not known why this state of affairs makes its appearance.

To make the relation between the various varieties which arise as visible as possible, we derive them from a common geometric object which "encodes" the group Σ_6: the tetrahedron in $\mathbb{P}^3$. We then show how to modify $\mathbb{P}^3$ so as to obtain a cubic hypersurface in $\mathbb{P}^4$, the Segre cubic $\mathcal{S}_3$, and also how to modify it to obtain a quartic hypersurface in $\mathbb{P}^4$, the Igusa quartic $\mathcal{I}_4$. We give complete proofs that $\mathcal{S}_3$ and $\mathcal{I}_4$ are arithmetic quotients – in fact Satake compactifications (or Baily-Borel embeddings, on the nose) of arithmetic quotients. This fact, together with Shimura's theory (Chapter 1), yields a *moduli interpretation* of each of the varieties, and a fascinating correlation occurs between the geometry, in terms of *subvarieties*, and moduli interpretation, in terms of *degenerations or specializations*. These questions will be taken up for all examples in this chapter, which include in addition to $\mathcal{S}_3$ and $\mathcal{I}_4$ the Hessians of these varieties (all threefolds), as well as the Coble variety $\mathcal{Y}$, a four-dimensional arithmetic quotient which also has a Σ_6-symmetry – it is the double cover of $\mathbb{P}^4$ branched along the Igusa quartic.

A lot has been written about $\mathcal{S}_3$ and $\mathcal{I}_4$, so necessarily we prove little which was not known. However, the point of view we take is novel, and simply sheds light on the matter, given a new perspective as well as exposing the mysterious relation mentioned above with the Hessian variety.

3.1 The tetrahedron in $\mathbb{P}^3$

3.1.1 Arrangements defined by Weyl groups

Let $\Phi(T,G) \subset \mathfrak{t}^*$ be a root system of a simple Lie group G (over $\mathbb{C}$). Using notations as in Bourbaki we have the roots (for those systems which will be of interest to us in the sequel)

$$\mathbf{A_n} \quad \pm(\varepsilon_i - \varepsilon_j),\ 1 \le i < j \le n+1;$$

$$\mathbf{B_n} \quad \pm(\varepsilon_i \pm \varepsilon_j),\ \pm\varepsilon_i, 1 \le i < j \le n;$$

$$\mathbf{C_n} \quad \pm(\varepsilon_i \pm \varepsilon_j),\ \pm 2\varepsilon_i, 1 \le i < j \le n;$$

$$\mathbf{D_n} \quad \pm(\varepsilon_i \pm \varepsilon_j),\ 1 \le i < j \le n; \qquad\qquad (3.1)$$

$$\mathbf{F_4} \quad \pm(\varepsilon_i \pm \varepsilon_j),\ \pm\varepsilon_k,\ \pm\tfrac{1}{2}(\varepsilon_1 \pm \varepsilon_2 \pm \varepsilon_3 \pm \varepsilon_4), 1 \le i < j \le 4,\ k = 1,\ldots,4;$$

$$\mathbf{E_6} \quad \pm(\varepsilon_i \pm \varepsilon_j), 1 \le i < j \le 5,\ \pm\tfrac{1}{2}(\varepsilon_1 \pm \varepsilon_2 \pm \varepsilon_3 \pm \varepsilon_4 \pm \varepsilon_5 - \varepsilon_6 - \varepsilon_7 + \varepsilon_8),\ \text{with an even number of "}-\text{" signs in the parenthesis;}$$

Each root α determines an orthogonal plane $\alpha^\perp$, and for any arrangement $\mathbf{X_n}$,

$$\mathcal{A}(\mathbf{X_n}) := \{\alpha^\perp \big| \alpha \text{ a root}\} \qquad\qquad (3.2)$$

is a central arrangement in $\mathbb{C}^n$, i.e., each of the planes passes through the origin. This induces a projective arrangement in $\mathbb{P}^{n-1}$, as follows. Blow up the origin of $\mathbb{C}^n$; the exceptional divisor is a $\mathbb{P}^{n-1}$. The *projective arrangement* is the union of the intersections $[H] \cap \mathbb{P}^{n-1}$ in the exceptional divisor, where $[H]$ is the proper transform of the hyperplane $H = \alpha^\perp$ under the blow up. The projective arrangements for $\mathbf{B_n}$ and $\mathbf{C_n}$ coincide, and these arrangements are given in $\mathbb{P}^{n-1}$ with projective coordinates $(x_1 : \ldots : x_n)$ as follows:

$\mathcal{A}(\mathbf{A_n})$: $\quad \{x_i = 0,\ i = 1,\ldots,n;\ x_i = x_j,\ 1 \le i < j \le n\};$
$\mathcal{A}(\mathbf{B_n})$: $\quad \{x_i = 0,\ i = 1,\ldots,n;\ x_i = \pm x_j,\ 1 \le i < j \le n\};$
$\mathcal{A}(\mathbf{D_n})$: $\quad \{x_i = \pm x_j,\ 1 \le i < j \le n\};$
$\mathcal{A}(\mathbf{F_4})$: $\quad \{x_i = 0,\ i = 1,\ldots,n;\ x_i = \pm x_j,\ 1 \le i < j \le 4,\ \tfrac{1}{2}(x_1 \pm x_2 \pm x_3 \pm x_4)\};$
$\mathcal{A}(\mathbf{E_6})$: $\quad \{x_i = \pm x_j,\ 1 \le i < j \le 5,\ \tfrac{1}{2}(x_1 \pm x_2 \pm x_3 \pm x_4 \pm x_5 + x_6)\}.$

$$(3.3)$$

For the arrangement of type $\mathbf{A_n}$ we have made the coordinate transformation $x_1 = \varepsilon_1 - \varepsilon_{n+1}, \ldots, x_n = \varepsilon_n - \varepsilon_{n+1}$, so $x_i - x_j = \varepsilon_i - \varepsilon_j$ for $1 \leq i < j \leq n$, and for $\mathbf{E_6}$ we have taken x_6 to replace $\varepsilon_8 - \varepsilon_7 - \varepsilon_6$.

The arrangements above are the arrangements defined by the projective reflection groups $PW(\mathbf{X_n})$. Each hyperplane is the reflection plane for the reflection on the corresponding root. From this point of view these arrangements are studied in [OS2].

3.1.2 Rank 4 arrangements

As described above, the groups $W(\mathbf{A_4})$, $W(\mathbf{B_4})$, $W(\mathbf{D_4})$ and $W(\mathbf{F_4})$ give rise to projective arrangements in $\mathbb{P}^3$. They consists of ten, 16, 12 and 24 planes, respectively. They may also be described as follows (see [GS]):

$\mathcal{A}(\mathbf{A_4})$:   four faces of a tetrahedron plus the six symmetry planes;

$\mathcal{A}(\mathbf{B_4})$:   six faces of a cube plus the nine symmetry planes plus the plane at infinity;

$\mathcal{A}(\mathbf{D_4})$:   six faces of a cube plus the six symmetry planes through two edges each, OR: eight faces of an octahedron plus three symmetry planes containing four vertices each plus the plane at infinity;

$\mathcal{A}(\mathbf{F_4})$:   "desmic figure": six faces of the cube, eight faces of an inscribed octahedron, nine symmetry planes and the plane at infinity; this is also determined by the regular 24-cell;

The combinatorial description of these arrangements can be encoded in numbers:

$$t_q(j) := \# \left\{ \begin{array}{l} \mathbb{P}^j\text{'s of the arrangement through which } q \text{ of the reflection} \\ \text{planes pass} \end{array} \right\}.$$

(3.4)

In the case of the above arrangements we have the following data ($t_q := t_q(0)$, the number of points):

$\mathcal{A}(\mathbf{A_4})$: $t_6 = 5$, $t_4 = 10$; $t_3(1) = 10$, $t_2(1) = 15$.

$\mathcal{A}(\mathbf{B_4})$: $t_9 = 4, t_6 = 8$, $t_5 = 12$, $t_4 = 16$; $t_4(1) = 6$, $t_3(1) = 16$, $t_2(1) = 36$.

$\mathcal{A}(\mathbf{D_4})$: $t_6 = 12$, $t_3 = 12$; $t_3(1) = 16$, $t_2(1) = 18$.

$\mathcal{A}(\mathbf{F_4})$: $t_9 = 24$, $t_4 = 96$; $t_4(1) = 18$, $t_3(1) = 32$, $t_2(1) = 72$.

(3.5)

Definition 3.1.1 An arrangement $\mathcal{A} \subset \mathbb{P}^n$ is said to be in (combinatorial) *general position*, if $t_q(j) = 0$ for all $q > n - j$. All $\mathbb{P}^j$'s $\subset \mathcal{A}$ for which $j > n - q$ holds are the *singularities* of the arrangement. The singularities are *genuine* if they are not the intersection of higher-dimensional singular loci with one of the planes of the arrangement. The union of all genuine singularities is the *singular locus*.

In the above arrangements we have the following singular loci:

$$
\begin{aligned}
\mathcal{A}(\mathbf{A_4}): &\quad \text{five singular points, ten singular lines;} \\
\mathcal{A}(\mathbf{B_4}): &\quad 12{=}4{+}8 \text{ (genuine) singular points, } 22{=}6{+}16 \text{ singular lines;} \\
\mathcal{A}(\mathbf{D_4}): &\quad 12 \text{ singular points, } 16 \text{ singular lines;} \\
\mathcal{A}(\mathbf{F_4}): &\quad 24 \text{ (genuine) singular points, } 50{=}18{+}32 \text{ singular lines.}
\end{aligned}
\tag{3.6}
$$

3.1.3 The tetrahedron

Consider now the arrangement $\mathcal{A}(\mathbf{A_4})$ in (3.5). By (3.6) the singular locus consists of five points and ten lines. We introduce the following notation: $P_1 = (1,0,0,0)$, $P_2 = (0,1,0,0)$, $P_3 = (0,0,1,0)$, $P_4 = (0,0,0,1)$, $P_5 = (1,1,1,1)$, and l_{ij} will denote the line joining P_i and P_j. Each line contains two of the five points, and at each of the points four of the ten lines meet. The arrangement is *resolved* by performing the following birational modification of $\mathbb{P}^3$:

    a) Blow up the five points, $\varrho_1 : \widehat{\mathbb{P}}^3 \longrightarrow \mathbb{P}^3$;

    b) Blow up the proper transforms of the ten lines, $\varrho_2 : \widetilde{\mathbb{P}}^3 \longrightarrow \widehat{\mathbb{P}}^3$, $\varrho : \widetilde{\mathbb{P}}^3 \longrightarrow \mathbb{P}^3$.
    (3.7)

In the resolution 15 exceptional divisors $E_1, \ldots, E_5$ and $L_{12}, \ldots, L_{45}$ are introduced. The E_i are the proper transforms of the exceptional divisors introduced under ϱ_1, and are isomorphic to $\mathbb{P}^2$ blown up in the four points $(1:0:0)$, $(0:1:0)$, $(0:0:1)$, $(1:1:1)$, as are the proper transforms H_i of the ten planes of the arrangement. The ten exceptional divisors L_{ij} are each isomorphic to $\mathbb{P}^1 \times \mathbb{P}^1$. The symmetry group of $\widetilde{\mathbb{P}}^3$ consists of projective linear transformations of $\mathbb{P}^3$ which preserve the arrangement $\mathcal{A}(\mathbf{A_4})$, together with certain *birational* transformations of $\mathbb{P}^3$ which are *regular* on $\widetilde{\mathbb{P}}^3$, i.e., which contain the singular locus (3.6) with simple multiplicity in their ramification locus. Hence the Weyl group itself, $W(\mathbf{A_4}) = \Sigma_5$ (symmetric group on five letters) is contained in the symmetry group. But in fact, Σ_6 is the symmetry group, and the extra generator is a permutation of one of the E_i and H_j, which clearly can be done *on* $\widetilde{\mathbb{P}}^3$.

3.1.4 A birational transformation

Note that since each of the ten lines in (3.6) contains two of the five points which are blown up under ϱ_1, the normal bundle of the proper transform of each line on $\widehat{\mathbb{P}}^3$ is $\mathcal{O}(1-2) \oplus \mathcal{O}(1-2) = \mathcal{O}(-1) \oplus \mathcal{O}(-1)$. By general results of threefold birational geometry, it follows that

    a) The divisors L_{ij} on $\widetilde{\mathbb{P}}^3$ may be blown down to an ordinary threefold rational point (node), i.e., a singularity given by the equation $x^2 + y^2 + z^2 + t^2 = 0$, OR:
    (3.8)

    b) The ten lines on $\widehat{\mathbb{P}}^3$ may be blown down to the nodes mentioned in a).

In other words, there is a threefold which we denote by T, which contains ten threefold nodes, with a birational triangle:

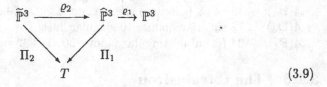

$$(3.9)$$

The map Π_2 blows down the union of ten disjoint "quadric surfaces" (i.e., divisors isomorphic to $\mathbb{P}^1 \times \mathbb{P}^1$) to ordinary nodes, while ϱ_2 blows these quadric surfaces down to ten disjoint lines, which Π_1 then blows down to the same ten isolated nodes. The 5+10 divisors E_i and H_j on $\widetilde{\mathbb{P}}^3$ have the following properties:

a) Each is isomorphic to $\mathbb{P}^2$ blown up in four points;

b) Each contains ten lines of intersection with the other 15, forming an arrangement in the blown up $\mathbb{P}^2$ of ten lines meeting in 15 points.

c) Under the birational map Π_2 each of the divisors E_i and H_j are blown down to a $\mathbb{P}^2$; the image of the ten lines of b) lie four at a time in each of these $\mathbb{P}^2$'s, as the four t_3-points of the following arrangement, which is the union of the intersections of the given $\mathbb{P}^2$ with the others:

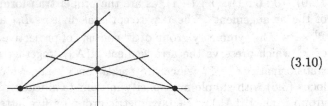

$$(3.10)$$

d) The composition $\Pi_2 \circ \varrho^{-1}$ restricted to each of the planes H_j is a usual Cremona transformation, blowing up three non-colinear points and blowing down the three joining lines. *Proof:* Take a face H_j of the tetrahedron; ϱ_1 blows up the three vertices it contains, so $\varrho_1^{-1}(H_j)$ (the proper transform of H_j) is $\mathbb{P}^2$ blown up in three points. Under ϱ_2, a fourth point of $\varrho_1^{-1}(H_j)$ is blown up, but it is blown down again under Π_2, as are the proper transforms of the three lines (in the plane H_j) joining the three vertices. By symmetry the same holds for all the H_j.

It follows that on T, the images $\widetilde{H}_j = \Pi_2(H_j)$ and $\widetilde{E}_i = \Pi_2(E_i)$ are copies of $\mathbb{P}^2$, each containing four of the ten nodes of T. Furthermore, in each of $\widetilde{H}_j$ and $\widetilde{E}_j$ we have the four t_3-points of the arrangement (3.10), which are these four nodes of T. Finally, since there are 15 $\mathbb{P}^2$'s, ten nodes and four of them in each of the 15 $\mathbb{P}^2$'s, there are five of these divisors passing through a given node. Explicitly, take the node n_{ij} corresponding to the line l_{ij} in (3.6). Then it meets the exceptional divisors $\widetilde{E}_i$, $\widetilde{E}_j$, as well as the three of the $\widetilde{H}_\nu$ for which H_ν contains the line l_{ij}.

3.1.5 Fermat covers associated with arrangements

Let $\mathcal{A} \subset \mathbb{P}^n$ be an arrangement of hyperplanes, i.e., a union $\mathcal{A} = \cup_{i=1}^k H_i$ of k hyperplanes, and let $d \geq 2$ be an integer. To the pair $(\mathcal{A}, d)$ there is an associated function field $\mathcal{L}(\mathcal{A}, d)$, an algebraic extension of the rational function field $\mathcal{M}(\mathbb{P}^n)$. It defines, in a unique way, a branched cover $Y(\mathcal{A}, d) \longrightarrow \mathbb{P}^n$, and a unique desingularization $\widetilde{Y}(\mathcal{A}, d)$. The function field is defined by:

$$\mathcal{L}(\mathcal{A}, d) = \mathbb{C}\left(\frac{x_1}{x_0}, \ldots, \frac{x_n}{x_0}\right)\left[\sqrt[d]{H_2/H_1}, \ldots, \sqrt[d]{H_k/H_1}\right], \qquad (3.11)$$

and the cover $Y(\mathcal{A}, d)$ is the so-called Fox closure of the étale cover over $\mathbb{P}^n - \mathcal{A}$ which is defined by (3.11). $Y(\mathcal{A}, d)$ is smooth outside of the *singular locus* of $\mathcal{A}$ (Definition 3.1.1), and the singularities of $Y(\mathcal{A}, d)$ are resolved by resolving the singularities of $\mathcal{A}$. This is done by first blowing up all (genuine, i.e., not near pencil) singular points, then all singular lines, and so forth. The resolution (3.7) is a typical example. This is described in more detail in the author's thesis; the desingularization $\widetilde{Y}(\mathcal{A}, d)$ is the fibre product in the following diagram:

$$
\begin{array}{ccc}
\widetilde{Y}(\mathcal{A}, d) & \longrightarrow & Y(\mathcal{A}, d) \\
\widetilde{\pi} \downarrow & & \downarrow \pi \\
\widetilde{\mathbb{P}}^n & \xrightarrow{\varrho} & \mathbb{P}^n
\end{array}
\qquad (3.12)
$$

where the horizontal arrows are modifications and the vertical arrows are Galois covers with Galois group $(\mathbb{Z}/d\mathbb{Z})^{k-1}$, which is the Galois group of the field extension of (3.11). ϱ is the modification of $\mathbb{P}^n$ which resolves the singularities of $\mathcal{A}$. For example, each singular point P on $Y(\mathcal{A}, d)$ is resolved by an *irreducible* divisor D_P, which itself is a Fermat cover $Y(\mathcal{A}', d)$, where $\mathcal{A}' \subset \mathbb{P}^{n-1}$ is the arrangement induced in the exceptional $\mathbb{P}^{n-1}$ which resolves the point $P' = \pi(P)$. It consists of k' planes, where k' is the number of the k hyperplanes which meet at the point P', and in the process of resolving $Y(\mathcal{A}, d)$, the cover $Y(\mathcal{A}', d)$ is resolved also. Hence on $\widetilde{Y}(\mathcal{A}, d)$ there is a *smooth* divisor D_P which resolves the singular point P of $Y(\mathcal{A}, d)$.

The singular covers $Y(\mathcal{A}, d)$ can also be realized as complete intersections, namely the intersections of $N = k - n$ Fermat hypersurfaces in $\mathbb{P}^{k-1}$:

$$
\begin{array}{ccl}
F_1 & = & a_{11}x_1^d + \cdots + a_{1k}x_k^d \\
\vdots & \vdots & \\
F_N & = & a_{N1}x_1^d + \cdots + a_{Nk}x_k^d
\end{array}
\qquad (3.13)
$$

where $a_{11}H_1 + \cdots + a_{1k}H_k, \ldots, a_{N1}H_1 + \cdots + a_{Nk}H_k$ are the $(k-n)$ linear relations among the k hyperplanes H_i. The map $Y(\mathcal{A},d) \longrightarrow \mathbb{P}^n$ is realized explicitly by the map $(x_1, \ldots, x_k) \mapsto (x_1^d, \ldots, x_k^d)$.

3.1.6 The hypergeometric differential equation

The Fermat covers for the arrangement $\mathcal{A}(\mathbf{A_4})$ are closely related to solutions of the hypergeometric differential equation on $\mathbb{P}^3$, which is an algebraic differential equation with regular singular points, whose singular locus *coincides* with the arrangement $\mathcal{A}(\mathbf{A_4})$, meaning that solutions are locally branched along the planes of the arrangement.

First we introduce a new notation for the 15 surfaces E_i, H_j. These can be numbered by pairs (i,j), $i < j \in \{0, \ldots, 5\}$, with $E_i = H_{0i}$ and

$$H_{ij} \cap H_{kl} \neq \emptyset \iff i \neq j \neq k \neq l. \tag{3.14}$$

We denote by $0i$ the point P_i in $\mathbb{P}^3$, and by $0ij$ the singular line joining $0i$ and $0j$ in $\mathbb{P}^3$. We then let L_{0ij} denote the exceptional divisor on $\widetilde{\mathbb{P}}^3$. We have (in $\mathbb{P}^3$)

$$H_{ij} \cap H_{kl} = 0mn \iff \{i,j,k,l\} \cap \{0,m,n\} = \emptyset. \tag{3.15}$$

We want to consider branched covers $Y \longrightarrow \widetilde{\mathbb{P}}^3$ (with $\widetilde{\mathbb{P}}^3$ as in (3.7)), which are branched along the H_{ij} and the L_{0ij}. Hence we let

$$n_{ij} := \text{ branching degree along } H_{ij}; \quad n_{0ij} := \text{ branching degree along } L_{0ij}, \tag{3.16}$$

and of course n_{ij}, $n_{0ij} \in \mathbb{Z} \cup \infty$. (It makes sense to allow negative branching degrees, as we will see below.)

To define the hypergeometric differential equation we may just as well work on $\mathbb{P}^n$ with homogenous coordinates $(x_0 : \ldots : x_n)$, and consider the arrangement $\mathcal{A}(\mathbf{A_n})$ of (3.3). Let $\lambda_i \in \mathbb{Q}$, $i = 0, \ldots, n+1, \infty$, $\sum_i \lambda_i = n+1$. The hypergeometric differential equation is (see [Te]):

$$\begin{cases} (x_i - x_j)\partial_i \partial_j F + (\lambda_i - 1)(\partial_i F - \partial_j F) = 0, \ 1 \le i < j \le n \\ x_i(x_i - 1)\partial_i^2 F + P_i(x,\lambda)\partial_i F + (\lambda_i - 1) \sum \frac{x_\alpha(x_\alpha-1)}{(x_i - x_\alpha)}\partial_\alpha F \\ \qquad + \lambda_\infty(1 - \lambda_i)F = 0, \ 1 \le i \le n. \end{cases} \tag{3.17}$$

where

$$P_i(x,\lambda) = x_i(x_i - 1)\sum \frac{1 - \lambda_\alpha}{x_i - x_\alpha} + \lambda_0 + \lambda_i - 3 - (2\lambda_i + \lambda_0 + \lambda_{n+1})x_i.$$

A solution of (3.17) turns out to be a period of an algebraic curve (the periods are many valued, as are the solutions of (3.17)). The curve is

$$y^\nu = x^{\mu_0}(x-1)^{\mu_{n+1}}(x - t_1)^{\mu_1} \cdots (x - t_n)^{\mu_n}, \tag{3.18}$$

where the μ_i, ν are related to the λ_i by the relation

$$\frac{\mu_i}{\nu} = 1 - \lambda_i. \tag{3.19}$$

The equation (3.17) has an $(n+1)$-dimensional solution space, spanned by $(n+1)$ periods of differentials of the curve (3.18):

$$\omega_i = \int_{\gamma_i} \frac{dx}{y}, \quad <\gamma_0,\ldots,\gamma_n> = H^1(C,\mathbb{Z}). \qquad (3.20)$$

Taking these gives a homogenous many valued map

$$(\omega_0,\ldots,\omega_n) : D \subset \widetilde{\mathbb{P}}^n \xrightarrow{\phi} \mathbb{P}^n, \qquad (3.21)$$

where D is some Zariski open set (see (3.23) below). The map is well-defined, since not all ω_i vanish simultaneously. For very special values of the parameters λ_i, the image of ϕ is the complex ball $\mathbb{B}_n \subset \mathbb{P}^n$ (this is just the Borel embedding of $\mathbb{B}_n$ in its compact dual). In fact, one has the following theorem:

Theorem 3.1.2 ([DM], [Te]) *If the following conditions are satisfied, then* $\phi(D) = \mathbb{B}_n$:

$$\sum \mu_i = 2, \quad \forall_{i,j} : (1 - \mu_i - \mu_j)^{-1} \in \mathbb{Z} \cup \infty.$$

In this case there exists a finite cover

$$Y \longrightarrow D$$

branched along the total transform of $\mathcal{A}(\mathbf{A_n})$, *which is a (compactification of a) quotient* $\Gamma\backslash\mathbb{B}_n$ *with* Γ *torsion free.*

The integers $n_{ij} := (1-\mu_i-\mu_j)^{-1}$ are then just the branching degrees of $Y \longrightarrow D$ along the divisor H_{ij}. In fact the numbering introduced in (3.14) can be done analogously for any n.

In the special case of $\mathcal{A}(\mathbf{A_4})$ on $\mathbb{P}^3$, the integers n_{0ij} of (3.16) are determined by the relation

$$n_{0ij} = 2\left(\frac{1}{n_{kl}} + \frac{1}{n_{lm}} + \frac{1}{n_{km}}\right)^{-1},$$

where the line $0ij$ is the intersection of H_{kl}, H_{lm}, H_{km}, and these together with the n_{ij} describe the branching degrees along the entire branch locus. The solutions of the equations in Theorem 3.1.2 are as follows:

1) $\frac{1}{3}, \frac{1}{3}, \frac{1}{3}, \frac{1}{3}, \frac{1}{3}, \frac{1}{3}$,

2) $\frac{1}{2}, \frac{1}{2}, \frac{1}{4}, \frac{1}{4}, \frac{1}{4}, \frac{1}{4}$

3) $\frac{3}{4}, \frac{1}{4}, \frac{1}{4}, \frac{1}{4}, \frac{1}{4}, \frac{1}{4}$

4) $\frac{1}{2}, \frac{1}{3}, \frac{1}{3}, \frac{1}{3}, \frac{1}{3}, \frac{1}{6}$ $\qquad (3.22)$

5) $\frac{3}{8}, \frac{3}{8}, \frac{3}{8}, \frac{3}{8}, \frac{3}{8}, \frac{1}{8}$

6) $\frac{5}{12}, \frac{5}{12}, \frac{5}{12}, \frac{1}{4}, \frac{1}{4}, \frac{1}{4}$

7) $\frac{7}{12}, \frac{5}{12}, \frac{1}{4}, \frac{1}{4}, \frac{1}{4}, \frac{1}{4}$

The set D of (3.21) is determined as the complement of

$$H_\infty = \{H_{ij} \big| n_{ij} = \infty; \ L_{0ij} \big| n_{0ij} = \infty\} \subset \widetilde{\mathbb{P}}^3. \qquad (3.23)$$

This is the locus which the uniformizing map (3.21) maps onto the *boundary* of $\mathbb{B}_3 \subset \mathbb{P}^3$, i.e., $\phi(D) = \mathbb{B}_n$, $\phi(H_\infty) \subset \partial \mathbb{B}_n$. This requires of course that the corresponding covers of the divisors on the cover $\widetilde{Y}$ be abelian varieties (as these are the compactification divisors on ball quotients). This can happen as follows

  (i) On one of the H_{ij}, this can occur if the branching degrees are: 2 for the six lines of (3.10), and -4 for the four exceptional curves.

  (ii) On L_{0ij}, this can happen if $\mu_k + \mu_l + \mu_m = 1$, $\mu_0 + \mu_i + \mu_j = 1$.
$$(3.24)$$
In the second case, the surface S_{0ij} covering L_{0ij} is of the form $C_1 \times C_2$, where $C_1 \longrightarrow \mathbb{P}^1$ (respectively $C_2 \longrightarrow \mathbb{P}^1$) is a cover, with branching determined by (μ_k, μ_l, μ_m) (respectively determined by (μ_0, μ_i, μ_j)). It is an abelian variety $\iff$ both curves C_i are elliptic. Note that $Y \longrightarrow \mathbb{P}^3$ will be a Fermat cover $\iff$ all n_{ij} coincide $\iff$ all μ_i coincide. In particular,

Proposition 3.1.3 *The only ball quotient in the list (3.22) which is a Fermat cover which is a ball quotient is the solution 1), namely $Y(\mathcal{A}(\mathbf{A}_4), 3)$ is a ball quotient.*

Remark 3.1.4 We will see later (see I3 following Lemma 6.4.2 below) that the solution 4) gives rise also to a Fermat cover which is a ball quotient, namely $Y(\mathcal{A}(\mathbf{D}_4), 3)$.

3.2 The Segre cubic $\mathcal{S}_3$

In this section we will show that the variety T of (3.9) has a projective embedding as a cubic hypersurface known as the Segre cubic, which we denote by $\mathcal{S}_3$.

3.2.1 Segre's cubic primal

In $\mathbb{P}^5$ with homogenous coordinates $(x_0 : \ldots : x_5)$ consider the locus

$$\mathcal{S}_3 := \{\sum_{i=0}^{5} x_i = 0; \ \sum_{i=0}^{5} x_i^3 = 0\}. \qquad (3.25)$$

As the first equation is linear, this shows that $\mathcal{S}_3$ is a hypersurface, i.e., $\mathcal{S}_3 \subset \mathbb{P}^4 = \{x \in \mathbb{P}^5 \big| \sum x_i = 0\}$. Using $(x_0 : \ldots : x_5)$ as projective coordinates, the relation $x_5 = -x_0 - \cdots - x_4$ gives the equation of $\mathcal{S}_3$ as a hypersurface; however, the equation in $\mathbb{P}^5$ shows that $\mathcal{S}_3$ is invariant under the symmetry group Σ_6, acting on $\mathbb{P}^5$ by permuting coordinates, which is not so immediate from the hypersurface equation.

It is known that for any degree d there is an upper bound on the number of ordinary double points which a hypersurface of degree d can have, the so-called Varchenko bound[1]. For cubic threefolds this number is ten, and it has been known since the last century that S_3 is the *unique* (up to isomorphism) cubic with ten nodes. The nodes on S_3 are given by the points of $\mathbb{P}^5$ for which three of the coordinates are 1 and the other three are -1. This is just the Σ_6-orbit of

$$(1, 1, 1, -1, -1, -1). \tag{3.26}$$

There is another interesting locus on S_3. Consider, in $\mathbb{P}^5$, the planes P_σ given by

$$P_\sigma = \{x_{\sigma(0)} + x_{\sigma(3)} = x_{\sigma(1)} + x_{\sigma(4)} = x_{\sigma(2)} + x_{\sigma(5)} = 0\}, \tag{3.27}$$

where $\sigma \in \Sigma_6$. There are 15 such P_σ's, the Σ_6-orbit of

$$P_{id} = \{x_0 + x_3 = x_1 + x_4 = x_3 + x_5 = 0\}. \tag{3.28}$$

One checks easily that each P_σ contains four of the double points; for example P_{id} contains the following:

$$(1, 1, -1, 1, -1, -1), \ (1, -1, 1, 1, -1, -1), \ (1, -1, -1, 1, 1, -1), \ (1, -1, -1, 1, -1, 1).$$

Furthermore, the intersection of P_{id} with the other P_σ is the line arrangement (3.10). It is easily checked that each P_σ is contained entirely in S_3. One can also argue as follows. Any line in $\mathbb{P}^5$ which contains two of the nodes of S_3 meets S_3 with multiplicity 4, hence is contained in S_3. Similarly, each P_σ meets S_3 in the six lines of the arrangement (3.10), hence is contained in S_3.

We just remark that the hyperplane sections $\{x_i = 0\}$ of S_3 are cubic surfaces with equation

$$S_3 = \{\sum_{i=0}^{4} x_i = \sum_{i=0}^{4} x_i^3 = 0\}. \tag{3.29}$$

This cubic surface is known as the Clebsch diagonal surface and will be studied later in Chapter 4. It is the unique cubic surface having Σ_5 as symmetry group. The relation between S_3 and the icosahedral group was studied by Hirzebruch. It turns out that S_3 is A_5-equivariantly birational to the Hilbert modular surface for $\mathcal{O}_{\mathbb{Q}(\sqrt{5})}$, of level $\sqrt{5}$.

Other interesting hyperplane sections are given by the hyperplanes $T_{ij} = \{x_i - x_j = 0\}$; indeed, T_{ij} also contains four of the ten nodes, hence $T_{ij} \cap S_3$ is a four-nodal cubic surface. This four-nodal cubic surface is projectively unique, and is called the Cayley cubic.

---

[1]There are also better bounds due to Miyaoka, Čmutov and others. See the first chapter of A. Kalkar, "Cubic fourfolds with fifteen ordinary double points", Phd. Thesis, Leiden 1986.

3.2.2   A birational transformation

Theorem 3.2.1 *The variety T of equation (3.9) is biregular to S_3; the isomorphism $\psi : T \longrightarrow S_3$ defined below is Σ_6-equivariant.*

Proof: Following Baker [Ba1], IV, p. 152, we define a birational map

$$\beta : \mathbb{P}^3 - - \rightarrow S_3.$$

Consider all quadric surfaces in $\mathbb{P}^3$ passing through the points P_i of (3.6). Generators of this linear system are given by the following degenerate quadrics. Let $(z_0 : \ldots : z_3)$ be homogenous coordinates on $\mathbb{P}^3$, and set

$$\begin{aligned}
\xi &= z_0(z_3 - z_1), & \eta &= z_1(z_3 - z_2), & \zeta &= z_2(z_3 - z_0); \\
\xi' &= z_1(z_3 - z_0), & \eta' &= z_2(z_3 - z_1), & \zeta' &= z_0(z_3 - z_2).
\end{aligned} \tag{3.30}$$

These quadrics satisfy the relations $\xi + \eta + \zeta = \xi' + \eta' + \zeta'$ and $\xi\eta\zeta = \xi'\eta'\zeta'$. Now change coordinates by setting

$$\begin{aligned}
\xi &= X + Y, & \eta &= Y + Z, & \zeta &= X + Z; \\
\xi' &= -(X' + Y'), & \eta' &= -(Y' + Z'), & \zeta' &= -(X' + Z').
\end{aligned} \tag{3.31}$$

Then the relations $\xi + \eta + \zeta = \xi' + \eta' + \zeta'$ and $\xi\eta\zeta = \xi'\eta'\zeta'$ become

$$\begin{aligned}
X + Y + Z + X' + Y' + Z' &= 0 \tag{3.32} \\
X^3 + Y^3 + Z^3 + (X')^3 + (Y')^3 + (Z')^3 &= 0.
\end{aligned}$$

One sees this is just equation (3.25) of the Segre cubic. This yields a rational map $\beta : \mathbb{P}^3 - - \rightarrow S_3$, $\beta(z_0 : z_1 : z_2 : z_3) = (X, Y, Z, X', Y', Z')$. The base locus of the linear system of quadrics defining β (3.30) is the five points of (3.6), as the quadrics all contain these points. It follows that β blows up all five points, the exceptional divisors $E_1, \ldots, E_5$ being projective planes. Now consider one of the ten lines of (3.6); for example, the one given by $z_2 = z_3 = 0$. Then $\eta = \zeta = \eta' = \zeta' = 0$ and $\xi = \xi' = -z_0 z_1$. In other words, β maps that line to the point $(1, 0, 0, 1, 0, 0)$ in the $(\xi, \eta, \zeta, \xi', \eta', \zeta')$ space, which is the point $(1, 1, -1, -1, -1, 1)$ in the (X, Y, Z, X', Y', Z') space. But that is just one of the ten nodes of S_3. From Σ_6-symmetry we conclude that $\beta : \mathbb{P}^3 - - \rightarrow S_3$ coincides with the map $\Pi = \Pi_1 \circ \varrho_1^{-1}$, with Π_1 as in (3.9) and ϱ_1 as in (3.7). In other words, $\beta = \Pi$ is the composition of morphisms

$$\begin{array}{ccc}
 & \widehat{\mathbb{P}}^3 & \\
\varrho_1 \swarrow & & \searrow \Pi_2 \\
\mathbb{P}^3 \,\, - - - - \!\!\! & \xrightarrow{\;\Pi\;} & T,
\end{array} \tag{3.33}$$

and since (3.32) states that $\beta = \Pi$ maps onto S_3, this gives an isomorphism $T \cong S_3$. Explicitly, $t \in T$, $t \mapsto (\varrho_1 \circ \Pi_2^{-1})(t) \mapsto \beta((\varrho_1 \circ \Pi_2^{-1})(t)) = \psi(t) \in S_3$

is the desired map. The Σ_6-equivariance follows from the fact that the whole diagram (3.33) is Σ_6-equivariant. □

Now consider the Picard group of $\mathcal{S}_3$. From the explicit form of birational map as given by Theorem 3.2.1 and (3.9), we see that $\mathrm{Pic}(\mathcal{S}_3)$ is generated by the image of the hyperplane class, call it H, and the five exceptional classes E_i. It follows that $\mathrm{Pic}(\mathcal{S}_3)$ has rank 6, and the primitive part $\mathrm{Pic}^0(\mathcal{S}_3)$, i.e., the complement of the hyperplane class, has rank 5. The 15 classes H_{ij} introduced in (3.14) (these are the 15 linear $\mathbb{P}^2$'s on $\mathcal{S}_3$ noted in (3.27)) give classes in $\mathrm{Pic}(\mathcal{S}_3)$ and in $\mathrm{Pic}^0(\mathcal{S}_3)$. The 15 hyperplanes

$$\mathcal{H}_{ij} = \{x_i + x_j = 0\}, \tag{3.34}$$

each of which meets $\mathcal{S}_3$ in three of the 15 $\mathbb{P}^2$'s, give 15 *relations* in $\mathrm{Pic}^0(\mathcal{S}_3)$: since $\mathcal{H}_{ij} \cap \mathcal{S}_3$ is a hyperplane section, the sum of the three $\mathbb{P}^2$'s cut out by $\mathcal{H}_{ij}$, i.e., $H_{i_1,j_1} + H_{i_2,j_2} + H_{i_3,j_3} = \mathcal{H}_{ij} \cap \mathcal{S}_3$, is linearly equivalent to the hyperplane class. This yields the following exact sequence of $\mathbb{Z}$-modules:

$$
\begin{array}{ccccccccc}
1 & \longrightarrow & K & \longrightarrow & \mathbb{Z}\{\mathcal{H}_{ij}\} & \longrightarrow & \mathbb{Z}\{H_{ij}\} & \longrightarrow & \mathrm{Pic}^0(\mathcal{S}_3) & \longrightarrow & 1 \\
 & & \| \wr & & \| \wr & & \| \wr & & \| \wr & & \\
1 & \longrightarrow & \mathbb{Z}^5 & \longrightarrow & \mathbb{Z}^{15} & \longrightarrow & \mathbb{Z}^{15} & \longrightarrow & \mathbb{Z}^5 & \longrightarrow & 1.
\end{array}
\tag{3.35}
$$

Lemma 3.2.2 *In the sequence (3.35), all $\mathbb{Z}$-modules are Σ_6-modules, i.e., the exact sequence is one of Σ_6-modules.*

Proof: This is visible for the right three entries of the first sequence in (3.35), and it then follows for K. □

Now consider a generic hyperplane section of $\mathcal{S}_3$; this is a smooth cubic surface. Let $\nu : S = \mathcal{S}_3 \cap H \hookrightarrow \mathcal{S}_3$ denote the inclusion of the section, and let $\nu^* : H^2(\mathcal{S}_3, \mathbb{Z}) \longrightarrow H^2(S, \mathbb{Z})$ be the induced map on cohomology. Then by the Lefschetz hyperplane theorem, this map is *injective*, and since both S and $\mathcal{S}_3$ have no holomorphic 2-forms, we may view this as an injective map of the Picard groups: $\mathrm{Pic}(\mathcal{S}_3) \hookrightarrow \mathrm{Pic}(S)$, and a corresponding inclusion on the primitive part. Recall also that we have on the cubic surface 27 generators (the 27 lines), 45 relations among these (the 45 tritangents), and an exact sequence on $\mathrm{Pic}^0(S)$ as in (4.12). All in all we get the following map of sequences as in (3.35):

$$
\begin{array}{ccccccccc}
1 & \longrightarrow & \mathbb{Z}^5 & \longrightarrow & \mathbb{Z}^{15} & \longrightarrow & \mathbb{Z}^{15} & \longrightarrow & \mathbb{Z}^5 & \longrightarrow & 1 \\
 & & \downarrow & & \downarrow & & \downarrow & & \downarrow & & \\
1 & \longrightarrow & \mathbb{Z}^{24} & \longrightarrow & \mathbb{Z}^{45} & \longrightarrow & \mathbb{Z}^{27} & \longrightarrow & \mathbb{Z}^6 & \longrightarrow & 1.
\end{array}
\tag{3.36}
$$

where the right hand groups are $\mathrm{Pic}^0(\mathcal{S}_3)$ and $\mathrm{Pic}^0(S)$, respectively, and the down arrows are inclusions (by Lefschetz). Note that this corresponds to a symmetry breaking. Indeed, on the first sequence there is a symmetry group Σ_6 acting, as already noted, while on the group $\mathrm{Pic}^0(S)$, in fact on the whole second sequence, the group $W(E_6)$ acts naturally, as is well-known.

Proposition 3.2.3 *The ideal $\mathfrak{I}(10)$ of the ten nodes is generated by the five quadrics $\mathcal{R}_\lambda$ of the Jacobian ideal of $\mathcal{S}_3$.*

Proof: The inclusion $Jac(\mathcal{S}_3) \subset \mathfrak{I}(10)$ is obvious, and the five elements of $Jac(\mathcal{S}_3)$ are clearly independent. The fact that $\mathfrak{I}(10)$ has rank 5 has been verified by standard basis computations (with the algebra program Macaulay).   □

Corollary 3.2.4 *The ideal of the ten nodes of $\mathcal{S}_3$, $\mathfrak{I}(10)$, is the Jacobian ideal of $\mathcal{S}_3$.*   □

3.2.3 Uniformization

In this section we will show that the Segre cubic $\mathcal{S}_3$ is actually the Satake compactification of a Picard modular variety. Let $K = \mathbb{Q}(\sqrt{-3})$ be the field of Eisenstein numbers, and consider the $\mathbb{Q}$-group G whose set of $\mathbb{Q}$-points is given by $G(\mathbb{Q}) = U(3,1;K)$, the unitary group of the standard hermitian form on K^4 with signature (3,1). Consider the arithmetic group $\Gamma := G_\mathbb{Z} = U(3,1;\mathcal{O}_K) \subset G(\mathbb{Q})$, where $\mathcal{O}_K$ denotes the ring of integers in K. It acts on the three-ball with non-compact quotient X_Γ. Consider the principal congruence subgroups $\Gamma(\sqrt{-3})$ and $\Gamma(3)$, as defined in (2.17). These determine a corresponding level structure in the sense of Definition 2.2.4. Now note the following well-known isomorphisms:

$$\Gamma/\Gamma(\sqrt{-3}) = \Sigma_6, \quad \Gamma(3)/\Gamma(\sqrt{-3}) = (\mathbb{Z}/3\mathbb{Z})^9. \tag{3.37}$$

It follows from this that the corresponding quotients $X(a) := X_{\Gamma(a)}$, $a = 1, \sqrt{-3}, 3$, yield Galois covers

$$X(3) \xrightarrow{(\mathbb{Z}/3\mathbb{Z})^9} X(\sqrt{-3}) \xrightarrow{\Sigma_6} X(1), \tag{3.38}$$

which explicitly describe the level structures involved. As usual let $X(a)^*$ denote the Satake compactification.

Theorem 3.2.5 *There is a commutative diagram*

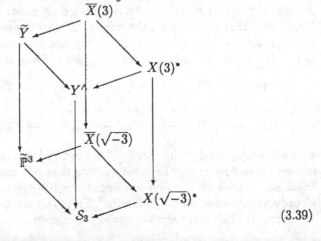

$$\tag{3.39}$$

where the horizontal maps from right to left are isomorphisms, those from left to right are birational, and the vertical maps are $(\mathbb{Z}/3\mathbb{Z})^9$ covers.

Proof: First we have, over $\mathbb{P}^3$, a singular cover T_{DM} defined by the solution 1) of (3.22). This is desingularized by blowing up the $\mathbb{P}^3$ along the singular locus of the arrangement, $\widetilde{\mathbb{P}}^3 \longleftarrow \widetilde{T}_{DM}$. From the fact that all $\mu_i = 1/3$, we see that all n_{ij} and n_{0ij} are equal to three, that is T_{DM} is the Fermat cover $Y(\mathcal{A}(\mathbf{A_4}), 3)$, and $\widetilde{T}_{DM}$ is its desingularization $\widetilde{Y} := \widetilde{Y}(\mathcal{A}(\mathbf{A_4}), 3)$ as in (3.12); see also Proposition 3.1.3. By Theorems 3.1.2 and 3.2.1, $\widetilde{Y}$ is the desingularization of the ball quotient $\Gamma'\backslash\mathbb{B}_3$, for some torsion free group Γ'. Blowing $\widetilde{Y}$ down from $\widetilde{\mathbb{P}}^3$ to $\mathcal{S}_3$ gives the singular variety $Y^\wedge$, which we will see in a minute is the Satake compactification of the ball quotient. Hence we only need to identify the groups and check the compactifications coincide. As to the first, we start with

Lemma 3.2.5.1 *Let $G_{\mathbb{Q}}$ be an isotropic $\mathbb{Q}$-form of $U(3,1)$, $G_{\mathbb{Q}} \sim U(3,1;L)$, L imaginary quadratic over $\mathbb{Q}$, and let $\Gamma \subset G_{\mathbb{Q}}$ a torsion free arithmetic subgroup with arithmetic quotient X_Γ, Baily-Borel compactification X_Γ^* and toroidal compactification $\overline{X}_\Gamma$. Then the isomorphism class of a single compactification divisor determines the field L, and hence $G_{\mathbb{Q}}$ up to isogeny.*

Proof: First note that for $U(3,1)$ the parabolic (there is only one conjugacy class of parabolics, as the $\mathbb{R}$-rank is one) takes on the particularly simple form

$$P \cong (\mathcal{R}\mathcal{K}) \rtimes \mathcal{Z}V, \quad \mathcal{R} \cong \mathbb{R}^\times, \quad \mathcal{K} = SU(2) \times U(1)$$
$$\mathcal{Z} = \mathbb{R}, \quad V = \mathbb{C}^2$$
$$(3.40)$$

For the $\mathbb{Q}$-form of P, it follows that $V_{\mathbb{Q}} \cong L^2$ for some imaginary quadratic field L, and for the arithmetic parabolic $\Gamma_P \subset P$, $\Gamma_P \cap V_{\mathbb{Q}} \subset (\mathcal{O}_L)^2$ is some lattice. Furthermore, the theory of toroidal embeddings shows that a compactification divisor of $\overline{X}_\Gamma$ is of the form $\mathbb{C}^2/(\Gamma_P \cap V_{\mathbb{Q}})$, which has complex multiplication by L, so its isomorphism class determines L, which was to be shown. $\square$

Now an easy calculation shows what the compactification divisors on $\widetilde{Y}$ are. Namely, these are the irreducible components of the inverse image in $\widetilde{Y}$ of the exceptional divisors $L_{0ij} \subset \widetilde{\mathbb{P}}^3$, $L_{0ij} \cong \mathbb{P}^1 \times \mathbb{P}^1$. The local geometry of the arrangement shows the branch locus in L_{0ij} is of the form $p_1^*(\mathcal{O}(3)) \otimes p_2^*(\mathcal{O}(3))$, i.e., of the form $\{0\} \times \mathbb{P}^1$, $\{1\} \times \mathbb{P}^1$, $\{\infty\} \times \mathbb{P}^1$ and $\mathbb{P}^1 \times \{0\}$, $\mathbb{P}^1 \times \{1\}$, $\mathbb{P}^1 \times \{\infty\}$. It is well-known that the Fermat cover $E \longrightarrow \mathbb{P}^1$, branched at $(0, 1, \infty)$ to degree 9, with Galois group $(\mathbb{Z}/3\mathbb{Z})^2$, is the elliptic curve $E_\varrho = \mathbb{C}/\mathbb{Z} \oplus \varrho\mathbb{Z}$, $\varrho = e^{2\pi i/3}$. From this it follows

Lemma 3.2.5.2 *The compactification divisors Δ_i of $\widetilde{Y}$ are products*

$$\Delta_i \cong E_\varrho \times E_\varrho,$$

where E_ϱ is the unique elliptic curve with $\mathbb{Z}/6\mathbb{Z}$ as automorphism group, i.e., $E_\varrho = \mathbb{C}/\mathbb{Z} \oplus \varrho\mathbb{Z} = \{x^3 + y^3 + z^3 = 0\}$ and $Aut(E_\varrho) = <\pm 1, \pm\varrho, \pm\varrho^2>$. $\square$

Note that the morphism $\widetilde{Y} \longrightarrow Y^\wedge$ blows down the Δ_i to singular points (just as $\widetilde{\mathbb{P}}^3 \longrightarrow S_3$ blows down the L_{0ij} to the nodes of S_3) which lie over the nodes of S_3. From this and the well-known fact that the Satake compactification of a ball quotient has only isolated, zero-dimensional singularities, which are resolved in a torus embedding by means of complex tori or quotients thereof, and the fact, known by [DM], that the group is arithmetic, we get the following

Corollary 3.2.5.1 *The variety $Y^\wedge$ is the Satake compactification of the quotient $\Gamma' \backslash \mathbb{B}_3$, with $\Gamma' \subset G_{\mathbb{Q}}$ and $G_{\mathbb{Q}} = PU(3,1;K)$, K the field of Eisenstein numbers as above.* $\qquad\square$

Now that it is established that Γ' is an arithmetic subgroup[2] of $PU(3,1;K)$, group-theoretic methods can be applied to determine the arithmetic subgroup. We utilize the method of [J], where it was applied to the levels $\sqrt{-3}$, 2 and $2\sqrt{-3}$. First, let y_b denote the number of cusps of a ball group, and y_m the number of subball quotients as in *loc. cit.*, Lemma 2.9.

Lemma 3.2.5.3 *The following table is correct*

group	y_b	y_m
$\Gamma(\sqrt{-3})$, Γ_{S_3}	10	15
$\Gamma(3)$, Γ'	810	405

where Γ_{S_3} is the ball group such that S_3 is the Satake compactification of $\Gamma_{S_3} \backslash \mathbb{B}_3$.

Proof: For $\Gamma(\sqrt{-3})$ this is contained in *loc. cit*, Lemma 2.9. For Γ_{S_3} we know the ten nodes of S_3 are the cusps, hence $y_b = 10$. For the 15 $\mathbb{P}^2$'s on S_3 we know by *loc. cit*, Proposition 3.9 that these are the subball quotients on S_3, hence $y_m = 15$. Next consider Γ' and the smooth Fermat cover $\pi : \widetilde{Y} \longrightarrow \widetilde{\mathbb{P}}^3$. By general facts of Fermat covers, the degree of π is 3^9, while each irreducible component over a subball (a $\mathbb{P}^2$) is a Fermat cover of degree 3^5; since π is ramified along these $\mathbb{P}^2$'s, the total degree over each is 3^8, hence there are $3^{8-5} = 27$ irreducible components over each. It follows that for Γ', $y_m = 27 \cdot 15 = 405$. Similarly, the cusps on S_3 are the images of the lines denoted $0mn$ in (3.15) on $\mathbb{P}^3$, and the number of irreducible components over each is $3^{9-3-2} = 81$. It follows for Γ' that $y_b = 10 \cdot 81 = 810$. For the group $\Gamma(3)$ it is easiest to determine y_m and y_b again by considering the cover $(\Gamma(3) \backslash \mathbb{B}_3)^* \longrightarrow (\Gamma(\sqrt{-3}) \backslash \mathbb{B}_3)^*$ (since 3 is not a prime in K, the ring $\mathcal{O}_K / 3\mathcal{O}_K$ is not a field and the description in terms of finite geometry as in Lemma 2.4.3 is not so convenient). Again this cover is of degree 3^9 and is branched along subball quotients and cusps. The local branching is easily checked to be identical to the cover $Y^\wedge \longrightarrow S_3$, and the arguments above apply to verify the second row of the table for $\Gamma(3)$. $\qquad\square$

We have already seen that Γ' is commensurable to $\Gamma(1) = U(3,1;\mathcal{O}_K)$.

---

[2]It is not completely obvious at this point that the unitary group is with respect to the *standard* form. It holds because the group in question arises from the family of curves (3.18) for the weights 1) in (3.22); the corresponding abelian varieties are Jacobians, hence principally polarized, so the form is standard by Proposition 2.2.3 and (1.9).

Lemma 3.2.5.4 Γ' *is conjugate to a subgroup of finite index of* $\Gamma(1)$.

Proof: The Lemmas 3.14 and 3.16 of *loc. cit.* apply to Γ', and it follows that Γ' is generated by subgroups, each of which is conjugate to a subgroup of $\Gamma_2(1) = U(2,1;\mathcal{O}_K)$. Furthermore, the intersections $N_i \cap N_j$ of these subgroups are conjugate, by the same elements as in the previous sentence, to subgroups of $\Gamma_1(1) = U(1,1;\mathcal{O}_K)$. Now conjugate Γ' by some element g such that for one of the N_i, say N_0 we have $N_0^g \subset \Gamma_2(1)$, where the latter group is the subgroup corresponding to deleting one of the anisotropic eigenvectors of the hermitian form. Since by *loc. cit.* 3.16 Γ' is generated by a parabolic P and the N_i which are incident to it, it is sufficient to show that the elements of $\Gamma'(1)$ which conjugate the N_0 to the other N_i are in $\Gamma(1)$, where $\Gamma'(1)$ is the group containing Γ' as a normal subgroup with quotient $(\mathbb{Z}/3\mathbb{Z})^9 \times \Sigma_6$. For more detailed arguments leading to this conclusion see Lemmas 5.6.7 and 5.6.8 below, which apply also to the present situation. So consider the elements of $\Gamma'(1)$ which conjugate N_0 to N_i, and consider this as an action on the compactification divisor $E_\varrho \times E_\varrho$ (Lemma 3.2.5.2). It is clear from the geometry of the situation that these elements are translations in the lattice $\Lambda(\infty)$ in the parabolic P (see the discussion following (5.39) below), hence are indeed in $\Gamma(1)$, and consequently Γ' is generated by subgroups P, N_0, each of which is conjugate (by a common element) to subgroups of $\Gamma(1)$, and conjugates of this conjugate of N_0 by elements of $\Gamma(1)$. Therefore Γ' is conjugate to a subgroup of $\Gamma(1)$. $\qquad\qquad\square$

We henceforth assume $\Gamma' \subset \Gamma(1)$. Then we have sequences (see *loc. cit.* p. 541 and 551)

$$1 \longrightarrow \Gamma' \longrightarrow \Gamma(1) \longrightarrow (\mathbb{Z}/3\mathbb{Z})^9 \times \Sigma_6 \longrightarrow 1$$
$$\| \qquad\qquad \|$$
$$1 \longrightarrow \Gamma(3) \longrightarrow \Gamma(1) \longrightarrow (\mathbb{Z}/3\mathbb{Z})^9 \times \Sigma_6 \longrightarrow 1.$$

Let P_3, $N_{3,0}$, $N_{3,i}$ denote the subgroups of $\Gamma(3)$ corresponding to P, N_0 and the N_i. We know that $P = P_3, N_0 = N_{3,0}$, and the same argument as in *loc. cit.* 3.15 implies that $N_{3,i} = N_i$ for all i, hence, as in *loc. cit.* 3.17 we have

Lemma 3.2.5.5 $\Gamma' = \Gamma(3)$.

Note that from Lemma 3.2.5.3 the modular and compactification divisors of $\widetilde{Y}$ are in one-to-one correspondence with those of $(\Gamma(3)\backslash\mathbb{B}_3)^\sim$, while by Lemma 3.2.5.2 and *loc. cit.*, Proposition 3.9 all components are isomorphic. Let Y_b (resp. $Y_{3,b}$) be the union of the 810 compactification divisors of $\widetilde{Y}$ (resp. $(\Gamma(3)\backslash\mathbb{B}_3)^\sim$) and Y_m (resp. $Y_{3,m}$) the union of the 405 modular subvarieties of $\widetilde{Y}$ (resp $(\Gamma(3)\backslash\mathbb{B}_3)^\sim$), and set

$$\mathcal{E} = Y_b \cup Y_m, \qquad \mathcal{E}_3 = Y_{3,b} \cup Y_{3,m}.$$

Lemma 3.2.5.6 *The normal crossings divisors* $\mathcal{E}$ *and* $\mathcal{E}_3$ *are isomorphic.*

Proof: This follows as in *loc. cit.*, Corollary 3.22: all components are isomorphic, the pairwise common intersections are also isomorphic, and the relevant graphs are isomorphic. $\qquad\qquad\square$

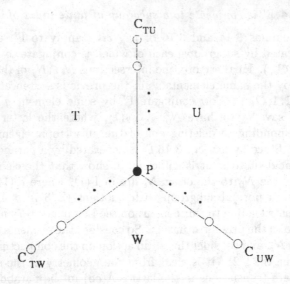

Figure 3.1: A fixed point for the action of $(\mathbb{Z}/3\mathbb{Z})^9$

We now consider the action of $(\mathbb{Z}/3\mathbb{Z})^9$ on $\tilde{Y}$ and on $(\Gamma(3)\backslash\mathbb{B}_3)^\sim$, (which we know by Lemma 3.2.5.5 and Corollary 3.2.5.1 are isomorphic).

Lemma 3.2.5.7 $\Gamma_{\mathcal{S}_3} = \Gamma(\sqrt{-3})$.

Proof: To show that $\Gamma_{\mathcal{S}_3} = \Gamma(\sqrt{-3})$ it will suffice to show that (under the isomorphism $\tilde{Y} \cong (\Gamma(3)\backslash\mathbb{B}_3)^\sim$) the action of $(\mathbb{Z}/3\mathbb{Z})^9$ on $\tilde{Y}$ coincides with that of $(\mathbb{Z}/3\mathbb{Z})^9$ on $(\Gamma(3)\backslash\mathbb{B}_3)^\sim$. As in *loc. cit.*, Corollary 3.11 we use the principle that two automorphisms of finite order on a smooth variety agree if they agree on some open set (in the complex topology), and they agree on an open set containing a fixed point it their derivatives agree on the tangent space to that point. As in *loc. cit.*, proof of Theorem 4.1, $\overline{X(1)} \xrightarrow{\sim} \overline{Y(1)}$, we consider a fixed point which is the intersection of three of the modular subvarieties. Compared with *loc. cit.*, Figure 6, we here have the Figure 3.1 where T, U, W are three of the modular surfaces and C_{TU}, C_{TW}, C_{UW} are the curves of intersection (themselves modular curves on each modular surface), and where there are a total of ninie points on each of these curves (only three are shown in Figure 3.1). In order to show that the two actions of $(\mathbb{Z}/3\mathbb{Z})^9$ agree on $\tilde{Y}$, it will suffice to show that the action on each of the curves C_{TU}, C_{TW}, C_{UW} coincide, as this action determines the derivative of each element at P. But for the Fermat cover $\tilde{Y}$, the action on each of the curves is just the action of $(\mathbb{Z}/3\mathbb{Z})^2$ on the Fermat curve $C \longrightarrow \mathbb{P}^1$ of degree $d = 3^2$ branched at three points. Thus it will suffice to show that for the cover on these curves induced by the cover $X(3) \longrightarrow X(\sqrt{-3})$, the cover is just the above mentioned Fermat cover. This is easy and left to the reader. □

To complete the proof of the Theorem it will suffice to check that the compactifications coincide; but this follows from Lemma 3.2.5.6.   □

3.2.4 Moduli interpretation

Now applying Shimura's theory we get the following moduli description of $\mathcal{S}_3$ (see [J], §2 for details).

Theorem 3.2.6 *Any point $x \in \mathcal{S}_3 - \{ten\ nodes\}$ determines a unique isomorphism class of principally polarized abelian fourfolds with the following PEL-data: $\{(K, \Phi, *), (Q, T, \mathcal{M}), y_1, \dots, y_8\}$,, where $Q = K^4$, T is the skew-hermitian form iJ, where J is the form of (5.38), $\mathcal{M} = \mathcal{O}_K^4$, and the points $y_1, \dots, y_8$ are the points of a level $\sqrt{-3}$ structure in the sense of Definition 2.2.4, i.e., the points of the kernel of multiplication by $\sqrt{-3}$. Any point*

$$x \in Y^\wedge - \{inverse\ image\ under\ \phi\ of\ (3.39)\ of\ the\ ten\ nodes\}$$

determines a unique isomorphism class of abelian fourfolds as above with a level 3 structure.

Moreover, the moduli interpretation of the 15 $\mathbb{P}^2$'s on $\mathcal{S}_3$ is given in [J].

Proposition 3.2.7 *The 15 $\mathbb{P}^2$'s on $\mathcal{S}_3$ are compactifications of two-dimensional ball quotients which are moduli spaces of those abelian fourfolds A_x^4 as above which split:*

$$A_x^4 \cong A_x^3 \times E_\varrho.$$

The intersections of the 15 planes determine moduli points of A_x^4 which further decompose, i.e., A_x^3 splits.

Remark 3.2.8 It is natural to ask whether, given a point $x \in \mathcal{S}_3$, one can give the equations defining the abelian variety A_x occuring in Theorem 3.2.6. In some sense one can. First it turns out the A_x is the Jacobian of an algebraic curve, as described by the hypergeometric equation as in equation (3.18). Since the parameters are by Proposition 3.1.3 the set 1) in (3.22), these curves have the form:

$$C_\tau = \{y^3 = \prod_{i=1}^{6} (x - t_i(\tau))\}; \tag{3.41}$$

C_τ obviously has an automorphism of order 3, given by $y \mapsto \varrho y$ with the third root of unity ϱ. This yields an automorphism of the Jacobian of C_τ. Without much difficulty one finds

(i) $\mathrm{Jac}(C_\tau) = A_\tau$ has complex multiplication by $\mathcal{O}_K$, the signature is (3,1).

(ii) The automorphism group is $\mathcal{O}_K^*$, and is given by multiplication by $\pm\varrho$ in $\mathcal{O}_K$.

The most direct way to see this is to write down the Jacobian of the curve (3.41) and show that its periods have the complex multiplication. A basis of the (1,0) differentials on C_τ written in the normal form

$$y^3 = x(x-1)(x-t_1)(x-t_2)(x-t_3) \qquad (3.42)$$

is given by

$$\int \frac{dx}{\sqrt[3]{x(x-1)(x-t_1)(x-t_2)(x-t_3)}}; \qquad (3.43)$$

choosing a basis of $H_1(C_\tau, \mathbb{Z})$ and taking the integrals over the elements of that basis gives the Jacobian; the multiplication by ϱ is then evident. Hence one may invoke Shimura's theory to conclude:

Lemma 3.2.9 *The isomorphism classes of the Jacobians of the curves (3.42) are given as the points of the arithmetic quotient* $PU(3, 1; \mathcal{O}_K)\backslash\mathbb{B}_3$. *Putting a* $\sqrt{-3}$ *level structure on the Jacobians yields the moduli space* $P\Gamma(\sqrt{-3})\backslash\mathbb{B}_3$.

The latter space has already been identified with the open subset of smooth points on $\mathcal{S}_3$.

The precise relation between the moduli *point* $\tau \in \mathbb{B}_3$ and the *values* of the t_i has been derived for surfaces, i.e., for τ in one of the subballs covering one of the 15 $\mathbb{P}^2$'s on $\mathcal{S}_3$, by Holzapfel. The result is: there are automorphic forms G_2, G_3 and G_4 of indicated weights on $\mathbb{B}_2$ such that

$$C_\tau = \{y^3 = x^4 - G_2(\tau)x^2 - G_3(\tau)x - G_4(\tau)\}, \qquad (3.44)$$

much akin to the Weierstraß equation for an elliptic curve. (The variable x in (3.44) is of course different than that in (3.42)). There is no doubt a similar expression for $\tau \in \mathbb{B}_3$.

3.3   The Igusa quartic $\mathcal{I}_4$

This variety has been known since the last century, and it is related to the configuration in $\mathbb{P}^4$ which is dual to the 15 hyperplanes of (3.34) and the 15 planes of (3.27) which they cut out on $\mathcal{S}_3$, and in fact $\mathcal{I}_4$ is just the dual variety of $\mathcal{S}_3$. It was also known in the last century that the tangent hyperplane sections of $\mathcal{I}_4$ are Kummer surfaces, giving $\mathcal{I}_4$ a moduli interpretation. Igusa, in the 1960's, made this rigorous and showed that $\mathcal{I}_4$ is the Satake compactification of $\Gamma(2)\backslash\mathbb{S}_2$, the Siegel modular threefold of level 2. We begin by discussing the projective variety, then turn to Igusa's results.

3.3.1   The quartic locus associated to a configuration of 15 lines

Let l_σ be the line dual in $\mathbb{P}^4$ to the P_σ of (3.27), and let h_{ij} denote the point dual to $\mathcal{H}_{ij}$ of (3.34). Then these 15 lines meet at the 15 points h_{ij}, and three of the 15 lines meet at each, corresponding to the three $\mathbb{P}^2$'s which are contained in each $\mathcal{H}_{ij}$. Furthermore, each of the 15 lines contains three of the 15 points, as each $\mathbb{P}^2$ is contained in three of the $\mathcal{H}_{ij}$. It is useful to introduce the following notation: each line is given a notation (ij), and two such lines (ij), (kl) meet if and only if the sets (ij), (kl) are disjoint. Hence the 15 points are numbered by *synthemes* (ij, kl, mn) and the three lines meeting each point are the indicated *duads* (pairs) (ij), (kl), (mn). Then there are ten sets such as 23, 31, 12 and 56, 64, 45 with the property that the first and last three do not meet, but each of the first meets each of the last. Therefore the six lines are generators of a quadric surface

$$Q_{ijk} = \begin{matrix} \text{quadric with } (ij), (jk), (ik) & \text{in one ruling and} \\ (lm), (mn), (ln) & \text{in the other ruling} \end{matrix} \qquad (3.45)$$

Then Q_{ijk} lies in a $\mathbb{P}^3$, and there are ten such, corresponding to the ways of dividing the six numbers into two *triads* (triples). Let us denote the corresponding $\mathbb{P}^3$ by K_{ijk}, so

$$Q_{ijk} \subset K_{ijk}. \qquad (3.46)$$

Then each of the 15 lines is contained in four of the K_{ijk}, and six of the K_{ijk} meet at each of the 15 points.

Consider now a set of four mutually skew of the 15 lines, for example 12, 23, 24, 25. Then there will be a two-dimensional space of $\mathbb{P}^2$'s which meet all four lines (as we are in $\mathbb{P}^4$, generically a plane and a line will not intersect). Of all of these planes, there are exactly two passing through a given point of space $x \in \mathbb{P}^5$. The locus we are interested in is:

$$Q := \left\{ x \in \mathbb{P}^5 \left| \begin{matrix} \text{the two planes meeting four skew lines} \\ \text{of the 15 } (ij) \text{ and passing through } x \\ \text{coincide} \end{matrix} \right. \right\}. \qquad (3.47)$$

If we take coordinates $\xi, \eta, \zeta, \xi', \eta', \zeta'$ satisfying $\xi + \eta + \zeta + \xi' + \eta' + \zeta' = 0$ as coordinates[3] on $\mathbb{P}^4$, then the condition (3.47) yields a locus with equation ([Ba1], p. 125):

$$\sqrt{(\eta - \zeta')(\eta' - \zeta)} + \sqrt{(\zeta - \xi')(\zeta' - \xi)} + \sqrt{(\xi - \eta')(\xi' - \eta)} = 0. \qquad (3.48)$$

After rationalizing, this is a quartic expression in the variables $\xi, \ldots, \zeta'$.

To find the dual variety of the locus Q, Baker does the following. Letting a, b, c be variables, $a' = (1 - a)$, $b' = (1 - b)$, $c' = (1 - c)$, consider the six points which are the vertices of a coordinate simplex in $\mathbb{P}^5$, and call them A, B, C, A', B', C'. Then any point of our $\mathbb{P}^4$ can be written as $x = A/bc' + B/ca' + C/ab' + A'/b'c + B'/c'a + C'/a'b$. Calculating the tangent plane of Q at a point $x \in Q$ which satisfies (3.48), in terms of the coordinates used in (3.48), one gets:

$$bc'\xi + ca'\eta + ab'\zeta - b'c\xi' - c'a\eta' - a'b\zeta' = 0. \qquad (3.49)$$

---

[3]These coordinates are the same as in (3.30) up to a sign

Now putting $u = bc', v = ca', w = ab', u' = -b'c, v' = -c'a, w' = -a'b$, the equation becomes

$$u\xi + v\eta + w\zeta + u'\xi' + v'\eta' + w'\zeta' = 0, \qquad (3.50)$$

with the two identities

$$u + v + w + u' + v' + w' = 0, \quad uvw + u'v'w' = 0. \qquad (3.51)$$

Since the identities (3.51) do not depend on the point, it follows that these equations define the dual variety. Now comparing with (3.30), we have

Proposition 3.3.1 *The dual variety of the quartic locus Q is the Segre cubic S_3.*

It is easy to see that Q is singular along the 15 lines. It was also noted classically that a tangent hyperplane section of Q is a Kummer quartic surface, with 16 nodes, 15 from the intersections with the 15 singular lines, and one from the point of tangency.

3.3.2   Igusa's results

The relation to the Kummer quartic surfaces is correctly understood by studying theta constants for the theta functions with 1/2-characteristics. This was done by Igusa in [I], and we now recall some of his results.

3.3.2.1   Theta functions

Let $\tau \in \mathbb{S}_g = \{M \in M_g(\mathbb{C}) | \tau = {}^t(\tau),\, \mathrm{Im}(\tau) \text{ positive definite}\}$, $z \in \mathbb{C}^g$, and $m = (m', m'') \in \mathbb{Q}^{2g}$. Note that $\mathbb{S}_g$ is a hermitian symmetric space of type $\mathrm{III}_\mathbf{g}$.

Definition 3.3.2 The *theta function of degree g and characteristic m* is defined by the power series

$$\theta_m(\tau, z) = \sum_{n \in \mathbb{Z}^g} \exp\left(\frac{1}{2}{}^t(n + m')\tau(n + m') + {}^t(n + m')(z + m'')\right).$$

As a function of τ the series θ_m converges precisely for $\tau \in \mathbb{S}_g$, while as functions of z by fixed τ these are theta functions on $A_\tau = \mathbb{C}^g / (\mathbb{Z}^g + \tau\mathbb{Z}^g)$. As such the zeros on A_τ are determined by the characteristic m. The corresponding *theta constant* is

$$\theta_m(\tau) := \theta_m(\tau, 0). \qquad (3.52)$$

Igusa has studied in [I] these theta constants, in particular the theta functions with characteristics $m \in \frac{1}{2^n}\mathbb{Z}$. Some of his results are the following.

Lemma 3.3.3 $\theta_m(\tau) \equiv 0 \iff m \bmod(1)$ *satisfies* $\exp(4\pi i({}^tm')m'') = -1$.

The Siegel modular group $\Gamma_g(1) = Sp(2g, \mathbb{Z})$ acts on the arguments (τ, z) as follows:

$$M = \begin{pmatrix} A & B \\ C & D \end{pmatrix}, \quad M(\tau, z) = \left((A\tau + B)(C\tau + D)^{-1}, (C\tau + D)^{-1} z \right), \quad (3.53)$$

and on the characteristic itself by

$$M(m) = \begin{pmatrix} D & -C \\ -B & A \end{pmatrix} \cdot m + \frac{1}{2} \begin{pmatrix} \mathrm{diag}(C^t D) \\ \mathrm{diag}(A^t B) \end{pmatrix}. \quad (3.54)$$

The behavior of the theta functions under M is given by

Lemma 3.3.4 *Let $M \in \Gamma_g(1)$ act on (τ, z) as in (3.53) and on the characteristic m as in (3.54). Then the theta functions transform according to the rule:*

$$\begin{aligned} \theta_{M(m)}(M(\tau, z)) &= \kappa(M) \exp(2\pi i \phi_m(M)) \det(C\tau + D)^{1/2} \times \\ &\quad \exp(\pi i\, {}^t z (C\tau + D)^{-1} C z) \theta_m(\tau, z), \end{aligned}$$

where $\kappa(M)$ is some eighth root of unity and $\phi_m(M)$ is defined by the formula

$$\phi_m(M) = -\frac{1}{2}{}^t m' BD m' + {}^t m''\, {}^t ACm'' - 2\, {}^t m''\, {}^t BCm'' - {}^t \mathrm{diag}(A^t B)(Dm' - Cm'').$$

In particular for the theta constants the formula becomes

$$\theta_{M(m)}(M\tau) = \kappa(M) \exp(2\pi i \phi_m(M)) \det(C\tau + D)^{1/2} \theta_m(\tau). \quad (3.55)$$

What the equation (3.55) says for $g = 2$ is that up to an eighth root of unity, non-vanishing theta constants with $1/2$-characteristics are automorphic forms of weight $1/2$ for the main congruence subgroup of level 2 in $Sp(4, \mathbb{Z})$. Indeed, for $M \in \Gamma(2)$, it holds that $e^{2\pi i \phi_m(M)} = 1$, as Igusa shows. There are 16 characterstics m; six are *odd* (i.e., $\theta_m(\tau, z) = -\theta_m(\tau, -z)$) so give rise to vanishing theta constants, while ten are even. The fourth powers θ_m^4 are genuine automorphic forms for $\Gamma(2)$, and determine a morphism

$$f : \Gamma(2) \backslash \mathbb{S}_2 \longrightarrow \mathbb{P}^9 = (\theta_{m_1}^4 : \cdots : \theta_{m_{10}}^4), \quad (3.56)$$

where $m_1, \ldots, m_{10}$ are the ten even characteristics.

3.3.2.2 The ring of automorphic forms

Among the ten coordinate theta functions there are five linear relations, the Riemann relations. This implies that the map f in (3.56) maps into a $\mathbb{P}^4$, displaying the quotient $X_{\Gamma(2)}$ as a hypersurface. In fact, since this is an embedding by means of automorphic functions whose closure $X_{\Gamma(2)}^* \subset \mathbb{P}^4$ is normal (see below), it follows that f gives a Baily-Borel embedding of the arithmetic quotient. The proof that f is an embedding given by Igusa is quite deep, involving showing that the ring of modular forms of $\Gamma(2)$ is the integral closure of the ring generated by the said theta functions. More precisely, his result is

Theorem 3.3.5 ([I],p. 397) *Take as coordinates in $\mathbb{P}^4$ the following theta constants:*

$$y_0 = \theta^4_{(0110)}(\tau), \ y_1 = \theta^4_{(0100)}(\tau), \ y_2 = \theta^4_{(0000)}(\tau),$$

$$y_3 = \theta^4_{(1000)}(\tau) - \theta^4_{(0000)}(\tau), \ y_4 = -\theta^4_{(1100)}(\tau) - \theta^4_{(0000)}(\tau),$$

where we let $(ijkl)$ denote the characteristic $(\frac{i}{2}\frac{j}{2}\frac{k}{2}\frac{l}{2})$. Set also

$$\chi_{10} = \prod_{even\ m} \theta^2_m.$$

Then the ring of modular forms of $\Gamma(2)$ is given by:

$$R(\Gamma(2)) = \mathbb{C}[y_0, \ldots, y_4, \chi_{10}]/\mathcal{E},$$

where $\mathcal{E}$ is the ideal generated by the following two relations:

$$\mathcal{E} = \begin{cases} R_1 &= (y_0y_1 + y_0y_2 + y_1y_2 - y_3y_4)^2 - 4y_0y_1y_2(\sum y_i) \\ R_2 &= \chi^2_{10} - \frac{1}{4}s(y_0, \ldots, y_4), \ s \text{ homogenous of degree 5} \end{cases}$$

However, the formula R_1 relating the theta functions was known long before Igusa. Since the five linear relations determining the image $\mathbb{P}^4$ of f are known, it is sufficient to give a single relation of minimal degree among the θ^4_m to determine the image. This relation can be found as early as in the 1887 paper of Maschke [Ma1], p. 505[4]. In terms of the theta constants above, this equation is

$$\left(\sum \theta^8_m\right)^2 - 4\left(\sum \theta^{16}_m\right) = 0, \tag{3.57}$$

which, as can be checked, is the same quartic as that given by R_1 in 3.3.5, as well as that given by (3.48). For this it is useful to have the equations of the ten hyperplanes $\theta_m = 0$ in the coordinates used in equation (3.48). These are (cf. [Ba1], p. 129)

$$\xi - \xi' = 0; \quad \eta - \eta' = 0; \quad \zeta - \zeta' = 0; \quad \xi + \eta + \zeta - \xi' - \eta' - \zeta' = 0$$

$$\eta - \zeta' = 0; \quad \zeta - \xi' = 0; \quad \xi - \eta' = 0; \quad \eta' - \zeta = 0; \quad \zeta' - \xi = 0; \quad \xi' - \eta = 0.$$

Definition 3.3.6 The *Igusa quartic* $\mathcal{I}_4$ is the quartic threefold defined in $\mathbb{P}^4$ by the relation R_1 of Theorem 3.3.5 or the equation (3.57).

As a corollary we have

Corollary 3.3.7 *The Igusa quartic $\mathcal{I}_4$ and the quartic locus $\mathcal{Q}$ of (3.48) coincide, and this quartic is the Satake compactification of $X_{\Gamma(2)}$.*

Hence we have described $X^*_{\Gamma(2)}$ as a singular quartic hypersurface in $\mathbb{P}^4$. There are the two interesting loci:

---

[4]the equation is somewhat hidden: "...daß dagegen die symmetrische Funktion vierter Dimension sich bis auf einen Zahlenfactor als das Quadrat der zweiten Dimension erweist."

(i) the singular locus, which is the boundary of the Baily-Borel embedding of $X_{\Gamma(2)}$;

(ii) the intersection of $\mathcal{I}_4$ with the coordinate hyperplanes in $\mathbb{P}^9$, which are the modular subvarieties $\overline{Y}_m(2)$ of [J], Thm. 3.19; these are quotients of symmetric subdomains isomorphic to a product of discs.

As already mentioned, the singular locus of $\mathcal{I}_4$ consists of 15 lines; this can be directly calculated from the equation. Alternatively, applying general formula for the number of cusps (see for example [Ya]) we see that $X_{\Gamma(2)}$ has 15 one-dimensional boundary components and 15 zero-dimensional boundary components; by 3.3.7 this is then the singular locus of $\mathcal{I}_4$. (That these boundary components are rational curves is obvious ($\Gamma(2)\backslash\mathbb{S}_1$ is rational); that they are actually *lines* is not so obvious, but an easy calculation). This line of reasoning also requires the result, also due to Igusa, that, although $\Gamma(2)$ is not torsion-free, there are nonetheless no singularities on $X_{\Gamma(2)}$.

3.3.3 Moduli interpretation

The embedding (3.56) of $X^*_{\Gamma(2)}$ as the quartic $\mathcal{I}_4$ shows that $\mathcal{I}_4$ has a moduli interpretation. In fact, $X_{\Gamma(2)}$ is a coarse moduli space of principally polarized complex abelian surfaces with a level 2 structure. In other words, $\mathcal{I}_4$ is the Baily-Borel embedding of the coarse moduli space for the functor $\mathbf{A}^*_{2,2}$ of section 1.2.2, 3. However, $\Gamma(2)$ contains torsion, namely the element -1, so $X_{\Gamma(2)}$ is *not* a fine moduli variety. This corresponds to the fact that the automorphism $z \mapsto -z$ of A_τ preserves the level 2 structure, hence the actual *object* which is parameterized by $X_{\Gamma(2)}$ is the *quotient* $A_\tau/(z \mapsto -z)$. This is just the Kummer quartic surface which already occured above. The precise relation is given by

Theorem 3.3.8 *For a point* $x \in \mathcal{I}_4 - \{$*intersections of* $\mathcal{I}_4$ *with the ten coordinate planes in (3.56)*$\}$*, the corresponding Kummer quartic surface* $K_x = A_\tau/\{\pm 1\}$*, where* $x = p(\tau)$ *for the natural projection* $p : \mathbb{S}_2 \longrightarrow X_{\Gamma(2)}$*, is the intersection of* $\mathcal{I}_4$ *with the tangent hyperplane at* x*,* $T_x\mathcal{I}_4$*:*

$$K_x = \mathcal{I}_4 \cap T_x\mathcal{I}_4.$$

In other words, $\mathcal{I}_4$ *is a compactification of the fine moduli space for the functor* $\mathbf{K}_2$ *of section 1.2.2, 3.*

This statement can be found for example in [Ba1], p. 138-139. It amounts to the fact, true in any dimension, that for $n \geq 3$ the theta functions with characteristics $\in \mathbb{Z}/n\mathbb{Z}$ on a fixed A_τ give an embedding of A_τ, while for $n = 2$ they map onto the Kummer variety.

The reason one must exclude the ten hyperplane sections in Theorem 3.3.8 is the following result.

Proposition 3.3.9 *The ten hyperplane sections* $\{\theta^4_m = 0\} \cap \mathcal{I}_4$ *are tangent hyperplane sections, i.e., the intersection is of degree 2 and multiplicity 2.*

A proof, based only on the equation of $\mathcal{I}_4$, can be found in [Ba1], p. 129. To understand the meaning of this, note that a general hyperplane section meets $\mathcal{I}_4$ in a quartic surface, while the intersections here are quadric surfaces, hence to preserve degree must be counted twice (i.e., multiplicity 2). Consider the symmetric subdomain $\mathbb{S}_1 \times \mathbb{S}_1 \subset \mathbb{S}_2$, which in this case is the set of reducible matrices:

$$\mathbb{S}_1 \times \mathbb{S}_1 = \left\{ \begin{pmatrix} \tau_1 & 0 \\ 0 & \tau_2 \end{pmatrix} \right\} \subset \left\{ \begin{pmatrix} \tau_1 & \tau_{12} \\ \tau_{12} & \tau_2 \end{pmatrix} \right\} = \mathbb{S}_2.$$

Then an easy calculation shows that the theta function of Definition 3.3.2 is a *product* of two theta functions of a single variable (i.e., $z \in \mathbb{C}$). This is equivalent to the fact that for reducible $\tau \in \mathbb{S}_2$, the abelian surface A_τ is a product of two elliptic curves, $A_\tau = E_1 \times E_2$. In this case, the map given onto the "product Kummer" variety is a map $s : E_1 \times E_2 \longrightarrow E_1/\{\pm 1\} \times E_2/\{\pm 1\} = \mathbb{P}^1 \times \mathbb{P}^1$, and this $\mathbb{P}^1 \times \mathbb{P}^1$ is the quadric surface occuring in 3.3.9. Since $\mathbb{P}^1 \times \mathbb{P}^1$ has no moduli, we see that *formally* the statement of Theorem 3.3.8 remains true for all $x \in \mathcal{I}_4 - \{15 \text{ singular lines}\}$, if we consider product Kummer varieties instead of the usual ones, and the hyperplane section is the quadric surface of Proposition 3.3.9. Note however, that this quadric surface, being a modular subvariety, can also be described as:

$$E_1/\{\pm 1\} \times E_2/\{\pm 1\} \cong (\Gamma_1(2)\backslash \mathbb{S}_1)^* \times (\Gamma_1(2)\backslash \mathbb{S}_1)^*, \qquad (3.58)$$

describing the product Kummer surface of a reducible abelian surface as a compactification of an arithmetic quotient, that is, as a Janus-like variety. We then get the following moduli interpretation of the quadric surfaces.

Proposition 3.3.10 *The ten quadric surfaces of Proposition 3.3.9 are modular subvarieties which correspond to abelian surfaces which split. More precisely, for any x on one of the quadric surfaces, but not on any of the singular lines (there are six such singular lines on each quadric surface, see (3.45)), determines a smooth abelian surface which splits, with a level 2 structure.*

Finally we note that this geometry can be described, as discussed already in [J] and many other places, in terms of the finite geometry

$$V = (\mathbb{Z}/2\mathbb{Z})^4. \qquad (3.59)$$

Let $< , >$ denote the induced symplectic form on V; every vector $v \in V$ is isotropic with respect to $< , >$. Since there are 15 non-zero vectors, there are 15 one-dimensional boundary components. Similarly, there are 15 isotropic planes in V, giving 15 zero-dimensional boundary components. The modular subvarieties of Proposition 3.3.10 correspond in this setting to non-singular pairs $\{\delta, \delta^\perp\}$, where δ is a two-dimensional subspace of V on which $< , >$ is non degenerate, and $\delta^\perp$ denotes the orthocomplement with respect to $< , >$. Of these there are exactly ten, as is easily checked. We leave further details to the reader.

3.3.4 Birational transformations

We have seen above in Proposition 3.3.1 that $\mathcal{S}_3$ and $\mathcal{I}_4$ are dual varieties. It follows from general theory that they are then in fact *birational*. In this section we describe the ensuing birational map explicitly. We consider the following modifications of $\mathbb{P}^4$.

a) Blow up the ten nodes (3.26) of $\mathcal{S}_3$; denote this by $\varrho_1 : \widehat{\mathbb{P}}^4 \longrightarrow \mathbb{P}^4$. There are ten exceptional $\mathbb{P}^3$'s, each with normal bundle $\mathcal{O}_{\mathbb{P}^3}(-1)$. Consider one of the 15 hyperplanes $\mathcal{H}_{ij}$ of (3.34). Since each hyperplane contains $4 + 2 + 1 = 7$ nodes, its proper transform on $\widehat{\mathbb{P}}^4$ is a $\mathbb{P}^3$ blown up in those seven points; each of the 15 $\mathbb{P}^2$'s of (3.27) lying on $\mathcal{S}_3$ contains four of the nodes, so their proper transforms are copies of $\mathbb{P}^2$ blown up in four points. Finally, let $\widehat{\mathcal{S}}_3 \subset \widehat{\mathbb{P}}^4$ denote the proper transform of $\mathcal{S}_3$ in $\widehat{\mathbb{P}}^4$; $\widehat{\mathcal{S}}_3$ is smooth, and $\varrho_{1|\widehat{\mathcal{S}}_3} : \widehat{\mathcal{S}}_3 \longrightarrow \mathcal{S}_3$ is a desingularization of $\mathcal{S}_3$, replacing each node with a quadric surface $\cong \mathbb{P}^1 \times \mathbb{P}^1$.

b) Let $\mathfrak{J}(15)$ denote the ideal of the 15 singular lines of $\mathcal{I}_4$; blow up $\mathfrak{J}(15)$, and let $\varrho_2 : \widehat{\mathbb{P}}^4 \longrightarrow \mathbb{P}^4$ denote this modification. Under ϱ_2, each of the lines is replaced by a $\mathbb{P}^2$-bundle over that line, and each point is replaced by a union of $\mathbb{P}^1$'s, one each for each *pair* (l_1, l_2) of *lines* meeting at the point; this mentioned $\mathbb{P}^1$ is then the intersection of the fibre $\mathbb{P}^2$ of ϱ_2 at that point with the (two) exceptional $\mathbb{P}^2$-bundles over the lines l_1 and l_2. Note that the proper transforms of the ten quadrics of Proposition 3.3.9 on $\mathcal{I}_4$ are still biregular to $\mathbb{P}^1 \times \mathbb{P}^1$, while the proper transforms of each of the lines turns out to be a *Kummer modular surface*, that is, $\mathbb{P}^2$ blown up in four points. Let $\widehat{\mathcal{I}}_4$ be the proper transform of $\mathcal{I}_4$ in $\widehat{\mathbb{P}}^4$; then $\varrho_{2|\widehat{\mathcal{I}}_4} : \widehat{\mathcal{I}}_4 \longrightarrow \mathcal{I}_4$ is a desingularization of $\mathcal{I}_4$.

Theorem 3.3.11 *The varieties $\widehat{\mathcal{S}}_3$ and $\widehat{\mathcal{I}}_4$ are biregular, and the explicit birational map $\varphi : \mathcal{S}_3 \dashrightarrow \mathcal{I}_4$ is the birational morphism completing the following diagram:*

$$
\begin{array}{ccc}
\widehat{\mathcal{S}}_3 & \xrightarrow{\widehat{\varphi}} & \widehat{\mathcal{I}}_4 \\
\varrho_1 \downarrow & & \downarrow \varrho_2 \\
\mathcal{S}_3 & \dashrightarrow{\varphi} & \mathcal{I}_4.
\end{array}
$$

Moreover, φ is Σ_6-equivariant.

Proof: As ϱ_1 and ϱ_2 are Σ_6-equivariant, the second statement follows from the first. Let $D \subset \mathcal{S}_3$ be the open set:

$$D = \mathcal{S}_3 - \{15 \text{ hyperplanes } P_\sigma \text{ of } (3.27)\}; \tag{3.60}$$

here we may take the regular map of D onto the set of tangent hyperplanes (now viewing $\mathcal{I}_4$ as the projective dual of $\mathcal{S}_3$), and set

$$
\begin{array}{cccc}
\varphi_{|D} : D & \longrightarrow & D' \subset \mathcal{I}_4 & \tag{3.61} \\
x & \mapsto & (\mathbb{P}^3)_x = \text{tangent hyperplane to } \mathcal{S}_3 \text{ at } x
\end{array}
$$

Lemma 3.3.11.1 *The subset $D' \subset \mathcal{I}_4$ is: $D' = \mathcal{I}_4 - \{10$ quadric surfaces of Proposition 3.3.10$\}$.*

Proof: Suppose $x \in D$; then $(\mathbb{P}^3)_x$ meets D in an irreducible cubic (the union of the P_σ are *all* the linear subspaces contained in $\mathcal{S}_3$, so outside of this locus $(\mathbb{P}^3)_x \cap \mathcal{S}_3$ cannot have a linear factor, so, being cubic, must be irreducible), while the ten quadric surfaces are the locus of the tangent hyperplanes meeting $\mathcal{S}_3$ in one of the nodes, all of which are excluded in D. $\qquad\square$

Now we glue D onto the rest of $\widehat{\mathcal{S}}_3$, and D' onto the rest of $\widehat{\mathcal{I}}_4$. The locus $\Lambda_1 = \widehat{\mathcal{S}}_3 - D$ coincides with $\Lambda_2 = \widehat{\mathcal{I}}_4 - D'$, as follows from the descriptions of the rational maps ϱ_1 and ϱ_2 above. Both the Λ_i consist of ten $\mathbb{P}^1 \times \mathbb{P}^1$'s and 15 rational surfaces, each isomorphic to $\mathbb{P}^2$ blown up in four points. Hence we can complete $\varphi_{|D}$ to a biregular isomorphism $\varphi : \widehat{\mathcal{S}}_3 \longrightarrow \widehat{\mathcal{I}}_4$, by fixing an isomorphism $\varphi_\Delta : \Lambda_1 \longrightarrow \Lambda_2$, and setting

$$\varphi(x) = \begin{cases} \varphi_{|D}(x), \text{ if } x \in D \\ \varphi_\Delta(x), \text{ if } x \in \Lambda_1 \end{cases}$$

completing the proof of Theorem 3.3.11. $\qquad\square$

The following description is more concrete. If x is one of the nodes of $\mathcal{S}_3$, there is a quadric cone of tangent (to $\mathcal{S}_3$) hyperplanes at x; so closing up φ maps x to the quadric surface over which the above is a cone, i.e., x is blown up. If x is *not* a node, then there is a unique tangent hyperplane $T_x\mathcal{S}_3$, determining a point of $\mathcal{I}_4$. Furthermore, $T_x\mathcal{S}_3$ and $T_y\mathcal{S}_3$ *coincide* for $x \neq y$, if and only if x and y are contained in a common Segre plane (3.27), and the line joining x and y in that Segre plane passes through one of the four nodes, say N, in that Segre plane. This is because $T_x\mathcal{S}_3 \cap \mathcal{S}_3 = \mathcal{H} \cup Q_x$, where Q_x is a residual quadric cone, and the quadric cone is the intersection of $T_x\mathcal{S}_3$ with the cone C_N which is the tangent cone of the node N in the Segre plane. So if x and y lie on a line through N, Q_x and Q_y coincide, so $T_x\mathcal{S}_3$ and $T_y\mathcal{S}_3$ coincide also.

Theorem 3.3.12 *The duality map $d : \mathcal{S}_3 - - \to \mathcal{I}_4$ is given by the linear system of quadrics 3.2.3, i.e., by the elements of the ideal $\mathcal{J}(10)$ of the ten nodes: $d = \varphi$.*

Proof: It suffices to check that d, viewed as a modification of $\mathcal{S}_3$, coincides with the birational map φ of Theorem 3.3.11. But this is easy. As the base locus is the set of nodes, these are blown up. As just explained, x and y in one of the Segre planes map to the same point on the image line precisely when the line joining them passes through one of the nodes in the Segre plane. As these lines are precisely what the map φ blows down, d certainly coincides with φ. $\qquad\square$

We also have the following analogue of Corollary 3.2.4.

Lemma 3.3.13 *The ideal $\mathcal{J}(15)$ of the 15 singular lines of the Igusa quartic coincides with the Jacobian ideal of $\mathcal{I}_4$.*

Proof: Once again the inclusion $\mathcal{J}ac(\mathcal{I}_4) \subset \mathcal{J}(15)$ is obvious, and the inverse inclusion can be verified by means of standard basis computations, namely that $\mathcal{J}(15)$ is generated by five cubics. $\qquad\square$

Along the same lines as Theorem 3.3.12 we then get

Theorem 3.3.14 *The duality map* $d : \mathcal{I}_4 \dashrightarrow S_3$ *is given by the system of cubics containing the 15 lines, i.e., by the Jacobian ideal of* $\mathcal{I}_4$: $d = \varphi^{-1}$.

Proof: As above, it suffices to show that d, viewed as a modification of $\mathcal{I}_4$, coincides with the map φ^{-1} of Theorem 3.3.11. This is readily verified, as the base locus, the 15 lines, are blown up, while the tangent planes for any two points x and y in a common quadric of $\mathcal{I}_4$ (of Proposition 3.3.9) coincide, blowing down the quadric surface to a node. □

3.3.5 The Siegel modular threefold of level 4

From the general theory of congruence subgroups, $X_{\Gamma(4)} \longrightarrow X_{\Gamma(2)}$ is a Galois cover, with Galois group $\Gamma(2)/\Gamma(4) \cong (\mathbb{Z}/2\mathbb{Z})^9$. Indentifying $X^*_{\Gamma(2)}$ with $\mathcal{I}_4$ and identifying $\mathcal{I}_4$ birationally with S_3, we can consider Fermat covers over $X^*_{\Gamma(2)}$, i.e., given by a diagram

$$
\begin{array}{ccccc}
Z(\mathbf{A_4}, n) & \overset{\overline{\varphi}^{-1}}{\dashrightarrow} & Y^\wedge(\mathbf{A_4}, n) & \longleftarrow & \widetilde{Y}(\mathbf{A_4}, n) \\
\downarrow & & \downarrow & & \downarrow \quad (\mathbb{Z}/2\mathbb{Z})^9 \qquad (3.62)\\
\mathcal{I}_4 & \overset{\varphi^{-1}}{\dashrightarrow} & S_3 & \longleftarrow & \widetilde{S}_3
\end{array}
$$

where $\overline{\varphi}^{-1}$ is *induced* by φ^{-1}, that is, (3.62) is a fibre square (cf. (3.39), where $\widetilde{S}_3$ is denoted $\widetilde{\mathbb{P}}^3$).

Theorem 3.3.15 *The Fermat cover* $Z(\mathbf{A_4}, 2)$ *is the Satake compactification of the Siegel modular threefold of level 4.*

Proof: It suffices to show that $\widetilde{Y}(\mathbf{A_4}, 2)$ is the induced cover over $\widehat{\mathcal{I}}_4$, where $\varrho_2 : \widehat{\mathcal{I}}_4 \longrightarrow \mathcal{I}_4$ is the desingularization of $\mathcal{I}_4$ of Theorem 3.3.11. Now the identification can be reduced to identifying what is in the branch locus of $\widetilde{Y}(\mathbf{A_4}, 2) \longrightarrow \widetilde{S}_3$. There are two kinds of components:

a) covers $\widetilde{Y}(\mathbf{A_3}, 2)$ of blown up $\mathbb{P}^2$'s, the H_{ij} of (3.14);

b) covers $\widetilde{Y}(\mathbf{A_2}, 2) \times \widetilde{Y}(\mathbf{A_2}, 2)$ of $\mathbb{P}^1 \times \mathbb{P}^1$'s, the L_{0ij}.

Lemma 3.3.15.1 $\widetilde{Y}(\mathbf{A_3}, 2) \cong S(4)$, *Shioda's elliptic modular surface of level 4.*

Proof: This is well-known. $\widetilde{Y}(\mathbf{A_3}, 2)$ is K3 since it is a Fermat cover branched over six lines. One constructs structures of fibre space $\widetilde{Y}(\mathbf{A_3}, 2) \longrightarrow \mathbb{P}^1$ with elliptic curves as fibres by taking the cover of the pencil of lines through a node (each such line meets four of the six lines outside the node, so the cover is branched at four points, i.e., is elliptic). The six fibres of type I_4 are readily identified, as are the 16 sections. □

Lemma 3.3.15.2 *The cover* $\widetilde{Y}(\mathbf{A_2}, 2) \longrightarrow \mathbb{P}^1$ *coincides with the cover*

$$(\Gamma_1(4) \backslash \mathbb{S}_1)^* \longrightarrow \mathbb{P}^1,$$

by which we mean the Galois actions coincide.

Proof: This is even more well-known.   □

The theorem now follows, provided we accept that $\widetilde{Y}(\mathbf{A_4}, 2)$ is a quotient of $\mathbb{S}_2$ at all, i.e., that the cover $\mathbb{S}_2 \longrightarrow \mathcal{I}_4^0$ factorizes (here $\mathcal{I}_4^0 = \mathcal{I}_4 - \{15 \text{ lines}\}$), $\widetilde{Y}(\mathbf{A_4}, 2)^0 := \widetilde{Y}(\mathbf{A_4}, 2) - q^{-1}(15 \text{ lines})$:

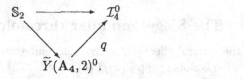

$$(3.63)$$

But there is an easy way to see that this is the case: we can, for any given $x \in \widetilde{Y}(\mathbf{A_4}, 2) - \Delta$ and $y = q(x)$, put a level 4 structure on A_y, such that the Galois group just permutes the level 4 over level 2 structures, that is, we make the identification $\Gamma(2) \backslash \Gamma(4) \cong (\mathbb{Z}/2\mathbb{Z})^9 \cong$ the Galois group of the cover. So $\widetilde{Y}(\mathbf{A_4}, 2)$, being a moduli space as in Shimura's theory, is a quotient of $\mathbb{S}_2$.   □

One could also imagine arguing with uniqueness of Galois covers, since we know the branch locus, branch degrees and Galois group. However there is in general no such uniqueness of covers, so we have to be careful. In our situation, there are two possible approaches to show uniqueness:

1) Since the modular subvarieties determine, on the group-theoretic side, generators of the corresponding arithmetic group, we could conclude, from the isomorphisms 3.3.15.1 and 3.3.15.2, the desired result.

2) Since the branch divisors are totally geodesic with respect to the Bergmann metric, on the cover the metric retains its symmetry property.

Method 1) has been applied in [J], and 2) can be carried out for ball quotients.

3.4   The Hessian varieties of $\mathcal{S}_3$ and $\mathcal{I}_4$

3.4.1   The Nieto quintic

Let $(x_0 : \ldots : x_5)$ be the projective coordinates on $\mathbb{P}^5$ used to define $\mathcal{S}_3$ in (3.25), and let $\sigma_i = \sigma_i(x_0, \ldots, x_5)$ be the i-th elementary symmetric function $\sigma_\lambda = \sum_{i_1 < \ldots < i_\lambda} x_{i_1} \cdots x_{i_\lambda}$ in $(x_0 : \ldots : x_5)$. Define the *Nieto quintic* $\mathcal{N}_5$ by the equations

$$\mathcal{N}_5 = \begin{cases} \sigma_1 = 0 \\ \sigma_5 = 0 \end{cases} \subset \mathbb{P}^4 = \{\sigma_1 = 0\} \subset \mathbb{P}^5. \qquad (3.64)$$

The symmetry of $\mathcal{N}_5$ under the symmetric group Σ_6 is evident from the equation. This quintic was discovered in the thesis [N] and further studied in [BN], which will be our general reference for this section. We just briefly describe the geometry of $\mathcal{N}_5$ without discussing details.

The singular locus is relatively easy to determine, just by calculating the Jacobian of (3.64). The result is

Proposition 3.4.1 ([BN], 3.1) $\mathcal{N}_5$ *has the following singular locus:*

(i) *20 lines* $L_{ijk} = \{x_i = x_j = x_k = 0 = \sum x_i\}$;

(ii) *ten isolated points, the* Σ_6-*orbit of* $(1, 1, 1, -1, -1, -1)$, *which are the points* $P_{ij} = (1, \pm 1, \ldots \pm 1)$, *with* $+1$ *in the i-th and j-th positions.*

We will give a different proof of this below, see the discussion following Proposition 6.3.2. Note that the ten points occuring in (ii) are just the ten nodes of $\mathcal{S}_3$ (see (3.26), cf. also Remark 3.4.8 below). Furthermore, a local calculation shows that the singularities of $\mathcal{N}_5$ along the lines of (i) are of the type $\{\text{disc}\} \times A_1$, and at the points of (ii) are ordinary double points. Hence the former are resolved by a $\mathbb{P}^1$-bundle over the line L_{ijk}, while the points are resolved, as with the case of $\mathcal{S}_3$, by quadric surfaces. The 20 lines L_{ijk} of 3.4.1 meet at the following 15 points:

$$Q_{ij} = (0, \ldots, 1, \ldots, -1, \ldots) = \{\Sigma_6 - \text{orbit of } Q_{56} = (0 : 0 : 0 : 0 : 1 : -1)\}.$$
$$(3.65)$$

Lemma 3.4.2 *The 20 lines* L_{ijk} *of Proposition 3.4.1 meet four at a time at the 15 points* Q_{ij}; *each line* L_{ijk} *contains three of the points, namely we have* $Q_{ij} \in L_{klm} \iff \{i, j\} \cap \{k, l, m\} = \emptyset$.

Proof: The line L_{123} contains the three points Q_{46}, Q_{45} and Q_{56}, so by Σ_6-invariance each line contains three of the Q_{ij}. The point Q_{56} is contained in the four lines L_{123}, L_{124}, L_{134} and L_{234}, so by Σ_6-invariance, each point is contained in four lines. $\square$

Also, $\mathcal{N}_5$ contains a finite number of linear planes.

Lemma 3.4.3 $\mathcal{N}_5$ *contains the following 30* $\mathbb{P}^2$ '*s:*

(i) *15 planes* $N_{ijkl} = \{x_i + x_j = x_k + x_l = x_m + x_n = 0\}$;

(ii) *15 planes* $N_{ij} = \{x_i = x_j = 0 = \sum_{k \neq i,j} x_k\}$.

Proof: It is immediately verified that these planes satisfy the equation (3.64). $\square$

Presumably these are in fact *all* the linear planes contained in $\mathcal{N}_5$. Note that the N_{ijkl} are just the 15 planes (3.27) lying on the Segre cubic $\mathcal{S}_3$.

Among the 15 planes N_{ijkl} the common intersections were described in the discussion of the planes P_σ on the Segre cubic (see (3.10)).

Lemma 3.4.4 *Each plane N_{ijkl} contains the following four of the ten points of 3.4.1, (ii):*

$$P_{km}, \ P_{kn}, \ P_{lm} \ and \ P_{ln};$$

it also contains the following three of the 15 points Q_{ij} of (3.65): Q_{ij}, Q_{kl} and Q_{mn}.

Proof: Consider N_{0123}; it contains the four nodes $(1:-1:1:-1:1:-1)$, $(1:-1:-1:1:1:-1)$, $(1:-1:1:-1:-1:1)$ and $(1:-1:-1:1:-1:1)$ which are the points P_{24}, P_{25}, P_{34} and P_{35}, which gives the first statement by Σ_6-symmetry (there is an asymmetry in the notation, since we may take $i < j, k < l$ in the notation for N_{ijkl}, and since the first coordinate of P_{ij} may be assumed to be $+1$). Similarly, N_{0123} contains the three points Q_{01}, Q_{23} and Q_{45}, giving the second statement by Σ_6-symmetry. □

We now note that these seven points lie in the plane N_{0123} as in Figure 3.2. This is in fact easily checked. Note that the lines in N_{0123}, i.e., the intersections

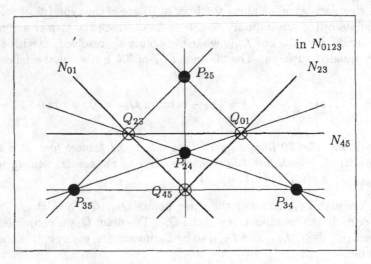

Figure 3.2: The plane N_{0123}

with the other N_{ijkl} are *not* the lines of Proposition 3.4.1; those lines have equations such as $x_0 = x_1 + x_2 = x_1 + x_3 = x_4 + x_5$. However, in the 15 planes N_{ij} of 3.4.3, several of the 20 singular lines L_{ijk} *do* lie. In fact, we have

Lemma 3.4.5 *Each N_{ij} contains the four lines L_{ijk}, L_{ijl}, L_{ijm} and L_{ijn}. There are three planes passing through L_{ijk}, namely N_{ij}, N_{ik} and N_{jk}. N_{ij} contains none of the nodes of 3.4.1, (i), but contains six of the points Q_{ij} of (3.65), namely Q_{kl}, Q_{km}, Q_{kn}, Q_{lm}, Q_{ln} and Q_{mn}. These six points lie three at a time on the L_{ijk} and form in each N_{ij} a configuration as shown in Figure 3.3. The three light lines are intersection of N_{01} with N_{ijkl} as indicated.*

Proof: This is once again easily verified. □

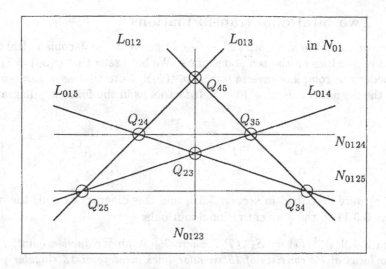

Figure 3.3: The plane N_{01}

Finally we note that there are hyperplanes in $\mathbb{P}^4$ cutting out these $\mathbb{P}^2$'s on $\mathcal{N}_5$.

Lemma 3.4.6 *The fifteen hyperplanes $\mathcal{H}_{ij} = \{x_i + x_j = 0\}$ meet $\mathcal{N}_5$ each in the union of the three planes N_{ijkl}, N_{ijkm} and N_{ijkn}, and a residual quadric; the six hyperplanes $x_i = 0$ meet $\mathcal{N}_5$ each in the union of five planes N_{ij}, N_{ik}, N_{il}, N_{im} and N_{in}.*

Proof: This is once again just a computation. □

Now let us consider the intersection of $\mathcal{S}_3$ and $\mathcal{N}_5$. As is obvious from the above description, they both contain the 15 planes N_{ijkl}, and, the intersection being of degree 15, this is the entire intersection. From general arguments on projective varieties, from the fact that the dual of $\mathcal{S}_3$, namely the Igusa quartic $\mathcal{I}_4$, is *normal*, it follows that the parabolic divisor on $\mathcal{S}_3$, which is the intersection of $\mathcal{S}_3$ with the *Hessian variety*, must get blown down under the duality map, i.e., the intersection $Hess(\mathcal{S}_3) \cap \mathcal{S}_3$ consists also of the 15 planes on $\mathcal{S}_3$! Since the Hessian has degree 5, this is again the entire intersection, and it is natural to ask whether $\mathcal{N}_5$ and $\mathbf{Hess}(\mathcal{S}_3)$ are related. In fact, we have

Lemma 3.4.7 *The Nieto quintic is the Hessian of $\mathcal{S}_3$, i.e., $\mathcal{N}_5 = \mathbf{Hess}(\mathcal{S}_3)$, with equality, not just isomorphism.*

Proof: This is an easy computation (at least for a computer). □

Remark 3.4.8 Since the Hessian variety $\mathbf{Hess}(V)$ aquires nodes where V has nodes, this "explains" the ten isolated singularities on $\mathcal{N}_5$. A completely different explanation is given in section 6.3.3.

3.4.2   Two birational transformations

We consider the rational map $\varphi : \mathbb{P}^4 -- \to \mathbb{P}^4$ given by the Jacobian ideal of $\mathcal{S}_3$, i.e., by five quadrics on the ten nodes of $\mathcal{S}_3$. We have seen that $\varphi(\mathcal{S}_3) = \mathcal{I}_4$, and we consider the complete inverse image $\varphi^{-1}(\mathcal{I}_4)$. Note that as φ has ten base points, the degree of φ is $2^4 - 10 = 6$. It factors as in the following diagram:

$$
\begin{array}{ccccccc}
\mathbb{P}^4 & \xleftarrow{\varrho_1} & \widehat{\mathbb{P}}^4 & \xrightarrow{\tilde{\alpha}} & \widehat{\widehat{\mathbb{P}}}^4 & \xrightarrow{\varrho_2} & \mathbb{P}^4 \\
\cup & & \cup & & \cup & & \cup \\
\mathcal{S}_3 & \longleftarrow & \widehat{\mathcal{S}}_3 & \xrightarrow{\sim} & \widehat{\mathcal{I}}_4 & \longrightarrow & \mathcal{I}_4
\end{array}
\qquad ; \qquad (3.66)
$$

ϱ_1 and ϱ_2 were described in section 3.3.4, and this diagram extends the one of Theorem 3.3.11 to the ambient rational fourfolds.

Lemma 3.4.9 $\varphi^{-1}(\mathcal{I}_4) = \mathcal{S}_3 \cup \mathcal{P}_5$, where $\mathcal{P}_5$ is an irreducible quintic. The singular locus of $\mathcal{P}_5$ consists of 15 singular lines meeting at 15 singular points, plus and additional 55 (counting multiplicities) points. Ten of the 55 points are double points and 15 further points have multiplicity three.

Proof: To get the complete inverse image, one inserts the partial derivatives of $\mathcal{S}_3$ into the equation for $\mathcal{I}_4$. The resulting expression is of degree 8, and dividing out the equation for $\mathcal{S}_3$, the resulting threefold is a quintic. The calulation of the singular locus was done with Macaulay. It turned out that the 15 singular lines meet in 15 points, and form a configuration isomorphic to the singular locus of $\mathcal{I}_4$. $\square$

Note that since $\mathcal{S}_3$ is generically 1:1 over $\mathcal{I}_4$, the cover $\mathcal{P}_5 \longrightarrow \mathcal{I}_4$ is generically 5:1. It follows that φ is not Galois. Using explicit calculations we have found the following facts about $\mathcal{P}_5$:

- Under φ, the entire singular locus of $\mathcal{P}_5$ maps to the singular locus of $\mathcal{I}_4$.

- The 15 singular lines of $\mathcal{P}_5$ are cut out by three hyperplanes of the type $\mathcal{T}_{ij} = \{x_i - x_j = 0\}$ (recall these hyperplanes contain four of the nodes of $\mathcal{S}_3$ and cut out on $\mathcal{S}_3$ a copy of the Cayley cubic), in fact $\mathcal{T}_{12} \cap \mathcal{T}_{34} \cap \mathcal{T}_{56}$ is such a line.

- The intersection $\mathcal{S}_3 \cap \mathcal{P}_5$ consists of the 15 Segre planes on $\mathcal{S}_3$ (as does the intersection $\mathcal{S}_3 \cap \mathcal{N}_5$).

- The 45 additional singular points of $\mathcal{P}_5$ lie on both $\mathcal{S}_3$ and $\mathcal{N}_5$.

We now consider a birational image of $\mathcal{N}_5$ which strongly resembles $\mathcal{P}_5$: the dual variety.

Proposition 3.4.10 The dual variety $\mathcal{N}_5^\vee$ of $\mathcal{N}_5$ contains the following singularities:

(i) 15 singular lines, the images of the 15 Segre planes, meeting at 15 points;

(ii) 15 singular points, the images of the 15 planes N_{ij}.

Moreover, there are ten bitangent hyperplane sections, the images of the ten nodes of $\mathcal{N}_5$.

Proof: The proof of (i) is exactly the same as in the case of the dual of S_3. To see (ii), note that each of the planes N_{ij} contains four of the singular lines (see Figure 3.3). Since the Jacobian is given by quartics, each vanishing on four lines in each N_{ij}, it follows that they all have the same value on N_{ij}, i.e., N_{ij} maps to a point. Note that when the lines are blown up, the N_{ij} get blown up in six points; the corresponding cubic surface is the Cayley cubic (blown up in four points), which subsequently gets blown down to a point (the normal bundle of the plane N_{ij} gets twisted by $\mathcal{O}(-1)$ for each line in it which is blown up). The statement on the bitangents follows again as for the Igusa quartic. □

Of course, the dual $\mathcal{N}_5^{\vee}$ may be singular in codimension 1. We also describe a slight variant of $\mathcal{N}_5^{\vee}$ (or rather of its normalization). Note that the four hyperplanes $\mathcal{H}_1, \ldots, \mathcal{H}_4$ ($\mathcal{H}_i = \{x_i = 0\}$ the hyperplane occuring in Lemma 3.4.6) contain all 20 singular lines. By Σ_6-symmetry the same is true for any such product of four of the $\mathcal{H}_i$. Adjoining the quartic $x_1 \cdot x_2 \cdot x_3 \cdot x_4$ to the Jacobian ideal we get the ideal $\mathcal{I}(20)$ of the 20 lines (explicit computation). The ideal $\mathcal{I}(20)$ determines a rational map $\beta : \mathbb{P}^4 - - \rightarrow \mathbb{P}^5$. The image $\beta(\mathcal{N}_5)$ still has the ten nodes of $\mathcal{N}_5$, and the Segre planes are not blown down (it is easy to verify that the ratios of the partial derivatives of $\mathcal{N}_5$ and of $x_1 \cdot x_2 \cdot x_3 \cdot x_4$ form a two dimensional image, so the image of the Segre plane is a surface).

Finally one can also consider the image of $\mathcal{N}_5$ under φ. This is a degree 20 subvariety of $\mathbb{P}^4$, and its singular locus consists of a surface of degree 100, and possibly also lower dimensional components (explicit computation). It also has the 20 singular lines of $\mathcal{N}_5$, since the lines are not blown up by φ.

The other birational transformation is the following.

a) Blow up the 15 points Q_{ij} of (3.65); let $p_1 : \widetilde{\mathcal{N}}_5 \longrightarrow \mathcal{N}_5$ denote this blow up.

b) As each of the lines L_{ijk} contains three points (see Lemma 3.4.2), each L_{ijk} can be blown down to an isolated singular point (the normal bundle is $\mathcal{O}(-2) \oplus \mathcal{O}(-2)$, cf. (3.8)). Let $p_2 : \widetilde{\mathcal{N}}_5 \longrightarrow \widehat{\mathcal{N}}_5$ denote this blow down.

(3.67)

The following is easy to see (see Figures 3.2 and 3.3).

Lemma 3.4.11 *The singular locus of $\widehat{\mathcal{N}}_5$ consists of the 20 isolated cusps from (3.67) b), and the ten cusps, the images of the singular points P_{ij} of Proposition 3.4.1, (ii). The proper transforms of the N_{ijkl} of Lemma 3.4.3, (i) on $\mathcal{N}_5$ are $\mathbb{P}^2$'s blown up in three points, a del Pezzo surface; the proper transforms of the N_{ij} of Lemma 3.4.3, (ii) are $\mathbb{P}^2$'s blown up in six points, then the L_{ijk} are blown down to four nodes, so this is the Cayley cubic.*

3.4.3 Moduli interpretation

The Nieto quintic was discovered as the solution of a certain moduli problem, and we briefly state the results of [BN] describing this.

The point of departure is the action of the Heisenberg group $H_{2,2}$ on $\mathbb{P}^3$, and the study of quartics which are invariant under the action. $H_{2,2}$ is a group of order 32 generated by the following linear transformations of $\mathbb{P}^3$ with coordinates $(z_0 : z_1 : z_2 : z_3)$:

$$
\begin{aligned}
\sigma_1 &: & (z_0 : z_1 : z_2 : z_3) &\mapsto & (z_2 : z_3 : z_0 : z_1) \\
\sigma_2 &: & (z_0 : z_1 : z_2 : z_3) &\mapsto & (z_1 : z_0 : z_3 : z_2) \\
\tau_1 &: & (z_0 : z_1 : z_2 : z_3) &\mapsto & (z_0 : z_1 : -z_2 : -z_3) \\
\tau_2 &: & (z_0 : z_1 : z_2 : z_3) &\mapsto & (z_0 : -z_1 : z_2 : -z_3)
\end{aligned}
\tag{3.69}
$$

The center of the group is ± 1 and $PH_{2,2} = H_{2,2}/ \pm 1$ has a nice interpretation:

$$
PH_{2,2} \cong (\mathbb{Z}/2\mathbb{Z})^4,
\tag{3.70}
$$

which carries, as in (3.59), an induced symplectic form. This means that one can speak of isotropic elements of the *group* $PH_{2,2}$. The normalizer of $H_{2,2}$ in $SL(4,\mathbb{C})$ maps surjectively to $\Sigma_6 \cong Sp(4,\mathbb{Z}/2\mathbb{Z})$, which acts transitively on diverse geometric loci of the sympectic form *inside* the group $PH_{2,2}$. These loci are:

a) 15 pairs of skew lines

b) 15 invariant tetrahedra

c) ten fundamental quadrics.
$$
\tag{3.71}
$$

The moduli problem considered is a special set of quartics which are invariant under (3.70). The set of *all* invariant quartics is just a $\mathbb{P}^4$, spanned for example by the five quartics:

$$
g_0 := z_0^4 + z_1^4 + z_2^4 + z_3^4
$$
$$
g_1 := 2(z_0^2 z_1^2 + z_2^2 z_3^2) \qquad g_2 = 2(z_0^2 z_2^2 + z_1^2 z_3^2) \qquad g_3 := 2(z_0^2 z_3^2 + z_1^2 z_2^2)
$$
$$
g_4 := 4z_0 z_1 z_2 z_3.
$$

Let (A, B, C, D, E) denote the coordinates of a particular quartic $Q_{(A,B,C,D,E)} = \{Ag_0 + Bg_1 + Cg_2 + Dg_3 + Eg_4 = 0\}$. The generic quartic $Q_{(A,B,C,D,E)}$ is smooth, and the locus of singular quartics can be determined as an equation in $(A, \ldots, E)$. Note that the $(A, \ldots, E)$ are functions of $(z_0 : \cdots : z_3)$, so the answer as to whether $Q_{(A,B,C,D,E)}$ is singular depends on the point $z \in \mathbb{P}^3$. This is discussed in detail in [BN]. The result is given in Table 3.1. As one sees, the first row of the table is equivalent to Theorem 3.3.8 above! The special class of quartics to be considered here is, however, a quite different set, consisting of generically smooth quartics. This is the set of Kummer surfaces of (1,3)-polarized abelian

Table 3.1: Singular Heisenberg invariant quartics

$z \in \mathbb{P}^3$	$\dim Q^{sing}$	$Q_{(A,B,C,D,E)}$	$S_{(A,B,C,D,E)}$
$\notin$ fix line	0	Kummer surface	Segre cubic
$\in$ one fix line	2	singular in four coordinate vertices	$A = 0$
$\in$ the intersection of two fixed lines	3	singular along two fixed lines	$A = B = 0$

Notations: Q^{sing} denotes the space of quartics singular at z, $S_{(A,B,C,D,E)}$ denotes the equation of the locus Q^{sing} in the coordinates (A, B, C, D, E).

surfaces, which, as it turns out, can be smoothly embedded in $\mathbb{P}^3$. This was first discovered by M. Traynard [Tr], Chapitre VIII, and another proof was given by L. Godeaux [Go]. This was subsequently rediscovered independently by Naruki and Nieto (see [Na] and [N]). The 16 exceptional $\mathbb{P}^1$'s resolving the 16 double points of the Kummer surface are 16 disjoint *lines* on the quartics. Also, by a result of Nikulin [Ni], the converse is true, i.e., any quartic containing 16 lines is a Kummer surface. Furthermore, the quartic being invariant under $PH_{2,2}$, if it contains one line, it contains all 16 transforms, so the moduli involved is the condition:

L is a line in $\mathbb{P}^3$ lying on a smooth Heisenberg invariant quartic surface

$$(3.72)$$

The equation describing this in the Grassmannian $\mathbb{G}(2,2) = \{x_0^2 + \cdots + x_5^2 = 0\}$ is calculated in [N]. It is

$$\mathcal{M}_{20} = \{\sigma_5(x_0^2, \ldots, x_5^2) = 0 = \sigma_1(x_0^2, \ldots, x_5^2)\}. \qquad (3.73)$$

Now one considers the natural 2-power map

$$m_2 : \mathbb{P}^5 \longrightarrow \mathbb{P}^5 \qquad (3.74)$$
$$(x_0, \ldots, x_5) \mapsto (x_0^2, \ldots, x_5^2) = (u_0, \ldots, u_5)$$

and the image of $\mathcal{M}_{20}$ in $\mathbb{P}^5$. Comparing the equations (3.64) and (3.73) we have

Lemma 3.4.15 $m_2(\mathcal{M}_{20}) = \mathcal{N}_5$.

The main results of [BN] can be described as follows. First we define a Zariski open subset $M^s \subset \mathcal{M}_{20}$. The following 15 quadric surfaces $q_{ij} \subset \mathbb{P}^5$ actually lie on $\mathcal{M}_{20}$, as is easily verified:

$$q_{ij} = \{x_i = x_j = 0 = \sum_{m \neq i,j} x_m^2\}. \qquad (3.75)$$

Under the squaring map m_2 (3.74) the quadric q_{ij} maps to the plane

$$N_{ij} = \{u_i = u_j = 0 = \sum_{m \neq i,j} u_m\}; \qquad (3.76)$$

so the image of $\mathbf{Q} := \cup_{i,j} q_{ij}$ is $\mathbf{N} := \cup_{i,j} N_{ij}$, and the planes N_{ij} are the 15 planes of Lemma 3.4.3 (ii). Furthermore the N_{ij} are contained in the branch locus of $m_{2|\mathcal{M}_{20}} : \mathcal{M}_{20} \longrightarrow \mathcal{N}_5$; this locus is *singular* on $\mathcal{M}_{20}$ because $\mathcal{N}_5$ is tangent to $u_i = 0$ and $u_j = 0$ in all of N_{ij}.

Next consider the inverse image under m_2 of the ten nodes; since the nodes lie on *none* of the branch planes $u_i = 0$, each node has $\deg(m_2) = 32$ inverse images, so $\mathcal{M}_{20}$ has 320 singular points (clearly also nodes), which are the Σ_6-orbit, call it $\mathbf{P}$, of the points

$$(\pm 1 : \pm 1 : \pm 1 : \pm i : \pm i : \pm i). \tag{3.77}$$

Finally consider the inverse images of the 15 Segre planes of Lemma 3.4.3 (i). This locus is given by the 15 equations which are the Σ_6-orbit of

$$x_0^2 + x_1^2 = x_2^2 + x_3^2 = x_4^2 + x_5^2 = 0. \tag{3.78}$$

Inspection shows that this degree 8 surface on $\mathcal{M}_{20}$ splits into eight planes, giving altogether 120=15.8 planes on $\mathcal{M}_{20}$; let $\mathbf{R}$ denote their union. Now define:

$$M^s := \mathcal{M}_{20} - \mathbf{Q} - \mathbf{P} - \mathbf{R}, \quad \mathcal{N}_5^s := m_2(M^s). \tag{3.79}$$

Then the statement proved in [BN] is

Theorem 3.4.16   a) *M^s is isomorphic to a Zariski open subset of the moduli space $\mathcal{A}^*_{(1,3)}(2)$ of complex abelian surfaces with a $(1,3)$ polarization and a level 2 structure;*

b) *There is a double cover $p : \widetilde{\mathcal{N}} \longrightarrow \mathcal{N}_5$ for which $p^{-1}(\mathcal{N}_5^s)$ is isomorphic to a Zariski open set of the moduli space $\mathcal{A}_{(2,6)}(2)$;*

c) *$\mathcal{N}_5^s$ is the moduli space of $PH_{2,2}$-invariant smooth quartic surfaces containing 16 skew lines, corresponding to the functor $K3_{PH_{2,2}}$ of section 1.2.2, 4.*

Since the varieties $\mathcal{M}_{20}$, $\widetilde{\mathcal{N}}$ and $\mathcal{N}_5$ are compactifications of the Zariski open sets of (3.79), we have the following:

Corollary 3.4.17 *There are birational equivalences:*

$$\mathcal{M}_{20} - - \to (\Gamma_{(1,3)}(2)\backslash \mathbb{S}_2)^*, \quad \widetilde{\mathcal{N}} - - \to (\Gamma_{(2,6)}(2)\backslash \mathbb{S}_2)^*, \quad \mathcal{N}_5 - - \to (\Gamma \backslash \mathbb{S}_2)^*,$$

where $\Gamma_{(1,3)} \subset \Gamma_{(2,6)} \subset \Gamma$, $[\Gamma : \Gamma_{(2,6)}(2)] = 2$.

As is shown in [BN], the map $\widetilde{\mathcal{N}} \longrightarrow \mathcal{N}_5$ is given in the following way. It just happens to turn out the *any* of the $PH_{2,2}$-invariant quartics with 16 skew lines actually contains 32 lines, the first skew set of 16 and a second set of 16 skew lines. The second set of sixteen is found as the image of the first set under the involution

$$(x_0 : \ldots : x_5) \mapsto \left(\frac{-1}{x_0} : \frac{1}{x_1} : \cdots : \frac{1}{x_5} \right), \tag{3.80}$$

which can be adjoined to the group $PH_{2,2}$ to form a group of order 32. Altogether the 32 lines have the following properties.

a) Each line of one set of 16 skew lines meets ten of the other set;

b) Each pair of one set of 16 skew lines meets six of the other set; (3.81)

c) The 32 lines form in 15 manners a quadruple of *double fours*, which consist of two sets of four skew lines.

A configuration with the properties (3.81) is called a (32_{10})-configuration.

From Nikulin's results just mentioned, it follows that the second set of 16 lines are also the images of blown-up torsion points on another abelian surface, so there are *two* abelian surfaces with $(2,6)$ polarization and level 2 structure giving rise to the *same* resolved Kummer surface, i.e., the map is given by

$$\widetilde{N} \longrightarrow \mathcal{N}_5 \qquad (3.82)$$
$$(A_{\tau_1}, A_{\tau_2}) \mapsto \overline{(A_{\tau_1}/\{\pm 1\})} \cong \overline{(A_{\tau_2}/\{\pm 1\})},$$

where the isomorphism permutes the two sets of 16 skew lines.

The next step is to identify the modular subvarieties on the arithmetic quotients of Corollary 3.4.17. From the structure of the periods we know that in terms of abelian surfaces, these modular subvarieties parameterize the abelian surfaces which split. These loci are described to some extent in [BN].

Theorem 3.4.18 *a) Points on $\mathcal{N}_5$ parameterize smooth quartic surfaces unless they lie on one of the 30 planes of Lemma 3.4.3;*

b) points on $\mathcal{N}_5$ parameterize quartic surfaces containing more that 32 lines if an only if the corresponding abelian surfaces are products. Furthermore, a line on a surface of this set of quartic surfaces has coordinates in $\mathbb{P}^5$ which is in the Σ_6-orbit of

$$x_0^4(x_1^2 + x_2^2) + x_1^4(x_2^2 + x_0^2) + x_2^4(x_0^2 + x_1^2) - 6x_0^2 x_1^2 x_2^2 = 0.$$

Unfortunately, these results do not allow us to explicitly describe the relation between the compactification $\mathcal{M}_{20}$ and compactifications of $\mathcal{A}_{(1,3)}(2)$, in particular the Baily-Borel embedding. This must be considered an interesting open problem.

3.4.4 A conjecture

To end this section we make a conjecture on one of the birational models of the variety $\mathcal{N}_5$. Consider the birational map $\mathcal{N}_5 \,-\,-\,\rightarrow \widehat{\mathcal{N}}_5$ of (3.68). Recalling now the Janus-like isomorphism between the Picard modular variety $Y(\sqrt{-3})$ and the Siegel modular variety $\overline{X(2)}$ (see Theorem 2.4.11), it is natural to ask about an analogue here, since the involved Siegel modular varieties of Corollary 3.4.17 all are related to level 2, albeit with different polarizations. So consider abelian fourfolds with complex multiplication by $\mathbb{Q}(\sqrt{-3})$ of signature (3,1), with a level $\sqrt{-3}$ structure, but with (1,1,1,3) polarizations.

Problem 3.4.19 Is $\widehat{\mathcal{N}_5}$ the Satake compactification of

$$Y_{(1,1,1,3)}(\sqrt{-3}) := \Gamma_{(1,1,1,3)}(\sqrt{-3})\backslash \mathbb{B}_3,$$

where $\Gamma_{(1,1,1,3)}(\sqrt{-3})$ denotes the arithmetic group giving equivalence of complex multiplication by $\mathbb{Q}(\sqrt{-3})$, signature (3, 1), with a level $\sqrt{-3}$ structure and a (1, 1, 1, 3)-polarization?

I conjecture that for *some* subgroup of $\Gamma_{(1,1,1,3)}(\sqrt{-3})$, this does in fact hold. Evidence for the conjecture:

i) The proper transforms of the 15 Segre planes are by Proposition 3.2.7 the moduli space of principally polarized abelian threefolds with complex multiplication by $\mathbb{Q}(\sqrt{-3})$, signature (2,1), with a level $\sqrt{-3}$ structure, (although these moduli spaces are blown up in three points on $\widehat{\mathcal{N}_5}$). These could parameterize abelian fourfolds with said complex multiplication, signature (3,1) with a level $\sqrt{-3}$ structure and polarization (1, 1, 1, 3) which split:

$$A_4 \cong A_3 \times A_1,$$

where A_3 has complex multiplication, signature (2,1), polarization (1,1,1), and A_1 has complex multiplication, but a polarization 3.

ii) The proper transforms of the 15 planes N_{ij} of Lemma 3.4.3, (ii), are four nodal cubic surfaces (Lemma 3.4.14). These surfaces occur also on the ball quotient S_3 above: pick any four of the nodes which are not coplanar; they determine a unique $\mathbb{P}^3$ in $\mathbb{P}^4$, and its intersection with S_3, a cubic surface, has four nodes in the four nodes of S_3 in that $\mathbb{P}^3$. (Note that there is a unique four-nodal cubic surface, as it is $\mathbb{P}^2$ blown up in the six intersection points of four (general) lines, a complete quadrilateral in $\mathbb{P}^2$, and any two such quadrilaterals are projectively equivalent. This cubic surface is usually called the Cayley cubic , mentioned above.)

iii) The singular locus consists of isolated singular points, resolved by quadric surfaces, so these singularities are rational. Recall that at each P_{ij}, six of the 15 Segre planes meet. At each Q_{ij} (the 15 points (3.65)), three of the Segre planes and six of the N_{ij} of Lemma 3.4.3, (ii) meet. In both cases, the exceptional $\mathbb{P}^1 \times \mathbb{P}^1$ can be covered equivariantly by a product $E_\varrho \times E_\varrho$ of the elliptic curve E_ϱ with branching only at the intersection with the proper transforms of the 30 planes above, as follows:

   – P_{ij}: $E_\varrho \times E_\varrho \longrightarrow \mathbb{P}^1 \times \mathbb{P}^1$ a Galois $\mathbb{Z}/3\mathbb{Z}$-quotient;
   – Q_{ij}: $E_\varrho \times E_\varrho \longrightarrow \mathbb{P}^1 \times \mathbb{P}^1$ is the product of two double covers branched at $0, 1, \varrho, \varrho^2$.

This supports by Lemma 3.2.5.1 the idea that this could be the compact-ification locus of $X_{(1,1,1,3)}(\sqrt{-3})$.

3.5 The Coble variety $\mathcal{Y}$

We now give an explicit projective description of the Baily-Borel compactifi-cation of the arithemetic quotient $\Gamma(2)\backslash \mathcal{D}$ of Proposition 2.3.3. All the facts presented here were proved originally by Coble [C1] or by Yoshida and his collab-orators in [MSY]. We have the four-dimensional family of K3-surfaces discussed in Chapter 2, defined in terms of a set of (ordered) six lines in the plane. Dual to the six lines are six points, and so the relation with the moduli space of cubic surfaces is evident. Let two ordered sets of six lines, $(l_1,\ldots,l_6)$, $(l'_1,\ldots,l'_6)$ be given.

Definition 3.5.1 The two sets of lines $(l_1,\ldots,l_6)$, $(l'_1,\ldots,l'_6)$ are said to be *associated*, if the following relation holds. Since the set $(l_1,\ldots,l_6)$ is ordered, we can form two triangles,

$$\Delta(l_1,l_2,l_3), \quad \Delta(l_4,l_5,l_6);$$

these two triangles have together six vertices, which come equipped with a num-bering, say $(p_1,\ldots,p_6)$, and these correspond dually (with respect to a conic) to another ordered set of six lines[5], $(l_{p_1},\ldots,l_{p_6})$. Then $(l_1,\ldots,l_6)$ and $(l'_1,\ldots,l'_6)$ are associated, if: $(l_{p_1},\ldots,l_{p_6}) = (l'_1,\ldots,l'_6)$, as a set of six ordered lines. It is easy to see that this definition does not depend on the conic used.

Of course, starting with two sets of ordered six points, one can define in the same way the notion of association. Since, as abstract moduli spaces, the space of ordered sets of six lines is the "same" (by duality) as the set of ordered sets of six points, we see that we are dealing here with the space of sets of six ordered points in $\mathbb{P}^2$. This problem was dealt with in the papers of Coble [C1], and has been given a modern treatment in [DO]. It can be described as follows. The relevant moduli space is easy to describe: let $(p_1,...,p_n)$ be a set of n points in $\mathbb{P}^k$; this is represented by M, the $n \times (k+1)$ matrix whose i^{th} column gives the coordinates of the point p_i. The moduli space is then the GIT quotient

$$\mathbf{P}_n^k = GL(k+1)\backslash M(n,k+1)/(\mathbb{C}^*)^n. \tag{3.83}$$

By taking the set of semistable points in $M(n,k+1)$ the above quotient is compact, although singular. It is classical that $\mathbf{P}_6^1$ is a threefold whose compact-ification can be identified with a cubic threefold in $\mathbb{P}^4$ with ten ordinary double points, which is just the Segre cubic $\mathcal{S}_3$. Note that the similar moduli problem, namely six points on a conic in $\mathbb{P}^2$, is realized by the Igusa quartic $\mathcal{I}_4$, so these are very closely related, but not identical, moduli problems.

---

[5]the line l_{p_i} is thus the first polar of p_i with respect to the conic in the sense of Definition B.1.6

Our interest here is in $\mathbf{P}_6^2$, a fourfold. In this case we may represent elements by matrices

$$\mathbf{P}_6^2 \ni M = \begin{bmatrix} 1 & 0 & 0 & 1 & x & w \\ 0 & 1 & 0 & 1 & y & z \\ 0 & 0 & 1 & 1 & u & u \end{bmatrix}, \qquad (3.84)$$

and as Coble shows, the map $\mathbf{P}_6^2 \dashrightarrow \mathbf{P}^4$, $M \mapsto [x : y : w : z : u]$ is a birational map (it is clear that $\mathbf{P}_6^2$ is rational, this map simply gives an explicit birationalization). The GIT theory here consists of finding G-invariant functions on $\mathbf{P}_6^2$, and these turn out to be generated by 3×3 minors of M.

In terms of the matrix M the process of association can be described as follows. Each such matrix M determines a second one: since the six points are ordered, one can define six lines by $l_{12} = \overline{p_1 p_2}$, $l_{13} = \overline{p_1 p_3}$, $l_{23} = \overline{p_2 p_3}$, $l_{45} = \overline{p_4 p_5}$, $l_{46} = \overline{p_4 p_6}$, $l_{56} = \overline{p_5 p_6}$; these six lines determine dually six points, whose coordinates are then brought into the normal form given above. It turns out that the entries of the second matrix are determined by the fact that the maximal minors are proportional to the maximal minors of the first. More precisely, if we let (ijk) denote the 3×3 minor of M which is given by the columns i, j, k, and if we let M' be the associated matrix, $(ijk)'$ the corresponding minor, then the minors of M and M' are related by:

$$(123)(145)(246)(356) = (124)'(135)'(236)'(456)'. \qquad (3.85)$$

Now association is an involution on $\mathbf{P}_6^2$, and one can take the *quotient* by this involution.

Definition 3.5.2 Let $\mathcal{Y}$ be the double cover of $\mathbb{P}^4$ branched along the Igusa quartic $\mathcal{I}_4$, $\pi : \mathcal{Y} \longrightarrow \mathbb{P}^4$.

Clearly $\mathcal{Y}$ will be singular precisely along the singular locus of $\mathcal{I}_4$, i.e.,

Lemma 3.5.3 *The singular locus of $\mathcal{Y}$ consists of 15 lines, the inverse images of the 15 singular lines of $\mathcal{I}_4$.*

Theorem 3.5.4 ([DO],Example 4, p. 37) *The moduli space of six ordered points in $\mathbb{P}^2$ is equal to the double cover $\mathcal{Y}$, and the double cover involution on $\mathcal{Y}$ coincides with the association involution on $\mathbf{P}_6^2$.*

In other words, a set $(p_1, \ldots, p_6)$ is *associated to itself*, if and only if the six points lie on a conic in $\mathbb{P}^2$.

Consider one of the hyperplanes H in $\mathbb{P}^4$, $H = \{\theta_m^4 = 0\}$ of Proposition 3.3.9. Since H is *tangent* to $\mathcal{I}_4$, the inverse image $\pi^{-1}(H)$ in $\mathcal{Y}$ will *split into two copies of $\mathbb{P}^3$*. In this way, we get a union of 20 $\mathbb{P}^3$'s on $\mathcal{Y}$,

Lemma 3.5.5 *The inverse images $\pi^{-1}(H)$ of the tangent hyperplanes $H = \{\theta_m^4 = 0\}$ consist of two copies each of $\mathbb{P}^3$, and these two $\mathbb{P}^3$'s on $\mathcal{Y}$ meet in the quadric surface which is the inverse image under π of the quadric on $\mathcal{I}_4$ to which H is tangent. This gives a total of 20 such $\mathbb{P}^3$'s on $\mathcal{Y}$.*

A resolution of singularities of $\mathbf{P}_6^2$ is affected by resolving the Igusa quartic by blowing up the ideal of the 15 lines; this is the map ϱ_2 of Theorem 3.3.11. Let $\widehat{\mathbf{P}_6^2}$ denote this desingularization $\widehat{\mathbf{P}_6^2} \longrightarrow \mathbf{P}_6^2$. On $\widehat{\mathbf{P}_6^2}$ we have a set of 36 divisors, the *discriminant locus*, the proper transforms of the Igusa quartic, the 20 $\mathbb{P}^3$'s and the 15 exceptional divisors of the blow up.

It is clear how this variety is the moduli space of cubic surfaces : blow up $\mathbb{P}^2$ in the six points, and embed by the linear system of cubic curves through the six points. The ordering of the six points of course determines a marking of the 27 lines in the well-known manner. We will discuss this in more detail in the next chapter. The symmetry group of $\widehat{\mathbf{P}_6^2}$ is $\Sigma_6 \times \mathbb{Z}/2\mathbb{Z}$; although the Weyl group $W(E_6)$ acts birationally on it, the action is not regular. For that it is neccessary to modify $\widehat{\mathbf{P}_6^2}$ even more. Dolgachev mentions in [DO] that he suspects it is sufficient to blow up $\widehat{\mathbf{P}_6^2}$ in the intersection of the 36 divisors.

One of the many things proved in [MSY] is the following.

Theorem 3.5.6 *The variety $\mathcal{Y}$ is the Baily-Borel compactification of the arithmetic quotient $\Gamma(2)\backslash \mathcal{D}$ of Proposition 2.3.3.*

The proof given in [MSY] of this fact simply (!) calculates the image of the period map, and in determining when the periods lie on the boundary of the period domain $\mathcal{D}$, the authors find that this locus coincides with the set of K3 surfaces whose set of six lines correspond to those singularities of Lemma 3.5.3 of $\mathcal{Y}$.

Corollary 3.5.7 *The Loci 5) and 7) of Table 2.1 are the inverse images on $\mathcal{Y}$ of the 15 singular lines and 15 singular points, respectively, of the branch locus $\mathcal{I}_4$. The Loci 3) of Table 2.1 are the inverse images of the ten special hyperplane sections of Lemma 3.5.5, i.e., the quadrics. The loci 2) of Table 2.1 are the 20 $\mathbb{P}^3$'s of Lemma 3.5.5, and Locus 1) is just the branch locus of the double cover.*

Chapter 4

The 27 lines on a cubic surface

Whole books have been written on the 27 lines on a general cubic surface, for example [Hen], [Seg]. Still this configuration retains its charm, and in a sense, the rest of the book is devoted to some of the geometry which canonically arises from this beautiful configuration. In fact, a lot of the geometry, including all of Chapter 6, is essentially new, giving a complement to well-known results.

Virtually all the geometry is closely tied up with the *automorphism group* of the lines, the Weyl group of E_6. The two fundamental roles played by $W(E_6)$ (respectively by its simple normal subgroup of index two, $G_{25,920}$) are:

- $W(E_6)$ (resp. $G_{25,920}$) is a reflection (resp. unitary reflection) group in $\mathbb{P}^5$ (resp. $\mathbb{P}^4$ and $\mathbb{P}^3$).

- $W(E_6)$ (resp. $G_{25,920}$) is the Galois group of an algebraic equation, namely the algebraic equation for the 27 lines on the cubic surface (resp. the algebraic equation which results after extracting the discriminant of the above).

We will consider the second role in this chapter and the first in the following chapters. Of course the two are interwoven, and in fact the latter problem is *reduced* to the former, and it is the former (a form problem) which can be solved (by transcendental methods).

In the first section we recall the normal forms for equations of cubic surfaces and the configurations related to the 27 lines. In the second we recall the general method developed in the last century to solve algebraic equations with simple Galois groups, giving the standard example of the quintic as developed in [Kl3]. In the third section we then recall the actions of $G_{25,920}$ on $\mathbb{P}^3$ and $\mathbb{P}^4$, and the invariants, etc., of these actions. Finally, in the last section we discuss the equation for the 27 lines, an equation with Galois group $W(E_6)$. There are two approaches to solving the equation. The first, due to Klein and Jordan, and carried out by Maschke and Burkhardt, uses foremost the representation of $G_{25,920}$ in $\mathbb{P}^3$ and its invariants. The second, due to Coble, uses the representation in

$\mathbb{P}^4$ and reduces the problem to the form problem for this group, and then to a *special* form problem, which is then a problem about the invariant of fourth degree J_4, the topic of the next chapter. We should remark that as far as solving the equation is concerned, all this work is more or less unnecessary. In fact, one has a standard method for giving the solution of *any* equation in terms of theta functions, as shown in the appendix to Mumford's Theta III. But the work is worth it as far as understanding the geometry of the situation is concerned.

4.1 Cubic surfaces

In this section we introduce a key configuration for this and the following examples. Although of course very well-known, we review many of the facts about the 27 lines on a general cubic surface. We begin with the classical approach, in terms of cubic forms, and then proceed to the more modern approach, in which the cubic surface is considered as a del Pezzo surface of degree 6.

4.1.1 The classical approach

Towards the middle of the last century algebraic geometry was being born. One of the first topics studied was algebraic surfaces in $\mathbb{P}^3(\mathbb{C})$. Irreducible quadrics have a familiar structure: either they are smooth, in which case they are isomorphic to $\mathbb{P}^1 \times \mathbb{P}^1$, or they have a double point, in which case they are a cone over a quadric curve. In either case there is a continuous family of lines lying on the quadric, which in those days was indicated by a statement "a smooth quadric has two ∞'s of lines, a singular quadric a single ∞". The next simplest surface is a cubic surface, the vanishing locus of a homogenous polynomial of degree 3. It was natural to wonder whether a smooth (or singular) cubic also has an ∞ of lines. It seems this question was first taken up by Arthur Caley, who reasoned as follows. Since a generic line in $\mathbb{P}^3$ intersects a cubic surface in three points, it will be contained in the cubic if it contains a fourth point lying on the cubic. In other words, the condition that a line lie on a cubic surface places four conditions on the line. On the other hand, there are ∞^4 lines in $\mathbb{P}^3$ (i.e., the Grassmann of lines in $\mathbb{P}^3$ is four-dimensional; it is the "Klein quadric" in $\mathbb{P}^5$). It then stands to reason that there should in fact be only finitely many lines on a cubic surface (not necessarily smooth, the reasoning applies also at least to cubic surfaces with isolated singularities). Arthur Cayley communicated this to George Salmon in 1849, who promtly replied that the finite number is 27 for a smooth cubic surface. His reasoning went as follows.

Consider hyperplane sections H_t passing through one of the lines. Since $H_t \cap S$ is a third degree plane curve, which already contains a line, it consists generically of a quadric and a line. Since this intersection contains two double points, it follows that *every* such plane is a bitangent, i.e., tangent to S in two points. For a finite number of hyperplanes H_t this intersection degenerates into three lines, as illustrated in Figure 4.1.

Such planes H_t which are tangent to S at three points (and contain three

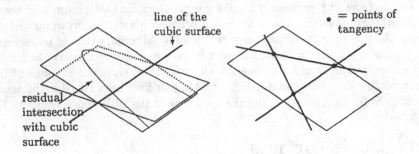

Figure 4.1: (a) generic (transversal) section   (b) tangent section

lines) are called accordingly *tritangent planes*. Now suppose a smooth cubic surface $\{f = 0\}$ together with a line L on it is given. Fix coordinates in such a way that $L = \{x_3 = x_4 = 0\}$ for homogenous coordinates $(x_1 : x_2 : x_3 : x_4)$ on $\mathbb{P}^3$. Then the equation $f = 0$ can be written $x_3 U + x_4 V = 0$ with U, V quadratic. To find the tritangent planes one puts $x_3 = \mu x_4$ into this relation, divides by x_4 and forms the discriminant of the ensuing equation. Viewing this as an equation in μ, one sees easily that it has degree 5, i.e., there are five values of μ for which the corresponding plane is a tritangent, or in other words the given line lies in five tritangent planes. Now Salmon counts all lines on the cubic surface as follows. Since there are five tritangent planes through L, there are 5 triangles on the cubic surface, all of which contain L. The same holds for any line. We see that a fixed tritangent plane meets a total of $4 \cdot 3 = 12$ other tritangents in one of its three lines, and each of these contains two more lines, hence we find a total of $24 + 3 = 27$ lines. In [Sal] this argument is used to prove the fact that a smooth cubic surface contains 27 lines, in particular, that there is a line L to which the argument may be applied.

We consider now the group of permutations, $\mathrm{Aut}(\mathcal{L})$, of the 27 lines (or of the 45 planes), by which we mean the permutations of the lines preserving the intersection behavior of the lines. For this it is useful to consider the famous double sixes and the notation for the 27 lines introduced by Schläfli. A *double six* is an array

$$N = \begin{bmatrix} a_1 & a_2 & a_3 & a_4 & a_5 & a_6 \\ b_1 & b_2 & b_3 & b_4 & b_5 & b_6 \end{bmatrix}$$

of 12 of the 27 lines with the property that two of these 12 meet if and only if they are in different rows and columns. (This notation distinguishes this particular set of 12, although any such double six is equivalent to it under $\mathrm{Aut}(\mathcal{L})$). The other lines are given by the $\binom{6}{2} = 15$ $c_{ij} = a_i b_j \cap a_j b_i$, where $a_i b_j$ denotes the tritangent spanned by those two lines. There are 36 double sixes, namely the N

above, 15 N_{ij} and 20 N_{ijk}:

$$N_{ij} = \begin{bmatrix} a_i & b_i & c_{jk} & c_{jl} & c_{jm} & c_{jn} \\ a_j & b_j & c_{ik} & c_{il} & c_{im} & c_{in} \end{bmatrix}$$

$$N_{ijk} = \begin{bmatrix} a_i & a_j & a_k & c_{mn} & c_{ln} & c_{lm} \\ c_{jk} & c_{ik} & c_{ij} & b_l & b_m & b_n \end{bmatrix}$$

(4.1)

Since a double six describes, by definition, the intersection behavior of the lines, we see immediately that Σ_6 (the symmetric group on six letters) acts by permutations on a double six and a $\mathbb{Z}_2$ acts by exchanging rows. Since there are 36 double sixes, we see $|\mathrm{Aut}(\mathcal{L})| = |\Sigma_6| \cdot 2 \cdot 36 = 51{,}840$. In fact $\mathrm{Aut}(\mathcal{L})$ is nothing but the Weyl group $W(E_6)$, and the 36 double sixes correspond to the positive roots. The 27 lines correspond to the 27 fundamental weights of E_6, and the many other sets of objects (lines, tritangents, etc.)correspond to natural sets of objects (roots, weights, etc.) of E_6. In the remainder of this section we will consider the sets of objects assoicated with the 27 lnes. Many of these will be translated below into the $W(E_6)$ formulation, in particular in Chapter 6. Consider two double sixes; a natural question arising here is: how many lines do they have in common? The answer is twofold:

- either: four (like a_1, a_2, b_1, b_2, which N and N_{12} have in common) which have the property of lying in pairs in planes, the pairs being however mutually disjoint;

- or: six (like a_1, a_2, a_3, b_1, b_2, b_3, which N and N_{123} have in common) which form two triples.

Following Sylvester one speaks accordingly of *syzygetic* and *azygetic* pairs of double sixes. A given double-six is syzygetic to 15 others and azygetic to 20 others. A pair of azygetic double sixes form through the 12 lines they do *not* have in common a third double-six, which is azygetic to both. There are 120 such triads of azygetic double sixes, and $36 \cdot 15 \div 2 = 270$ pairs of syzygetic double sixes.

A further geometric curiosity of the 45 planes is the following. Take two of the tritangents which do not meet in a line on the cubic surface, say α_1, α_2 (for example $a_1 b_2 c_{12}$ and $a_3 b_4 c_{34}$). These two planes determine three other planes, denoted β_1, β_2, β_3, by the property of containing each a line of α_1 and of α_2 (in the example above, $a_1 b_4 c_{14}$, $b_2 a_3 c_{23}$, $c_{12} c_{34} c_{56}$). The third line in each of β_i which is not one of α_i all lie in a common tritangent (for example here $c_{14} c_{23} c_{56}$). Then this is a unique third tritangent denoted α_3. The set of 6 tritangents $(\alpha_1, \alpha_2, \alpha_3)$, $(\beta_1, \beta_2, \beta_3)$ have the special property that the nine lines $\alpha_\mu \cap \beta_\nu$ together with a point (19 conditions) determine S. Such a set of six tritangents is called a *trihedral pair*, and there are 120 such; this implies that the equation of the cubic surface S can be written in 120 ways as

$$y_1 y_2 y_3 + z_1 z_2 z_3 = 0.$$

Table 4.1: Loci assoicated with the 27 lines on a smooth cubic surface

# objects	description of the objects
27	lines on a cubic surface
135	intersection points of two of the lines
216	pairs of skew lines
36	double sixes
45	tritangents
120	trihedral pairs, set of six tritangents containing nine lines
40	triples of trihedral pairs, set of 18 tritangents containing all 27 lines
120	triads of azygetic double sixes
270	pairs of syzygetic double sixes

Furthermore there are 40 *triads (triples) of trihedral pairs*, such that each such triad contains all 27 lines. The α's meet the β's in nine lines on the cubic surface and vica versa. One finds the following types:

$$
\begin{array}{|ccc|ccc|ccc|}
\hline
\multicolumn{3}{|c|}{(20)} & \multicolumn{3}{c|}{(10)} & \multicolumn{3}{c|}{(90)} \\
\hline
a_i & b_j & c_{ij} & c_{il} & c_{jm} & c_{kn} & a_i & b_j & c_{ij} \\
b_k & c_{jk} & a_j & c_{mn} & c_{ik} & c_{lj} & b_l & a_k & c_{kl} \\
c_{ik} & a_k & b_i & c_{jk} & c_{ln} & c_{im} & c_{il} & c_{jk} & c_{mn} \\
\hline
\end{array}, \qquad (4.2)
$$

the rows (of each box) giving α_1, α_2, α_3 and the columns giving β_1, β_2, β_3. This configuration is in some sense complementary to the double sixes: *Starting with nine lines lying in such a trihedral pair, the remaining 18 lines form a unique azygetic triple of double sixes and conversely, the nine lines not contained in a given azygetic triple of double sixes always lie in a trihedral pair.* We sum up the configuration in Table 4.1.

4.1.2　Equations

As far as the equation defining a cubic surface is concerned, we have the following result, claimed by Sylvester (1851) and Steiner, and proved by Clebsch (1861); it is the so-called *pentrahedral form* of the cubic. More recent proofs of this theorem may be found in [Ri] and [Sh].

Theorem 4.1.1 *A general cubic form $F(x_0 : x_1 : x_2 : x_3)$ can be put, in a unique way, in the form*

$$a_1 y_1^3 + a_2 y_2^3 + a_3 y_3^3 + a_4 y_4^3 + a_5 y_5^3 = 0,$$

where the coordinates y_i are linear in $(x_0 : \cdots : x_3)$ and satisfy

$$y_1 + \ldots + y_5 = 0.$$

The five planes $y_i = 0$ are the faces of a *Sylvester pentahedron*; they meet in pairs in ten lines and three at a time in ten points. These ten lines are contained in the Hessian of the cubic surface S (a quartic, see section B.5.3.2), and the ten points are *double points* of that Hessian. Viewing these ten points as hyperplanes (in the dual $\mathbb{P}^3$), the product gives a polynomial of degree ten, whose vanishing defines the ten points. Clebsch's proof of 4.1.1 was by showing that the degree ten equation has a resolvent of degree five (see Lemma 4.2.2 below), which can be solved by the methods described in the next section (see in particular Theorem 4.2.9 and Figures 4.8, 4.9 and 4.12 below). The five solutions of this equation then determine five linear forms y_i in the x_i, and the uniqueness[1] of the representation above is relatively clear from the construction. This form of the cubic equation is very convenient for the *invariant theory* (see section B.4 in the Appendix). We also mention that in the appendix we describe the Hessian, Steinerian and dual varieties of a cubic surface. Also there the notion of invariants of the cubic form is introduced, and the known invariants are listed.

A second special form of the equation, the *hexahedral form*, is given by the *polar hexagons*:

Theorem 4.1.2 *A general cubic form F can be put in a four-dimensional family of forms:*

$$z_1^3 + \cdots + z_6^3 = 0,$$

where the z_i are linear in the x_i.

Cremona (Math. Ann. XIII) recognized that there are 36 special equations of the form 4.1.2 – one associated with each double six. Indeed, the 15 lines *not* in the given double six lie in 15 tritangent planes, and from these one can construct ten trihedral pairs; the ten pairs of intersection points of the trihedra form the pairs of *opposite* vertices of the hexagon (six planes meet in $\binom{6}{2} = 15$ lines and $\binom{6}{3} = 20$ points). Furthermore, from the hexahedral equation one can get the pentahedral equation.

Proposition 4.1.3 (Cremona, 1840) *In each of the 36 hexagons associated with any of the double sixes of the cubic surface S, there is a space cubic curve inscribed, all of which have the same five ocsulating planes; these form the pentahedron of 4.1.1.*

Cremona also noted that for a cubic in hexahedral form, one can explicitly write down the equations for the 27 lines and 45 tritangents. We will return to this below. It is possible to pass from either of the above equations to the other; this is described in [Sal], §310 (p. 403). The hexahedral form determines immediately

---

[1]assuming F is generic, $\mathrm{Aut}(S) = \{1\}$. For cubics with non-trivial automorphisms, for example if $a_i = a_j$, uniqueness is lost, but there will still be finitely many pentahedral forms representing a given F.

the pentahedral form by the previous proposition, while the determination of the hexahedral form from a given pentahedral form requires the solution of a sextic equation. However, there is no purely (rationally) algebraic method of passing from a given quatenary cubic F to the pentahedral or hexahedral form, which follows from the fact that the Galois group of the problem is simple; indeed any algebraic reduction to the form 4.1.1 would imply solvability of the Galois group; this actually occurs for special cubic surfaces, but not in general.

Finally we briefly describe the "equation of 27th degree which determines the 27 lines" which will be discussed extensively below; see in particular section 4.3 for this. It was already known to Cayley that one can write the set of 27 lines as the intersection of a covariant of degree nine of the cubic with the cubic surface S, as described in (B.51) of the appendix. This is an equation of degree 27 in the variables of $\mathbb{P}^3$, and applying elimination theory reduces this (but not in a covariant manner) to a degree 27 equation in a single variable. It almost goes without saying that the Galois group of this equation is $\text{Aut}(\mathcal{L})$. It can be reduced to $G_{25,920}$ by extracting the square root of the discriminant (see, for example, [Kl1]).

4.1.3   Special cubic surfaces

There are two particular cubic surfaces which are very special and easy to study. We will do that in this section, where we will use several general results on cubic surfaces proved or stated later, to give a rather complete discussion of them. Let us list several aspects, about which one would generaly like to know as much as possible for any given cubic surface S:

1. the pentahedral equation 4.1.1;

2. the hexahedral form 4.1.2;

3. the 36 sets of six points in the plane which are defined by the double sixes of S;

4. invariants and linear covariants of S (B.49) and (B.50);

5. the equations of the lines and tritangents;

6. the Hessian variety, a symmetroid quartic surface (B.5.11);

7. the dual variety (B.37).

The equations were discussed above. The sets of six points are considered in general in the next section. Finding equations for the lines and tritangents is a kind of unifying theme of the rest of the book and occurs repeatedly from now on. The symmetroids are discussed in the appendix on quartic surfaces, and the equation of the dual variety is described in the appendix on cubic forms. As far as it is feasible we will discuss these issues for the two special cubic surfaces. These surfaces each have a comparitively large automorphism group, the symmetric groups Σ_4 and Σ_5, respectively, and correspondingly, each is

particularly symmetric. A general cubic surface will not be as symmetric as
these are, a fact the reader should bear in mind.

The Cayley cubic $\mathcal{C}$. There is a unique cubic surface with four nodes, usually
called the Cayley cubic surface, which we will denote by $\mathcal{C}$. If one takes the four
vertices

$$(1,0,0,0), (0,1,0,0), (0,0,1,0), (0,0,0,1)$$

to be those nodes (this does not restrict generality as any four points in $\mathbb{P}^3$ are
equivalent to these under $PGL(4)$), the equation can be written

$$\frac{1}{u_0} + \frac{1}{u_1} + \frac{1}{u_2} + \frac{1}{u_3} = 0, \text{ or } u_0 u_1 u_2 + u_0 u_1 u_3 + u_0 u_2 u_3 + u_1 u_2 u_3 = 0. \quad (4.3)$$

First we describe the pentahedral equation for $\mathcal{C}$.

Proposition 4.1.4 *The pentahedral form of the Cayley cubic (4.3) is*

$$\mathcal{C} = \{y_1^3 + y_2^3 + y_3^3 + y_4^3 + \frac{1}{4}y_5^3 = 0\}, \quad \sum y_i = 0.$$

Proof: We recall from the discussion of the Segre cubic S_3 that there were
special kinds of hyperplane sections; the sections $x_i + x_j$ cut S_3 in three $\mathbb{P}^2$'s, the
sections $x_i = 0$ cut out the diagonal surface, while the other sections of interest
were the 15

$$S_3 \cap \{T_{ij} = 0\}, \quad (4.4)$$

where $T_{ij} = x_i - x_j$ contains four of the nodes, for example $T_{56} = x_5 - x_6$ contains

$$(1,-1,-1,-1,1,1), \quad (1,1,1,-1,-1,-1), \quad (1,1,-1,1,-1,-1)$$

and

$$(1,-1,1,1,-1,-1).$$

Since as is easily checked the cubic surface (4.4) is irreducible, it has four nodes
and so does $\mathcal{C}$. Since $\mathcal{C}$ is the unique cubic surface with four nodes, this hyperplane section is isomorphic to $\mathcal{C}$. From this we can determine the pentahedral
equation: we have $\sum_1^6 x_i = 0$, $x_5 - x_6 = 0$, so the equation for S_3 becomes:

$$x_1^3 + x_2^3 + x_3^3 + x_4^3 + 2x_5^3 = 0, \quad 2x_5 = -x_1 - x_2 - x_3 - x_4.$$

So if we set:

$$y_i = x_i, \ i = 1,\ldots,4, \quad y_5 = 2x_5, \quad (4.5)$$

then we have $\sum_1^5 y_i = 0$ and the equation for $\mathcal{C}$ becomes

$$x_1^3 + \cdots + x_4^3 + 2x_5^3 = y_1^3 + \cdots + y_4^3 + \frac{1}{4}y_5^3 = 0,$$

which is the desired pentahedral form. □

Since the cubic surface $\mathcal{C}$ is *singular*, there are not 27 lines, 45 tritangents
and 36 double sixes. As to the double sixes: since $\mathcal{C}$ is a section of the Segre
cubic S_3, we get immediately its hexahedral form:

Proposition 4.1.5 *The 15 hexahedral equations of* $\mathcal{C}$ *are given by the* Σ_6-*orbit of*

$$(\bar{a},\ldots,\bar{f}) = (0,0,0,0,1,-1).$$

Proof: We have the coordinates $(a,\ldots,f) = (x_1,\ldots,x_6)$, and the equation of $\mathcal{T}_{56}$ defines the $(\bar{a},\ldots,\bar{f})$ as stated. $\square$

To count the lines, let us begin by describing the six points in the plane giving rise to $\mathcal{C}$.

Proposition 4.1.6 *The six points in* $\mathbb{P}^2$ *determining* $\mathcal{C}$ *are the six intersection points of a complete quadralateral, i.e., as pictured:*

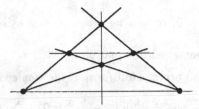

The thin lines are the three diagonals of the quadralateral. The set of six points is rigid; it cannot be deformed, hence defines a unique set of six points.

Proof: Blowing up the six points, each edge of the quadralateral contains three points, hence becomes a -2-curve and is blown down to an ordinary node, so the corresponding cubic surface has four nodes. As such a surface is unique, it is (isomorphic to) $\mathcal{C}$. $\square$

Now we can find all lines on $\mathcal{C}$. They are the images of the six points, the lines joining the six points in twos, of which there are now only three instead of 15, the diagonals of the quadlateral, and the conics on five points, all of which coincide with sets of two of the four lines (and which get blown down on the surface). This establishes the first statement of

Proposition 4.1.7 *The Cayley cubic* $\mathcal{C}$ *contains nine lines and 11 tritangents. The latter are given by:*

$$\begin{aligned} y_i + y_j, \quad 1 \leq i < j \leq 4 \quad &(6) \\ 2y_i + y_5 \quad\quad\quad &(4) \\ y_5 \quad\quad\quad &(1). \end{aligned}$$

Proof: The 15 Segre planes on $\mathcal{S}_3$, intersected with $\mathcal{C}$, give five lines. These 15 planes are cut out by the 15 hyperplanes $x_i + x_j, 1 \leq i < j \leq 6$, which from $2x_5 = y_5$ become the 11 listed planes. That there are no further tritangents follows from (4.14) below and the hexahedral form 4.1.5 for $\mathcal{C}$. Indeed, in this case $d_2 = 1$, while $\bar{a} = \bar{b} = \bar{c} = \bar{d} = 0$, $\bar{e} = 1, \bar{f} = -1$. So the equations (4.14) become, for example

$$(\bar{e}\bar{f} - 1)(e + f) \equiv (-2)(x_5 + x_6) = -4x_5 = -2y_5,$$

Figure 4.2: The Cayley cubic surface

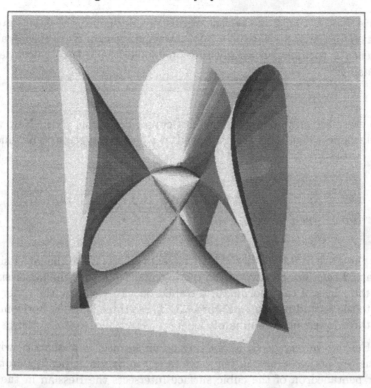

The affine equation used here is

$$(1 - 3x - 3y - 3z)(xy + xz + yz) + 6xyz = 0.$$

These affine coordinates are derived by sending the coordinate tetrahedron $\Delta = \{u_0 u_1 u_2 u_3 = 0\}$ to the simplex $\Delta' = \{xyz(x + y + z - 1/3) = 0\}$. As a plane at infinity one chooses $v = 3(u_0 + u_1 + u_2 + 2u_3)$, so that, in terms of the new coordinates u_0, u_1, u_2, v, we have $u_3 = \frac{1}{2}(1/3v - u_0 - u_1 - u_2)$, and this is inserted in the homogenous equation for $\mathcal{C}$. We get

$$\mathcal{C} = \{u_0 u_1 u_2 + (u_0 u_1 + u_0 u_2 + u_1 u_2)\frac{1}{2}(\frac{1}{3}v - u_0 - u_1 - u_2) = 0\}.$$

Now dehomogenizing by dividing by v^3 and setting $x = u_0/v$, $y = u_1/v$, $z = u_2/v$, we get the affine equation

$$\mathcal{C} = \left\{ xyz + \frac{1}{2}(xy + yz + xz)(\frac{1}{3} - x - y - z) = 0 \right\}$$

$$= \left\{ \frac{1}{6}(6xyz + (xy + yz + xz)(1 - 3x - 3y - 3z) = 0) \right\}.$$

so this plane has already been acounted for, and similarly for the others.   □

Remark 4.1.8 Actually, in the case of this singular cubic, the Galois group of the equation for the 27 lines reduces from the Weyl group of E_6 to the tetrahedral group, hence is solvable, so at any rate one should be able to calculate these tritangents.

Since there are only nine lines there of course no double sixes. Now we can easily determine which lines are contained in which tritangents. The nine lines are:

(3) $y_5 = 0 = y_i + y_j = y_k + y_l,\ \ \{i,j,k,l\} = \{1,2,3,4\},$

(6) $2y_i + y_5 = 2y_j + y_5,\ 1 \le i < j \le 4.$

The first three are contained in three tritangents apiece, while the latter six lines are contained each in only two of the tritangents. Note that the latter six lines are just the edges of the coordinate tetrahedron. To visualize the other three, we refer to the second picture in Figure 4.3. These three lines are horizontal and join the three lower local minima of $\mathcal{C}$.

The Hessian variety of $\mathcal{C}$, as for any cubic surface, is a quartic with ten nodes, the ten nodes being the ten vertices of the Sylvester pentahedron. The Sylvester pentahedron of the cubic surface intersects the Hessian in the union of ten lines on that quartic. A picture for the Cayley cubic is given in Figure 4.4. This quartic has in fact 14 nodes instead of just ten, which can be seen as follows. In the following set of pictures, two tetrahedra have been drawn into the view of the Hessian.

Note that the central teepee forms a tetrahedron: a "top" tetrahedron. Now locate the bottom face of that tetrahedron; at the center of that face is a node, connecting to the bottom branch of the surface, which goes vertically downwards. This node and the three other vertices of the face form another tetrahedron, a "bottom" tetrahedron. The three lower branches have two nodes apiece, one at the center of each face of the bottom tetrahedron, and one on the bottom branch of the surface.

Hence there are three planes[a], horizontal in the picture, each of which contains three nodes, accounting for nine nodes. In addition there is the node joining the bottom tetrahedron with the bottom branch, and the node joining the top tetrahedron with the top branch, so we see in all 11 nodes.

---

[a]Two of these planes are the horizontal planes of the Sylvester pentahedron. The third, lying above the other two, contains the three vertices of the common face of the top and bottom tetrahedron.

There are three further nodes in the plane at infinity, so this quartic has 14 ordinary double points (note that the two parallel planes of the Sylvester pentahedron meet in the plane at infinity, and this two-fold line meets each of the other three planes of the Sylvester pentahedron in a point which is a vertex of the Sylvester pentahedron, i.e., a point where three of the planes meet, hence a node of the Hessian). Of course the four additional nodes (compared with the Hessian of a smooth cubic surface) come from the nodes of the Cayley cubic, and with some imagination, one can superimpose the center tetrahedron of the Cayley cubic over the top tetrahedron of the Hessian and see the four additional nodes (the pictures are not at exactly the same scale and angle, but roughly). Looking at the picture of the Hessian with the planes of the Sylvester pentahedron, the four additional nodes are precisely those not lying on one of these five planes – these four points are the vertices of the bottom tetrahedron.

Note that in the picture with the Sylvester pentahedron, one of the five planes is *head on*, and one only sees a line. That plane contains four of the nodes which are finite.

Finally, we describe the transition from pentahedral to tetrahedral coordinates. As we have seen, the tetrahedron of reference is given by $\Delta = \{u_0 u_1 u_2 u_3 = 0\}$. In the y_i coordinates, we take the four tritangents $2y_i + y_5$ of Proposition 4.1.7, i.e., we set

$$u_0 = 2y_1 + y_5, \qquad u_1 = 2y_2 + y_5,$$
$$u_2 = 2y_3 + y_5, \qquad u_3 = 2y_4 + y_5, \tag{4.6}$$

and a direct verification shows that

$$u_0 u_1 u_2 + u_0 u_1 u_3 + u_0 u_2 u_3 + u_1 u_2 u_3 = \frac{8}{3} \left(\sum_1^4 y_i^3 + \frac{1}{4} y_5^3 \right).$$

Figure 4.3: The Cayley cubic with the coordinate tetrahedron and its Hessian variety

Figure 4.4: The Cayley cubic and its Hessian with the Sylvester pentahedron

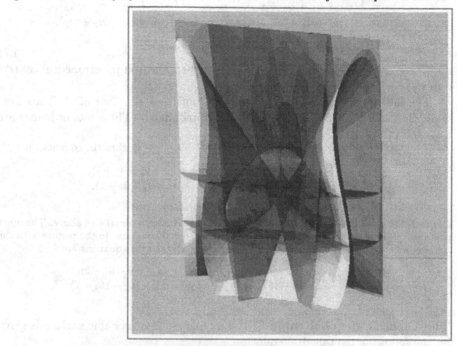

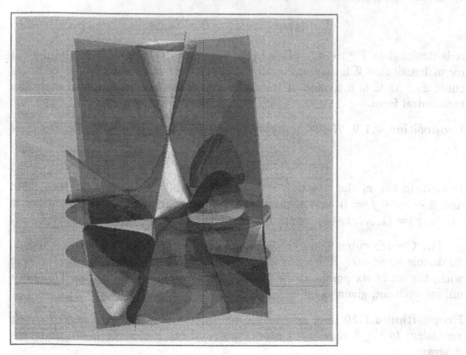

Inverting the equations we have

$$y_5 = \tfrac{1}{2}(u_0 + u_1 + u_2 + u_3), \quad y_1 = -\tfrac{1}{4}(-u_0 + u_1 + u_2 + u_3),$$
$$y_2 = -\tfrac{1}{4}(u_0 - u_1 + u_2 + u_3), \quad y_3 = -\tfrac{1}{4}(u_0 + u_1 - u_2 + u_3),$$
$$y_4 = -\tfrac{1}{4}(u_0 + u_1 + u_2 - u_3),$$

(4.7)

and these are the equations for the Sylvester pentahedron in tetrahedral coordinates.

The four planes of (4.6) are tritangents but the latter four of (4.7) are not, while the plane $y_5 = 0$ does turn out to be a tritangent. These sets of planes are depicted in Figure 4.3.

If we take the affinisation given by $u_3 = 1/2(1/3 - x - y - z)$, then the equations are:

$$y_1 = \tfrac{1}{6}(-9x + 3y + 3z + 1), \quad y_2 = \tfrac{1}{6}(3x - 9y + 3z + 1),$$
$$y_3 = \tfrac{1}{6}(3x + 3y - 9z + 1), \quad y_4 = \tfrac{1}{6}(9x + 9y + 9z - 1),$$
$$y_5 = \tfrac{1}{3}(1 + 3x + 3y + 3z).$$

Of these, the plane $(1 + 3x + 3y + 3z) = 0$ is the special tritangent mentioned above. The other pentahedral planes are, as one can see in the picture, not tritangents. In the picture with the Hessian we have instead used $u_3 = 1/2(2/5 - x - y - z)$; then the equations are

$$y_1 = \tfrac{1}{40}(15x - 5y - 5z - 2), \quad y_2 = \tfrac{1}{40}(-5x + 15y - 5z - 2),$$
$$y_3 = \tfrac{1}{40}(-5x - 5y + 15z - 2), \quad y_4 = \tfrac{1}{40}(-15x - 15y - 15z + 2),$$
$$y_5 = \tfrac{1}{20}(2 + 5x + 5y + 5z).$$

The Clebsch diagonal cubic $\mathfrak{C}$ The other very famous cubic surface is given by the following pentahedral equation:

$$\mathfrak{C} = \{y_1^3 + \cdots + y_5^3 = 0 = \sum y_i\};$$

(4.8)

it is depicted in Figure 4.5. This surface happens to be given in pentahedral form. Recall that $\mathfrak{C}$ is isomorphic to the hyperplane sections $x_i = 0$ of the Segre cubic $\mathcal{S}_3$. As $\mathfrak{C}$ is a section of the Segre cubic, again we get immediately the hexahedral form.

Proposition 4.1.9 *The 36 hexahedral forms for* $\mathfrak{C}$ *are given by the* $W(E_6)$*-orbit of*

$$(\bar{a}, \ldots, \bar{f}) = (1, \ldots, 1, -5).$$

Proof: In the x_i the equation is $\sum_1^6 x_i^3 = 0$, while $x_6 = 0$ on the one hand and $\bar{a} + \cdots + \bar{f} = 0$ on the other. It follows that $\bar{a} = \cdots = \bar{e}$, and hence $(\bar{a}, \ldots, \bar{f}) = (1, \ldots, 1, -5)$, as stated. $\square$

The Clebsch cubic $\mathfrak{C}$ is smooth, hence contains 27 lines, has 45 tritangents, 36 double sixes, etc. But these 27 lines are in very special position. To begin with, the set of six points in the plane must admit an action of Σ_5. There is a unique such set, given as follows:

Proposition 4.1.10 *Any set of six ponts in* $\mathbb{P}^2$ *giving rise to* $\mathfrak{C}$ *is projectively equivalent to the vertices of a pentagram, and the center, as in the following diagram.*

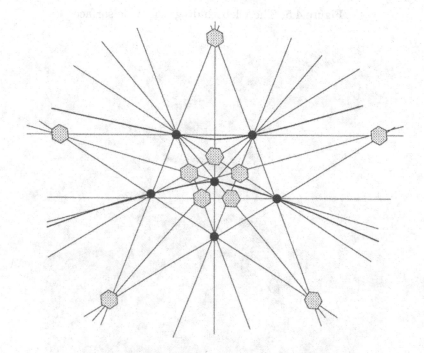

The ten points indicated by the hexagons are Eckard points, which are three-fold points of the arrangement; the six points are five-fold points of the arrangement. One of the conics through five of the six points is also drawn.

Proof: This is the only Σ_5-invariant set of six points[2], and indeed $\mathfrak{C}$ is the unique cubic surface whose automorphism group is Σ_5.   □

Given three of the c_{ij} in $\mathbb{P}^2$, the triangle they form in $\mathbb{P}^2$ becomes a triangle on the cubic surface, i.e., is the intersection of the cubic surface with a tritangent. In case c_{ij}, c_{kl}, c_{mn} all meet *in a single point*, (that is the triangle shrinks to a point), one refers to the point as an *Eckard point* of the surface[3], and the tritangent $< c_{ij}c_{kl}c_{mn} >$ is called an *Eckard plane*. From the diagram above we see that $\mathfrak{C}$ has ten such Eckard points, and again, $\mathfrak{C}$ is the unique cubic surface with ten Eckard points. Note that these points occur, when, writing the cubic surface in pentahedral form, two or more of the coefficients coincide; for $\mathfrak{C}$ there are five coefficients which coincide, hence $\binom{5}{2} = 10$ Eckard points.

Corollary 4.1.11 $\mathfrak{C}$ *is the unique cubic surface with ten Eckard points.*

Before turning to the equations for the lines and tritangents, we pause to explain the name "diagonal" cubic. The five planes $y_i = 0$ of the Sylvester pentahedron

---

[2]this arrangement is given by the natural projective action of the icosahedral group on $\mathbb{P}^2$, and the six points are the five-fold points of the arrangement and form a single orbit

[3]via Cremona transformations of $\mathbb{P}^2$, described below, any such triple intersection in a tritangent on a cubic surface is equivalent to an intersection of three lines c_{ij}

Figure 4.5: The Clebsch diagonal cubic surface

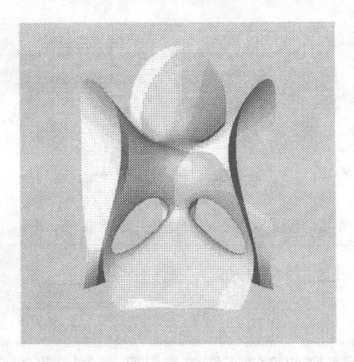

.The affine equation for the Clebsch cubic $\mathfrak{C}$ is derived as follows. Inserting the expressions in the u_i of (4.10) into the equation (4.8) for $\mathfrak{C}$, we get an expression in the u_i. Then divide by v^3, where $v = 2(u_0 + u_1 + u_2 + 4u_3)$ is the plane at infinity. The result is

$$\mathfrak{C} = \{81(x^3 + y^3 + z^3) - 189(x^2y + x^2z + xy^2 + xz^2 + y^2z + yz^2) +$$
$$54xyz + 126(xy + xz + yz) - 9(x^2 + y^2 + z^2) - 9(x + y + z) + 1 = 0\}.$$

---

meet in ten lines, which meet three at a time in ten points. In each such penta-hedral plane, the intersection with the other four form a complete quadralateral as in Proposition 4.1.6. The six intersection points are all Eckard points, so the ten vertices of the Sylvester pentahedron lie on the ten Eckard planes. The consequence of this is that the diagonals of the quadralaterals in each pentahedral plane are lines on the cubic $\mathfrak{C}$, as these lines join two Eckard points, hence meet six of the lines of the surface (and are consequently themselves lines on the surface). Now we turn to the equations for the 27 lines and 45 tritangents, which in this case are easily derived, as we have the hexahedral form for the surface.

Proposition 4.1.12 *15 tritangents of $\mathfrak{C}$ are given by the equations*

$$0 = y_i + y_j, \quad 1 \le i < j \le 5 \quad (10), \qquad y_i = 0, \quad i = 1, \ldots, 5 \quad (5).$$

Of these, the first ten are the Eckard planes of $\mathfrak{C}$. These 15 tritangents cut $\mathfrak{C}$ in 15 lines, namely the 15 lines complementary to the double six on $\mathfrak{C}$ used to get the hexahedral equation. The remaining tritangents are given by equations defined over $\mathbb{Q}(\sqrt{5})$, as described in (4.14). They are as follows:

$$\alpha_{ijkl} \quad 0 = (1 - 3\sqrt{5})(y_i + y_j) - (1 + 3\sqrt{5})(y_k + y_l), \qquad i \ne j \ne k \ne l \ne 5 \quad (6)$$

$$\beta_{ijk} \quad 0 = (1 - 3\sqrt{5})(y_i + y_j) - (3\sqrt{5} - 5)(y_k + y_5), \qquad i \ne j \ne k \ne 5 \quad (12)$$

$$\gamma_{ijk} \quad 0 = (-5 - 3\sqrt{5})(y_i + y_5) - (1 + 3\sqrt{5})(y_j + y_k), \qquad i \ne j \ne k \ne 5 \quad (12)$$

$$\delta_{ij} \quad 0 = (-5 - 3\sqrt{5})(y_i + y_5) - (-5 + 3\sqrt{5})(y_i + y_5), \quad i \ne j \ne 5 \quad (6).$$

The 27 lines are given by:

$$0 = y_i + y_j = y_k + y_l, \quad i \ne j \ne k \ne l \ne 5 \quad (3)$$

$$0 = y_i + y_j = y_k + y_5, \quad i \ne j \ne k \ne 5 \quad (12)$$

$$0 = \delta_{ij} = \gamma_{jkl}(= \beta_{kli}), \quad i \ne j \ne k \ne l \ne 5 \quad (12).$$

Proof: Just apply (4.14), noting that here $d_2^2 = 45$, so $d_2 = 3\sqrt{5}$. $\qquad\square$

To get a nice equation for this cubic in homogenous coordinates on $\mathbb{P}^3$, we utilize the tetrahedral coordinates used above. This time we would like four of the ten Eckhard planes to comprise this coordinate tetrahedron, i.e., we set

$$u_0 := y_1 + y_5, \; u_1 := y_2 + y_5, \; u_2 := y_3 + y_5, \; u_3 := y_4 + y_5, \qquad (4.9)$$

and since we have the linear relation $\sum y_i = 0$, the last coordinate can be expressed as $u_3 = -y_1 - y_2 - y_3$, and the sum of the u_i is $3y_5$, which is one of the Sylvester planes (they are the $y_i = 0$). To get an affine picture we this time set:

$$u_0 = x, \; u_1 = y, \; u_2 = z, \; u_3 = 1/4(1 - x - y - z),$$

i.e., this time we take the plane at infinity to be $u_0 + u_1 + u_2 + 4u_3 = 0$. Then, we must invert the equations (4.9), and insert these expressions into the equation (4.8). This gives:

$$y_5 = \tfrac{1}{3}(u_0 + u_1 + u_2 + u_3), \quad y_1 = -\tfrac{1}{3}(-2u_0 + u_1 + u_2 + u_3),$$

$$y_2 = -\tfrac{1}{3}(u_0 - 2u_1 + u_2 + u_3), \quad y_3 = -\tfrac{1}{3}(u_0 + u_1 - 2u_2 + u_3),$$

$$y_4 = -\tfrac{1}{3}(u_0 + u_1 + u_2 - 2u_3),$$

$$(4.10)$$

and this is inserted into (4.8), giving us the equation for $\mathfrak{C}$ in the u_i coordinates. With this, we can again derive the affine equation, and the result is pictured in Figure 4.5.

The equations for the lines in the u_i variables are convenient for the purpose of drawing them. We have

Figure 4.6: The Clebsch cubic with coordinate tetrahedron and Hessian

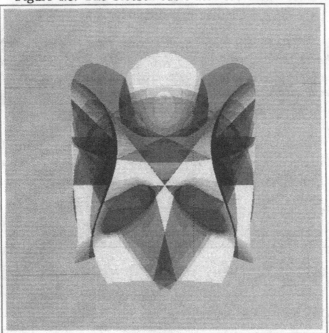

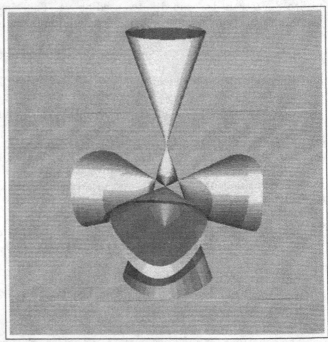

The equations for the faces of the coordinate tetrahedron are given by the hyperplanes of (4.9). These are all Eckard tritangents, which is quite visible.

Figure 4.7: The Clebsch cubic and its Hessian with the Sylvester pentahdron

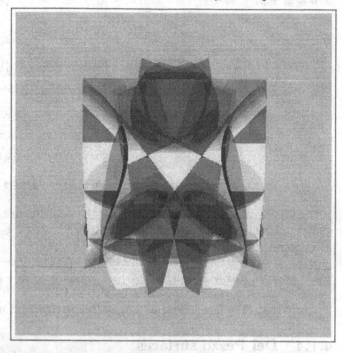

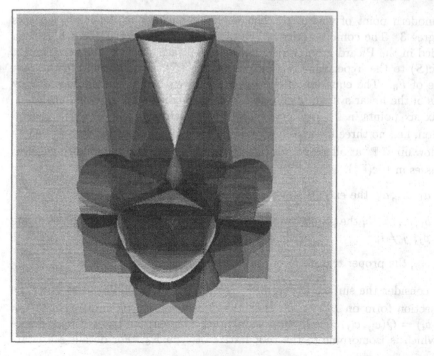

$$0 = y_i + y_j = y_k + y_l = -\frac{1}{3}(u_0 + u_1 + u_2 + u_3) + \frac{1}{2}(u_i + u_j)$$

$$= -\frac{1}{3}(u_0 + u_1 + u_2 + u_3) + \frac{1}{2}(u_k + u_l);$$

$$0 = y_i + y_j = y_k + y_5 = -\frac{1}{3}(u_0 + u_1 + u_2 + u_3) + \frac{1}{2}(u_i + u_j) = u_k;$$

$$0 = (-5 - 3\sqrt{5})(y_i + y_5) - (-5 + 3\sqrt{5})(y_i + y_5)$$

$$= (-5 - 3\sqrt{5})u_i - (-5 + 3\sqrt{5})u_j$$

$$= (-5 - 3\sqrt{5})(y_j + y_5) - (1 + 3\sqrt{5})(y_k + y_l) =$$

$$(-5 - 3\sqrt{5})u_j - (1 + 3\sqrt{5}) \cdot$$

$$(-\frac{1}{3}(u_0 + u_1 + u_2 + u_3) + \frac{1}{2}(u_k + u_l))$$

In Figures 4.6 and 4.7 we show both the coordinate tetrahedron of Eckard planes, as well as the Sylvester pentahedron for $\mathfrak{C}$. These are given, in affine coordinates, by

$$y_1 = -\tfrac{1}{18}(-9x + 3y + 3z + 1), \quad y_2 = -\tfrac{1}{18}(3x - 9y + 3z + 1),$$
$$y_3 = -\tfrac{1}{18}(3x + 3y - 9z + 1), \quad y_4 = -\tfrac{1}{18}(3x + 3y + 3z - 1),$$
$$y_5 = \tfrac{1}{9}(3x + 3y + 3z + 1).$$

Note that these are indeed tritangent planes (though not Eckhard planes).

4.1.4   Del Pezzo surfaces

The modern point of view is to consider cubic surfaces as del Pezzo surfaces of degree 3. The combinatorics of the 27 lines, as listed in Table 4.1, is then encoded in the Picard group of the del Pezzo surface. In fact, the complement in $\mathrm{Pic}(S)$ to the hyperplane section, call it $\mathrm{Pic}^0(S)$, is isomorphic to the root lattice of E_6. The equation of the surface is given by the embedding of S by means of the linear system of elliptic curves through the six given points.

Fix six points in $\mathbb{P}^2$, say $x = (p_1, \ldots, p_6)$, such that the p_i are in general position, i.e., no three lie on a line, and not all six lie on a conic. Let $\widehat{\mathbb{P}}^2_x$ denote the blow up of $\mathbb{P}^2$ at all six points, $\varrho_x : \widehat{\mathbb{P}}^2_x \longrightarrow \mathbb{P}^2$. Consider the following curves as classes in $\mathrm{Pic}(\widehat{\mathbb{P}}^2_x)$:

i) $a_1, \ldots, a_6$, the exceptional divisors over $(p_1, \ldots, p_6)$;

ii) $b_1, \ldots, b_6$, b_i the proper transform of the conic q_i passing through all points p_j, $j \neq i$;

iii) c_{ik}, the proper transform of the line $\overline{p_i p_k}$.

If we consider the surface $\widehat{\mathbb{P}}^2_x$, we have $H^2(\widehat{\mathbb{P}}^2_x, \mathbb{Z}) = [l]\mathbb{Z} \oplus_i \mathbb{Z}a_i$. Let Q be the intersection form on $H^2(\widehat{\mathbb{P}}^2_x, \mathbb{Z})$; then the classes a_i, b_i, c_{ij} fulfill $Q(a_i, a_i) = Q(b_i, b_i) = Q(c_{ij}, c_{ij}) = -1$. In a well-known manner one takes a rank 6 subset, which is isomorphic to the root lattice of type $\mathbf{E_6}$. For details see [Man]

or [DO] and references therein. Consider the orthocomplement of the canon-
ical class on $\widehat{\mathbb{P}}_x^2$, and denote this by $\mathrm{Pic}^0(\widehat{\mathbb{P}}_x^2)$. Recall that the anti-canonical
class is $3l + \sum_{i=1}^6 a_i$, and that the anti-canonical embedding of $\widehat{\mathbb{P}}_x^2$ is as a cubic
surface. Consequently we may view $\mathrm{Pic}^0(\widehat{\mathbb{P}}_x^2)$ as the orthocomplement of the
hyperplane section class of $\mathrm{Pic}(S_x)$, where S_x is the cubic surface which is the
anti-canonical embedding. The following elements λ with $Q(\lambda, \lambda) = -2$ form a
basis of $\mathrm{Pic}^0(S_x)$:

$$
\begin{aligned}
\alpha_0 &= l - a_1 - a_2 - a_3 \\
\alpha_1 &= a_1 - a_2 \\
\alpha_2 &= a_2 - a_3 \\
\alpha_3 &= a_3 - a_4 \\
\alpha_4 &= a_4 - a_5 \\
\alpha_5 &= a_5 - a_6
\end{aligned}
\tag{4.11}
$$

These also form a base of a root system of type $\mathbf{E_6}$, by taking $\alpha_1, \ldots, \alpha_5$ as the
sub-root system of type $\mathbf{A_5}$. Since the classes a_i, b_i, c_{ij} are exceptional, they
all represent elements of $\mathrm{Pic}^0(S_x)$. This leads to the following exact sequence of
$\mathbb{Z}$-modules

$$
0 \longrightarrow \mathbb{Z}^{24} \longrightarrow \mathbb{Z}^{45} \longrightarrow \mathbb{Z}^{27} \longrightarrow \mathrm{Pic}^0(S_x) \longrightarrow 0
\tag{4.12}
$$

which was already discussed in the section on the Segre cubic.

4.1.5 Coble's hexahedral form

Taking the embedding $\varepsilon : \widehat{\mathbb{P}}_x^2 \longrightarrow S_x \subset \mathbb{P}^3$ given by the linear system of cubic
curves through the six points, one sees without difficulty that a_i, b_i, c_{ij} are all
lines on S_x, and that their intersections are given by the 36 double sixes (4.1),
so we have reproduced the situation studied above. To give an explicit form of
the equation of S_x, it seems that Coble's hexahedral form is most convenient.
This is given by

Theorem 4.1.13 ([C1], see also [DO]) *Let $x = (p_1, \ldots, p_6)$ be a set of six
points in general position in $\mathbb{P}^2$, and let $\widehat{\mathbb{P}}_x^2$ be the corresponding del Pezzo surface
of degree 6. There exist six cubics $(a, b, c, d, e, f) \in H^0(\mathbb{P}^2, \mathcal{O}(3))$ on the six
points, such that there are coefficients $(\bar{a}, \bar{b}, \bar{c}, \bar{d}, \bar{e}, \bar{f})$, satisfying the following
conditions*

i) $a + \cdots + f = 0;$

ii) $\bar{a} + \cdots + \bar{f} = 0;$

iii) $\bar{a}a + \cdots + \bar{f}f = 0,$

and such that the equation defining the cubic surface $\varepsilon(\widehat{\mathbb{P}}_x^2)$ is

$$
a^3 + \cdots + f^3 = 0.
$$

We note that the equation iii) is just the equation of a hyperplane in the symmetric $\mathbb{P}^4$, and the equation for the cubic surface shows this is just a hyperplane section of the Segre cubic. Hence we have

Theorem 4.1.14 *Given* $x = (p_1, \ldots, p_6)$ *as in 4.1.13 and the coordinates*

$$(a, \ldots, f),$$

there exists a hyperplane $H_{(\bar{a}, \ldots, \bar{f})} \in (\mathbb{P}^4)^\vee$, *given by*

$$H_{(\bar{a}, \ldots, \bar{f})} = \{\bar{a}a + \cdots + \bar{f}f = 0\},$$

such that

$$S_x = S_3 \cap H_{(\bar{a}, \ldots, \bar{f})}$$

is a cubic surface isomorphic to $\varepsilon(\widehat{\mathbb{P}}_x^2)$. *If one fixes an ordering of the six points* $x = (p_1, \ldots, p_6)$, *then any choice of double six on* $\widehat{\mathbb{P}}_x^2$ *gives rise to a marked cubic surface, and for this marking the choice of* $(a, \ldots, f)$ *is unique.* $\qquad\square$

So in the hexahedral form, a marked cubic surface is given by a hyperplane section of the Segre cubic. The exceptional locus (of such hyperplane sections) can be determined from this description: any hyperplane containing one of the 15 Segre planes (3.27) meets S_3 in a *degenerate* cubic surface.

As already mentioned, Cremona had noticed that the equations of the lines and tritangents can be directly expressed for a cubic surface given in the hexahedral form of Theorem 4.1.14. In terms of the coordinates $a, \ldots, f$ used above, they can be given as follows.

i) The 15 Segre planes (3.27) on the Segre cubic intersect the hyperplane sections $H_{(\bar{a}, \ldots, \bar{f})}$ in the 15 lines of the type

$$a + d = b + e = c + f = 0 \qquad (4.13)$$

(which are lines on $S_x = S_3 \cap H_{(\bar{a}, \ldots, \bar{f})}$ if $\bar{a}a + \cdots + \bar{f}f = 0$); taking any two of these equations defines a tritangent plane.

ii) The 12 other lines can be given by

$$\begin{cases} (be - d_2)(b + e) - (cf + d_2)(c + f) &= 0 \\ (cf - d_2)(c + f) - (ad + d_2)(a + d) &= 0 \\ (ad - d_2)(a + d) - (be + d_2)(b + e) &= 0, \end{cases} \qquad (4.14)$$

where $d_2^2 = \sigma_2(\bar{a}, \ldots, \bar{f})^2 - 4\sigma_4(\bar{a}, \ldots, \bar{f}) = 0$ is the condition on the six points that they lie on a conic in $\mathbb{P}^2$.

Any two equations of the left hand side of (4.14) define the remaining 30 tritangents, by permutations of the letters, such that for an odd permutation, the sign of d_2 in (4.14) is changed accordingly (see [C1], I, p. 173).

If one considers the six points in $\mathbb{P}^2$ up to permutation, then the invariant algebra is given by

Theorem 4.1.15 ([C1], I, p. 176) *The invariants $\sigma_2, \ldots, \sigma_6, d_2\sqrt{d}$ consistute a rational and integral complete system of $(\Sigma_6\text{-})$ invariants of $\mathbf{P}_6^2$ (see (3.84)), where*

$$d = \prod(\bar{a} - \bar{b})^2.$$

$d = 0$ is the condition that three lines c_{ij}, c_{jk}, c_{ik} meet at a point (i.e., determine an Eckard point on the cubic surface).

The actual symmetry group of the 27 lines – $W(E_6)$ – contains this Σ_6 (acting regularly on the cubics). It is extended to $W(E_6)$ by a *Cremona transformation* in the plane: the quadratic transformation given by blowing up three of the points a_4, a_5, a_6 and then blowing down the joining lines c_{45}, c_{46}, c_{56} yields the switch of double sixes:

$$N = \begin{bmatrix} a_1 & a_2 & a_3 & a_4 & a_5 & a_6 \\ b_1 & b_2 & b_3 & b_4 & b_5 & b_6 \end{bmatrix} \mapsto \begin{bmatrix} a_1 & a_2 & a_3 & c_{56} & c_{46} & c_{45} \\ c_{23} & c_{13} & c_{12} & b_4 & b_5 & b_6 \end{bmatrix} = N_{123},$$

which, together with Σ_6, generates $W(E_6)$. This gives a *rational* action of $W(E_6)$ on $\mathbf{P}_6^2$. Hence one can ask for the set of invariants under the $G_{6,2}$[4] action.

   In [C1], III, Coble determines invariants for $G_{6,2}$, as expressions in the coordinates $\bar{a}, \ldots, \bar{f}$. These, in turn, may be expressed in terms of x, y, z, t, u, where the six points in $\mathbb{P}^2$ are brought into the form

$$\begin{pmatrix} 1 & 0 & 0 & 1 & x & z \\ 0 & 1 & 0 & 1 & y & t \\ 0 & 0 & 1 & 1 & u & u \end{pmatrix}. \tag{4.15}$$

The basic idea is to apply the Clebsch transference principle (see section B.2.4) to the surface $S_{(\bar{a}, \ldots, \bar{f})}$, which by the discussion above is a hyperplane section of the Segre cubic. From this it follows that the invariants of the quatenary cubic form (B.49) determine corresponding invariants of the hexahedral form as derived in section B.4.2 in the Appendix. Passing on to expressions in the (x, y, z, t, u), note that the $(\bar{a}, \ldots, \bar{f})$ are quadratic in the (x, y, z, t, u) (see [C1] I, p. 196).

$$\begin{aligned} 3a &= \rho - 3(ux + ut), & 3d &= \rho - 3(uy + uz), \\ 3b &= \rho - 3(ux + yz), & 3e &= \rho - 3(uy + xt), \\ 3c &= \rho - 3(ut + yz), & 3f &= \rho - 3(uz + xt), \end{aligned} \tag{4.16}$$

$$\rho = u(x + y + z + t) + xt + yz.$$

Then we have

Theorem 4.1.16 *A complete system of invariants is given by:*

---

[4]we are using Coble's notation for the Cremona groups; as abstract groups $G_{6,2} \cong W(E_6)$ and $\Gamma_{6,2} \cong G_{25,920}$

- *for the quatenary cubic: $i_8, i_{16}, i_{24}, i_{32}, i_{40}$ and i_{100} of degrees indicated and of weights 6, 12, 18, 24, 30 and 75 (see (B.49), where they are denoted by I_k instead of i_k).*

- *for a hexahedral cubic: $I_6, I_{12}, I_{18}, I_{24}, I_{30}$ and I_{75} of the degrees indicated in $(\bar{a}, \ldots, \bar{f})$ (see (B.56), (B.57) and (B.58) for these).*

- *for $G_{6,2}$: $j_{10}, j_{20}, j_{30}, j_{40}, j_{50}$ and j_{125} of the degrees indicated in (x, y, z, t, u).*

Of course these systems are related, but, and that is the whole point, in a very non-trivial manner. In this respect one defines the following related problems.

Definition 4.1.17 The *equation problem* for $G_{6,2}$ is: given numerical values of $j_{10}, \ldots, j_{125}$, calculate the coordinates (x, y, z, t, u) of a point $P \in \mathbf{P}_6^2$ for which the j_k take the assigned values.

The next problem concerns the subgroup $\Gamma_{6,2} \subset G_{6,2}$. Since (semi-)invariants of $G_{6,2}$ which change sign under odd permutations are genuine invariants of $\Gamma_{6,2}$, the invariants of the latter can be deduced from those of the former (cf. section B.1.2.1). As Coble shows, the invariant j_{40} may be taken to be the *discriminant* Δ of the cubic surface (in fact, i_{32}, I_{24} and j_{40} may *all* be represented by the discriminant, each expressed in the appropriate variables), and $\sqrt{\Delta}$ can be taken as an invariant for $\Gamma_{6,2}$. To see how this occurs, note that the discriminant has rational factors corresponding to the conditions (on the six points in the plane) that the so defined cubic surface is *singular*: either all six points lie on a conic, three lie on a line, or two points coincide in some direction. Hence, we have "$\Delta = \delta \prod \delta_{ijk}$", where δ_{ijk} expresses the condition that three points (labeled p_i, p_j, p_k) lie on a line. This expression, however, is not correct, as in order to express Δ rationally in terms of $\bar{a}, \ldots, \bar{f}$, we must utilize the relation (here $\delta = d_2$, where d_2 was defined above):

$$\delta^2 = \sigma_2^2 - 4\sigma_4,$$

which means the discriminant will divide δ^2, hence by symmetry, will also divide δ_{ijk}^2, so what we actually have is

$$\Delta = \delta^2 \prod \delta_{ijk}^2.$$

This expression is now of degree 24 in $\bar{a}, \ldots, \bar{f}$, but in these variables it is a square, and up to sign it makes sense to speak of $\sqrt{\Delta} = \delta \prod \delta_{ijk}$. Note that

$$\delta_{123}\delta_{456} = (\bar{a} + \bar{b} + \bar{c}) = -(\bar{d} + \bar{e} + \bar{f}),$$

etc. A definite sign can be given to $\sqrt{\Delta}$ by the assignment

$$\sqrt{\Delta} = d_2 \prod_{i,j \in \{\bar{b}, \ldots, \bar{f}\}} (\bar{a} + i + j), \qquad (4.17)$$

and d_2 is given the sign

$$d_2 = \begin{vmatrix} (341)(561) & (531)(461) \\ (342)(562) & (532)(462) \end{vmatrix}, \qquad (4.18)$$

where $(ijk) =$ the 3×3 minor of (4.15) given by the $(i,j,k)^{th}$ columns. Then

Lemma 4.1.18 *A complete system of invariants for* $\Gamma_{6,2}$ *is given by*

$$j_{10}, j_{20}, j_{30}, \sqrt{\Delta}, j_{50} \text{ and } j_{125}.$$

The skew invariant i_{100} or I_{75} or j_{125} has the same behavior under $G_{6,2}$ and $\Gamma_{6,2}$ as $\sqrt{\Delta}$: it is invariant under $\Gamma_{6,2}$ and changes sign under all odd elements of $G_{6,2}$. The invariant is a product over 45 factors, each expressing the condition that a given tritangent plane contains an Eckard point.

Definition 4.1.19 The *form problem* for $\Gamma_{6,2}$ is: given numerical values for $j_{10}, j_{20}, j_{30}, \sqrt{\Delta}, j_{50}$, calculate the ratios of the coordinates (x, y, z, t, u) of a point $p \in \mathbf{P}_6^2$ for which the j_k take the assigned values.

Finally, it is important to know whether the three sets of invariants of Theorem 4.1.16 are *covariantly* equivalent. This means that they change in the same way upon affecting some permutation of the 27 lines. For the invariants I_k this can be achieved by using the Clebsch transference principle, since this preserves the covariant relations, and is described in the appendix.

4.2 Solving algebraic equations by means of theta functions

In this section we review the general procedure, developed in the last century, for solving algebraic equations by means of transcendental functions, in particular theta functions, presenting explicitly the example of the general quintic equation, the Galois group of which is Σ_5. Part of the reason for presenting this "old" theory in such detail is because it forms, in some sense, the beginnings of the theory of arithmetic quotients and their geometry. This will be much extended in the example of the equation for the 27 lines of a general cubic surface, where much more geometry comes into play. But still much is based on this case, which then, in addition to being part of the history of our subject, serves also as a pattern for a more interesting example.

4.2.1 Algebraic equations

To start we recall a few general facts about the roots of an algebraic equation. These relate the solution of a given equation to solutions of other equations, so-called resolvents. As a first step, there is a general method to simplify a general degree n equation

$$P(x) = x^n + a_{n-1}x^{n-1} + \cdots + a_1 x + a_0 = 0. \qquad (4.19)$$

It follows from Galois theory that (4.19) can be solved by means of *radicals*, if and only if the Galois group is *solvable*. Hence, supposing the Galois group to be simple, this method does not yield a solution. Letting $(\alpha_1, \ldots, \alpha_n)$ denote the roots of (4.19), the Galois group of the equation (4.19) is the Galois group of the field extension

$$K = k(\alpha_1, \ldots, \alpha_n), \tag{4.20}$$

where the coefficients a_i of (4.19) are assumed to lie in k. If the equation is general, then the Galois group is Σ_n. (It is not so easy to explicitly write down a "general" equation; see for example Lang's book "*Algebra*" for examples of general quintics (a criterion due to Artin)). Let Δ denote the discriminant of (4.19); then the Galois group $Gal(K|k(\sqrt{\Delta})) \cong A_n$ is simple.

Now let $\psi \in k[x]$ be a polynomial of degree $\leq n - 1$, and set

$$Q(Y) = \tau_\psi(P)(Y) := \prod(Y - \psi(\alpha_i)) \in k[Y]. \tag{4.21}$$

The polynomial $\tau_\psi(P)$ is called the *Tschirnhaus transformation* of P by ψ. Assume that P and Q have no multiple roots (which of course holds if they are general). Then we have

Lemma 4.2.1 *The roots of P can be rationally calculated in terms of the roots of Q.*

Proof: Let $\beta_1, \ldots, \beta_n$ be the roots of Q, $\beta_i = \psi(\alpha_i)$. Set

$$\lambda(Y) = Q(Y) \cdot \left(\frac{\alpha_1}{Y - \beta_1} + \cdots + \frac{\alpha_n}{Y - \beta_n} \right).$$

$\lambda(Y)$ can be written $\lambda(Y) = \sum_\mu \prod_{i \neq \mu} (Y - \beta_i) \alpha_\mu$, and the coefficients, being rationally expressible in terms of the elementary symmetric functions of the roots, are in k. Hence $\lambda(\beta_k) = \prod_{i \neq k} (\beta_k - \beta_i) \alpha_k$, and consequently

$$\alpha_k = \frac{\lambda(\beta_k)}{\prod_{i \neq k} (\beta_k - \beta_i)} = \frac{\lambda(\beta_k)}{Q'(\beta_k)}, \tag{4.22}$$

giving the desired expression. □

4.2.2 Resolvents

Similarly, let $y \in K$ be given, and let $y_1 = y, y_2, \ldots, y_m$ be the conjugates of y under the action of the Galois group $G = Gal(K|k)$. Then

$$R(x) := \prod_{i=1}^{m} (x - y_i) \tag{4.23}$$

is a polynomial with coefficients in k. It is called a *resolvent* of $P(x)$. In the particular case that the stabilizer of y_i in G is the identity element, we have $m = |G|$ and R is called a *Galois resolvent*. As above, set

$$\lambda(x) = R(x) \left(\frac{\sigma_1(\alpha_1)}{x - y_1} + \cdots + \frac{\sigma_m(\alpha_1)}{x - y_m} \right),$$

where $y_i = \sigma_i(y_1)$ for $\sigma_i \in G$. Then the same argument as above shows

$$\sigma_k(\alpha_1) =: \alpha_k = \frac{\lambda(y_k)}{R'(y_k)},$$

so again

Lemma 4.2.2 *Let $R(x)$ be a resolvent of $P(x)$. Then the roots of $P(x)$ can be rationally calculated in terms of the roots of $R(x)$.*

Theorem 4.2.3 (Reduction of equations, Tschirnhaus 1683) *By a linear tranformation, or by a quadratic Tschirnhaus transformation whose coefficients involve a single square root, any equation*

$$x^n + c_{n-1}x^{n-1} + \cdots + c_1 x + c_0 = 0$$

can be reduced to a principal *equation*

$$x^n + c_{n-3}x^{n-3} + \cdots + c_1 x + c_0 = 0.$$

The square root which must be introduce above is called an *accessory irrationality*, as it is only introduced to put the given equation is some normal form. Going one step further,

Theorem 4.2.4 *By adjunction of a cube root and three square roots any equation of degree n can be reduced to the form*

$$x^n + c_{n-4}x^{n-4} + \cdots + c_1 x + c_0 = 0.$$

For $n = 5$ this is due to Bring (1786), and the Bring form of a quintic equation is (cf. (B.25))

$$y^5 + 5by + c = 0. \tag{4.24}$$

The general method of solving algebraic equations amounts to finding resolvents of a given equation, which one knows how to solve for some reason or another. For example, a "pure" equation is one which can be reduced to the form $x^n = a$, which can then be solved by $x = \exp(\frac{1}{n}\ln a)$. Many equations of interest can be reduced to equations involving theta functions, of which the modular equation, to be discussed next, is the prototype.

4.2.3 The modular equation

Let $E = \mathbb{C}/\Lambda$ be an elliptic curve, given in the Weierstraß form

$$y^2 = 4x^3 - g_2 x - g_3, \tag{4.25}$$

with $g_2 = 60 \sum_{\substack{\omega \in \Lambda \\ \omega \neq 0}} \frac{1}{\omega^4}$, $g_3 = 140 \sum_{\substack{\omega \in \Lambda \\ \omega \neq 0}} \frac{1}{\omega^6}$ and discriminant $\Delta = g_2^3 - 27g_3^2$. The J-invariant is $J = g_2^3/\Delta$, and as Λ varies, J becomes a modular function (for

$PSL(2, \mathbb{Z}))$ on the upper half plane $\mathbb{S}_1 = \{\tau \in \mathbb{C} | \mathrm{Im}(\tau) > 0\}$ (where $\Lambda = \mathbb{Z} \oplus \tau \mathbb{Z}$). The *modular equation* is: given $u \in \mathbb{C}$, find $\tau \in \mathbb{S}_1$, such that

$$J(\tau) = u. \qquad (4.26)$$

A solution τ of (4.26) can be constructed as follows. Given $u \in \mathbb{C}$,

$$E_u: \quad y^2 = 4x^3 - \frac{27u}{u-1}(x+1) \qquad (4.27)$$

is the Weierstraß equation of an elliptic curve E_u with J-invariant equal to u. To find $\tau \in \mathbb{S}_1$ with $E_u = \mathbb{C}/\mathbb{Z} \oplus \tau \mathbb{Z}$, one calculates the *periods* of E_u,

$$\omega_i = \int_{\gamma_i} \frac{dx}{y} = \int_{\gamma_i} \frac{dx}{\sqrt{4x^3 - \frac{27u}{u-1}(x+1)}}, \qquad (4.28)$$

where (γ_1, γ_2) is a $\mathbb{Z}$-basis of $H_1(E_u, \mathbb{Z})$. Then the *solution* of (4.26) is

$$\tau_u = \omega_1/\omega_2; \quad J(\tau_u) = u. \qquad (4.29)$$

Geometrically, solving (4.26) is equivalent to finding an inverse image of u under

$$J : \mathbb{S}_1 \longrightarrow \Gamma \backslash \mathbb{S}_1, \quad \Gamma = SL(2, \mathbb{Z}).$$

A solution $\tau_u \in \mathbb{S}_1$ is determined up to the action of Γ, corresponding to the choice of a basis (γ_1, γ_2) in (4.28).

The division problem Consider the two elliptic curves E_τ and $E_{n\tau}$. Since the fundamental parallelogram of $E_{n\tau}$ consists of n copies of the fundamental paralellogram of E_τ, one sees that there is a map $E_\tau \xrightarrow{\phi_n} E_{n\tau}$ given by the diagram

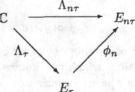

$$(4.30)$$

and since $[\Lambda_\tau : \Lambda_{n\tau}] = n$, ϕ_n is a finite map of degree n. The kernel is finite and consists of points $k\tau$, $1 \le k \le n$. A surjective morphism

$$\phi : E \longrightarrow E'$$

of elliptic curves is called an *isogeny*, if $|\mathrm{Ker}(\phi)| < \infty$.

Division problem: Given an isogeny $\phi : E_\tau \longrightarrow E'_{\tau'}$, determine the relation between the values $\xi(\tau)$ and $\xi(\tau')$ for any modular form ξ.

Theorem 4.2.5 *Given an isogeny $\phi : E \longrightarrow E'$, there are algebraic relations between the elliptic functions on E and on E'.*

Proof: Let $K(E)$ and $K(E')$ be the function fields; each is generated by the elliptic functions. Then from the isogeny we get that $K(E)$ is an algebraic extension of the field $K(E')$, and this implies that there are algebraic relations between the elliptic functions on E and E'. □

Corollary 4.2.6 *The same holds for invariants of $K(E)$ (for example g_2, g_3, Δ, etc.).*

In fact, for the isogenies ϕ_n the extension of function fields is even Galois, with Galois group $\mathbb{Z}/n\mathbb{Z}$.

With isogenous curves one gets other equations related to the modular equation (4.26). For this, consider the principal congruence subgroup $\Gamma(N) \subset SL(2, \mathbb{Z})$ and the corresponding factorization

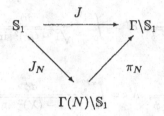

$$(4.31)$$

The function J_N is called the level N modular function (Hauptmodul der Stufe N). Note that since $\Gamma(N)$ is a normal subgroup of $SL(2, \mathbb{Z})$, the finite cover

$$\Gamma(N)\backslash \mathbb{S}_1 \xrightarrow{\pi_N} \Gamma\backslash \mathbb{S}_1 \qquad (4.32)$$

is a Galois cover, with Galois group $P\Gamma/P\Gamma(N)$. In particular, for $N = 5$, we have the identification

$$P\Gamma/P\Gamma(5) \cong PSL(2, \mathbb{Z}/5\mathbb{Z}) \cong A_5, \qquad (4.33)$$

the alternating group on five letters, which is the Galois group $Gal(K|k(\sqrt{\Delta}))$ of a general equation of degree 5, i.e., for K as in (4.20) with $n = 5$.

The solutions to these modular equations can be found somewhat more directly than with periods as above in terms of hypergeometric functions. These functions arise as solutions of the Picard-Fuchs equations for the corresponding family of elliptic curves with level N structures, that is pairs (E_u, L), where E_u is an isomorphism class of elliptic curves as in (4.27) and L is a level N structure on E_u. These Picard-Fuchs equations for $N = 2, 3, 4$ or 5 take the form (cf. [Kl3], (38), p. 80):

$$y'' + \frac{y'}{u} + \frac{y}{4(u-1)^2 u^2} \cdot \left\{ -\frac{1}{\nu_2^2} + u\left(\frac{1}{\nu_2^2} + \frac{1}{\nu_3^2} - \frac{1}{\nu_1^2} + 1\right) - \frac{u^2}{\nu_3^2} \right\} = 0, \quad (4.34)$$

for triples $(\nu_1, \nu_2, \nu_3) = (2, 2, 3), (2, 3, 3), (2, 3, 4), (2, 3, 5)$ for $N = 2, 3, 4$ and 5, respectively. Solutions to the modular equations can be given in terms of

Riemann P-functions ([Kl3], p. 81)

$$P \left(\begin{array}{ccc} \frac{1}{2\nu_2} & \frac{1}{2\nu_3} & \frac{1}{4} \\ -\frac{1}{2\nu_2} & -\frac{1}{2\nu_3} & \frac{3}{4} \end{array}, u \right).$$

(4.35)

The only singularities of (4.34) occur at $u = J(\tau) = 0, 1, \infty$, corresponding to finite torsion in $SL(2, \mathbb{Z})$ at i, $\varrho = e^{2\pi i/3}$ and the parabolic at ∞. In this way, the periods (4.28) can be calculated in terms of hypergeometric functions.

Jacobi's equation (1829)

The following lines are taken directly from Jacobi's *"Fundamenta nova theoriae functionum ellipticarum"*, which appeared in 1829.

Let

$$K := \int_0^{\pi/2} \frac{d\phi}{\sqrt{1 - \kappa^2 \sin^2 \phi}}, \quad K' := \int_0^{\pi/2} \frac{d\phi}{\sqrt{1 - (\kappa')^2 \sin^2 \phi}}, \quad \kappa^2 + (\kappa')^2 = 1.$$

If

$$\Lambda := \int_0^{\pi/2} \frac{d\phi}{\sqrt{1 - \lambda^2 \sin^2 \phi}}, \quad \Lambda' := \int_0^{\pi/2} \frac{d\phi}{\sqrt{1 - (\lambda')^2 \sin^2 \phi}}, \quad \lambda^2 + (\lambda')^2 = 1$$

and

$$\left(\begin{array}{c} \Lambda \\ \Lambda' \end{array} \right) = A \left(\begin{array}{c} K \\ K' \end{array} \right) \quad \text{with } A = \left(\begin{array}{cc} a & b \\ c & d \end{array} \right), \quad ad - bc = n,$$

then, setting $u = \sqrt[4]{\kappa}$, $v = \sqrt[4]{\lambda}$ we get for prime n equations of degree $(n + 1)$ between u and v, for example

$n = 3$: $\qquad\qquad\qquad\qquad u^4 - v^4 - 2uv(1 - u^2 v^2) = 0$
$n = 5$: $\qquad\qquad\qquad\qquad u^6 = v^6 + 5u^2 v^2(u^2 - v^2) + 4uv(1 - u^4 v^4).$

(4.36)

The latter equation is "Jacobi's 6th degree equation for the transformations of order five of the elliptic functions." Putting the above into more modern language we have: K, K' are nothing but (up to a factor) the periods ω_1 and ω_2, and Λ, Λ' are the periods corresponding to the elliptic curve $E_{n\tau}$ (see (4.30)). The functions κ and λ are rational functions of $J(\tau)$ and $J(n\tau)$:

$$J(\tau) = 1728 \cdot 16 \frac{(\kappa^4 + 14\kappa^2 + 1)^3}{\kappa^2 (\kappa^2 - 1)^4}, \quad J(n\tau) = \text{same with } \kappa \text{ replaced by } \lambda.$$

(4.37)

The function κ (resp. λ) corresponds to the Legendre normal form:

$$y^2 = (1 - x^2)(1 - \kappa^2 x^2) \quad (\text{resp. } y^2 = (1 - x^2)(1 - \lambda^2 x^2))$$

(4.38)

for which all branch points of the double cover $E_\tau \longrightarrow \mathbb{P}^1$ (resp. for which all branch points of $E_{n\tau} \longrightarrow \mathbb{P}^1$) are finite (as opposed with the Weierstraß form, which has one of the branch points at infinity). If $J_1, \ldots, J_{s_n}$ denotes the J-invariants of all curves n-isogenous to E_τ, then these values satisfy a modular equation

$$\Phi_n(x, J) = 0 \qquad (4.39)$$

with $\Phi_n \in \mathbb{Z}[X, Y]$, and ($n$ a prime) Φ_n has degree $(n + 1)$ in both X and Y. For more details on the following, see [L], Chapter 5.

Lemma 4.2.7 *For general E and prime n, the Galois group of $\Phi_n(x, J)$ over $\mathbb{Q}(\varepsilon, J)$ (ε an nth root of unity) is isomorphic to $PSL(2, \mathbb{Z}/n\mathbb{Z})$ acting on the $(n + 1)$ roots of Φ_n in the J variable, $J_0, \ldots, J_n$ as on the points of $\mathbb{P}^1(\mathbb{F}_n)$.*

The equation for $n = 5$, when J is replaced by κ by means of (4.37), is nothing but Jacobi's equation of the sixth degree.

The important fact about the equation (4.36) is, by considering u as a *parameter*, it is an equation of 6th degree (in the variable v) with one parameter, *which can be solved by means of elliptic functions*.

4.2.4   Hermite, Kronecker and Brioschi

In a paper "Sur la résolution de l'équation du cinquième degré" published in 1858, Hermite found the resolvent of fifth degree which derives from Jacobi's equation of sixth degree. This arises, because, as just mentioned, the modular equation has Galois group A_5, so there exists a resolvent of fifth degree (there exists a resolvent of any degree which divides 60), corresponding to a root with stabilizer A_4 which is of index five in A_5. He did this as follows: set $y = (v_\infty - v_0)(v_1 - v_4)(v_2 - v_3)$, where v_i are the *roots* of Jacobi's equation. The corresponding resolvent (4.23) is

$$y^5 - 2^4 \cdot 5^3 u^4 (1 - u^8)^2 \cdot y - 2^6 \sqrt{5^5} \cdot u^3 (1 - u^8)^2 (1 + u^8) = 0. \qquad (4.40)$$

This takes the form of a "Bring equation" $t^5 - t - A = 0$ (see (4.24) and (B.25)) by setting

$$y = 2\sqrt[4]{5^3} u \sqrt{1 - u^8} t$$
$$A = \frac{2}{\sqrt[4]{5^5}} \cdot \frac{1 + u^8}{u^2 \sqrt{1 - u^8}}. \qquad (4.41)$$

Hence, since we have seen by a Tschirnhaus transformation any quintic equation can be brought into the Bring form (by adjoining a cube root and three square roots, see Theorem 4.2.4), which can be solved by the elliptic functions for the transformations of fifth order, this gives a solution of the quintic equation. A flow chart for this solution is given in Figure 4.8.

In the mentioned work of Jacobi, he applied Tschirnhaus transformations to the equation of sixth degree (4.36) (respectively of $(n+1)^{st}$ degree) to get other equations of sixth (resp. $(n + 1)^{st}$) degree. These had the following amazing property:

Figure 4.8: Hermite's solution

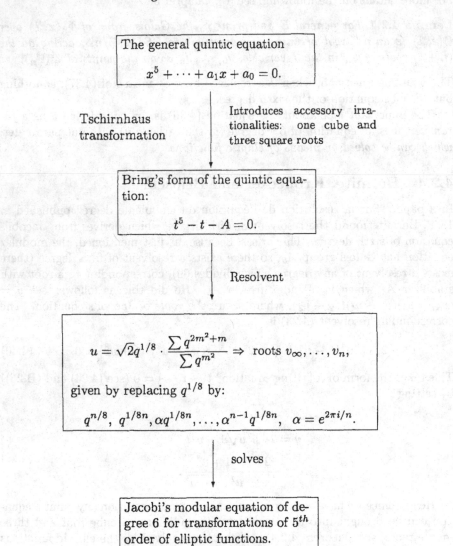

The general quintic equation

$$x^5 + \cdots + a_1 x + a_0 = 0.$$

Tschirnhaus transformation

Introduces accessory irrationalities: one cube and three square roots

Bring's form of the quintic equation:

$$t^5 - t - A = 0.$$

Resolvent

$$u = \sqrt{2}q^{1/8} \cdot \frac{\sum q^{2m^2+m}}{\sum q^{m^2}} \Rightarrow \text{ roots } v_\infty, \ldots, v_n,$$

given by replacing $q^{1/8}$ by:

$$q^{n/8}, \; q^{1/8n}, \alpha q^{1/8n}, \ldots, \alpha^{n-1} q^{1/8n}, \; \alpha = e^{2\pi i/n}.$$

solves

Jacobi's modular equation of degree 6 for transformations of 5th order of elliptic functions.

Theorem 4.2.8 (Jacobi, 1829) *The given set of equations satisfy: the square roots of the roots $z_\infty, z_0, \ldots, z_{n-1}$ can be* linearly *combined from $\frac{n+1}{2}$ elements $\mathbf{A}_0, \ldots, \mathbf{A}_{\frac{n-1}{2}}$:*

$$\begin{cases} \sqrt{z_\infty} = \sqrt{(-1)^{\frac{n-1}{2}} n \mathbf{A}_0} \\ \sqrt{z_\nu} = \mathbf{A}_0 + \varepsilon_\nu \mathbf{A}_1 + \cdots + \varepsilon^{\left(\frac{n-1}{2}\right)^2} \mathbf{A}_{\frac{n-1}{2}}, \end{cases}$$

where $\nu = 1, \ldots, n-1$, $\varepsilon = e^{\frac{2\pi i}{n}}$, and there are choices of the square roots so that the given relations hold.

Brioschi then showed how to form a "general" Jacobi equation of sixth degree: there are homogenous polynomials A, B, C of the $\mathbf{A}_i$ of degrees 2,6 and 10, such that

$$(z - A)^6 - 4A(z - A)^5 + 10B(z - A) + (5B^2 - AC) = 0 \qquad (4.42)$$

is the most general form for a sextic such that the (square roots of the) roots can be linearly expressed in terms of three elements. Furthermore, he finds the resolvent of fifth degree for this equation:

$$x^5 + 10Bx^3 + 5(9B^2 - AC)B^2 x - \sqrt[4]{\frac{\Pi}{5^5}} = 0, \qquad (4.43)$$

where Π is the discriminant of the equation (4.42). For $B = 0$ this is a Bring equation, and for $A = 0$, of the form

$$x^5 + 10Bx^3 + 45B^2 x - \sqrt[4]{\frac{\Pi}{5^5}} = 0. \qquad (4.44)$$

Kronecker found that the sextic (4.42) with $A = 0$ could be solved by means of elliptic functions:

$$\zeta^6 - 10Z\zeta^3 + 12Z^2\zeta + 5Z^2 = 0 \qquad (4.45)$$

has a root

$$\zeta_0 = \frac{-g_2(\tau) \sqrt[12]{\Delta(5\tau)}}{\sqrt[12]{\Delta(\tau)^5}} \qquad (4.46)$$

where

$$Z = J(\tau). \qquad (4.47)$$

Furthermore he proved (most of)

Theorem 4.2.9 *A general quintic equation (after adjoining the square root of the determinant and hence reducing the Galois group to A_5) gives rise to a sextic resolvent which is a Jacobi equation (and hence can be solved by means of elliptic functions). By adjoining an additional square root it can be transformed into the form $A = 0$ above, but without such an irrationality cannot be reduced to a one-parameter equation.*

Kronecker claimed but did not prove the statement on the necessity of adjoining an accessory irrationality. Klein was to give a proof of this later. We sketch Kronecker's solution in the flow chart given in Figure 4.9.

Figure 4.9: Kronecker's solution

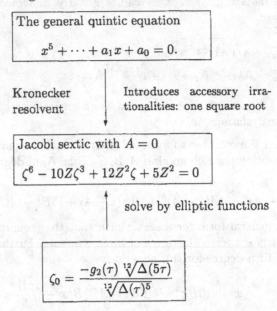

4.2.5   The geometric description

Let

$$P(x) = x^5 + a_4 x^4 + \cdots + a_1 x + a_0 = 0 \qquad (4.48)$$

be a general quintic equation and let $x_1, \ldots, x_5$ denote its *roots*. Then one can view $P(x)$ as determining a point

$$P(x): \quad (x_1 : \ldots : x_5) \in \mathbb{P}^4; \qquad (4.49)$$

but since different ordering of the roots $x_1, \ldots, x_5$ give rise to the same polynomial $P(x)$, one can view the space of quintic polynomials as $\mathbb{P}^4/\Sigma_5$, where Σ_5 acts on $\mathbb{P}^4$ by permuting coordinates. Recall that the coefficients of the equation (4.48) are ($\pm$) the elementary symmetric functions of the roots. Hence, the Tschirnhaus transformation which transforms (4.48) into a quintic with $a_4 = 0$ corresponds to restricting attention to polynomials for which $\sum x_i = 0$. Since this is symmetric in the x_i, it follows that we have a well defined action of Σ_5 on the $\mathbb{P}^3$ given by that relation,

$$\mathbb{P}^3 = \{(x_1 : \ldots : x_5) \in \mathbb{P}^4 \mid \sum x_i = 0\}.$$

Now we can describe the different Tschirnhaus transformations in terms of certain Σ_5-*invariant surfaces or curves* in this $\mathbb{P}^3$. For example the "Hauptgleichung" of Felix Klein is

$$y^5 + \alpha y^2 + \beta y + \gamma = 0; \qquad (4.50)$$

Figure 4.10: The icosahedron inscribed in the sphere

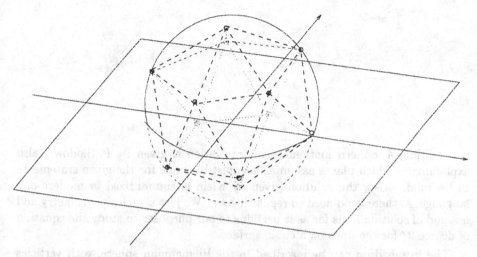

it corresponds to $\sum x_i^2 = 0$, i.e., to the quadric surface

$$Q = \{(x_1 : \ldots : x_5) \in \mathbb{P}^4 \mid \sum x_i = \sum x_i^2 = 0\} \subset \mathbb{P}^3. \tag{4.51}$$

The "quintic resolvent with $A = 0$" (4.43) corresponds to $\sum x_i^3 = 0$, i.e., to the Clebsch diagonal cubic surface (see (4.8))

$$\mathfrak{C} = \{(x_1 : \ldots : x_5) \in \mathbb{P}^4 \mid \sum x_i = \sum x_i^3 = 0\} \subset \mathbb{P}^3. \tag{4.52}$$

Finally the "Bring equation" (4.40) corresponds to $\sum x_i^2 = \sum x_i^3 = 0$, i.e. to the Bring curve

$$B = \{(x_1 : \ldots : x_5) \in \mathbb{P}^4 \mid \sum x_i = \sum x_i^2 = \sum x_i^3 = 0\} \subset \mathbb{P}^3. \tag{4.53}$$

All of the above are invariant under the action of Σ_5. Recall also that the first two cases require an accessory square root, while in the third case one must adjoin a cube root and three square roots (see Theorem 4.2.3).

4.2.6 The icosahedron

There is a universal Galois resolvent for general quintic equations: the icosahedral equation. In studying this resolvent one can apply the geometry of the icosahedron, making the algebra very geometric. The icosahedral equation is easily solved in terms of transcendental functions, giving a "universal solution" to the problem of finding roots of quintic equations. Although studied by many authors as described above, it was Felix Klein who built the entire theory of quintic equations on these foundations, laid down in the classic [Kl3]. In the

Figure 4.11: An inscribed octahedron in the icosahedron

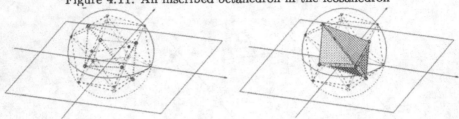

new edition, a modern mathematical translation is given by P. Slodowy, also explaining at which places assumptions must be made for the given statements to be valid. Also, the solution given by Klein is summarized in modern-day language, so there is no need to repeat it here. We just sketch the geometry and method of solution in as far as it pertains to our purpose: to study the equation of degree 27 for the lines on a cubic surface.

The icosahedron can be inscribed in the Riemannian sphere, with verticies given by (see Figure 4.10):

$$z = 0, \infty, \varepsilon^\nu(\varepsilon + \varepsilon^4), \varepsilon^\nu(\varepsilon^2 + \varepsilon^3), \ \varepsilon = e^{2\pi i/5}. \tag{4.54}$$

There are 12 vertices, 30 edges and 20 faces. As each symmetry of the icosahedron yields a linear transformation of $\mathbb{P}^1$, the icosahedral group $\mathcal{I}$ is naturally a (finite) subgroup of $PGL(2, \mathbb{C})$. But in fact, since the vertices are defined over the field $\mathbb{Q}(\varepsilon)$,

$$\mathcal{I} \subset PSL(2, \mathbb{Q}(\varepsilon)). \tag{4.55}$$

One can inscribe five octahedra in the icosahedron I: the 30 edges fall into five sets of six edges, the midpoints of which are the vertices of the octahedron. One such set is pictured in Figure 4.11.

It is easy to see that each symmetry of I affects an *even* permutation of the five octahedra, and hence

Lemma 4.2.10 $\mathcal{I} \cong A_5$, *the alternating group on 5 letters.*

Since $\mathcal{I}$ is a finite group of automorphisms of $\mathbb{P}^1$, one can find a finite set of generators of the ring of invariant functions. Clearly, if v_i denote the 12 vertices, w_j the 30 midpoints of the edges and u_k the 20 centerpoints of the faces, then we have invariants of degrees 12, 20 and 30:

$$f(z_0 : z_1) = \prod_{i=1}^{12}(z_0(v_i)_1 - z_1(v_i)_0) = z_0 z_1(z_0^{10} + 11z_0^5 z_1^5 - z_1^{10}),$$

$$
\begin{aligned}
H(z_0 : z_1) &= \prod_{k=1}^{20}(z_0(u_k)_1 - z_1(u_k)_0) = \mathbf{Hess}(f) = \left| \frac{\partial^2 f}{\partial z_i \partial z_j} \right| \\
&= -(z_0^{20} + z_1^{20}) + 228(z_0^{15}z_1^5 - z_0^5 z_1^{15}) - 494 z_0^{10} z_1^{10}, \\
T(z_0 : z_1) &= \prod_{j=1}^{30}(z_0(w_j)_1 - z_1(w_j)_0) = \mathcal{J}(f, H) \\
&= (z_0^{30} + z_1^{30}) + 522(z_0^{25}z_1^5 - z_0^5 z_1^{25}) \\
&= -10005(z_0^{20}z_1^{10} + z_0^{10}z_1^{20}).
\end{aligned}
\tag{4.56}
$$

Here we have used the Hessian variety and Jacobian, see Definitions B.1.8 and B.1.7, respectively. Since there are only two coordinates, there is a relation between these forms, which is:

$$
T^2 = -H^3 + 1728 f^5.
\tag{4.57}
$$

As a subgroup of $PGL(\mathbb{C})$, $\mathcal{I}$ acts on $\mathbb{P}^1$; let

$$
q : \mathbb{P}^1 \longrightarrow \mathbb{P}^1/\mathcal{I} \cong \mathbb{P}^1
\tag{4.58}
$$

denote the projection onto the quotient. An easy argument shows that

$$
q(z_0 : z_1) = \frac{H^3(z_0 : z_1)}{1728 f^5(z_0 : z_1)},
\tag{4.59}
$$

where f, H are the degree 12 and degree 20 invariants above. Let $u \in \mathbb{P}^1$ be an affine coordinate on the quotient; then the degree of q is 60, with the following branching at the points $u = 0, 1, \infty$:

u	$q^{-1}(u)$	$stab(z), z \in q^{-1}(u)$
0	20 points	$\mathbb{Z}/3\mathbb{Z}$
1	30 points	$\mathbb{Z}/2\mathbb{Z}$
∞	12 points	$\mathbb{Z}/5\mathbb{Z}$

By Lemma 4.2.10 and (4.33) we have

Lemma 4.2.11 $\mathcal{I} \cong PSL(2, \mathbb{Z}/5\mathbb{Z}) \cong A_5$.

Moreover, the *natural* action of $PSL(2, \mathbb{Z}/5\mathbb{Z})$ on $\mathbb{P}^1$, with quotient morphism π_5 as in (4.32), can be identified with the *natural* action of $\mathcal{I}$ induced by the symmetries of the icosahedron:

Proposition 4.2.12 *The cover* $\mathbb{P}^1 \longrightarrow \mathbb{P}^1/\mathcal{I}$ *can be naturally identified with the cover* $\pi_5 : (\Gamma(5)\backslash \mathbb{S}_1)^* \cong \mathbb{P}^1 \longrightarrow (\Gamma\backslash \mathbb{S}_1)^* \cong \mathbb{P}^1$, *where* $(-)^*$ *denotes compactification.*

Furthermore, since $u = \infty$ corresponds to $J(\infty)$ (on $\Gamma \backslash \mathbb{S}_1$), we know that the 12 vertices of the icosahedron are *the 12 cusps* of $\Gamma(5)$.

The *icosahedral equation* is just $u = q(z_0 : z_1)$ (see (4.59)), or, after dehomogenizing,

$$\left((z^{20} + 1) - 228(z^{15} - z^5) + 494z^{10}\right)^3 + 1728uz^5(z^{10} + 11z^5 - 1)^5 = 0, \quad (4.60)$$

and it follows from the fact that the Galois group of q is A_5 that

Lemma 4.2.13 *The icosahedral equation (4.60) is its own Galois resolvent.*

Geometrically we have by Proposition 4.2.12 the following diagram

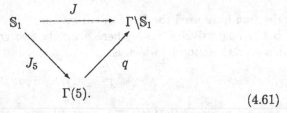

$$\Gamma(5). \tag{4.61}$$

Here J_5 is the "Hauptmodul fünfter Stufe", and can be written in terms of τ as

$$J_5(\tau) = q^{2/5} \cdot \frac{\sum_{-\infty}^{\infty}(-1)^n q^{5n^2 - 3n}}{\sum_{-\infty}^{\infty}(-1)^n q^{5n^2 - n}} = q^{2/5}\frac{\theta_1(2\tau, q^5)}{\theta_1(\tau, q^5)}, \tag{4.62}$$

where $q = e^{2\pi i \tau}$ and θ_1 is the theta function with characteristic $(1, 0)$. The diagram (4.61) also gives a *solution* of the icosahedral equation: given $u \in \mathbb{P}^1$, set

$$z = J_5(\tau), \tag{4.63}$$

where τ is a solution of the modular equation (4.26). Then from $q \circ J_5 = J$, we have $q(z) = q(J_5(\tau)) = J(\tau) = u$, i.e., $q(z) = u$. A solution of (4.62) can also be given directly in terms of hypergeometric functions; we refer to [Kl3] and the references therein (in the new addition, references provided by Slodowy) for details and formulas.

4.2.7   Klein's solution

The solution suggested by Klein had an algebraic and a transcendental part. We sketch this briefly, as these are typical steps for any resolution.

Algebraic Part

Classical: Reduction of equation with solvable Galois group to a "pure equation"	*Icosahedral*: Reduction of equation of fifth degree to solution of the "icosehedral equation"
$$x^n = X$$	$$\frac{H^3(x)}{1728 f^5(x)} = X$$

Actually, Klein's method for dealing with this reduction is almost completely geometric, which was his objective in writing [Kl3]. First of all, if we are given a general quintic $P(x)$ as in (4.49), then by means of a Tschirnhaus transformation it can be put in the form (4.50), determining a point on the quadric Q (4.51). Let P_P denote this point; P_P determines two fibres of $Q \cong \mathbb{P}^1 \times \mathbb{P}^1$, one in each ruling. The action of Σ_5 on Q induces an action of A_5 on each factor, the two being related by an outer automorphism, and the orbit of P_P consists of 120 points (in $\mathbb{P}^3$). Let L_P, M_P denote the two lines in $\mathbb{P}^3$ which are the fibres of Q which meet at P_P, on each of which we have an action of A_5 on $\mathbb{P}^1$. We may consider the quotient of $\mathbb{P}^1$ by A_5 on each copy; each quotient morphism is a copy of q as in (4.58), which we denote by q_1 and q_2. Let z_1, z_2 be affine coordinates on the first and second copy of $\mathbb{P}^1$; for $p \in \mathbb{P}^4$, we have two icosahedral equations (cf. *loc. cit.* p. 184):

$$u_1(p) = \frac{H^3(z_1)}{1728 f^5(z_1)}, \quad u_2(p) = \frac{H^3(z_2)}{1728 f^5(z_2)}, \qquad (4.64)$$

where the "icosahedral solvents" u are determined rational algebraically in terms of the coefficients of the given Hauptgleichung (4.50), α, β, γ and $\nabla = \sqrt{\Delta}$. This is done as follows.

Let $\mathcal{O}_\nu, \nu = 1, \dots, 5$ be the five inscribed octahedra; let

$$f_\nu = \prod (z - \{\text{vertices of } \mathcal{O}_\nu\}),$$

$$W_\nu = \mathbf{Hess}(f_\nu) = \prod (z - \{\text{vertices of dual cube}\})$$

be the invariant forms (for the νth octahedral subgroup of $\mathcal{I}$) of degrees six and eight, respectively. Then, supposing $(y_1, \dots, y_5)$ are the five roots of P, upon permutation of the roots, permuting the octahedra should result, i.e., (after some calculations), letting $z = z_0/z_1$ be an affine coordinate on one of the $\mathbb{P}^1$'s of Q,

$$y_\nu = mT(z)W_\nu(z) + 12n f^2(z) f_\nu(z) W_\nu(z), \qquad (4.65)$$

for affine parameters m, n and z being acted on by the icosahedral group. Also, letting $y_1, \dots, y_5$ be the five roots of a quintic equation, lying on the quadric Q, it is possible to express the parameters z_1, z_2 of (4.64) in terms of $y_1, \dots, y_5$: set (these are the expressions of Lagrange)

$$
\begin{aligned}
p_1 &= y_0 + \varepsilon y_1 + \varepsilon^2 y_2 + \varepsilon^3 y_3 + \varepsilon^4 y_4, \\
p_2 &= y_0 + \varepsilon^2 y_1 + \varepsilon^4 y_2 + \varepsilon y_3 + \varepsilon^3 y_4, \\
p_3 &= y_0 + \varepsilon^3 y_1 + \varepsilon y_2 + \varepsilon^4 y_3 + \varepsilon^2 y_4, \\
p_5 &= y_0 + \varepsilon^4 y_1 + \varepsilon^3 y_2 + \varepsilon^2 y_3 + \varepsilon y_4.
\end{aligned}
$$

Then, as Klein shows, we have the relations (here the line geometry on the Grassmann $\mathbf{G}(2,4) \subset \mathbb{P}^5$, the Klein quadric, is applied)

$$z_1 = -\frac{p_1}{p_2} = \frac{p_3}{p_4}, \quad z_2 = -\frac{p_2}{p_4} = \frac{p_1}{p_3}. \qquad (4.66)$$

Given a Hauptgleichung (4.50), the equations (4.64) and (4.65) may be combined, and utilizing furthermore the expression of z in terms of line coordinates (4.66) for $q(z) = u$, the icosahedral equation associated to the given set of roots is derived. Since now the elementary symmetric functions of the y_ν are the coefficients α, β and γ of the equation (and determine ∇), one gets a system of equations which can be solved rationally for m, n and u (as functions of $\alpha, \beta, \gamma, \nabla$). The results are:

$$
\begin{aligned}
m &= \frac{(11\alpha^3\beta + 2\beta^2\gamma - \alpha\gamma^2) \pm \alpha\nabla}{24(\alpha^4 - \beta^3 + \alpha\beta\gamma)} \\
u &= \frac{(48\alpha m^2 - 12\beta m - \gamma)^3}{64\alpha^2(12(\alpha\gamma - \beta^2)m - \beta\gamma)} \\
n &= -\frac{96\alpha m^3 + 72\beta m^2 + 6\gamma m - 12\alpha^2 u}{144\alpha m^2 + 12\beta m + \gamma}.
\end{aligned}
\tag{4.67}
$$

The second equation gives the parameter u for an icosahedral equation $q(z_0 : z_1) = u$ (the two possible values correspond to the two icosahedral equations (4.64)); inserting a solution $z = z_0/z_1$ of (4.64) into (4.65), as well as the values of m, n from (4.67), gives an explicit formula for the roots of the given Hauptgleichung.

Transcendental Part

Classical: Solution of the "pure equation" by means of logarithims: $$x = e^{\frac{1}{n}\log X}$$	*Icosahedral:* Solution of the icosahedral equation by means of elliptic functions: $$x = q^{\frac{2}{5}}\frac{\theta_1(\frac{2\pi i K'}{K}, q^5)}{\theta_1(\frac{\pi i K'}{K}, q^5)}$$

After determining the parameter $u = u(\alpha, \beta, \gamma, \nabla)$ for the given quintic equation, the corresponding icosahedral equation $q(z) = u$ can be solved by (4.63); the solution z (there are actually two solutions, corresponding to the two icosahedral equations (4.64)), together with values for m, n, u from (4.67), are inserted in (4.65). This gives the roots as sketched above. We summarize Klein's solution in the flow chart in Figure 4.12.

4.3   Solving the equation of 27th degree for the 27 lines

In [Kl1], Klein sketches in a letter to Jordan how one could use a similar process to solve the equation of 27th degree determining the 27 lines on a smooth cubic surface. This was the starting point of investigations of Maschke [Ma2] and Burkhardt [Bu] in Göttingen. They were students of Klein, who, following

Figure 4.12: Klein's solution

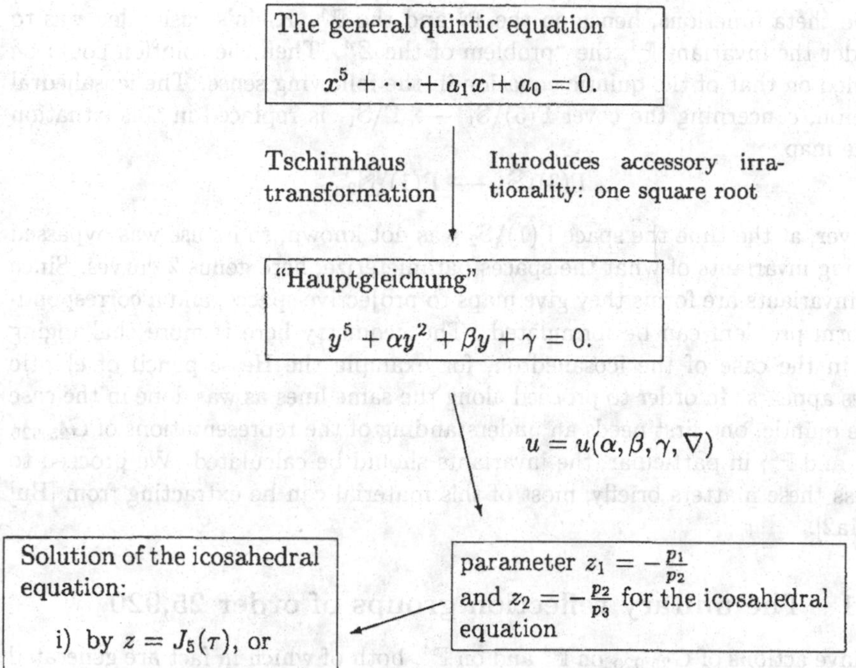

> The general quintic equation
> $$x^5 + \cdots + a_1 x + a_0 = 0.$$

Tschirnhaus transformation | Introduces accessory irrationality: one square root

> "Hauptgleichung"
> $$y^5 + \alpha y^2 + \beta y + \gamma = 0.$$

$$u = u(\alpha, \beta, \gamma, \nabla)$$

> Solution of the icosahedral equation:
>
> i) by $z = J_5(\tau)$, or
>
> ii) by means of hypergeometric functions (4.35).

> parameter $z_1 = -\frac{p_1}{p_2}$ and $z_2 = -\frac{p_2}{p_3}$ for the icosahedral equation

solves

> roots
> $$y_\nu = (4.65).$$

suggestions of [Kl1] and Jordan, worked on the possibility of solving the equation of 27th degree for the lines of the cubic surface by using hyperelliptic functions (nowadays theta functions of two variables). Hence these theta functions replace the theta functions of one variable used in the case of a quintic equation.

Start with the following nine theta functions with characteristics (see Definition 3.3.2):

$$X_{\alpha\beta} = \Theta \begin{bmatrix} 0 & 0 \\ \frac{\alpha}{3} & \frac{\beta}{3} \end{bmatrix} (\tau, z), \quad \tau \in \mathbb{S}_2, \ z \in \mathbb{C}^2, \ \alpha, \ \beta \in \mathbb{Z}/3\mathbb{Z}. \tag{4.68}$$

The $X_{\alpha\beta}$ have the property that transformation of τ (z being held fixed) by an element $\gamma \in P\Gamma(3)$ (the principal congruence subgroup of level three in $PSp(4, \mathbb{Z})$) induces a *linear* transformation of the $X_{\alpha\beta}$, yielding a linear action of $G_{25,920}$ on $\mathbb{P}^8$. This action splits into two invariant subspaces, namely:

$$\begin{aligned} \mathbb{P}^3 &= \{Z_{\alpha\beta} = \tfrac{1}{2}(X_{\alpha\beta} - X_{-\alpha-\beta})\}, \\ \mathbb{P}^4 &= \{Y_{\alpha\beta} = \tfrac{1}{2}(X_{\alpha\beta} + X_{-\alpha-\beta})\}. \end{aligned} \tag{4.69}$$

We get a covariant action of the group $G_{25,920}$: let $\alpha \in G_{25,920}$ be given.

Then as $\alpha \in PSp(4, \mathbb{Z}/3\mathbb{Z})$, a representative $\tilde{\alpha}$ of α acts on $\tau \in \mathbb{S}_2$ (which is independent of the representative modulo $\Gamma(3)$), while at the same time α acts on the theta functions, hence on the $\mathbb{P}^3$ and the $\mathbb{P}^4$. Klein's basic idea was to consider the invariant $\mathbb{P}^3$, the "problem of the Z". Then the solution could be modeled on that of the quintic equation in the following sense. The icosahedral equation, concerning the cover $\Gamma(5)\backslash\mathbb{S}_1 \longrightarrow \Gamma\backslash\mathbb{S}_1$, is replaced in this situation by the map

$$\Gamma(3)\backslash\mathbb{S}_2 \longrightarrow \Gamma(1)\backslash\mathbb{S}_2.$$

However, at the time the space $\Gamma(1)\backslash\mathbb{S}_2$ was not known, so its use was bypassed by using invariants of what the spaces parameterize, here genus 2 curves. Since such invariants are forms they give maps to projective spaces, and a corresponding form problem can be formulated. The geometry here is more challanging than in the case of the icosahedron, for example the Hesse pencil of elliptic curves appears. In order to proceed along the same lines as was done in the case of the quintic, one first needs an understanding of the representations of $G_{25,920}$ in $\mathbb{P}^3$ and $\mathbb{P}^4$; in particular, the invariants should be calculated. We proceed to discuss these matters briefly, most of this material can be extracting from [Bu] or [Ma2].

4.3.1   The unitary reflection groups of order 25,920

We have actions of $G_{25,920}$ on $\mathbb{P}^3$ and on $\mathbb{P}^4$, both of which in fact are generated by unitary reflections. Such actions have *invariant forms* (or subvarieties) $f_i(x)$, where the variables x are homogenous variables of the corresponding projective space, the determination of which is the area of invariant theory. The *form problem* now is as follows. Given ratios for the *values* of the invariant forms f_i, determine explicitly ratios for the values x_i of the variables (i.e., a point in the projective space). See the appendix, section B.1.2 for more details from a general point of view. Klein and his students studied the two particular representations in the 1890's. Maschke considered the action on $\mathbb{P}^3$; Burkhardt worked on $\mathbb{P}^4$. Tables for these arrangements, taken from [OS1], are given in Table 4.2 and Table 4.3. The first rows of these tables give the numbers of each type of subspace; the other rows give the inclusions, where we are using the notations of (3.4) for the subspaces.

The arrangement in $\mathbb{P}^3$: The arrangement induced in each of the 40 planes of this arrangement is the *extended Hesse pencil*, the arrangement of 21 lines which are the 12 lines of the Hesse pencil together with the nine lines joining corners of the four triangles. The 12 lines of the Hesse pencil are the 12 two-fold lines lying in the plane, the nine other lines are the four-fold lines lying in the plane. We might also remark here that the five-fold points of the arrangement split into two different types of singular points in the planes, namely 36 two-fold points and nine four-fold points which are the base points of the Hesse pencil. The 12-fold points of the arrangement lying in one of the planes are five-fold points of the induced arrangement. For each 12-fold point we can also speak of the induced arrangement, by blowing up the point in $\mathbb{P}^3$ and considering the

Table 4.2: The arrangement in $\mathbb{P}^3$ defined by $G_{25,920}$

space	k	$t_4(1)$	$t_2(1)$	t_5	t_{12}
#	40	90	240	360	40
$t(2)$	1	9	12	45	12
$t_4(1)$		1	0	4	4
$t_2(1)$			1	6	2

proper transforms of the 12 planes intersecting the point in the exceptional $\mathbb{P}^2$. In our case we get the Hesse pencil itself, i.e., the 12 lines of the four degenerate cubics.

The arrangement in $\mathbb{P}^4$: This arrangement is exactly the dual of the arrangement of the 45 Jordan primes which we will be discussing in the next chapter, of relevance to the Burkhardt quartic. For example, the 40 18-fold points correspond to the 40 Steiner primes containing 18 of the 45 nodes, the 45 12-fold points correspond to the 45 Jordan primes, each containing 12 of the nodes, etc. We will be discussing this arrangement in great detail in the next chapter.

We now describe briefly the invariant forms under these representations. The invariant forms under the action of $G_{25,920}$ on $\mathbb{P}^3$ were calculated by Maschke in [Ma2]. This is done essentially by reducing the problem to that of the invariants of the Hesse group of order 648 acting on $\mathbb{P}^2$, as follows. The 40 planes of the arrangement in $\mathbb{P}^3$ are given explicitly in homogenous coordinates $(z_0 : ... : z_3)$:

$$\begin{aligned}
(4) \quad & z_i = 0 \quad (i = 0, ..., 3) \\
(9) \quad & (z_1^3 + z_2^3 + z_3^3)^3 - 27z_1^3 z_2^3 z_3^3 = 0 \\
(9) \quad & (z_0^3 + z_1^3 + z_3^3)^3 - 27z_0^3 z_1^3 z_3^3 = 0 \\
(9) \quad & (z_0^3 + z_1^3 + z_2^3)^3 - 27z_0^3 z_1^3 z_2^3 = 0 \\
(9) \quad & (z_0^3 + z_2^3 + z_3^3)^3 - 27z_0^3 z_2^3 z_3^3 = 0
\end{aligned} \qquad (4.70)$$

In the $\mathbb{P}^2$ given by $z_0 = 0$ with homogenous coordinates (z_1, z_2, z_3) the action of G_{648} is generated by five collineations $A, B, C, D, E : z \mapsto z'$,

	A	B	C	D	E	
$z_1' =$	z_2	z_1	z_1	z_1	$z_1 + z_2 + z_3$	
$z_2' =$	z_3	z_3	ϱz_2	ϱz_2	$z_1 + \varrho z_2 + \varrho^2 z_3$	(4.71)
$z_3' =$	z_1	z_2	$\varrho^2 z_3$	ϱz_3	$z_1 + \varrho^2 z_2 + \varrho z_3$	

A complete system of invariants for G_{648} has degrees 6, 9, 12, 12 and 18 and can

Table 4.3: The arrangement in $\mathbb{P}^4$ defined by $G_{25,920}$

space	k	$t_2(2)$	$t_3(2)$	$t_3(1)$	$t_4(1)$	$t_6(1)$	$t_9(1)$	t_4	t_{10}	t_{12}	t_{18}
#	45	270	240	270	720	540	40	540	216	45	40
$t(3)$	1	12	16	18	64	72	8	84	48	12	16
$t_2(2)$		1	0	3	8	6	0	18	12	3	4
$t_3(2)$			1	0	3	9	2	9	9	3	7
$t_3(1)$				1	0	0	0	6	0	2	0
$t_4(1)$					1	0	0	3	3	0	1
$t_6(1)$						1	0	1	2	1	2
$t_9(1)$							1	0	0	0	4

be given explicitly by the following forms:

$$
\begin{aligned}
C_6 &= z_1^6 + z_2^6 + z_3^6 - 10(z_1^3 z_2^3 + z_2^3 z_3^3 + z_3^3 z_1^3), \\
C_9 &= (z_1^3 - z_2^3)(z_2^3 - z_3^3)(z_3^3 - z_1^3), \\
C_{12} &= (z_1^3 + z_2^3 + z_3^3)[(z_1^3 + z_2^3 + z_3^3)^3 + 216 z_1^3 z_2^3 z_3^3], \\
\mathfrak{C}_{12} &= z_1 z_2 z_3 [27 z_1^3 z_2^3 z_3^3 - (z_1^3 + z_2^3 + z_3^3)^3], \\
C_{18} &= ((z_1^3 + z_2^3 + z_3^3)^6 - 540 z_1^3 z_2^3 z_3^3 ((z_1^3 + z_2^3 + z_3^3)^3) - 5832 z_1^6 z_2^6 z_3^6.
\end{aligned}
\tag{4.72}
$$

C_{18} is the functional determinant of C_{12} and $\mathfrak{C}_{12}$, $\mathfrak{C}_{12}$ is just the product of the 12 lines of the Hesse arrangement. C_{12} is the Hessian of $\mathfrak{C}_{12}$ and vise versa. C_9 is the so-called difference product of z_1^3, z_2^3, z_3^3.

The two relations among these invariants are:

$$
\begin{aligned}
432 C_9^2 &= C_6^3 - 3 C_6 C_{12} + 2 C_{18}, \\
1728 \mathfrak{C}_{12}^3 &= C_{18}^2 - C_{12}^3.
\end{aligned}
\tag{4.73}
$$

Maschke proves that the action of $G_{25,920}$ in $\mathbb{P}^3$ is generated by the action of a G_{648} acting in one of the 40 planes and a tetrahedral group of order 24 consisting of the permutation group acting on the z_i. This tetrahedral group stabilizes the tetrahedron consisting of the first four forms in (4.70), and $G_{25,920}$ is generated by it and the G_{648} acting on $z_0 = 0$ as above. From this one deduces that each invariant form of $G_{25,920}$ can be written as a polynomial in z_0 with coefficients which are polynomial expressions in the invariants of G_{648}. The result is as follows. There are invariant forms of degrees 12, 18, 24, 30 and 40. F_{40} is just the product of the 40 planes defining the arrangement in $\mathbb{P}^3$. For example F_{12} is:

$$
\begin{aligned}
F_{12} &= 6 z_0^{12} + 6 \cdot 22 C_6 z_0^6 + 6 \cdot 220 C_9 z_0^3 + C_6^2 + 5 C_{12} \tag{4.74} \\
&= 6\{ \Sigma z_i^{12} + 22 \Sigma z_i^6 z_j^6 + 220 \Sigma \pm z_i^6 z_j^3 z_k^3 \},
\end{aligned}
$$

where there are rules for determining which sign $\pm$ is applied. The forms F_{12}, F_{18}, F_{24}, F_{30} and F_{40} satisfy the (surprisingly complicated) relation,

$$2^{28} \cdot 3^{15} \cdot 5^{15} \cdot F_{40}^3 =$$

$$\begin{vmatrix} \Phi_{30} & 2\Phi_{24}^2 & F_{18}\Phi_{24} & F_{12}\Phi_{24} \\ 2\Phi_{24} & 27F_{12}\Phi_{30} - 11F_{18}\Phi_{24} & 3F_{12}\Phi_{24} - 4F_{18}^2 & 3\Phi_{30} - 4F_{12}F_{18} \\ F_{18} & 3F_{12}\Phi_{24} - 4F_{18}^2 & 13F_{12}F_{18} - 3\Phi_{30} & 9F_{12}^2 - 2\Phi_{24} \\ F_{12} & 3\Phi_{30} - 4F_{12}F_{18} & 9F_{12}^2 - 2\Phi_{24} & F_{18} \end{vmatrix} \quad (4.75)$$

where $4\Phi_{24} = 25F_{24} - 9F_{12}^2$, $6\Phi_{30} = 25F_{30} - F_{12}F_{18}$. There are two simpler relations involving F_{40}. First, F_{40} is the Hessian determinant of F_{12}, and secondly the Jacobian determinant $\mathcal{J}(F_{12}, F_{18}, F_{24}, F_{30}) = \left|\frac{\partial F_j}{\partial z_i}\right|$ satisfies (see Theorem B.1.14, (iii)):

$$\mathcal{J}(F_{12}, F_{18}, F_{24}, F_{30}) = 2^{26} \cdot 3^{15} \cdot 5^6 \cdot F_{40}^2.$$

Burkhardt proceeded similarly to determine the invariants of $G_{25,920}$ acting on $\mathbb{P}^4$. Let the coordinates be given as

$$y_0 = Y_{00}, \; y_1 = Y_{10}, \; y_2 = Y_{01}, \; y_3 = Y_{11}, \; y_4 = Y_{12} \quad (4.76)$$

where the $Y_{\alpha\beta}$ are the theta functions of (4.69). Using the isomorphism $G_{25,920} \cong PSp(4, \mathbb{Z}/3\mathbb{Z})$, it suffices to use generators of the symplectic group $PSp(4, \mathbb{Z}/3\mathbb{Z})$ to get generators of the action of $G_{25,920}$ on $\mathbb{P}^4$. However, Burkhardt used instead the corresponding hyperelliptic curves, and a certain *Weierstraß form* for them to describe the level 3 structure. At any rate, generators of the group are transformations B, C, D and S_2, which act as in the following table:

	B	C	D	S_2
$y_0' =$	$\frac{1}{\sqrt{-3}}(y_0 + 2y_1)$	y_0	$-y_0$	y_0
$y_1' =$	$\frac{1}{\sqrt{-3}}(y_0 - y_1)$	y_1	$-y_2$	$\varrho^2 y_1$
$y_2' =$	$\frac{1}{\sqrt{-3}}(y_2 + y_3 + y_4)$	y_4	$-y_1$	y_2
$y_3' =$	$\frac{1}{\sqrt{-3}}(y_2 + \varrho y_3 + \varrho^2 y_4)$	y_2	$-y_3$	$\varrho^2 y_3$
$y_4' =$	$\frac{1}{\sqrt{-3}}(y_2 + \varrho^2 y_3 + \varrho y_4)$	y_3	$-y_4$	$\varrho^2 y_4$

$$(4.77)$$

Consider the hyperplane $\mathcal{S}$ given by $\mathcal{S} = \{y_0 = 0\}$ and the plane $\mathcal{J} = \{y_0 = y_1 = 0\}$. The stabiliser of each is a subgroup of order 648, but these two subgroups of order 648 are *not* conjugate to each other; indeed,

$$N(\mathcal{S}) = \langle C, D, S_2 \rangle, \quad N(\mathcal{J}) = \langle B, C, S_2 \rangle. \quad (4.78)$$

Hence, just as above, $G_{25,920}$ is generated by the subgroup of order 648 acting on $\mathcal{J}$, generated by B, C and S_2, and by the centraliser of $\mathcal{J}$. Burkhardt shows this to be a homogenous tetrahedral group, and its invariants are:

$$\begin{aligned} \Phi &= y_0^4 + 8y_0 y_1^3; \\ \Psi &= y_0^3 y_1 - y_1^4; \\ t &= y_0^6 - 20y_0^3 y_1^3 - 8y_1^6; \end{aligned} \quad (4.79)$$

which satisfy the relation

$$\Phi^3 - 64\Psi^3 = t^2. \tag{4.80}$$

Furthermore, letting $\varphi = y_2 y_3 y_4$, $\psi = y_2^3 + y_3^3 + y_4^3$ and $u = y_0\psi + 6y_1\varphi$, all invariants of $G_{25,920}$ can be written as linear combinations of the following: Φ, $u;, t$, Ψ_1, C_6; C_9; Φ_3; t_3, C_{12}; C_{18}, where the expressions C_m are the invariants (4.72), the Φ, u, and t were just defined, and

$$\begin{aligned}
\Psi_1 &= \psi(-y_0^3 + 4y_1^3) + 18\varphi y_0^2 y_1; & (4.81)\\
\Phi_3 &= -\psi^3 y_0 + 18\varphi\psi^2 + 108\varphi^3 y_0;\\
t_3 &= \psi^3(y_0^3 + 8y_1^3) - 54\varphi\psi^2 y_0^2 y_1 + 324\varphi^2\psi y_0 y_1^2.
\end{aligned}$$

Burkhardt then calculates the following expressions. There are invariants of degrees 4, 6, 10, 12, 18 and 45. The invariant of degree 45 is just the product of the 45 reflection hyperplanes defining the arrangement in $\mathbb{P}^4$. As to the others, Burkhardt gives the following expressions:

$$\begin{aligned}
J_4 &= \Phi + 8u & (4.82)\\
J_6 &= t + 20\Psi_1 - 8C_6 & (4.83)\\
J_{10} &= \frac{1}{24}(\Phi\Psi_1 + ut + 2\Phi C_6 + 2u\Psi_1 - 2\Phi_3 - 2uC_6) & (4.84)\\
J_{12} &= \frac{1}{24}(3t\Psi_1 + 3u\Phi^2 + 19\Psi_1^2 - 9u^2\Phi - 10C_6 t - 11t_3 + 9u^3 & (4.85)\\
& \quad -2C_6\Psi_1 - 4C_{12} + 4C_6^2)
\end{aligned}$$

and a similar expression for J_{18}. There is a relation between these, which takes the form $J_{45}^2 =$ rational expression in the other invariants, which Burkhardt does not explicitly calculate. There is also an invariant of degree 40: the product of the 40 Steiner primes. Denoting this by J_{40}, the following relation holds:

$$3^{33} J_{40} = [J_4^2(2^9 J_{12} - J_4^3) - 3.2^{18} J_{10}^2]^2 - 2^{19}[J_4 J_6 - 3.2^8 J_{10}][J_4^3 J_{18} - 2^{11} J_{10}^3]. \tag{4.86}$$

Coble mentions in [C1], III, p. 350 that "the expressions in y for the invariants J_4, J_6, J_{10}, J_{12}, and J_{18} given by Burkhardt seem open to suspicion", and he gives other formulas for the invariants. He uses the fact, already noticed by Burkhardt, that the 40 *squares* of the linear forms defining the Steiner primes are permuted among each other, so that taking the sum of the kth powers of these 40 squares is either an invariant of degree $2k$ or zero. Coble then shows that for $k = 2, 3, 5, 6, 9$ the sum of the kth powers does not vanish, and hence, this yields a system of invariants. In fact, letting ξ_i be the 40 forms defining the 40 Steiner primes, it is clear that $I_k := \sum_{i=1}^{40} \xi_i^{2k}$ is an invariant if it does not vanish. In order for I_k, $k = 2, 3, 5, 6, 9$ to be non-vanishing it is sufficient to show that the Jacobian determinant does not vanish. This is what Coble shows, a calculation he simplifies by taking a convenient choice of coordinates.

4.3.2   Klein's suggestion

We sketch the method suggested by Klein, which was in many respects completed by Burkhardt in [Bu], III, briefly here.

Algebraic Part: Reduction to the "Maschke form problem":

$$\frac{F_{24}(z)}{F_{12}^2(z)} = \alpha, \quad \frac{F_{30}(z)}{F_{12}(z)F_{18}(z)} = \beta, \quad \frac{F_{18}^2(z)}{F_{12}^3(z)} = \gamma.$$

Step 1: First, by (B.51) of the appendix we have the equation of a covariant surface, whose intersection with the given cubic is the set of 27 lines. By elimination theory this is reduced to a degree 27 equation in one variable, with Galois group $W(E_6)$. By adjoining the square root of the discriminant of the cubic, the group is reduced to $G_{25,920}$. A slightly different way of viewing this is as follows. The 27 lines determine 27 points on the Grasmannian $G(2,4)$, which is a quadric hypersurface in $\mathbb{P}^5$. Letting A, B be two generic $\mathbb{P}^4$'s in the $\mathbb{P}^5$, each intersects the quadric in a divisor, and letting λ, μ be parameters, the solution to the equation

$$\lambda A + \mu B = 0$$

for a point on the quadric may be viewed as a solution $(\lambda : \mu) \in \mathbb{P}^1$ of the equation in one variable mentioned above. The elimination theory corresponds to projecting the 27 points onto a line, by which the covariance is lost.

Step 2: The action of the Galois group on the 27 roots, say $\alpha_1, \ldots, \alpha_{27}$, as well as the action of $W(E_6)$ (or $G_{25,920}$) on $\mathbb{C}^6$ (and consequently on $\mathbb{P}^5$), are fixed. In order to *covariantly* associate to a given equation the roots to a point in $\mathbb{P}^5$, we need equivariance, i.e., for $\gamma \in W(E_6)$ (or $G_{25,920}$), $\zeta(\alpha_i)$ the point in $\mathbb{C}^6$ associated with α_i,

$$\zeta(\gamma\alpha_i) = \gamma\zeta(\alpha_i).$$

Now consider the roots $\alpha_1, \ldots, \alpha_{27}$ as variables; the Galois group then acts on the space $\mathbb{C}^{27} = \langle \alpha_1, \ldots, \alpha_{27} \rangle$. On the other hand, $W(E_6)$ acts linearly on $\mathbb{C}^6$; recall the 27 linear functions a_i, b_i, c_{ij}, $i < j, i = 1, \ldots, 6$ of (6.7) below. It is possible to find a linear map

$$\xi : \mathbb{C}^{27} \longrightarrow \mathbb{C}^6 \subset \mathbb{C}^{27} \tag{4.87}$$
$$\langle \alpha_1, \ldots, \alpha_{27} \rangle \mapsto \xi(\alpha_1, \ldots, \alpha_{27}) = (\xi_1, \ldots, \xi_{27})$$

such that $\xi_i \equiv a_i$ $(i = 1, \ldots, 6)$, $\xi_{i+6} \equiv b_i$, $(i = 1, \ldots, 6)$ and $\xi_{k+12} \equiv c_{ij}$, $(k = 1, \ldots, 15)$ by setting ([Bu], §75):

$$\xi_i(\alpha_1, \ldots, \alpha_{27}) = 4\alpha_i - 2C_i + N_i, \tag{4.88}$$

where C_i is the sum of roots corresponding to the ten lines meeting the given α_i while N_i is the sum of the other 16 roots (so in particular, $C_i + N_i = \sum_{j=1}^{27} \alpha_j - \alpha_i$ for all $i = 1, \ldots, 27$). One sees in fact without difficulty that ξ has image in the

$\mathbb{C}^6 \subset \mathbb{C}^{27}$ defined by the 45 relations $a_i + b_j + c_{ij} = 0$, $c_{ij} + c_{kl} + c_{mn} = 0$ given by the tritangents (see the relation (6.10) below). Furthermore, ξ is clearly equivariant with respect to the action of $W(E_6)$.

Step 3: A linear change of coordinates now passes from the a_i, b_i, c_{ij} to the "line coordinates" a_{ij} ([Bu], p. 324):

$$\begin{aligned}
a_1 &\equiv \varrho a_{12} - a_{13} - \varrho^2 a_{34} + a_{42}, \\
a_2 &\equiv \varrho a_{12} - \varrho a_{13} - \varrho^2 a_{34} + \varrho^2 a_{42}, \\
a_3 &\equiv \varrho a_{12} - \varrho^2 a_{13} - \varrho^2 a_{34} + \varrho a_{42}, \\
a_4 &\equiv a_{14} - \varrho^2 a_{12} - a_{23} + \varrho a_{34}, \\
a_5 &\equiv \varrho a_{14} - \varrho^2 a_{12} - \varrho^2 a_{23} + \varrho a_{34}, \\
a_6 &\equiv \varrho^2 a_{14} - \varrho^2 a_{12} - \varrho a_{23} + \varrho a_{34}.
\end{aligned}$$

and the corresponding inverse relations

$$a_{12} = \frac{i}{3\sqrt{3}}(\varrho a_1 + \varrho a_2 + \varrho a_3 + \varrho^2 a_4 + \varrho^2 a_5 + \varrho^2 a_6),$$

$$a_{13} = -\frac{1}{3}(a_1 + \varrho^2 a_2 + \varrho a_3), \quad a_{14} = \frac{1}{3}(a_4 + \varrho^2 a_5 + \varrho a_6),$$

$$a_{34} = \frac{i}{3\sqrt{3}}(\varrho^2 a_1 + \varrho^2 a_2 + \varrho^2 a_3 + \varrho a_4 + \varrho a_5 + \varrho a_6),$$

$$a_{42} = \frac{1}{3}(a_1 + \varrho a_2 + \varrho^2 a_3), \quad a_{23} = -\frac{1}{3}(a_4 + \varrho a_5 + \varrho^2 a_6).$$

By means of these one now gets *rational functions of the roots* $\alpha_1, \ldots, \alpha_{27}$ *which are a_{ij}'s.* This requires taking the a_{ij} as homogeneous coordinates, and further, assuming that

$$a_{12}a_{34} + a_{13}a_{42} + a_{14}a_{23} = 0,$$

i.e., the point $\zeta(\alpha_1, \ldots, \alpha_{27}) \in \mathbb{P}^5$ defined by ϕ (for a fixed equation of degree 27) actualy lies on the Klein quadric, see [Bu], p. 326 for details.

Step 4: Reduction from the problem of the a_{ik} to that of the $z \in \mathbb{P}^3$. There is a general method to do this, developed by Klein for his studies of the equations of seventh and eighth degree. The idea is to use the fact that the line coordinates a_{ik} describe variable lines in $\mathbb{P}^3$, so, given two such lines which intersect, we get a point in $\mathbb{P}^3$. So suppose we could, from the given point in $\mathbb{P}^5$, get the equations for two covariant lines which *intersect*; then the point of intersection is a covariantly associated point. To do this explicitly, one uses the following fact: the most general kind of expression in $\mathbb{P}^5$ which is covariant is of the following form (let ζ_i denote the ith coordinate of $\zeta(\alpha_1, \ldots, \alpha_{27}) \in \mathbb{P}^5$, i=1,..., 6):

$$X_i = \lambda_1 \zeta_i + \lambda_2 \left(\zeta_i^2 - \frac{s_2}{6}\right) + \cdots + \lambda_6 \left(\zeta_i^6 - \frac{s_6}{6}\right), \tag{4.89}$$

where X_i is the covariant expression in the roots α_i, s_k is given by the sum of the kth powers of the ζ_i and we are assuming $s_1 = 0$. Now for this to describe a variable *line*, this point must lie on the Grassmannian, in other words must fulfill a quadratic relation. Given two such, say X_i' and X_i'', they will determine lines which *meet*, if, in addition the relation $\sum X_i' X_i'' = 0$ is fulfilled. So we set X_i' and X_i'' as in (4.89), where X_i' will have coefficients λ_i', X_i'' will have coefficients λ_i'', and require:

$$\sum X_i'^2 = 0, \quad \sum X_i' X_i'' = 0, \quad \sum X_i''^2 = 0; \qquad (4.90)$$

the first and last equations are equations for variable lines (i.e., that the covariant points lie on the Grassmannian), while the second expresses the condition that these lines meet. Inserting now the expressions (4.89) for X_i' and X_i'' into (4.90), one gets *quadratic equations* for the coefficients λ_i' and λ_i'', which can be solved. These expressions are then accessory irrationalities, necessary for the solution. At any rate we now have the variable covariant points X_i' and X_i''. To get the corresponding z (given by the intersection of the lines which X_i' and X_i'' define), note that a general line ξ will meet both X_i' and X_i'' (and hence also their intersection point) if

$$\sum X_i' \xi = \sum X_i'' \xi = 0. \qquad (4.91)$$

If we now express that general line ξ in terms of the a_{ik}, then the equation (4.91) becomes a system of linear equations in the a_{ik}, and if one utilizes finally the relation

$$a_{ij} = z_i z_j' - z_i' z_j$$

we get a system of equations linear in two sets of variables, z_i and z_i'. The *solution* is then the point z for which (4.91) holds *for all* z_i'. For more details see the Abh. LVIII, §§8-11 in Klein's collected papers.

The solution z from the last step gives the *values* for the Maschke form problem, as a rational expression of the coefficients and two accessory square roots introduced in Step 4, and the reduction to this problem has been done.

Transcendental Part: Solution of the Maschke equation in terms of the hyperelliptic functions $Z_{\alpha,\beta}$ of (4.69)

$$Z_{\alpha\beta} = \frac{1}{2}(X_{\alpha\beta} - X_{-\alpha-\beta}).$$

Once again there are several steps involved.

Step 5: The first observation is that the invariants of the Maschke group, $F_{12}, \ldots, F_{40}$, are *invariants for a corresponding binary sextic*, the roots of which define a genus 2 curve, hence an abelian surface as Jacobian, whose theta functions will give the solution. Furthermore, by expressing the $Z_{\alpha\beta}$ in terms of $\tau \in \mathbb{S}_2$ (the Siegel space of degree two), one also gets expressions for the $F_{12}, \ldots, F_{40}$ in terms of τ. If one uses the Fourier expansions as usual, then setting $p := \exp(2\pi i \tau_{11})$, $q := \exp(2\pi i \tau_{12})$, $r := \exp(2\pi i \tau_{22})$, where we are writing

$$\tau = \begin{pmatrix} \tau_{11} & \tau_{12} \\ \tau_{12} & \tau_{22} \end{pmatrix}, \text{ and setting } s := q - q^{-1}, \text{ Burkhardt finds ([Bu] III, p. 328)}$$

$$
\begin{aligned}
p_{12}^6 \cdot F_{12} &= 6p\{1 - 12p^2 + 176r^2 + \cdots\} \\
p_{12}^9 \cdot F_{18} &= 54p^{3/2}\{-1 + 18p^2 + 136r^2 + \cdots\}, \\
p_{12}^{12} \cdot F_{24} &= 1728p^2\{8r^2 + \cdots\}, \\
p_{12}^{15} \cdot F_{30} &= 2592p^{5/2}\{-8r^2 + \cdots\} \\
p_{12}^{20} \cdot F_{40} &= 8p^4r^4\{-2 - 4rs - 38p^2 - 2r^2 - s^2 + \cdots\},
\end{aligned}
\tag{4.92}
$$

where the expressions $\cdots$ are of third or higher degree. Here

$$p_{12} = \tau_{11}\tau_{22} - \tau_{12}^2$$

is the determinant of the period.

Step 6: Given a binary sextic $f(x_1, x_2) = 0$, a level 2 structure on the genus 2 (hyperelliptic) curve described by it amounts to an ordering of the branchpoints (see 2.4.7). The sextic may be described in ten ways $f = \phi_3\psi_3$ as the product of two cubics, corresponding to the ten even thetas with $\frac{1}{2}$-characteristics, and in six ways $f = \phi_1\psi_5$ as the product of a linear and quintic term, corresponding to the six odd characteristics. As the $Z_{\alpha\beta}$ are *even* functions for *odd* characteristics (see [Bu] II, p. 184), for the problem of the Z's one considers the latter type of decomposition:

$$f = \phi(x_1, x_2)\psi_5(x_1, x_2). \tag{4.93}$$

Then invariants of f are *simultaneous invariants* (see Definition B.1.4) of ϕ and ψ_5. In order to express the $F_{12}, \ldots, F_{40}$ in terms of the coefficients of f in (4.93), Burkhardt determines the corresponding degrees, using the fact that the complete system of invariants for sextics is known (see (B.29)). Using the power series expansion (4.92) and the discriminant D of the quintic ψ_5 in (4.93), he finds

$$
\begin{aligned}
F_{12} &= D^{1/2} \cdot J_{6,2}, \\
F_{18} &= D^{3/4} \cdot J_{9,3}, \\
F_{24} &= D \cdot J_{12,4}, \\
F_{30} &= D^{5/4} \cdot J_{15,5}, \\
F_{40} &= D^2 \cdot J_{20,8}.
\end{aligned}
$$

Here $J_{\kappa,\lambda}$ is a common invariant of ϕ and ψ_5, of degree κ in the coefficients of ϕ and degree λ in the coefficients of ψ_5. Without much difficulty he determines the $J_{\kappa,\lambda}$:

$$
\begin{aligned}
J_{6,2} &= \text{Hessian of } \psi_5 = [H], \\
J_{9,3} &= \mathfrak{J}(J_{6,2}, \psi_5) = [T],
\end{aligned}
$$

$$J_{12,4} = \psi_5^2 \cdot (\psi_5\psi_5)^4 = [\psi_5^2 i] \quad \begin{pmatrix} (\psi_5\psi_5)^4 =: i \text{ is the fourth} \\ \text{transvection of } \psi_5 \text{ over itself, see} \\ \text{(B.15)} \end{pmatrix} \quad (4.94)$$

$$J_{15,5} = \psi_5^2 \cdot ((\psi_5\psi_5)^4\psi_5)^1 = [\psi_5^2(i\psi_5)^1] \quad (\ )^1 \text{ is again the transvection}$$

$$J_{20,8} = [\psi_5^4].$$

In each case the bracket [ ] indicates that the variable in the expressions on the right hand side are replaced by the coordinates of the zero of ϕ. In particular, for the so-called Weierstraß form

$$\phi = x_2, \quad \psi_5 = 4x_1^4 - g_2 x_1^3 x_2^2 - g_3 x_1^2 x_2^2 - g_4 x_1 x_2^4 - g_5 x_2^5, \quad (4.95)$$

one has

$$[H] = -\frac{2}{5}g_2,$$

$$[T] = -\frac{8}{5}g_3,$$

$$[i] = -\frac{4}{5}g_4 + \frac{3}{100}g_2^2, \quad (4.96)$$

$$[(i\psi_5)] = -16g_5 + \frac{2}{25}g_2g_3,$$

$$[\psi_5] = 4.$$

Hence, once we are given the *values* of the Maschke form problem, the coefficients of an explicit sextic are given by (4.96).

Step 7: From the equation $f(x_1, x_2) = 0$, the non-normalized period Ω is given is given by:

$$\Omega = \begin{pmatrix} \int_{\alpha_1} \frac{x\,dx}{\sqrt{f}} & \int_{\alpha_2} \frac{x\,dx}{\sqrt{f}} & \int_{\beta_1} \frac{x\,dx}{\sqrt{f}} & \int_{\beta_2} \frac{x\,dx}{\sqrt{f}} \\ \int_{\alpha_1} \frac{dx}{\sqrt{f}} & \int_{\alpha_2} \frac{dx}{\sqrt{f}} & \int_{\beta_1} \frac{dx}{\sqrt{f}} & \int_{\beta_2} \frac{dx}{\sqrt{f}} \end{pmatrix}, \quad (4.97)$$

where as usual $\alpha_1, \alpha_2, \beta_1, \beta_2$ is a base of $H_1(C, \mathbb{Z})$, C the genus 2 curve defined by f. The normalized period τ defined by Ω is a solution of the Maschke form problem, i.e.,

$$\frac{F_{24}(z)}{F_{12}^2(z)} = \alpha, \quad \frac{F_{30}(z)}{F_{12}(z)F_{18}(z)} = \beta, \quad \frac{F_{18}^2(z)}{F_{12}^3(z)} = \gamma.$$

take on the given values.

This approach of Klein uses the representation of $G_{25,920}$ on $\mathbb{P}^3$, and not so much that on $\mathbb{P}^4$. But in fact one *can* use the representation in $\mathbb{P}^4$ instead of the above to solve the equation, a fact which was recognized somewhat later by Coble [C1]. Although this solution is by far more complicated, it is more canonical, and, moreover, it uses the action of $G_{25,920}$ on $\mathbb{P}^4$, which we also wish to study, so we will also sketch Coble's method.

4.3.3 Coble's solution

Coble's solution also consists of an algebraic and a trancendental part. We will sketch the algebraic part in this section, and the transcendental part will be discussed in the next chapter.

Algebraic part: Reduction of the problem of finding the equations for the 27 lines to the Burkhardt form problem (see Definition B.1.15 in the appendix) for the group $G_{25,920}$ acting on $\mathbb{P}^4$, which we will also refer to as "the Burkhardt group".

This reduction consists of several steps, one of which is quite complicated, one of which is quite easy.

Step 1: This is the easy part: reduction to the equation problem for $G_{6,2}$ acting on $\mathbf{P}_6^2$, see Definition 4.1.17.

At this point we note the $G_{6,2}$ acts only birationally on $\mathbf{P}_6^2$, so what we do here is only valid on a Zariski open set of the moduli space of cubic surfaces; but on the complement of this Zariski open set, the Galois group of the problem should shrink to a solvable subgroup (as was the case with the Cayley cubic). For the following, let $S \subset \mathbb{P}^3$ be given, say by $P = 0$, and we assume that $i_{100}(S) = I_{75}(S) \neq 0$ (if this does not hold, then the Galois group reduces to $W(F_4)$, which is solvable). *If we know both the quatenary cubic $P(u)$, $u = (u_0, u_1, u_2, u_3)$ homogenous coordinates on $\mathbb{P}^3$, and the hexahedral form $\bar{a}, \ldots, \bar{f}$ of S,* then the relation (B.55) implies that the coordinates u in $\mathbb{P}^3$ and $(a, \ldots, f)$ in hexahedral form are related by:

$$(l_{11}u) = (L_8 a), \quad (l_{19}u) = (L_{14}a), \tag{4.98}$$
$$(l_{27}u) = (L_{20}a), \quad (l_{43}u) = (L_{32}a),$$

where the notation $(L_j a)$ means L_j is a linear function of $(a, \ldots, f)$. Also, for the sum $a + \cdots + f$ we will use the abbreviation $\sum a$, and similarly for other expressions as well as for the coefficients $\bar{a}, \ldots, \bar{f}$. For example, $\sum \bar{a}^3$ will mean $\bar{a}^3 + \cdots + \bar{f}^3$.

Given the cubic surface S, the quatenary invariants and the hexahedral invariants are related by equalities like $i_8(S) = I_6(S)$, etc. Consequently the following steps can be carried out and complete the reduction of Step 1.

1° For the given S the invariants $i_8(S), \ldots, i_{40}(S)$ and the linear covariants $(l_{11}u), \ldots, (l_{43}u)$ are calculated.

2° The values $i_8(S), \ldots, i_{40}(S)$ furnish the values for $I_6, \ldots, I_{30}$ of the equation problem for $G_{6,2}$. Assuming this problem has been solved, we have the coordinates (x, y, z, t, u) of a point in $\mathbf{P}_6^2$.

3° These coordinates can be inserted into (4.15), and the hexahedral form for the surface $(\bar{a}, \ldots, \bar{f})$ is derived from (4.16), while the value of d_2 is derived from (4.18).

4° The coefficients of the linear covariants $(L_8 a), \ldots, (L_{32} a)$ are determined by the $\bar{a}, \ldots, \bar{f}$, so from the values calculated in 3°, the linear covariants can be calculated. Also, from the equations (4.13) and (4.14) the equations for the tritangents and lines are calculated, in terms of the $\bar{a}$.

5° By 1° we have the linear covariants $(l_j u)$, and by 4° we have the linear covariants $(L_j a)$. Hence, we can solve (4.98) for the coordinates a in terms of the u_i, assuming that the determinant of the four linear forms does not vanish. Thus we have the coordinates $a, \ldots, f$ which display S as a section of the Segre cubic, and the equations for the lines and tritangents in the u_i coordinates are derived by setting the expressions obtained for the $a, \ldots, f$ as linear expressions in the u_i in the equations for the lines and tritangents (4.13) and (4.14).

If the determinant of the four covariants used in 5° vanishes, this means that $I_{75} = 0$, which we have excluded. This completes the reduction to the equation problem for $G_{6,2}$.

Step 2: Translation of the *rational* action of $G_{6,2}$ on the hexahedral variables $a, \ldots, f$ into a *linear* action, that of the Burkhardt group.

Here one wants to give a relationship between coordinates on the space of coefficients $\bar{a}, \ldots, \bar{f}$ on the one hand, and coordinates y_i for the $\mathbb{P}^4$ with action of Burkhardt's group on the other. As we now have so many different coordinates, we pause to fix their notations.

Notations 4.3.1

- $\mathbf{P}_6^2$ is the invariant space of six points in $\mathbb{P}^2$, birational to $\mathbb{P}^4$ via the map

$$\mathbb{P}_6^2 \longrightarrow \mathbb{P}^4$$

$$p \mapsto \begin{pmatrix} 1 & 0 & 0 & 1 & x & z \\ 0 & 1 & 0 & 1 & y & t \\ 0 & 0 & 1 & 1 & u & u \end{pmatrix}, \quad (x, y, z, t, u) \in \mathbb{P}^4,$$

see (4.15).

- $\mathbb{P}^4$ with coordinates $(a, \ldots, f)$, $\sum a = 0$, with a linear action of Σ_6 and a rational action of $W(E_6)$. The dual coordinates are $\bar{a}, \ldots, \bar{f}$, and the usual relation $\sum \bar{a} a = 0$ holds (see (B.6)).

- $\mathbb{P}^4$ with coordinates $(x_a, \ldots, x_f)$, $\sum x_a = 0$, with a linear action of $G_{25,920}$ (Burkhardt's group). There are contragredient variables $(u_a, \ldots, u_f)$, $\sum u_a = 0$. Here the sum $\sum x_a u_a$ is an invariant (see (4.108)).

- the same $\mathbb{P}^4$, but with coordinates y_i as in (4.76). Here the dual coordinates will be denoted v_i. In these variables there is once again an action of Burkhardt's group.

Remark 4.3.2 The transformation from the (x_a, u_a) to the (y_i, v_i) are given in (5.2) and (5.3) below.

Note in particular that the notations $x_1, \ldots, x_6$ used in our description of the Segre cubic are *denoted* here $a, \ldots, f$. Now recall the Cremona transformations defined by the Jacobian ideals of S_3 and $\mathcal{I}_4$, which are the quadrics on ten points (see Corollary 3.2.4) and the cubics on 15 lines (see Lemma 3.3.13):

$$\varphi : \mathbb{P}^4 \longrightarrow \mathbb{P}^4 \text{ given by the Jacobian ideal of } S_3 \qquad (4.99)$$

$$d : \mathbb{P}^4 \longrightarrow \mathbb{P}^4 \text{ given by the Jacobian ideal of } \mathcal{I}_4. \qquad (4.100)$$

In our situation, from the rational action of $G_{6,2}$ on $\mathbb{P}^4$ with coordinates (x, y, z, t, u), we get a rational action of $G_{6,2}$ on $\mathbb{P}^4$ with coordinates $a, \ldots, f$, $\sum a = 0$. Coble then shows how one can, guided by the invariants associated with the 27 lines, turn this into a linear action (of $G_{25,920}$) on $\mathbb{P}^4$ with coordinates $x_a, \ldots, x_f$. This is given by

Theorem 4.3.3 ([C1], III §3) *Under the duality map d of (4.100), the rational action of $G_{25,920}$ is transformed into the linear action of the Burkhardt group on $\mathbb{P}^4$.*

Before we sketch the proof let us pause a moment to see why this is natural. Recall that $\mathbf{P}_6^2$ could be identified with the double cover of $\mathbb{P}^4$ branched along the Igusa quartic $\mathcal{I}_4$ (Theorem 3.5.4), hence also with 15 singular lines. On the other hand, $\mathcal{I}_4$ itself represented the condition on six points that they lie on a conic, while the 15 lines represented the condition that three of the points are on a line, so that under a *regular* action of $W(E_6)$, these loci must be exchangeable with the Igusa quartic, i.e., must be divisors isomorphic to (the proper transform of) $\mathcal{I}_4$ on some modification of $\mathbf{P}_6^2$. This means that to get a regular action we should blow up the 15 lines, and that is precisely what the map d does.

Remark 4.3.4 There is another rather remarkable interpretation of the maps d and φ, also due to Coble (see [C2]). For this we first describe two normal forms for binary sextics (compare also (4.95)).

1) The Maschke normal form:

$$\Phi = \prod(y - \Phi_i) = y^6 - 6F_8 y^4 + 4F_{12} y^3 + 9F_8^2 y^2 - 12F_{20} y + 4F_{24} = 0.$$

2) The Joubert normal form, given in (B.36):

$$S = \prod(X - Q_i) = X^6 + 15q_2 X^4 + 15q_4 X^2 + 6q_5 X + q_6 = 0.$$

In the situation at hand, the sextic in Joubert normal form is of the following special type:

$$T = t^6 - 3 \cdot 5 \cdot 2^7 A_0 t^4 + 3 \cdot 5(5 \cdot 4^7 A_0^2 - 9B_0)t^2$$
$$+ \frac{\sqrt{\Delta_0}}{8} t - 10(5^2 \cdot 4^{10} A_0^3 - 5 \cdot 12^3 A_0 B_0 - 3^3 C_0) = 0.$$

This is related to the Maschke normal form by the relations

$$A_0 = \frac{\rho^2}{2\cdot 5\cdot 2^7}\frac{F_{12}^2 - F_{24}}{F_8 F_{12} - F_{20}}, \quad B_0 = \frac{\rho^4}{3^4\cdot 5}F_8, \quad C_0 = \frac{-\rho^6}{3^6\cdot 5}F_{12},$$

$$\Delta_0 = \frac{2^8\cdot\rho^{10}}{3^4}(F_8 F_{12} - F_{20}), \quad \rho = \frac{(2\pi i)^2}{p_{12}}.$$

Here p_{12} is again the determinant of the period of the genus 2 curve defined by the sextic, and $F_4,\ldots,F_{20}$ are invariants defined by Maschke (which we have not defined). In fact, there is a Tschirnhaus transformation between the two normal forms. For a sextic in Joubert form a quadratic Tschirnhaus transformation leads to a sextic in Maschke normal form, and conversly, a sextic in Maschke normal form can be transformed by a cubic Tschirnhaus transformation into the Joubert normal form,

$$Q_i = -\Phi_i^3 + 3F_8\Phi_i - 2F_{12}, \tag{4.101}$$

where Φ_i are the roots of Φ and Q_i are the roots of the sextic in Joubert form. Then the incredible geometric interpretation is the following

Theorem 4.3.5 ([C2], (33)) *The quadratic and cubic Tschirnhaus tranformations just mentioned are inverses, mapping the Joubert normal form to the Maschke normal form and vica versa. Geometrically these are the birational maps φ and d given by the Jacobian ideals of the Segre cubic S_3 and the Igusa quartic $\mathcal{I}_4$.*

That is, thinking of Q_i (respectively Φ_i) as homogenous coordinates on $\mathbb{P}^5$ with $\sum Q_i = 0$ (respectively on $(\mathbb{P}^5)^\vee$ with $\sum \Phi_i = 0$), then the birational maps φ and d are given by the Tschirnhaus transformations, φ by a quadratic one and d by the cubic one (4.101).

Sketch of proof of 4.3.3 Let $\sigma_i(\bar{a})$ be the elementary symmetric functions of the coefficients $\bar{a},\ldots,\bar{f}$, and consider the Tschirnhaus transformation (4.101). Here this is:

$$\bar{\alpha} = \sigma_3(\bar{a}) - \sigma_2(\bar{a})\bar{a} - 2\bar{a}^3,$$
$$\vdots \quad \vdots \quad \vdots \tag{4.102}$$
$$\bar{\zeta} = \sigma_3(\bar{a}) - \sigma_2(\bar{a})\bar{f} - 2\bar{f}^3,$$

Then the usual relation $\bar{\alpha}+\bar{\beta}+\bar{\gamma}+\bar{\delta}+\bar{\varepsilon}+\bar{\zeta} = 0$ holds. Coble defines 40 irrational invariants of the cubic surface corresponding to the 40 complexes of triples of azygetic double sixes as described in (4.2) and (6.2). "Irrational invariant" means it is defined by homogenous polynomials with coefficients in the splitting field of the Galois extension defining the equation for the 27 lines and not in the field over which the cubic surface is defined. In terms of the six points given by (x,y,z,t,u) as in 4.3.1, these are given by expressions like (the notation (ijk) for the 3×3 minors of (3.84) was explained there)

$$\gamma_{123,456} = d_2(123)(456) = d_2(xt - yz + u(y - t - x + z)), \tag{4.103}$$

and so is of degree 3 in the coordinates of each point p_i. This yields $\frac{1}{2}\binom{6}{3} = 10$ such invariants. There are 30 others given by expressions

$$\gamma_{ij,kl,mn} = (ikl)(jkl)(kmn)(lmn)(mij)(nij),$$

once again cubic in the variables. For these one has

$$\gamma_{12,34,56} + \gamma_{12,56,34} = \overline{\alpha} + \overline{\delta},$$

and permutations. The important change of variables, now *linear*, is given by

$$
\begin{aligned}
x_a &= \overline{\alpha} + (\rho - \rho^2)d_2\overline{a}, \ldots, x_f = \overline{\zeta} + (\rho - \rho^2)d_2\overline{f}, \\
u_a &= \overline{\alpha} + (\rho^2 - \rho)d_2\overline{a}, \ldots, u_f = \overline{\zeta} + (\rho^2 - \rho)d_2\overline{f},
\end{aligned}
\tag{4.104}
$$

These again fulfill $x_a + \ldots + x_f = 0 = u_a + \ldots + u_f$, and the 40 invariants above take the form

$$
\begin{aligned}
2(\rho - \rho^2)\gamma_{123,456} &= x_a + x_b + x_c - u_a - u_b - u_c, \\
2(\rho - \rho^2)\gamma_{12,34,56} &= \rho x_a - \rho^2 x_d - \rho^2 u_a + \rho u_d, \\
2(\rho - \rho^2)\gamma_{12,56,34} &= -\rho^2 x_a + \rho x_d + \rho u_a - \rho^2 u_d,
\end{aligned}
\tag{4.105}
$$

and permutations of this. Moreover, he defines 45 *tritangent invariants*, which have the property of vanishing when the surface defined by the six points aquires an Eckard point. They are given in terms of the x_a and u_a as follows:

$$
\begin{aligned}
2t_{12,34,56} &= x_a - x_d, \\
2\tau_{12,34,56} &= u_a - u_d, \\
2t_{1256,4} &= -\rho(x_b + x_c) + \rho^2(x_e + x_f), \\
2\tau_{1256,4} &= -\rho^2(u_b + u_c) + \rho(u_e + u_f).
\end{aligned}
\tag{4.106}
$$

Since the variables $(x_a, \ldots, x_f)$ and $(u_a, \ldots, u_f)$ are invariant under permutations, to check that these coordinates describe the action of Burkhardt's group it suffices to determine what the action of the Cremona transformation A_{123} (which exchanges the double sixes N and N_{123}) is on these variables. It is easily checked that A_{123} acts on these variables as:

$$
\begin{aligned}
6x'_a &= (-3\rho^2 + \rho)u_a + (3\rho^2 + \rho)(u_b + u_c) + (u_d + u_e + u_f), \\
6x'_b &= (-3\rho^2 + \rho)u_b + (3\rho^2 + \rho)(u_a + u_c) + (u_d + u_e + u_f), \\
6x'_c &= (-3\rho^2 + \rho)u_c + (3\rho^2 + \rho)(u_a + u_b) + (u_d + u_e + u_f), \\
6x'_d &= (u_a + u_b + u_c) + (-3\rho + \rho)^2 u_d + (3\rho + \rho^2)(u_e + u_f), \\
6x'_e &= (u_a + u_b + u_c) + (-3\rho + \rho)^2 u_e + (3\rho + \rho^2)(u_d + u_f), \\
6x'_f &= (u_a + u_b + u_c) + (-3\rho + \rho)^2 u_f + (3\rho + \rho^2)(u_d + u_e),
\end{aligned}
\tag{4.107}
$$

Similarly, the images under A_{123} of the u_a can be expressed in terms of the x_a, with coefficients which are the conjugates of the coefficients above. As one sees from this, under the action of $W(E_6)$ on the space $\mathbb{P}^9 = \{x_a, u_a, \sum x_a = 0 =$

$\sum u_a\}$, the element A_{123} exchanges the x and the u coordinates, so the two skew $\mathbb{P}^4$'s given by

$$\mathbb{P}^4 = \{u_a = \cdots = u_f = 0\}, \quad (\mathbb{P}^4)^\wedge = \{x_a = \cdots = x_f = 0\}$$

are projective spaces which are exchanged under *odd* elements of $W(E_6)$, but fixed by $G_{25,920}$. Furthermore, a calculation shows that

$$\sum_{40} \gamma_{123,456}^2 = 2 \sum x_a u_a = 4I_6, \tag{4.108}$$

where I_6 is the invariant (B.56) of the hexahedral form which corresponds to the invariant i_8 of the quatenary cubic. From this it follows that under $G_{25,920}$, the variables x_a and u_a are contragredient, and odd elements of $W(E_6)$ exchange $\mathbb{P}^4$ and $(\mathbb{P}^4)^\wedge$, but have the invariant (4.108). This state of affairs is often referred to by saying $W(E_6)$ is a correlation group acting on $\mathbb{P}^4$, with $G_{25,920}$ as the invariant collineation group. The collineation group is now easily identified with the Burkhardt group, and 4.3.3 is established. □

The Burkhardt group will be studied more thoroughly in the next chapter. We just remark that in the coordiantes x_a, the Burkhardt invariant J_4 (4.82) becomes

$$J_4 = \sum x_a x_b x_c x_d, \quad \sum x_a = 0. \tag{4.109}$$

It is now a straightforward matter to relate the hexahedral invariants to the invariants of the Burkhardt group. This is done by utilizing (4.103) and (4.104) to change variables from the x_a to the $\bar{a}, \ldots, \bar{f}$. The result is

Theorem 4.3.6 ([C1], III (42)) *Any invariant of the Burkhardt group of total degree $2k$ in x and u, or the sum or difference of two dual invariants of total degree $2k$, is an invariant of the hexahedral form of degree $6k$ in $\bar{a}, \ldots, \bar{f}$, which is rational in $I_6, \ldots, I_{30}$.*

Step 3: Reduction to the Burkhardt form problem.

Now one must show how a solution of the Burkhardt form problem (see Definition B.1.15) gives a solution of the equation problem for $G_{6,2}$ (Definition 4.1.17). Let $p = (p_a, \ldots, p_f)$ be a solution, in the coordinates x_a, of the Burkhardt form problem. Then the factor of proportionality λ in the x_a is determined by the value of J_6/J_4, at least up to sign. It remains to transfer this solution to one in $\mathbf{P}_6^2$, i.e., to find the corresponding point (x, y, z, t, u). For this one requires linear covariants, as in (4.98) above. Let $J_{0,4}$ be the dual form of $J_4 = J_{4,0}$, where here the two indices indicate degree in the x_a and in the u_a. This is the invariant of the Burkhardt group acting on the u variables of degree four; explicit formula in the y_i variables (see 4.3.1) and their dual coordinates v_i are

$$J_{4,0} = y_0^4 + 8y_0(y_1^3 + \cdots + y_4) + 48y_1 y_2 y_3 y_4,$$

$$J_{0,4} = v_0^4 + v_0(v_1^3 + \cdots v_4^3) + 3v_1 v_2 v_3 v_4.$$

Let $\mathcal{P}_{3,1} := \Delta_v^3 J_{0,4}$ be the cubic polar (Definition B.1.6, see also Lemma B.2.2). Then acting with $\mathcal{P}_{3,1}$ on any invariant $J_{r,0}$ yields a linear covariant:

$$\mathcal{P}_{3,1} \vdash J_{r,0} = J_{r-3,1}. \tag{4.110}$$

Proposition 4.3.7 ([C1], (45)) *The invariants $J_4,\ldots,J_{18}$ of the Burkhardt group yield by (4.110) linear covariants, linearly independent in v, and with determinant J_{45}.*

Combining this with Theorem 4.3.6, we get, from the invariants $J_{r,0}$ and covariants $J_{r-3,1}$, the following sets of invariants of the hexahedral form (i.e., in the coordinates $\bar{a},\ldots,\bar{f}$):

- $I_{12}', I_{18}', I_{30}', I_{36}', I_{54}',$
- $I_6'', I_{24}'', I_{30}'', I_{42}'', I_{48}''.$

$$\tag{4.111}$$

Consider the following system of equations, linear in the u_a once values for $I_6'',\ldots,I_{48}''$ have been choosen:

$$J_{1,1} = I_6'', \quad J_{7,1} = I_{24}'', \quad \ldots, \quad J_{15,1} = I_{48}''; \tag{4.112}$$

together with the relation $\sum u_a = 0$, the system (4.112) can be *solved* for u_a, up to the same factor λ as occured in the ratios of the x_a. The reduction from the equation problem for $G_{6,2}$ to the Burkhardt form problem is as follows:

1) The given values of $I_6,\ldots,I_{30}$ (the equation problem for $G_{6,2}$) are the values for $I_{12}',\ldots,I_{54}'$ of (4.111), and these are the values of $J_4,\ldots,J_{18}$ for the Burkhardt form problem.

2) A solution is a point $(x_a,\ldots,x_f) \in \mathbb{P}^4$, a factor of proportionality λ is fixed by the ratio J_6/J_4.

3) The values of $I_6'',\ldots,I_{48}''$ are the values of the linear covariants of the Burkhardt group, determined as in (4.110), and the values of $u_a,\ldots,u_f$ are found as the solution of (4.112), up to the same factor of proportionality λ.

4) From the equations (4.104) we get equations

$$x_a - u_a = 2(\rho - \rho^2)d_2\bar{a}, \ \ldots, x_f - u_f = 2(\rho - \rho^2)d_2\bar{f},$$

$$x_a + u_a = 2\bar{a}, \ \ldots, x_f + u_f = 2\bar{\zeta}.$$

The first equations determine $\bar{a},\ldots,\bar{f}$ up to a factor μ, and the second set of equations then determines the value of μ^3 rationally. Moreover, $d_2\mu$ is rationally determined (from (4.17) upon replacing $\bar{a}$ by $\mu\bar{a}$, etc.). But μ need not be determined, as it drops out in the next step.

5) The sought for (x, y, z, t, u) is given by replacing $\bar{a}, \ldots, \bar{f}$ by $\mu\bar{a}, \ldots, \mu\bar{f}$ and ρ by $\mu\rho$ in the following set of equations,

$$6ux = \rho + 3(-\bar{a} - \bar{b} + \bar{c}),$$
$$6uy = \rho + 3(-\bar{d} - \bar{e} + \bar{f}),$$
$$6uz = \rho + 3(-\bar{d} + \bar{e} - \bar{f}),$$
$$6ut = \rho + 3(\bar{a} + \bar{b} - \bar{c}),$$
$$6uu = \frac{(\rho + 3(-\bar{d} + \bar{e} - \bar{f}))(\rho + 3(-\bar{d} - \bar{e} + \bar{f}))}{(\rho + 3(\bar{a} - \bar{b} - \bar{c}))},$$
$$\rho = 6\frac{(\bar{b}\bar{c} + \bar{c}\bar{a} + \bar{a}\bar{b} - \bar{e}\bar{f} - \bar{f}\bar{d} - \bar{d}\bar{e}) + d_2}{(\bar{a} + \bar{b} + \bar{c} - \bar{d} - \bar{e} - \bar{f})}.$$

These equations are a solution of the form problem for Σ_6.

At this point it is legitimate to ask whether anybody in his or her right mind is going to start the bewildering calculations involved in the reduction, requiring not only the x_a variables but also the u_a. Probably not, but in this age of computer algebra, it wouldn't be all that difficult. However, the reduction of the problem to the Burkhardt form problem is what we were after, as this will give an alternative solution which will allow us to associate to a given cubic surface *not only* a genus 2 curve (as was the case with Klein's solution), but also the local geometry of the Burkhardt quartic at the moduli point. So the equation for the 27 lines has led us naturally to the object of the next chapter, the unique degree 4 invariant of $G_{25,920}$ acting on $\mathbb{P}^4$.

Chapter 5

The Burkhardt quartic

In this chapter we take a glimpse at the geometry of a very special Janus-like variety (the variety $\overline{X(3)}$ or $\overline{Y(2)}$ of section 2.4), which possesses an embedding as a *hypersurface* in $\mathbb{P}^4$ of degree $d = 4$, the Burkhardt quartic $\mathcal{B}_4$, whose equation was written down for the first time by Burkhardt [Bu] in 1888, but which has recently drawn a lot of attention from algebraic geometers for a number of reasons. As we shall see, in this case the *boundary varieties* (of the Baily-Borel compactification $X(3)^*$ or $Y(2)^*$) and the *modular subvarieties* are given essentially by special hyperplane sections. Some of the very special properties of $\mathcal{B}_4$ (which will also be denoted $\mathcal{B}$ in the sequel) are:

- $\mathcal{B}_4$ is the unique invariant of degree $d = 4$ for the unitary reflection group $G_{25,920}$ acting on $\mathbb{P}^4$. Hence the symmetry group of $\mathcal{B}_4$ is $G_{25,920}$.

- $\mathcal{B}_4$ has 45 ordinary double points, which characterize $\mathcal{B}_4$ among nodal quartic hypersurfaces of $\mathbb{P}^4$, 45 being the maximal number of ordinary double points which a quartic hypersurface in $\mathbb{P}^4$ can have (Varchenko bound).

- The 45 nodes of $\mathcal{B}_4$ are in covariant (with respect to $G_{25,920}$) relation to the 45 tritangents of a smooth cubic surface.

- There are 40 special hyperplane sections $H \cap \mathcal{B}_4$ which split into a tetrahedron of hyperplanes (such a tetrahedron is also a quartic), yielding 40 linear $\mathbb{P}^2$'s contained in $\mathcal{B}_4$. The dual variety $\check{\mathcal{B}}_4$ has 40 singular lines, dual to the 40 $\mathbb{P}^2$'s, and 45 *tropes*, dual to the 45 double points.

- The Hessian variety $\mathbf{Hess}(\mathcal{B})$, of degree $d = 10$, meets $\mathcal{B}_4$ in the union of the 40 $\mathbb{P}^2$'s lying on $\mathcal{B}_4$.

- $\mathcal{B}_4$ is a self-Steinerian variety, that is, $\mathcal{B}_4$ is its own Steinerian (see Definition B.1.10), hence there is a finite map $\mathbf{Hess}(\mathcal{B}) \longrightarrow \mathcal{B}_4$, which has degree ten.

- There are 45 hyperplane sections which are "most singular" K3-surfaces – they are desmic surfaces, the Kummer surface of $E_\varrho \times E_\varrho$, where E_ϱ is the elliptic curve with modulus $\varrho = e^{\frac{2\pi i}{3}}$.

- There are 216 hyperplane sections which are copies of the Hessian variety of the Clebsch cubic diagonal surface $\mathfrak{C}$ (see Figure 4.6), which has symmetry group Σ_5; $\mathbf{Hess}(\mathfrak{C})$ is a special *symmetroid*, and it is the unique symmetroid with symmetry group Σ_5.

We will see how these subloci and properties are tied in with the structure of $\mathcal{B}_4$ as arithmetic quotient in the two guises: a ball quotient on the one hand and a Siegel space quotient on the other. The 40 special hyperplane sections turn out to be modular subvarieties in the one guise and boundary components in the other, and the moduli interpretation of some of the special sections is derived.

All of this geometry has a bearing on the solution of the equation of degree 27 for the lines on a cubic surface, a relationship discovered by Coble. In the last chapter we discussed the reduction to the form problem for the Burkhardt group, and in this chapter this problem will be solved by means of the moduli interpretation in terms of genus 2 hyperelliptic curves.

The other intention of this chapter is to give a complete proof of the second interpretation of $\mathcal{B}_4$ as an arithmetic quotient, namely as the Satake compactification $Y(2)^*$, a fact which has been known since 1990. A proof using theta functions was given in [Ge], while we here give a direct proof without theta functions, which incidently was also the original proof. We start just with the variety $\mathcal{B}_4$, together with all the geometry (subvarieties, singular locus, etc.), and using Yau's theorem we show that $\mathcal{B}_4$ is the compactification of a ball quotient, for some arithmetic group $\Gamma \subset U(3,1)$ with torsion. More precisely, we show that it suffices to prove certain numerical results on the modular subvarieties to be able to conclude that $\mathcal{B}_4$ itself is a ball quotient, and this is what we do. Then we identify the group Γ as $\Gamma_{\sqrt{-3}}(2) \subset U(3,1;\mathcal{O}_K)$ by applying the results of [J] on the arithmetic quotient $Y(2)$.

5.1 Burkhardt's quartic primal

Our interest in this chapter is with the invariant J_4 of degree 4 (4.82), the famous *Burkhardt quartic*. Calculating J_4 in terms of the variables $[y_0 : .. : y_4]$, which are related to Burkhardt's variables Y_i (which we denoted above by y_i) by $(y_0, y_1, y_2, y_3, y_4) = (-Y_0, 2Y_1, 2Y_2, 2Y_3, 2Y_4)$, J_4 is given by the equation

$$\mathcal{B} := \{y_0^4 - y_0(y_1^3 + y_2^3 + y_3^3 + y_4^3) + 3y_1y_2y_3y_4 = 0\}. \qquad (5.1)$$

This form of the equation is not completely symmetric: it singles out the hyperplane $\mathcal{S} = \{y_0 = 0\}$ (see (4.78)) in $\mathbb{P}^4$, and is the equation written with respect to this hyperplane. A much more symmetric form of the equation can be given as follows ([Ba2], p.39): set

$$
\begin{aligned}
x_0 &= y_0 + y_1 + y_4 & -x_3 &= y_0 + y_2 + y_3 \\
x_1 &= y_0 + \varrho y_1 + \varrho^2 y_4 & -x_4 &= y_0 + \varrho y_2 + \varrho^2 y_3 \,. \\
x_2 &= y_0 + \varrho^2 y_1 + \varrho y_4 & -x_5 &= y_0 + \varrho^2 y_2 + \varrho y_3
\end{aligned}
\tag{5.2}
$$

Here $(x_0 : ... : x_5)$ are homogenous coordinates in $\mathbb{P}^5$, and obviously $\sum x_i = 0$. To get the equation in terms of the x_i note that

$$
\begin{aligned}
3y_0 &= x_0 + x_1 + x_2, & -3y_0 &= x_3 + x_4 + x_5, \\
x_0 x_1 x_2 &= y_0^3 + y_1^3 + y_4^3 - 3y_0 y_1 y_4, & -x_3 x_4 x_5 &= y_0^3 + y_2^3 + y_3^3 - 3y_0 y_2 y_3, \\
x_1 x_2 &+ x_2 x_0 + x_0 x_1 = 3(y_0^2 - y_1 y_4), \\
x_4 x_5 &+ x_5 x_3 + x_3 x_4 = 3(y_0^2 - y_2 y_3);
\end{aligned}
\tag{5.3}
$$

hence

$$
\begin{aligned}
y_0^4 - y_0(y_1^3 + y_2^3 + y_3^3 + y_4^3) + 3 y_1 y_2 y_3 y_4 &= -y_0(y_0^3 + y_1^3 + y_4^3 - 3 y_0 y_1 y_4) \\
&= -y_0(y_0^3 + y_2^3 + y_3^3 - 3 y_0 y_2 y_3) \\
&\quad + 3(y_0^2 - y_1 y_4)(y_0^2 - y_2 y_3) \\
&= \frac{1}{3}(x_3 + x_4 + x_5) x_0 x_1 x_2 + \frac{1}{3}(x_0 + x_1 + x_2) x_3 x_4 x_5 \\
&\quad + \frac{1}{3}(x_1 x_2 + x_2 x_0 + x_0 x_1) \cdot (x_4 x_5 + x_5 x_3 + x_3 x_4) \\
&= \frac{1}{3} \sigma_4(x_0, .., x_5)
\end{aligned}
$$

where σ_4 is the 4th elementary symmetric polynomial in the x_i. Hence $\mathcal{B}$ is defined in $\mathbb{P}^5$ by the completely symmetric equations (cf. (4.109)):

$$
\mathcal{B} = \left\{ \begin{array}{l} \sigma_1(x_0, ..., x_5) = 0 \\ \sigma_4(x_0, ..., x_5) = 0 \end{array} \right\}.
\tag{5.4}
$$

Writing the equations in this form one "sees" the action of Σ_6, which is itself a subgroup of $G_{25,920}$, on $\mathcal{B}$: by permutation of coordinates. Let us now sketch some of the beautiful geometry of $\mathcal{B}$.

First of all $\mathcal{B}$ has 45 nodes (ordinary double points), for which we shall use the following notation ($\varrho = e^{\frac{2\pi i}{3}}$):

15 nodes $(ij) = (x_1, ..., x_6)$; $x_i = 1$, $x_j = -1$, $x_k = 0$, $k \neq i, j$.
30 nodes $(ij.kl.mn) = (x_1, ..., x_6)$: $x_i = x_j = 1$, $x_k = x_l = \quad$ (5.5)
ϱ, $x_m = x_n = \varrho^2$.

In this notation $(ij.kl.mn) = (ji.kl.mn) = (kl.mn.ij)$, etc., $(ij) = (ji)$. The most fundamental subsets of these 45 nodes are, as we are working in $\mathbb{P}^4$, sets of five nodes not all lying in a $\mathbb{P}^3$, i.e., spanning $\mathbb{P}^4$. There are in fact 27 such sets, called *Jordan pentahedra*, and as the numbers already suggest these correspond to the 27 lines, and similarly, the 45 double points correspond to the 45 tritangents, of a smooth cubic surface. The precise relationship is as follows.

The 45 nodes are joined by lines containing two (respectively three) of them, which are denoted ε-lines, each containing two nodes (respectively κ-lines, each containing three nodes). One sees from the coordinates for the nodes (5.5) that no more than three can lie on a line. The 27 Jordan pentahedra are spanned by sets of five nodes, all of which are joined by ε-lines, such as the set:

$$(ij), \quad (kl), \quad (mn), \quad (ij.kl.mn), \quad (ij.mn.kl). \tag{5.6}$$

Thus these five nodes are the vertices of a coordinate simplex, and the ten ε-lines joining them in pairs are the edges (one-dimensional faces) of the simplex. The two-dimensional faces of the pentahedron, containing three nodes and three ε-lines, are called *f-planes* and intersect $\mathcal{B}$ in a three-nodal (rational) quartic curve:

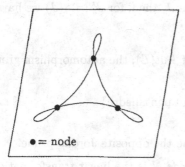 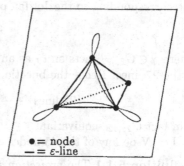

Figure 5.1: (a) f-plane   (b) coordinate tetrahedron in Jordan Prime

The five sides (three-dimensional faces) of the simplex are so-called *Jordan primes*, containing four of the five nodes, six of the ε-lines, and four of the f-planes of the pentahedron of reference. For each of the nodes of the simplex there is a Jordan prime which is the opposite side of the simplex. These are given by the equations:

node=(ij), opposite face: $J_{ij} = \{x_i - x_j = 0\}$,
node=$(ij.kl.mn)$, opposite face: $J_{(ij.kl.mn)} = \{x_i + x_j + \varrho^2(x_k + \quad$ (5.7)
$x_l) + \varrho(x_m + x_n) = 0\}$.

The intersection of this Jordan prime with $\mathcal{B}$ contains, in addition to these four, two other sets of four nodes, each itself the set of vertices of a (different) coordinate simplex. The intersection $\mathcal{B} \cap J_{ij}$ is therefore a 12-nodal quartic surface in $\mathbb{P}^3$, and is in fact one of the famous "desmic surfaces", so called because of the triad of desmic tetrahedra whose vertices are the nodes (see Figure 5.4). See section B.5.2.1 for some facts about these surfaces. We will have more to say about this surface later.

The relation to the 27 lines and 45 tritangents of the cubic surface can now be formulated: The 45 nodes correspond to the 45 tritangents in such a way that:

> *two nodes are joined by an ε−line (i.e., are the vertices of a*
> *Jordan pentahedron)*
> $\Longleftrightarrow$ (5.8)
> *the corresponding tritangents meet in one of the 27 lines.*

Hence the five nodes spanning a Jordan pentahedron correspond to the five tritangents passing through a fixed line.

Now that the relation to the 27 lines is known we can consider covariance. The finite group $W(E_6)$ acts on the set of 27 lines, hence so does $G_{25,920}$, and this action coincides with the action of $G_{25,920}$ on $\mathcal{B}$ under the above correspondence. Let a Jordan pentahedron J be given, and let $\gamma \in G_{25,920}$ be any element. Then $\gamma(J)$ is another of the Jordan pentahedra, and if ℓ_J denotes the line corresponding to the Jordan pentahedron J, then, for $J' := \gamma(J)$ we have

$$\gamma(\ell_J) = \ell_{J'},$$

where $\gamma \in G_{25,920}$ acts on ℓ_J as an element of $\mathrm{Aut}(\mathcal{L})$, the automorphism group of the 27 lines. Hence the bijection

$$\{27 \text{ lines}\} \longleftrightarrow \{27 \text{ Jordan pentahedra}\}$$

is in fact $G_{25,920}$-equivariant.

Let N be any of the 45 nodes and let J be the opposite Jordan prime.

Definition 5.1.1 The *projection from the node N* is the linear transformation

$$p(N) : x_i \mapsto x_i - \frac{1}{3} n_i \cdot J(x_1, \dots, x_6),$$

where $N = (n_1, \dots, n_6)$ in the coordinates $(x_1, \dots, x_6)$, $\sum x_i = 0$ on $\mathbb{P}^4$.

It is readily verified that for all nodes N, $p(N) \in G_{25,920}$, and they in fact generate $G_{25,920}$. The transformation $p(N)$ has the following effect:

1) $p(N)$ preserves N and J.

2) For any κ−line through N, let N_1, N_2 denote the other two nodes on the κ−line. Then $p(N)(N_1) = N_2$ and $p(N)(N_2) = N_1$.

These properties also determine a unique symmetry σ, such that $\sigma \in G_{25,920}$ satisfies 1) and 2) (for a fixed N).

5.2   Subloci on $\mathcal{B}_4$

5.2.1   The configuration of the 45 nodes

As mentioned above, we are concerned with the *dual* of the arrangement given in Table 4.3, which we call a *configuration*. It consists of a finite number of points together with the set of linear subspaces of dimension $j-1$ containing $\geq j$ of the points. Hence referring to Table 4.3, we have (a beautifully written description is also given in the first section of [Td1]):

- points: 45, dual to the 45 $t(3)$. We shall also refer to these 45 points as the *nodes* of the configuration.

- lines: 270 $\varepsilon-lines$, dual to the $t_2(2)$, each containing two of the nodes; 240 $\kappa-lines$, dual to the $t_3(2)$, each containing three of the nodes.

- planes: 270 $f-planes$, dual to the $t_3(1)$, each containing three of the nodes; 720 $d-planes$, dual to the $t_4(1)$, each containing four of the nodes; 540 $c-planes$, dual to the $t_6(1)$, each containing six of the nodes; 40 $j-planes$, dual to the $t_9(1)$, each containing nine of the nodes.

- primes[1]: 540 $x-primes$, dual to the t_7, each containing seven of the nodes; 216 $n-primes$, dual to the t_{10}, each containing 10 of the nodes; 45 *Jordan primes*, dual to the t_{12}, each containing 12 of the nodes; 40 *Steiner primes*, dual to the t_{18}, each containing 18 of the nodes.

Our first purpose will be to describe the geometry of each of the loci above.

I **Lines.** The 270 $\varepsilon-$lines are the one-dimensional simplices of the Jordan pentahedra introduced above. The 240 $\kappa-$lines, on the other hand, contain each three nodes of *different* Jordan pentahedra. In this way, one can define the notion of *polar* $\kappa-$line.

Definition 5.2.1 Let L be a given $\kappa-$line, containing the three nodes a_1, a_2, a_3. Let J_{a_i} denote the Jordan prime opposite the node a_i. Then the line

$$L^\kappa := J_{a_1} \cap J_{a_2} \cap J_{a_3}$$

is the *polar* $\kappa-$line, i.e., L and L^κ are polar.

Of course L and L^κ are skew.

If a node N is given, corresponding to a tritangent T, then other 44 nodes split up into $12+32$, the 12 corresponding to the tritangents which intersect T in a *line on the cubic surface*, the 32 corresponding to the other cases, which meet T in a line *not* on the cubic surface. In the situation here each of the former 12 lies on a unique $\varepsilon-$line through N, while the other 32 lie two at a time on 16 $\kappa-$lines through N. Thus there are 12 $\varepsilon-$lines and 16 $\kappa-$lines through N. These numbers are those of the second column, second row of Table 4.3. N is dual to a prime P which contains 12 (resp. 16) of the $t_2(2)$ (resp. of the $t_3(2)$).

---

[1] the word *prime* is a rather old-fashioned term for hyperplane, and *primal* is similarly an old-fashioned term for hypersurface

II **Planes.** The 270 f–planes are the two-dimensional simplices of the Jordan pentahedra above; each contains three nodes and the three ε–lines joining these in pairs. The 720 d–planes each connect a κ–line with a node a_i on its polar L^κ; each contains four nodes, a κ–line and the three ε–lines joining a node on the κ–line with the fourth node. These two planes are pictured below, where the ε–lines are dashed and the κ–line is solid.

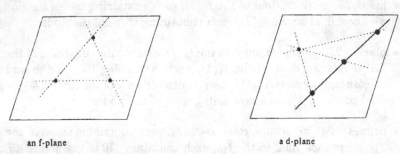

an f-plane   a d-plane

The 540 c–planes each contain six nodes, which lie in threes on four κ–lines which form a complete quadrilateral in the c–plane; in addition each contains three ε–lines, the "diagonals" of the quadralateral. This is depicted in Figure 5.2. The 40 j–planes each contain a copy of the Hesse-configuration: the nine nodes contained in each j–plane are the base points of the Hesse pencil. There are therefore 12 lines in this plane, each of which contains three of the nine nodes, i.e., is a κ–line. The κ–lines meet in twos in 12 further points, and four at a time at the nodes, forming the well known figure of four triangles which are the degenerate cubics of the Hesse pencil splitting into three lines apiece. Schematically one has the Figure 5.3, where three of the κ–lines are drawn together with the nine nodes lying on them, and the other nine κ–lines have to be imagined, three through each node, each line containing three nodes.

If a node N is given, the number of planes which contain N is given by the numbers in the third column and second row of Table 4.3; there are 18 f–planes, 64 d–planes, 72 c–planes and 8 j–planes containing N. The 8 j–planes in particular split into two sets of four, each set consisting of *skew* planes (intersecting only at the node), members of different sets intersecting in *lines* through the node.

III **Primes.** Consider a fixed c–plane; we will see below that it is the intersection of a unique Jordan prime J and two Steiner primes, a fact which can also be deduced from the last column and seventh row of Table 4.3, which states that each $t_6(1)$ contains a unique t_{12} and two t_{18}. Let N_J be the node opposite to J. Then the prime spanned by the fixed c–plane and the node N_J is an x–prime, containing seven of the nodes, six in the c–plane and the seventh being N_J. The c–plane contains three ε–lines, and the x–prime contains these as well as the six which join each node of the c–plane with N_J. For each of the κ–lines in the c–plane, the plane in

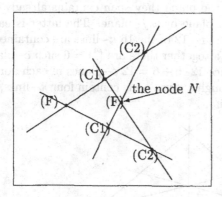

Each node is marked by the tetra-hedron F, $C1$ or $C2$ of the three desmic tetrahedra which contains it;
(F) is the tetrahedron of reference. The "diagonals" of the complete quadrilateral are the lines joining vertices with similar markings.

Figure 5.2: A c-plane

the x−prime joining it with N_J is a d−plane, and for each ε−line in the c−plane, the plane joining it with N_J is a f−plane. Hence an x−prime contains 7 nodes, 4 κ−lines, 9 ε−lines, 3 f−planes, 4 d−planes and 1 c−plane. The 216 n−primes each contain ten nodes; these lie on ten κ−lines, and these form the vertices and edges of a pentahedron of c−planes. We recognize this configuration as a Sylvester pentahedron (see Theorem 4.1.1), where the c−planes are the planes $y_i = 0$ for the pentahedral form of a cubic surface. We will also see later that the stabilizer, in $G_{25,920}$, of a n−prime is the icosahedral group of order 120, and it consequently acts on this pentahedron. The conclusion is that the $5 \cdot 3$ diagonals of the quadralaterals in the c−planes are the 15 lines lying on the Clebsch cubic, which give rise to its nick-name "diagonal cubic" (see the remark following Proposition 4.1.11). Hence a n−prime contains 10 nodes, 15 ε−lines, 10 κ−lines, 10 d−planes and 5 c−planes. An example is given by the prime $x_6 = 0$ in the coordinates above, which contains the ten nodes

$$(12)\ (13)\ (14)\ (15)\ (23)\ (24)\ (25)\ (34)\ (35)\ (45).$$

The 45 Jordan primes already described above each contain 12 nodes, which are the vertices of a triad of desmic tetrahedra (see section B.5.2.1). In fact, each of these tetrahedra gives rise to a Jordan pentahedron: the fifth vertex is the node which is opposite the given Jordan prime. The $3 \cdot 4 = 12$ faces of the tetrahedra are f−planes; in each we have three ε−lines, and through each such two of the f−planes pass, so there are $12 \cdot 3 \div 2 = 18$ ε−lines in the Jordan prime. Finally, there are κ−lines which join three nodes, each a vertex of one of the tetrahedra; of the 16 κ−lines through each of the nodes (these are the 16 lines of (B.59) 2)), four lie in the given Jordan prime, hence there are $4 \cdot 12 \div 3 = 16$ κ−lines altogether in the Jordan prime. If we fix one of the 12 nodes and choose two of the κ−lines through it, the plane they span contains already two κ−lines, so must be either a c−plane or a j−plane. The latter is easily excluded (as in the Jordan prime, no 12 of the 16 κ−lines are contained in a plane), this is a c−plane, and altogether there are $\binom{4}{2} = 6$ such c−planes passing through each node, hence $12 \cdot 6 \div 6 = 12$ c−planes in each Jordan prime. As the six c−planes through a given node contain four κ−lines, the local arrangement at the node is:

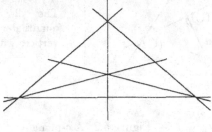

$$(5.9)$$

A Jordan prime contains 12 nodes, 18 ε−lines, 16 κ−lines, 12 f−planes and 12 c−planes.

The 40 Steiner primes each contain 18 of the nodes; these lie on four j−planes in the prime which form a tetrahedron (see Figure 5.3). The edges of the tetrahedron are κ−lines, and opposite edges are in fact *polar* to each other. As each j−plane contains 12 κ−lines, this accounts for $4 \cdot 12 = 48$ such, but the six forming the edges of the tetrahedron have been counted twice, so there are 42 κ−lines in each Steiner prime. There are also 27 ε−lines in such a prime, which can be seen by the following elegant argument due to Baker ([Ba2], p. 29). The three nodes on each κ−line are the third roots of -1. Hence, if $\Delta = XYZT = 0$ is the tetrahedron of j−planes in the Steiner prime, then on each edge, given by the vanishing of two of X, Y, Z, T, say $X = Y = 0$, the nodes fulfill the equation $Z^3 + T^3 = 0$; it follows that the nodes all lie on the cubic surface

$$X^3 + Y^3 + Z^3 + T^3 = 0,$$

and the 27 lines of this cubic surface then join the nodes in pairs; these 27 lines are ε−lines joining the nodes of any of the edges with the three nodes on the opposite edge. These lie three at a time in planes which then contain one κ−line and three ε−lines, i.e., is a d−plane; of these there are also 18, three for each of the edges of the tetrahedron. Consider a plane containing three nodes, which are on three different faces of the tetrahedron; these clearly intersect each of the j−planes containing the given edges in a κ−line, hence each such plane contains three κ−lines, hence must also contain a fourth and is a c−plane. There will be one such for each of the 27 ε−lines, in fact orthogonal to it. Consequently, each Steiner prime contains 18 nodes, 27 ε−lines, 42 κ−lines, 18 d−planes, 27 c−planes and 4 j−planes.

Fixing a node N, the number of primes passing through the node can be read off the last column and third row of Table 4.3; in particular there are 12 Jordan primes and 18 Steiner primes through each node. The Jordan primes intersect two at a time in the 270 f–planes, as already mentioned, and three at a time in the 240 planes which are dual to the 240 κ–lines. There are also two kinds of intersections among the Steiner primes; the intersection is either a j–plane or a c–plane. The Steiner primes meet *four at a time* in the j–planes: each of the four triangles of κ–lines in a j–plane is the face of a tetrahedron of j–planes, hence determines a Steiner prime. They meet *two at a time* in the c–planes: fixing a Steiner prime S, if $n_1, \ldots, n_{18}$ denote the 18 nodes in S, there are $27 (= 45 - 18)$ nodes *not* in S. These 27 nodes determine 27 opposite Jordan primes, and each of these meets S in a c–plane. The 18 Jordan primes opposite to $n_1, \ldots, n_{18}$ on the other hand meet S in a d–plane. All these statements are again easily read off of Table 4.3.

We now recall a few facts (df. [Td1]) on the stabilizers and centralizers of the linear spaces of the configuration.

Locus L	f–plane	d–plane	c–plane	j–plane
$N_G(L)$	Σ_4	$\Sigma_3 \times \mathbb{Z}_2$	$\Sigma_4 \times \mathbb{Z}_2$	G_{648}
$Z_G(L)$	$\mathbb{Z}_2 \times \mathbb{Z}_2$	1	$\mathbb{Z}_2$	$\mathbb{Z}_3$

Locus L	x–prime	n–prime	J–prime	S–prime
$N_G(L)$	$\Sigma_4 \times \mathbb{Z}_2$	Σ_5	G_{576}	G_{648}
$Z_G(L)$	1	1	$\mathbb{Z}_6$	1

This is known for the j–planes, J–primes and S–primes; these are the subgroups denoted G_4, G_5 and G_3, respectively, in [Td1], §7. The centralizers are the groups given by multiplication by roots of unity, see [Bu], p. 195 and p. 191 for this. The case of the n–primes is considered in [Td1], p. 339, the case of x–primes on *loc. cit.*, p. 338. Only the cases of the other planes have not been considered, so we must verify the statements for these. In the case of c–planes, it is clear that the permutation group of the four κ–lines it contains, isomorphic to Σ_4, is a subgroup of the normalizer, and as explained in *loc. cit.*, p. 338, the projection from the node N opposite to the Jordan prime which contains the c–plane fixes the c–plane and the node N. In the case of a d–plane, the symmetric group Σ_3 acts by permutations on the three nodes lying on the κ–line in the d–plane together with the projection from the fourth node (*loc. cit.*, p. 333, 334). The case of f–plane is not so immediate; as Todd shows, the group generated by the projections on the three nodes lying in the f–plane is $(\mathbb{Z}_2)^3$. However, as indicated by Figure 5.4, viewing the f–plane as a face of one of the tetrahedra, there are four κ–lines through the node opposite to the face, and any permutation of these also preserves the face (the f–plane). Note that G_{576} is the projective group of $W(F_4)$; as a reflection group in $\mathbb{P}^3$ it defines an arrangement $\mathcal{A}(F_4)$ (see (3.3)) consisting of two sets of 12 planes, each of which is an arrangement for an inclusion $W(D_4) \subset W(F_4)$. These are of course the 12 c-planes and the 12 f-planes.

5.2.2   Curves on $\mathcal{B}$

We consider now the intersection of $\mathcal{B}$ with the planes mentioned above. For the f-planes this was already described above; the intersection consists of a three-nodal quartic curve. Next note that the κ-lines are all contained in $\mathcal{B}$; indeed, they meet $\mathcal{B}$ in three points of multiplicity 2, hence with total multiplicity 6, hence, since $\mathcal{B}$ has degree 4, they lie on $\mathcal{B}$. For the d-planes, it is clear that the κ-line is part of the intersection, and the remainder must be a cubic. Since this cubic must pass through the fourth node in the d-plane, it must be nodal there. We have the picture

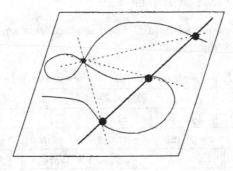

a d-plane

For the c-planes the intersection consists simply of the union of the four κ-lines contained in it, and the j-planes are wholly contained in $\mathcal{B}$, by an argument similar to the above. Each j-plane contains 12 κ-lines, which are all contained in $\mathcal{B}$, hence, letting j denote a j-plane, $j \cap \mathcal{B}$ contains 12 lines, so $j \cap \mathcal{B}$ has degree 12, while, since j is a plane and $\mathcal{B}$ is a quartic, it has degree 4. Hence $j \subset \mathcal{B}$.

We will now consider the primes individually.

5.2.3   Steiner primes

As we have already described, each of the Steiner primes contains a tetrahedron of j-planes. We can see this easily by utilizing the equation (5.1): setting $y_0 = 0$ (this is the equation of a Steiner prime), we have $\{y_0 = 0\} \cap \mathcal{B} = \{y_1 y_2 y_3 y_4 = 0\}$, and this is a tetrahedron of j-planes $\{y_0 = y_i = 0\}$, $i = 1, \ldots, 4$. We have seen above that the 27 ε-lines in the Steiner prime all lie on the cubic surface $C = \{y_1^3 + y_2^3 + y_3^3 + y_4^3 = 0\}$. Note that $\mathrm{Hess}(C) = 6^4 y_1 y_2 y_3 y_4$, and the tetrahedron which is the intersection of $\mathcal{B}$ with $\{y_0 = 0\}$ is the Hessian of the Fermat cubic surface C. So we have

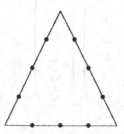 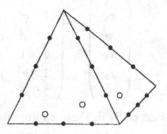

Figure 5.3: (a) 9 nodes in a Jacobi plane (b) 4 faces of tetrahedron of Jacobi planes in Steiner prime

Proposition 5.2.2 *The 40 hyperplane sections of $\mathcal{B}_4$ by a Steiner prime, consisting of a tetrahedron of j-planes, are the Hessian surfaces of the cubic surface in that Steiner prime whose 27 lines are the 27 ε-lines contained in the Steiner prime.*

5.2.4 Jordan primes

Here we have the most interesting case, already alluded to above, in which the hyperplane section of $\mathcal{B}_4$ by a Jordan prime J is a *desmic surface* (see section B.5.2.1). First we show that $J \cap \mathcal{B}$ is in fact a desmic surface. To see this is the case, note that $J \cap \mathcal{B}$ will have 12 nodes at the 12 nodes of $\mathcal{B}_4$ contained in J; these form the vertices of a triad of desmic tetrahedra, and consequently $J \cap \mathcal{B}$ is desmic. Note that the 16 lines, which an arbitrary desmic surface contains, are here the 16 κ-lines in J: since they are on $\mathcal{B}$, they are also contained in $J \cap \mathcal{B}$. To get a nice equation for this surface, we make another change of coordinates from the $x_1, \ldots, x_6$ to $\xi_1, \ldots, \xi_5$ such that $\xi_i = 0$ is the equation of a Jordan prime (see [Ba2], p. 49):

$$
\begin{pmatrix} 0 \\ \xi_1 \\ \xi_2 \\ \xi_3 \\ \xi_4 \\ \xi_5 \end{pmatrix} = \begin{pmatrix} 1 & 1 & 1 & 1 & 1 & 1 \\ 1 & \varrho & \varrho^2 & 1 & \varrho & \varrho^2 \\ 1 & \varrho^2 & \varrho & \varrho^2 & \varrho & 1 \\ 1 & \varrho & 1 & \varrho^2 & \varrho^2 & \varrho \\ 1 & 1 & \varrho & \varrho & \varrho^2 & \varrho^2 \\ 1 & \varrho^2 & \varrho^2 & \varrho & 1 & \varrho \end{pmatrix} \begin{pmatrix} x_1 \\ x_2 \\ x_3 \\ x_4 \\ x_5 \\ x_6 \end{pmatrix};
$$

$$6\begin{pmatrix} x_1 \\ x_2 \\ x_3 \\ x_4 \\ x_5 \\ x_6 \end{pmatrix} = \begin{pmatrix} 1 & 1 & 1 & 1 & 1 & 1 \\ 1 & \varrho^2 & \varrho & \varrho^2 & 1 & \varrho \\ 1 & \varrho & \varrho^2 & 1 & \varrho^2 & \varrho \\ 1 & 1 & \varrho & \varrho & \varrho^2 & \varrho^2 \\ 1 & \varrho^2 & \varrho^2 & \varrho & \varrho & 1 \\ 1 & \varrho & 1 & \varrho^2 & \varrho & \varrho^2 \end{pmatrix} \begin{pmatrix} 0 \\ \xi_1 \\ \xi_2 \\ \xi_3 \\ \xi_4 \\ \xi_5 \end{pmatrix}.$$

The equation of $\mathcal{B}_4$ in these coordinates is

$$\xi_1^2 \xi_2^2 + \xi_2^2 \xi_4^2 + \xi_4^2 \xi_5^2 + \xi_5^2 \xi_3^2 + \xi_3^2 \xi_1^2 + \varrho(\xi_1^2 \xi_4^2 + \xi_2^2 \xi_5^2 + \xi_4^2 \xi_3^2 + \xi_5^2 \xi_1^2 + \xi_3^2 \xi_2^2) = 0. \quad (5.10)$$

Setting $\xi_i = 0$ in this equation gives us the equation of the desmic surface, which we shall denote henceforth by S_{12} (note that all sections $J \cap B$, J a Jordan prime, are isomorphic). So set $\xi_5 = 0$; the equation of the desmic surface in this Jordan prime is then

$$S_{12} = \{\xi_1^2 \xi_2^2 + \xi_1^2 \xi_3^2 + \xi_2^2 \xi_4^2 + \varrho(\xi_1^2 \xi_4^2 + \xi_2^2 \xi_3^2 + \xi_3^2 \xi_4^2) = 0\} \quad (5.11)$$

Incidently, this also gives us a nice equation for the three-nodal quartic which is the intersection $P \cap B$ for any f–plane P (again all such sections are isomorphic), given by, say $\xi_4 = \xi_5 = 0$,

$$P \cap S_{12} = \{\xi_1^2 \xi_2^2 + \xi_1^2 \xi_3^2 + \varrho(\xi_2^2 \xi_3^2) = 0\}. \quad (5.12)$$

To describe more precisely "which" desmic surface (recall the set of desmic surfaces form a pencil) S_{12} is, we consider the geometry near one of the nodes. Note that the following slight change of variable:

$$x = \varrho^2 \xi_1, \quad y = \varrho^2 \xi_2, \quad z = \xi_3, \quad t = \xi_4, \quad u = \xi_5$$

transforms the equation of B to

$$(y^2 z^2 + x^2 t^2) + \varrho(z^2 x^2 + y^2 t^2) + \varrho^2(x^2 y^2 + z^2 t^2) + u(x^2 + y^2 + z^2 + t^2) = 0,$$

and the desmic surface in the prime $u = 0(= \xi_5)$ is just the equation (B.64) with $(\lambda, \mu, \nu) = (\varrho, \varrho^2, 1)$. By Theorem B.5.6, S_{12} is the Kummer surface of an abelian surface $E_\tau \times E_\tau$. By Corollary B.5.7, the modulus τ is determined by the four points of tangency of a conic and a quartic on the conic, a $\mathbb{P}^1$. To clarify this issue, we again indulge in a bit of geometry – this time of the desmic surface in a fixed Jordan prime, $S_{12} \subset J$.

To get a feeling for the "shape" of S_{12} we project from a node $N \in S_{12}$ onto a $\mathbb{P}^2$ (everything in the Jordan prime). Consider the coordinate simplex N belongs to, call it Σ. Each vertex of Σ is joined by κ–lines to other pairs of vertices belonging to other tetrahedra, which are switched by the projection from N. Hence the entire simplex of reference is left fixed by this projection. The face of the tetrahedron opposite N, say P, is left fixed (on each κ–line there is one "point at infinity", which is also fixed under the involution; this point coincides

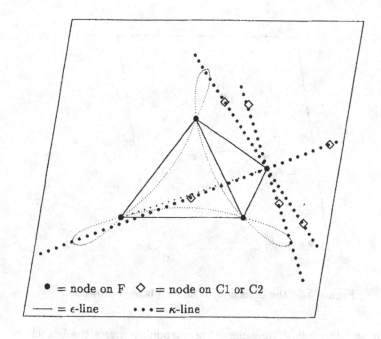

Figure 5.4: The 12 nodes of the desmic surface S_{12}

with the intersection of the κ–line with the face P). Thus P serves as a plane at infinity onto which S_{12} is projected.

We now have a double cover $S_{12} \longrightarrow \mathbb{P}^2$ and we can readily determine the branch locus. Since F is fixed under the involution its intersection with $\mathcal{B}$ (the three-nodal quartic curve (5.12)) is in the branch locus of $S_{12} \longrightarrow \mathbb{P}^2$. Since S_{12} is a K3 surface the total degree of the ramification locus is 6, so there is a conic factor (see Figure 5.5). Stated more rigorously, the cubic polar of N with respect to $\mathcal{B}_4$ divides the polar quadric, which can be checked algebraically (see [Ba2], §7).

Furthermore, the conic and the quartic are tangent at the four points which correspond to the κ–lines through N, and the three double points correspond to the three ε–lines through N. This gives four singularities of type A_3 and three of type A_1 of the branch locus. Note that the conic and quartic in ∞ have precisely the same symmetry as the famous line configuration (see (5.9)) which is here in an exceptional $\mathbb{P}^2$, the blow up of N. The six lines are the proper transforms of the intersection of N with the six c-planes which pass through N. So we see an A_4-symmetry of the branch locus, and this determines it among all such four-tangential (three-nodal-)quartic-conic configurations (a one-dimensional family).

We now describe the elliptic fibration of S_{12} corresponding to the pencil of lines $l_x \in \mathbb{P}^2$ through one of the A_3-points, say p. I am indebted to D. v. Straten

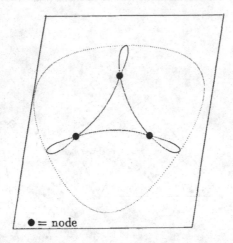

Figure 5.5: the branch locus in the plane of projection

for help in deriving this. The general line through p meets the branch locus in four other points, and after resolution of the A_3 the line doesn't ramify there, so the double cover is indeed an elliptic curve. The singular fibres correspond to:

1. the $\mathbb{P}^1$ tangent to the quartic at p,

2. the three lines joining p with an A_1 point, and

3. the four bitangents through p.

The elliptic fibering therefore has the singular fibres as depicted in Figure 5.6.

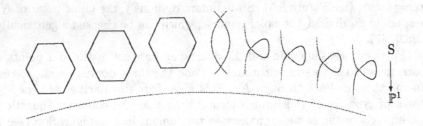

Figure 5.6: Elliptic fiber structure of the desmic surface

This elliptic fibering can also be described as follows: consider the set of planes passing through one of the κ-lines of the desmic surface containing N, say κ_1. There are four such κ-lines passing through N in S_{12}. Choose one of these other than κ_1, κ_2 say. These two determine a c-plane: let, in addition to

F, C_1 and C_2 denote the other two tetrahedra in the Jordan prime (so F, C_1 and C_2 are desmically related). Then since vertices of C_1 and C_2, respectively, lie on κ–lines, they therefore meet in a vertex of F and we have the situation of Figure 5.2.

Now what we have done with κ_2 we can do with the other two κ–lines, so we get $\binom{4}{2}=6$ c–planes through a node (a fact which was already mentioned) and we have a 12_6 configuration, i.e., 12 planes and 12 nodes, each plane containing six nodes, each node lying in six planes. Through κ_1 there are three such c–planes, and we recognize that the 12 c–planes form an arrangement of type $\mathcal{A}(W(\mathbf{D_4}))$; the 16 κ–lines are the $t_3(1)$, and the 18 ε–lines are the $t_2(1)$. Taking the (generic) planes through κ_1, since each contains a κ–line, its residual intersection with S_{12} splits into a line and a cubic, giving an elliptic fibration. The three c–planes through the κ–line yield the three singular fibres of type I_6. Through the κ–line there is one plane tangent to S_{12} there, i.e., whose residual intersection consists of a conic and the κ–line (a so-called "trope"), which yields the singular fibre of type I_2. Finally there are four other planes which are tangent to S_{12}, whose residual intersection is a cubic curve with a node, yielding the four singular fibres of type I_1.

We can now answer the question of the modulus τ.

Proposition 5.2.3 *The desmic surface S_{12} is the Kummer surface of $E_\varrho \times E_\varrho$.*

Proof: Recall that the double cover $E_\varrho \longrightarrow \mathbb{P}^1$ can be characterized by the fact that the four branch points on $\mathbb{P}^1$ are the vertices of an inscribed tetrahedron, hence the cover $E_\varrho \longrightarrow \mathbb{P}^1$ is invariant under a $\Sigma_4 \times \mathbb{Z}_2$, and there is in fact a $\Sigma_4 \times \mathbb{Z}_2 \times \mathbb{Z}_2$ acting on E_ϱ (these are of course not all automorphisms of the abelian variety E_ϱ). This distinguishes the curve E_ϱ among all elliptic curves. We note that our desmic surface has this symmetry after projecting from a node; the four points of tangency of the quartic and the conic in the $\mathbb{P}^2$ correspond to the four points of intersection of the four κ–lines through the opposite node in the Jordan prime. $\qquad\qquad\square$

We will give an independent proof of this result later (Lemma 5.7.6).

5.2.5 n–primes

We now come to a very beautiful case. Recall that the n–primes contain each ten nodes and five c–planes intersecting three at a time at the nodes. A typical example of n–prime is given by $\{x_i = 0\}$ for the coordinates $(x_1, \ldots, x_6)$, $\sum x_i = 0$ used above; from this the Σ_5-symmetry of the prime is evident. It follows from this that the same holds for the section $H_i \cap \mathcal{B}_4$, $H_i = \{x_i = 0\}$, which is a quartic surface in the $\mathbb{P}^3$ given (i.e., in H_i). Let us denote this surface by S_{10} – it has ten nodes at the nodes of $\mathcal{B}_4$ contained in H (H a n–prime, all such sections are isomorphic). Then S_{10} admits Σ_5 as symmetry group, and this will help identify the surface precisely.

We first show that S_{10} is a symmetroid which is the Hessian variety of a cubic surface, or equivalently (by Theorem B.5.17), $S_{10} = \Sigma$, where Σ denotes the Jacobian determinant. This now follows simply from the fact that the symmetroid S_{10} contains ten lines, namely the ten κ−lines in the n−prime. Hence Σ contains ten nodes and Σ and S_{10} coincide by the arguments in the proof of Theorem B.5.17.

This establishes $S_{10} = \mathbf{Hess}(C)$ for some cubic surface C, and since the Hessian is a covariant, C aquires the Σ_5-symmetry, and it follows by Proposition 4.1.10 that in fact $C = \mathfrak{C}$ is the Clebsch cubic.

Proposition 5.2.4 *The 216 hyperplane sections $H \cap \mathcal{B}_4$, H an n−prime, are all isomorphic to* $\mathbf{Hess}(\mathfrak{C})$*, the Hessian variety of the Clebsch cubic $\mathfrak{C}$. The 15 ε−lines in H are 15 of the 27 lines on $\mathfrak{C}$, which are the first 15 listed in Proposition 4.1.12.*

Proof: The first statement has already been verified. The 15 ε−lines are clearly contained in $\mathfrak{C}$, and these are the 15 lines mentioned.   □

Compare this result with Proposition 5.2.2.

5.2.6 x−primes

We now come to the last case of special hyperplane sections of $\mathcal{B}_4$: recall that an x−prime is spanned by a c−plane and the node N opposite the unique Jordan prime containing the given c−plane. It contains seven nodes. To determine more precisely which quartic this is, we again project, this time from N. By arguments as above, the branch locus of the induced double cover $B^\vee \longrightarrow \mathbb{P}^2$ (where B is the section $B = L \cap \mathcal{B}_4$, L an x−prime, and $B^\vee$ denotes B blown up at the node N) contains the four κ−lines in the c−plane (which is identified with the plane of projection). It is of degree 6, so contains a quadric factor in addition. Furthermore it is invariant under Σ_4, and tangent to each of the four κ−lines, and these properties uniquely determine the conic.

5.2.7 $\mathcal{B}_4$ is rational

The Burkhardt quartic is rational. This was first proved by Todd in 1936, and Baker gave in 1942 an explicit rationalization. This rationalization has recently been verified by Finkelnberg in his thesis [Fi], where an explicit modification of $\mathbb{P}^3$ onto $\mathcal{B}_4$ is given. The naked fact that $\mathcal{B}$ is rational is very easy to see. As was pointed out above, of the eight j−planes which pass through a node, there are four in each of the two rulings of a quadric cone. Taking one j−plane and the three κ−lines in it forming one of the triangles, it is easy to show that the three other j−planes through these three κ−lines do not meet each other, i.e. belong pairwise to the same ruling of the tangent cones at the nodes they contain. These three skew planes afford a rationalization: in $\mathbb{P}^4$ there is a unique line through a given point P which meets each of three skew planes. Fixing a $\mathbb{P}^3 \subset \mathbb{P}^4$, for $P \in \mathcal{B}$ the line will meet $\mathbb{P}^3$ in a well-defined point, yielding the rationalization.

The explicit rationalization exhibited by Baker is given by the system of quartics:

$$y_0 = \eta_3[\eta_0(\eta_1^2 + \eta_1\eta_2 + \eta_2^2) + \eta_1\eta_2(\eta_1 + \eta_2) - \eta_3^3]$$
$$y_1 = -\eta_3[(1 - \varrho)\eta_0^2(\eta_1 - \varrho^2\eta_2) - \varrho\eta_0(\eta_1 - \varrho^2\eta_2)^2 - \varrho\eta_1\eta_2(\eta_1 + \varrho\eta_2) + \eta_3^3]$$
$$y_2 = -\eta_1^2\eta_2^2 - \eta_0\eta_1\eta_2(\eta_1 + \eta_2) + (\eta_0 + \eta_1 + \eta_2)\eta_3^3$$
$$y_3 = -\varrho\eta_0^2(\eta_1 - \varrho^2\eta_2)^2 + (1 - \varrho)\eta_0\eta_1\eta_2(\eta_1 - \varrho^2\eta_2) + \eta_1^2\eta_2^2 - \varrho(\eta_1 + \varrho\eta_2)\eta_3^3$$
$$y_4 = \eta_0[\eta_0(\eta_1^2 + \eta_1\eta_2 + \eta_2^2) + \eta_1\eta_2(\eta_1 + \eta_2) - \eta_3^3],$$

where $(y_0 : \dots : y_4)$ are homogenous coordinates on $\mathbb{P}^4$, and $(\eta_0 : \dots : \eta_3)$ are homogenous coordinates on $\mathbb{P}^3$. The modification of $\mathbb{P}^3$ is as follows: Take a set $\mathcal{L}$ of nine lines in $\mathbb{P}^3$ with the following properties:

1) $\mathcal{L}$ consists of three triples of lines;

2) each triple consists of three lines in a pencil;

3) the three planes containing the pencils 2) are distinct; and

4) the nine lines form one closed polygon.

Then, as Finkelnberg shows, $\mathcal{L}$ is the base locus of a linear system of quartics in $\mathbb{P}^3$ of projective dimension 4. Blow up this base locus: $\widehat{\mathbb{P}_1^3} \longrightarrow \mathbb{P}^3$. $\widehat{\mathbb{P}_1^3}$ has the following singularities:

- nine nodes

- three D_4-points.

Now blow up the D_4-points: $\widehat{\mathbb{P}_2^3} \longrightarrow \widehat{\mathbb{P}_1^3}$. Then $\widehat{\mathbb{P}_2^3}$ has 18 nodes (three over each such D_4-point + the original nine), and there is a surjective map:

$$\widehat{\mathbb{P}_2^3} \longrightarrow \mathcal{B}$$

which blows down 27 $\mathbb{P}^1$'s on $\widehat{\mathbb{P}_2^3}$ to nodes. This is the explicit birational map from $\mathbb{P}^3$ to $\mathcal{B}$. We should remark that this birational map is *not* $G_{25,920}$-equivariant.

5.3 $\mathcal{B}_4$ is self-Steinerian

Given a hypersurface $V \subset \mathbb{P}^n$, one has the notion of Hessian variety as well as the closely related notion of Steinerian variety. These are described in the appendix (see Definitions B.1.8 and B.1.10 and the discussion there). The Hessian variety is the locus of points p for which the polar quadric of p with respect to V is *singular*, hence generically a cone with vertex at some point s_p, and the Steinerian is the locus of the points s_p. From this description it follows that the Hessian and Steinerian are birationally equivalent, however the actual *geometry* of the two is very subtle. The degree of the Hessian is determined by the degree of V, but the degree of the Steinerian may vary. It can in fact happen that the variety V *is its own* Steinerian, in which case we say V is *self-Steinerian*. It is rather mysterious what, geometrically, self-Steinerian means, but it is an extremely unusual phenomenon. It turns out that both $\mathcal{B}_4$ and the quintic $\mathcal{I}_5$ considered in the next chapter are self-Steinerian, but it is not clear what is implied by this.

The most straightforward way to show that $\mathcal{B}_4$ is self-Steinerian is to verify relation (B.5) in the appendix, i.e., to show that $f(\bigwedge^n H(x))$ is 0 modulo the Hessian, or

$$ f\left((\bigwedge^n H)_1(x), ..., (\bigwedge^n H)_{n+1}(x) \right) = G(x) \cdot Hess(V)(x). \qquad (5.13) $$

This relation can be checked on a computer (and has been checked for $V = \mathcal{B}$ using Macaulay).

There are also other ways of showing that $\mathcal{B}_4$ is self-Steinerian. For example, since $\mathcal{B}$ is invariant under $G_{25,920}$ and the Steinerian variety $\mathcal{ST}(\mathcal{B})$ is a covariant (of $\mathcal{B}$), $\mathcal{ST}(\mathcal{B})$ is also an invariant of $G_{25,920}$. Also $\mathbf{Hess}(\mathcal{B})$ is an invariant, of degree 10. In general one has a map

$$ \alpha : \mathbf{Hess}(V) \longrightarrow \mathcal{ST}(V), $$

sending a point $x \in \mathbf{Hess}(V)$ to the vertex of the cone C_x which is the quadric polar of x with respect to V. If one understands α sufficiently well, then since the degree of $\mathbf{Hess}(\mathcal{B})$ is 10, one gets an upper bound on the degree of $\mathcal{ST}(\mathcal{B})$. If one can in fact show that the degree of $\mathcal{ST}(\mathcal{B})$ is less than or equal to 5, then, since $\mathcal{B}$ is the unique invariant of degree $d \leq 5$, we get $\mathcal{ST}(\mathcal{B}) = \mathcal{B}$. A related method, developed by Coble, uses more of the very special geometry of the situation to show the self-Steinerian property for $\mathcal{B}$. We will sketch in the next two sections the following steps:

i) In section 4.3.2 we described the hyperelliptic functions giving rise to the unitary reflection groups, and considering the $X_{\alpha\beta}$ leads to an eight-dimensional representation, which contains an invariant $\mathbb{P}^3$ (with variables $Z_{\alpha\beta}$) and an invariant $\mathbb{P}^4$ (with variables $Y_{\alpha\beta}$) as in (4.69).

ii) These theta functions define in the usual manner embeddings of abelian surfaces (with level 3 structure in this case) given as common zeroes of quadrics.

iii) Let A_s be such an abelian surface, $A_s \subset \mathbb{P}^8$ as a degree 18 variety. Then $A_s \cap \mathbb{P}^3 = 6$ odd two-torsion points, $A_s \cap \mathbb{P}^4 = 10$ even two-torsion points.

iv) $A_s \cap \mathbb{P}^4 \subset \mathbf{Hess}(\mathcal{B})$ yielding a correspondence $s \longleftrightarrow x \in \mathbf{Hess}(\mathcal{B})$.

v) Mapping $A_s \cap \mathbb{P}^3 \mapsto \alpha(x) \in \mathcal{ST}(\mathcal{B})$ defines a 6:1 (rational) map of $\mathbb{P}^3$ (varying s) onto $\mathcal{ST}(\mathcal{B})$. This map is given by quartics and has 40 base points (the 40 t_{12} points of the arrangement defined in $\mathbb{P}^3$ by $G_{25,920}$, see Table 4.2). Hence the degree of the image is $\leq (64 - 40) \div 6 = 4 \Rightarrow \mathcal{ST}(\mathcal{B}) = \mathcal{B}$, since $\mathcal{B}$ is the unique degree 4 invariant.

5.3.1   The Witting configuration in $\mathbb{P}^3$

The action of $G_{25,920}$ on $\mathbb{P}^3$ in terms of the hyperelliptic functions $Z_{\alpha\beta}$ was first treated by Witting in his Göttingen dissertation. That is why the following configuration, essentially the arrangement defined by the action of $G_{25,920}$ on $\mathbb{P}^3$, is called a "Witting configuration". Most of what follows can be deduced from Table 4.2. There are 40 planes and 40 "poles", points polar to the planes, such that in each of the 40 planes there are 21 lines of the extended Hesse pencil, 12 of the 40 poles, and through each of the 40 points 12 of the planes pass. Each plane is associated to a unique pole; each plane is the face, its pole a vertex, of four tetrahedra of the 40 planes (coming from the four triangles in each plane). Hence there are 40 tetrahedra, which contain 240 edges (the $t_2(1)$'s), which form 120 skew pairs. Also the 90 $t_4(1)$'s fall into 45 pairs of skew lines. Fixing one such pair the remaining 44 fall into 32 (not skew to the given one) and 12 (skew to the given one). Of these 12 pairs there are four which are mutually skew, yielding *sets of five pairs of skew lines* $t_4(1)$'s. There are 27 such sets of five.

It is clear that in some way such a skew pair corresponds to a node of B, and we would like to express this by an explicit rational map. And, in fact, this is not too difficult, as was first pointed out by Burkhardt [Bu], III, §73, and computed in detail by Coble [C1], III, §5. The set of 40 points mentioned above is given explicitly by:

$$\text{(4)} \quad z_i = 1, \quad z_j = 0, \quad (i \neq j, i, j = 1, ..., 4)$$
$$\text{(9)} \quad z_1 = 0, \quad z_2^3 = z_3^3 = z_4^3;$$
$$\text{(9)} \quad z_2 = 0, \quad z_3^3 = -z_4^3 = z_1^3;$$
$$\text{(9)} \quad z_3 = 0, \quad z_4^3 = z_1^3 = -z_2^3;$$
$$\text{(9)} \quad z_4 = 0, \quad z_1^3 = z_2^3 = -z_3^3;$$

and it is easy to see that the family of quartics in $\mathbb{P}^3$ with these 40 points as base points is given explicitly by

$$
\begin{aligned}
\alpha_0 &= 6z_1 z_2 z_3 z_4 \\
\alpha_1 &= -z_1(z_2^3 + z_3^3 + z_4^3), \\
\alpha_2 &= z_2(z_1^3 + z_3^3 + z_4^3), \\
\alpha_3 &= z_3(z_1^3 - z_2^3 + z_4^3), \\
\alpha_4 &= z_4(z_1^3 + z_2^3 - z_3^3).
\end{aligned}
\qquad (5.14)
$$

so in particular, is a linear system of dimension 4. The α's must fulfill an algebraic equation, and one can check directly that $J_4(\alpha) = 0$, i.e., this gives a map of $\mathbb{P}^3$ onto B! There is, however, a more elegant way of seeing this fact, utilising the fact that the coordinates $z_i(= Z_{\alpha\beta})$ are theta functions of two variables. Let us just sketch this argument, which we learned from W. Barth.

We already mentioned above the transformation properties of the $Z_{\alpha\beta}(\tau, z)$ *for fixed z*, but one can of course also fix τ and consider the $Z_{\alpha\beta}$ as functions of z (i.e., as functions on a fixed abelian surface A_τ). As is well known these functions $X_{\alpha\beta}(\tau, z)$ yield, for fixed τ, an embedding of the abelian surface A_τ in $\mathbb{P}^8$. Now there are certain linear transformations of the z which map the embedded A_τ onto itself; these form the *Heisenberg group*. In our context, one takes any one of 81 three-division points P and considers the following transformation: $z \mapsto z + P$, which yields *linear* transformations of the $X_{\alpha\beta}$ and which maps A_τ onto itself. In addition there is an involution $I : z \mapsto -z$ which also induces a linear transformation of the $X_{\alpha\beta}$. These together define the Heisenberg group H of order $2 \cdot 81$, a central extension

$$1 \longrightarrow \mathbb{Z}_2 \longrightarrow H \longrightarrow \mathbb{Z}_3^2 \times \mathbb{Z}_3^2 \longrightarrow 1.$$

The involution I can be extended to an involution on $\mathbb{P}^8$ and has the same two invariant subspaces: $\mathbb{P}^3 = \{Z_{\alpha\beta}\}$ and $\mathbb{P}^4 = \{X_{\alpha\beta}\}$. It is obvious that the 16 half-periods (two-division points) are invariant under I, hence these 16 points constitute the intersection $A_\tau \cap (\mathbb{P}^3 \cup \mathbb{P}^4)$. Furthermore,

$$\begin{aligned} A_\tau \cap \mathbb{P}^3 &= 6 \text{ odd two-torsion points} \\ A_\tau \cap \mathbb{P}^4 &= 10 \text{ even two-torsion points.} \end{aligned} \tag{5.15}$$

Remark 5.3.1 One can of course project A_τ into the $\mathbb{P}^3$ and $\mathbb{P}^4$, respectively. Let us denote the projections by $p_4 : \mathbb{P}^8 \longrightarrow \mathbb{P}^4$ (projection *from* the $\mathbb{P}^3$), $p_3 : \mathbb{P}^8 \longrightarrow \mathbb{P}^3$ (projection *from* the $\mathbb{P}^4$). Then the result is:

- $p_4(A_\tau)=$ complete intersection of a cubic and a quadric, a K3-surface of degree 6 with ten nodes,

- $p_3(A_\tau)=$ quartic K3-surface with six nodes (a so-called Weddle surface).

Getting back to the embedding $A_\tau \subset \mathbb{P}^8$, it follows from general theory that (generically) A_τ is cut out by quadrics and cubics. But it turns out in this case that in fact quadrics suffice. Coble lists explicitly a set of nine quadrics whose intersection he claims is A_τ. He does not prove this, but it has recently been verified by W. Barth. One arrives at this by counting the number of quadrics. In $\mathbb{P}^8$ the family of quadrics $H^0(\mathbb{P}^8, \mathcal{O}(2))$ is 45-dimensional, while the set of quadrics restricted to A_τ (sections of $\mathcal{L}^{\otimes 2}$ if $\mathcal{L}$ gives the embedding) is 36-dimensional (since $\mathcal{L} = 3\Theta$ where Θ is the principal polarization). Therefore there *must* be nine relations, like

$$\alpha_0 X_{00}^2 + 2\alpha_1 X_{01} X_{02} + 2\alpha_2 X_{10} X_{20} + 2\alpha_3 X_{11} X_{22} + 2\alpha_4 X_{12} X_{21} = 0 \tag{5.16}$$
$$+ \text{ eight others.}$$

But one can say more: the Heisenberg group H acts on $H^0(\mathbb{P}^8, \mathcal{O}(2))$, so from one such relation one gets the eight others by applying H. The implication is then that *the coefficients $\alpha_0, ..., \alpha_4$ of the first relation are, in fact, coefficients for all nine relations*, i.e., from the above relation the other eight are determined by replacing $X_{00}^2, ..., X_{12}X_{21}$, respectively, by the rows of:

$$
\begin{array}{ccccc}
X_{01}^2 & X_{02}X_{00} & X_{11}X_{21} & X_{12}X_{20} & X_{10}X_{22} \\
X_{02}^2 & X_{00}X_{01} & X_{12}X_{22} & X_{10}X_{21} & X_{11}X_{20} \\
X_{10}^2 & X_{11}X_{12} & X_{20}X_{00} & X_{21}X_{02} & X_{22}X_{01} \\
X_{11}^2 & X_{12}X_{10} & X_{21}X_{01} & X_{22}X_{00} & X_{20}X_{02} \\
X_{12}^2 & X_{10}X_{12} & X_{22}X_{02} & X_{20}X_{01} & X_{21}X_{00} \\
X_{20}^2 & X_{21}X_{22} & X_{00}X_{10} & X_{01}X_{12} & X_{02}X_{11} \\
X_{21}^2 & X_{22}X_{20} & X_{01}X_{11} & X_{02}X_{10} & X_{00}X_{12} \\
X_{22}^2 & X_{20}X_{21} & X_{02}X_{12} & X_{00}X_{11} & X_{01}X_{10}
\end{array}
\tag{5.17}
$$

Now simply change coordinates from $X_{\alpha\beta}$ to the $Y_{\alpha\beta}$ and $Z_{\alpha\beta}$. If we denote these coordinates by $y_0, \ldots, y_4$ and $z_1, \ldots, z_4$, then the above nine quadrics are of the form:

$$
\begin{aligned}
&\alpha_0 y_0^2 + 2\alpha_1 y_1^2 + 2\alpha_2 y_2^2 + 2\alpha_3 y_3^2 + \alpha_4 y_4^2 \\
&-2\alpha_1 z_1^2 - 2\alpha_2 z_2^2 - 2\alpha_3 z_3^2 - 2\alpha_4 z_4^2 = 0, \\
&\alpha_0 y_1^2 + 2\alpha_1 y_0 y_1 + 2\alpha_2 y_3 y_4 + 2\alpha_3 y_2 y_4 + 2\alpha_4 y_2 y_3 \\
&+\alpha_0 z_1^2 - 2\alpha_2 z_3 z_4 - 2\alpha_3 z_2 z_4 - 2\alpha_4 z_2 z_3 = 0, \\
&\alpha_0 y_2^2 + 2\alpha_1 y_3 y_4 + 2\alpha_2 y_0 y_2 + 2\alpha_3 y_1 y_4 + 2\alpha_4 y_1 y_3 \\
&+\alpha_0 z_2^2 + 2\alpha_1 z_3 z_4 + 2\alpha_3 z_1 z_4 - 2\alpha_4 z_1 z_3 = 0, \\
&\alpha_0 y_3^2 + 2\alpha_1 y_2 y_4 + 2\alpha_2 y_1 y_4 + 2\alpha_3 y_0 y_3 + 2\alpha_4 y_1 y_2 \\
&+\alpha_0 z_3^2 + 2\alpha_1 z_2 z_4 - 2\alpha_2 z_1 z_4 + 2\alpha_4 z_1 z_2 = 0, \\
&\alpha_0 y_4^2 + 2\alpha_1 y_2 y_3 + 2\alpha_2 y_1 y_3 + 2\alpha_3 y_1 y_2 + 2\alpha_4 y_0 y_4 \\
&+\alpha_0 z_4^2 + 2\alpha_1 z_2 z_3 + 2\alpha_2 z_1 z_3 - 2\alpha_3 z_1 z_2 = 0, \\
&z_1 \pi_{01} + z_2 \pi_{43} + z_3 \pi_{24} + z_4 \pi_{32} = 0, \\
&z_1 \pi_{43} + z_2 \pi_{02} + z_3 \pi_{14} + z_4 \pi_{13} = 0, \\
&z_1 \pi_{24} + z_2 \pi_{14} + z_3 \pi_{03} + z_4 \pi_{12} = 0, \\
&z_1 \pi_{32} + z_2 \pi_{13} + z_3 \pi_{12} + z_4 \pi_{04} = 0,
\end{aligned}
\tag{5.18}
$$

where $\pi_{ij} = \alpha_i y_j - \alpha_j y_i$. Now, setting $Z_{\alpha\beta} = 0$ (i.e., looking at the invariant $\mathbb{P}^3$) one finds:

$$
[Hess(\mathcal{B})] \cdot \begin{pmatrix} \alpha_0 \\ \alpha_1 \\ \alpha_2 \\ \alpha_3 \\ \alpha_4 \end{pmatrix} = 0.
\tag{5.19}
$$

Therefore, by very definition, the parameters $(\alpha_0, ..., \alpha_4)$ map onto the Steinerian of $\mathcal{B}$. Note that this proves the following

Theorem 5.3.2 *There is a Zariski open subset of $\mathcal{ST}(\mathcal{B})$ which parameterizes abelian surfaces with a $(3,3)$ polarization, or equivalently, principally polarized abelian surfaces with a level 3 structure.*

Now we can sketch Coble's elegant proof that $ST(\mathcal{B}) = \mathcal{B}$. In the equation (5.18), set $Y_{\alpha\beta} = 0$; one finds

$$
\begin{aligned}
\alpha_0 z_1^2 & & -2\alpha_2 z_3 z_4 & & -2\alpha_3 z_2 z_4 & & -2\alpha_4 z_3 z_2 & = 0, \\
\alpha_0 z_2^2 & +2\alpha_1 z_3 z_4 & & & +2\alpha_3 z_1 z_4 & & -2\alpha_4 z_3 z_1 & = 0, \\
\alpha_0 z_3^2 & +2\alpha_1 z_2 z_4 & -2\alpha_2 z_1 z_4 & & & & +2\alpha_4 z_1 z_2 & = 0, \\
\alpha_0 z_4^2 & +2\alpha_1 z_3 z_2 & +2\alpha_2 z_1 z_3 & -2\alpha_3 z_1 z_2 & & & & = 0,
\end{aligned}
\tag{5.20}
$$

which is a system of equations of rank 4 in five variables and can therefore be solved for the α_i; doing this yields the quartic expressions (5.14) in the z_i above. Now it is clear that the six two-torsion points of $A_\tau \cap \mathbb{P}^3$ get mapped onto the same α, or in other words $\pi : \mathbb{P}^3 \longrightarrow ST(\mathcal{B})$ is a 6:1 map. Since there are 40 base points, the degree of the image is $(\frac{1}{6})(64 - 40) = 4$, and since π is $G_{25,920}$-equivariant, $ST(\mathcal{B})$ is a degree 4 invariant, and it follows that $ST(\mathcal{B}) = \mathcal{B}$:

Theorem 5.3.3 *The Burkhardt quartic $\mathcal{B}_4$ is self-Steinerian.*

As a corollary of this and Theorem 5.3.2 we have:

Corollary 5.3.4 *There is a Zariski open subset of $\mathcal{B}$ which is $G_{25,920}$-equivariantly biregular to a Zariski open subset of the moduli space of principally polarized abelian surfaces with a full level 3 structure.*

We will see later that the Zariski open subset of $\mathcal{B}$ is the complement of the 40 j-planes on $\mathcal{B}$ (see Lemma 5.7.1).

5.3.2   The map $\mathrm{Hess}(\mathcal{B}) \longrightarrow ST(\mathcal{B}) = \mathcal{B}$

The intersection of $\mathrm{Hess}(\mathcal{B})$ and $\mathcal{B}$ is of degree 40, and is in fact just the union of the 40 j-planes on $\mathcal{B}$. To see this, use the equation of $\mathcal{B}$ in the y_i's and consider the j-plane given by $\{y_0 = y_1 = 0\}$. Writing down the Hessian matrix of J_4 one sees immediately that $y_0 = y_1 = 0$ is identically satisfied by the determinant, i.e., the j-plane lies on the variety $\mathrm{Hess}(\mathcal{B})$. Now, since with $\mathcal{B}$ also $\mathrm{Hess}(\mathcal{B})$ is a covariant of $G_{25,920}$, all 40 j-planes, being an orbit under the group, lie on $\mathrm{Hess}(\mathcal{B})$, and the union of these 40 planes is then clearly the intersection of the two varieties.

It was already explained above that for a generic $\alpha \in \mathcal{B}$ the ten even division points of the abelian surface A_α lie on the Hessian variety $\mathrm{Hess}(\mathcal{B})$ and constitute the fibre of $\mathrm{Hess}(\mathcal{B}) \longrightarrow ST(\mathcal{B}) = \mathcal{B}$, hence $\mathrm{Hess}(\mathcal{B}) \longrightarrow ST(\mathcal{B}) = \mathcal{B}$ is a 10:1 cover. The term generic $\alpha \in \mathcal{B}$ means in this situation α not lying on one of the j-planes, and one sees easily that this union constitutes the branch locus of $\mathrm{Hess}(\mathcal{B}) \longrightarrow ST(\mathcal{B}) = \mathcal{B}$. Recall that the Hessian and Steinerian varieties are birational. Hence, birationally, this is a self map of degree 10 of $\mathcal{B}$.

5.3.3   Level 3 structures on Kummer surfaces

Recall from Remark 5.3.1 that projecting the $(3, 3)$ polarized abelian surface A_τ in the $\mathbb{P}^8$ from the invariant $\mathbb{P}^4$ (respectively from the invariant $\mathbb{P}^3$) displays A_τ

as a double cover of a quartic hypersurface in $\mathbb{P}^3$ with six nodes at the six two-torsion points of A_τ lying in the invariant $\mathbb{P}^3$ (respectively a sextic K3-surface in $\mathbb{P}^4$ with ten nodes at the ten even two-torsion points lying in the invariant $\mathbb{P}^4$). For the first, consider the four quadrics (5.20) and the Jacobian surface (Definition B.1.7 and (B.69)) they define ($\alpha \in \mathcal{B}$):

$$\mathcal{J}_\alpha = \mathcal{J}(Q_1, Q_2, Q_3, Q_4). \qquad (5.21)$$

Lemma 5.3.5 $\mathcal{J}_\alpha$ *is a Weddle surface, the corresponding symmetroid is a Kummer surface (see Lemma B.5.19)*

Proof: This follows from the fact that the quadrics Q_i in the invariant $\mathbb{P}^3$ all contain the six odd two-torsion points, so the Jacobian has six nodes and the symmetroid is Kummer by Lemma B.5.15. $\qquad \square$

Let π be the 6:1 $G_{25,920}$-equivariant map of $\mathbb{P}^3$ onto $\mathcal{ST}(\mathcal{B}) = \mathcal{B}$:

$$\pi : \mathbb{P}^3 \xrightarrow{\ 6:1\ } \mathcal{B}. \qquad (5.22)$$

We can consider the image of $\mathcal{J}_\alpha$ under π; this will be a divisor on $\mathcal{B}$. To see what it is, we note that π can is given explicitly by the system of quartics (5.14), as already mentioned there. The following result is also due to Coble.

Theorem 5.3.6 ([C1] III, (60)) *The image $\pi(\mathcal{J}_\alpha)$ of the Weddle surface $\mathcal{J}_\alpha$ is the tangent hyperplane section of $\mathcal{B}$ at $\alpha \in \mathcal{B}$. So each such hyperplane section is a Weddle surface in the system (5.14), and for each point $p \in \mathbb{P}^3$ there is such a Weddle surface with a node at p.*

Proof: Since the space of Weddle surfaces is three-dimensional, this could be proved by showing that a generic Weddle surface is projectively equivalent to a quartic in the system (5.14), because that would mean the generic Weddle maps under π in (5.22) onto a *hyperplane* section of $\mathcal{B}$. Since the six nodes of the Weddle surface map to $\alpha \in \mathcal{B}$ (by definition of π), this hyperplane section has a node at $\alpha \in \mathcal{B}$ for all α in the Zariski open set of Theorem 5.3.4, so must be a tangent hyperplane section.

Instead, let us, again following Coble, show this by utilizing the symbolic method. The Jacobian determinant of the quadrics (5.21) is easily calculated, and can be written as follows:

$$\begin{aligned}
\mathcal{J}_\alpha \ = \ & [\alpha_0^3 + 2(\alpha_1^3 + \cdots + \alpha_4^3)]6z_1 z_2 z_3 z_4 \\
& +6(\alpha_0\alpha_1^2 + 2\alpha_2\alpha_3\alpha_4)[-z_1(z_2^3 + z_3^3 + z_4^3)] \\
& + \ \ldots\ldots\ + 6(\alpha_0\alpha_4^2 + 2\alpha_1\alpha_2\alpha_3)[z_4(z_1^3 + z_2^3 - z_3^3)],
\end{aligned}$$

which, if we write J_4 is symbolic form $J_4 = (ay)^4 = (by)^4 = \ldots$, takes the simple form:

$$\mathcal{J}_\alpha = (a\alpha)^3 (a\alpha'),$$

where α' is the quartic expression in the z_i of (5.14). This symbolic expression tells us that under the map π, which correspond here to replacing the z_i by the

α', the expression $(a\alpha)^3(a\alpha')$ describes the tangent hyperplane to $\mathcal{B}$ at $\alpha \in \mathcal{B}$, and as $\alpha' \in \mathcal{B}$, the expression describes exactly the tangent hyperplane section $T_\alpha \mathcal{B} \cap \mathcal{B}$, as was to be shown.   □

We now turn to the other K3-surface, the sextic in $\mathbb{P}^4$ which we will denote by S_α. Since the $\mathbb{P}^4$ intersects A_τ in the ten even two-torsion points, S_α has ten nodes. We would like to locate S_α more precisely. Before proceeding, let us again fix notations for our coordinates: $(z_1, \ldots, z_4)$ coordinates on the "Maschke" $\mathbb{P}^3$; $(y_0, \ldots, y_4)$ coordinates on the "Burkhardt" $\mathbb{P}^4$; $(\alpha_0, \ldots, \alpha_4) = \pi(z_1, \ldots, z_4)$, π as in (5.22). For $\alpha \in \mathcal{B}$, consider the last four of the quadratic relations (5.18). For a general point in (y, z) on A_τ and fixed $\alpha \in \mathcal{B}$, the last four relations of (5.18) hold and z is non zero, so that on $p_4(A_\tau)$ the y's must satisfy

$$\mathcal{K}_\alpha = \begin{vmatrix} \pi_{01} & \pi_{43} & \pi_{24} & \pi_{32} \\ \pi_{43} & \pi_{02} & \pi_{14} & \pi_{13} \\ \pi_{24} & \pi_{14} & \pi_{03} & \pi_{12} \\ \pi_{32} & \pi_{13} & \pi_{12} & \pi_{04} \end{vmatrix} = 0. \tag{5.23}$$

Note that in three space, a determinant of the form (5.23) defines a quartic surface (a symmetroid, see section B.5.3.1), so that (5.23) defines a quartic *cone*, i.e., a cone over a symmetroid. In other words,

Lemma 5.3.7 *The sextic K3-surface S_α lies on the quartic cone $\mathcal{K}_\alpha$.*

One can deduce that $\mathcal{K}_\alpha$ is in fact a *Kummer* cone, i.e., that the symmetroid is a Kummer surface. These cones arise geometrically as the cones, in $\mathbb{P}^4$, over the tangent hyperplane sections of the Igusa quartic (see Theorem 3.3.8). If $H_p \cap \mathcal{I}_4 = h_p$ is the tangent hyperplane section for $p \in \mathcal{I}_4$, and if $C(h_p)$ denotes the cone in $\mathbb{P}^4$ over h_p with vertex at the point q_p corresponding under duality to h_p, this is also the enveloping cone of the Segre cubic from the point corresponding under duality to the given hyperplane section of $\mathcal{I}_4$ (compare Proposition 3.3.1 and Corollary 3.3.7). Moreover, we have

Lemma 5.3.8 *The polar cubic $\Delta_\alpha^1 J_4$ is a copy of the Segre cubic, with the ten nodes at the ten inverse images of α lying on the Hessian variety of $\mathcal{B}$.*

Proof: This follows from the fact that $\Delta_\alpha^1 J_4$ has a node for all $p \in \text{Hess}(\mathcal{B}) \cap st^{-1}(\alpha)$, where $st : \text{Hess}(\mathcal{B}) \longrightarrow \mathcal{B}$ is the Steinerian map (see the discussion following Definition B.1.10), and since all ten nodes in $st^{-1}\alpha$ lie on the Hessian variety, $Hess(J_4)$ vanishes at each and $\Delta_\alpha^1 J_4$ is a cubic hypersurface with ten nodes. It follows from uniqueness of S_3 that $\Delta_\alpha^1 J_4 \cong S_3$.   □

Lemma 5.3.9 $\mathcal{K}_\alpha$ *is the enveloping cone of $\Delta_\alpha^1 J_4$ from α.*

Proof: Again we give Coble's proof, which utilizes the symbolic expression for the cone $\mathcal{K}_\alpha$ (with α as variable):

$$48\mathcal{K}_\alpha = (a\alpha)^4 (ay)^4 - 4(a\alpha)^3 (ay) \cdot (a\alpha)(ay)^3 + 3[(a\alpha)^2 (ay)^2]^2.$$

In particular, for $\alpha \in \mathcal{B}$, this reduces to the last two terms, as $(a\alpha)^4 = 0$. On the other hand, the symbolic expression for the enveloping cone of the cubic polar $(a\alpha)(ay)^3$ is

$$3[(a\alpha)^2(ay)^2]^2 - 4(a\alpha)^3(ay) \cdot (a\alpha)(ay)^3,$$

so this coincides with $\mathcal{K}_\alpha$ for $\alpha \in \mathcal{B}$. $\square$

Lemma 5.3.10 $S_\alpha = \mathcal{K}_\alpha \cap \{\Delta_\alpha^1 J_4 = 0\} \cap \{\Delta_\alpha^2 J_4 = 0\}$, and $\mathcal{K}_\alpha$ is the cone of projection of S_α from α.

Proof: S_α is determined by eliminating the z_i from the system of quadrics (5.18), as this amounts to projecting to the $\mathbb{P}^4$. The remaining relation then takes the form (gotten by multiplying the first five relations by $\alpha_0, 2\alpha_1, \ldots, 2\alpha_4$ and adding)

$$\alpha_0^2 y_0^2 + 8y_0(\alpha_1 y_1^2 + \cdots + \alpha_4 y_4^2) + 8(\alpha_1\alpha_2 y_3 y_4 + \cdots + \alpha_3\alpha_4 y_1 y_2)$$

or symbolically

$$(a\alpha)^2 \cdot (ay)^2 = 0.$$

It can also be directly verified that the quadric above is the polar quadric of α with respect to $\mathcal{B}$. What this means is that any $y \in S_\alpha$ must satisfy the above quadratic relation, and hence $S_\alpha \subset \{\Delta_\alpha^2 J_4 = 0\}$. Now we may utilize the symbolic expression of Lemma 5.3.9 for $\mathcal{K}_\alpha$: if $\alpha \in \mathcal{B}$, so that $(a\alpha)^4 = 0$, and in addition $(a\alpha)^2(ay)^2 = 0$ and $\mathcal{K}_\alpha = 0$ (so y is on the cone), then either $(a\alpha)^3(ay) = 0$ or $(a\alpha)(ay)^3 = 0$, i.e., y is also on either the polar cubic or on the polar plane. Hence the intersection $\mathcal{K}_\alpha \cap \{\Delta_\alpha^2 J_4 = 0\}$ splits:

$$\mathcal{K}_\alpha \cap \{\Delta_\alpha^2 J_4 = 0\} = (\{\Delta_\alpha^3 J_4 = 0\} \cap \{\Delta_\alpha^2 J_4 = 0\}) \cap (\{\Delta_\alpha^1 J_4 = 0\} \cap \{\Delta_\alpha^2 J_4 = 0\}).$$

The first factor is a quadric cone in the polar hyperplane, while the second is a degree 6 surface which contains S_α and therefore coincides with it. The second statement then follows from Lemma 5.3.9. $\square$

Now we have the following K3-surfaces:

- The Kummer surface $K_{\alpha(\tau)} = A_\tau / \pm 1$.

- The sextic K3 S_α.

- The Weddle surface $\mathcal{J}_\alpha$.

First it is clear that $\mathcal{J}_\alpha$ projects onto K_α under the natural map of Lemma B.5.20, and all three of K_α, S_α and $\mathcal{J}_\alpha$ are birational. The cone $\mathcal{K}_\alpha$ is the cone on α in $\mathbb{P}^4$ over K_α (the tangent hyperplane section of $\mathcal{I}_4$), and S_α lies on $\mathcal{K}_\alpha$, thus also defining a natural rational map $S_\alpha \longrightarrow K_\alpha$. From the fact that $\mathcal{B}$ is the moduli space of $(3,3)$ polarized abelian surfaces, we see that such a triple $(K_\alpha, S_\alpha, \mathcal{J}_\alpha)$ determines a level 3 structure on A_τ. It would be interesting to derive this directly from the triple. We might formulate this as follows. "Locally" the moduli space $\mathcal{B}$ coincides with that of a level 2 structure, but with additional markings. We give a schematic representation of this in Figure 5.7.

Figure 5.7: The level 3 structure on Kummer surfaces

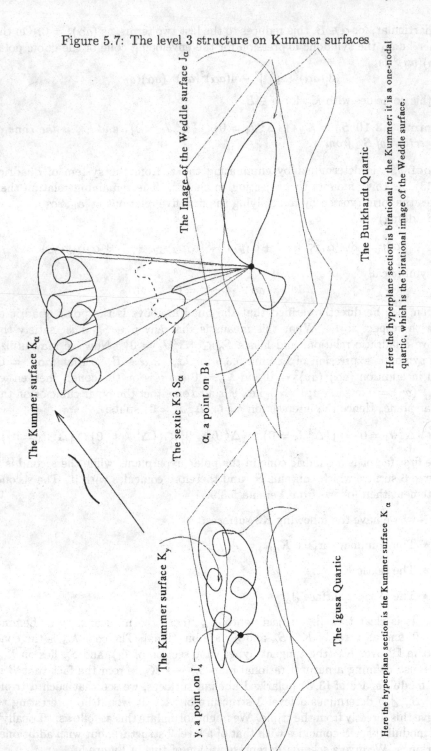

The Image of the Weddle surface J_α

The Burkhardt Quartic

Here the hyperplane section is birational to the Kummer; it is a one-nodal quartic, which is the birational image of the Weddle surface.

The Kummer surface K_α

The sextic K3 S_α

α, a point on B_4

The Kummer surface K_y

The Igusa Quartic

y, a point on I_4

Here the hyperplane section is the Kummer surface K_α

5.4 The solution of the Burkhardt form problem

In this section we sketch Coble's solution of the Burkhardt form problem.

Definition 5.4.1 Let ratios $(\lambda_1 : \ldots : \lambda_5)$ be given; the *Burkhardt form problem* is to find $y \in \mathbb{P}^4$, such that $(J_4(y) : \ldots : J_{18}(y)) = (\lambda_1 : \ldots : \lambda_5)$ (see Definition B.1.15). The *restricted Burkhardt form problem* is: given (λ, μ, ν), find $\alpha \in \mathcal{B}$ such that

$$(\lambda, \mu, \nu) = (J_{12}(\alpha)/J_6^2(\alpha), J_{18}(\alpha)/J_6^3(\alpha), J_{10}^3(\alpha)/J_6^5(\alpha)).$$

The reduction of the general form problem to the restricted one is algebraic, and follows the method of Klein ([Kl3], II,5,§2, p. 241), see [C1], III §7. This introduces, however, a quartic accessory irrationality. The transcendental part of the solution is the restricted form problem, and this solution can be given in the following steps:

1° Given $\alpha \in \mathcal{B}$, there is a binary sextic covariantly associated with α.

2° The absolute invariants of the sextic are λ, μ, ν.

3° The equation of the sextic is expressed in terms of these invariants by adding an accessory square root.

4° Given the sextic P_α, the periods are caculated, giving a point $\tau(\alpha) \in \mathbb{S}_2$, and the solution of the form problem is given by the theta functions $Y_{\alpha\beta}(\tau(\alpha))$.

Note that 3° and 4° are nothing but the last steps involved in the usual solution to the sextic equation, so are in this sense well-known. Of interest to us will be 1° and 2°.

1°: Recall that for $\alpha \in \mathcal{B}$ we have the Kummer cone $\mathcal{K}_\alpha$, the cone over the Kummer surface K_α. For this Kummer surface it is well known how to find the corresponding genus 2 curve (Lemma B.5.1). It is the genus 2 curve defined by the sextic on the plane conic in a trope of the Kummer, defined by the six nodes in that trope (the conic passes through all six points). This is the binary sextic we are looking for. To get the equation, first note that, from the symbolic expression for $\mathcal{K}_\alpha$ used in Lemma 5.3.9, one sees immediately that the trope of $\mathcal{K}_\alpha$ is given by the intersection with the polar plane of α with respect to $\mathcal{B}$, whose equation is $(a\alpha)^3(ay) = 0$, since then the equation reduces to

$$[(a\alpha)^2(ay)^2]^2 = 0,$$

which is a perfect square, i.e., the trope. Also this trope lies on the quadric polar, as its equation is $(a\alpha)^2(ay)^2$. Furthermore, the six double lines of $\mathcal{K}_\alpha$ over the six nodes in the trope are given by the equation

$$(a\alpha)(ay)^3 = 0,$$

which is the equation for the polar cubic, i.e., the copy of the Segre cubic mentioned above. Consequently,

Lemma 5.4.2 *The cubic, quadric and linear polars of α with respect to J_4 meet in six lines of a quadric cone in the polar hyperplane $\mathbb{P}^3 = \{(a\alpha)^3(ay) = 0\}$. The quadric cone is the intersection of the quadric and linear polars.*

Now using the fact that six lines on a quadric cone in $\mathbb{P}^3$ determine six points on a plane conic, we get the binary sextic explicitly.

2°: As a first step we will show that the sextic is projective to a sextic in $\mathbb{P}^3$ on a twisted cubic curve through six points $p_1, \ldots, p_5, p$. Indeed, the six lines of 1° lie entirely on the Segre cubic, here identified with the polar cubic of α with respect to $\mathcal{B}$. As we have seen in Theorem 3.2.1, this cubic threefold S_3 is birational to $\mathbb{P}^3$ by the six quadrics $q_a, \ldots, q_f$ on the five points $p_1, \ldots, p_5$ of the arrangement, and

$$S_3 = q_a^3 + \cdots + q_f^3 = 0, \quad q_a + \cdots + q_f = 0, \tag{5.24}$$

(these are the quadrics denoted $X, \ldots, Z'$ in (3.31)). Let us now temporarily introduce the notations S for the polar cubic of α and Q for the polar quadric of α, and finally $\mathcal{H}$ for the polar plane. As we have seen above, $S \cap Q$ is the sextic K3 S_α; $Q \cap \mathcal{H}$ is the quadric cone upon which the six lines lie, and the six lines are in fact on S, so letting $\mathcal{L}$ denote their union, we have as in Lemma 5.4.2

$$\mathcal{L} = \mathcal{H} \cap Q \cap S.$$

Now consider the Weddle surface $\mathcal{J}_\alpha$; the image $\pi(\mathcal{J}_\alpha)$ under the map π of (5.22) is contained in the tangent plane at α by Theorem 5.3.6. Consider the point $p \in \mathbb{P}^3$, which under the map given by the quadrics $q_a, \ldots, q_f, q : \mathbb{P}^3 \longrightarrow S$ maps to $\alpha = (\bar{q}_a, \ldots, \bar{q}_f) \in S$, $q(p) = \alpha$. We may ask about the inverse image in $\mathbb{P}^3$ under q of the sextic K3 $S_\alpha \subset S$.

Lemma 5.4.3 $q^{-1}(S_\alpha) \cong \mathcal{J}_\alpha$ *is the Weddle surface with nodes at $p_1, \ldots, p_5$ (the vertices of the coordinate tetrahedron and the center) and at p.*

Proof: $q^{-1}(S_\alpha)$ is of course birational to S_α, hence a quartic, with nodes at p and at the base points of the system of quadrics $q_a, \ldots, q_f$, which are $p_1, \ldots, p_5$, and no further nodes. The only surface fulfilling these properties is the Weddle $\mathcal{J}_\alpha$. □

Lemma 5.4.4 *Under the map $q : \mathbb{P}^3 \longrightarrow S$, the proper transform of the six lines determining the sextic are*

- *the five lines on W_α joining p and p_i, $i = 1, \ldots, 5$.*

- *the cubic space curve on the six points $p_1, \ldots, p_5, p$.*

Proof: The inverse image of the quadric cone $\mathcal{H} \cap Q$ under q is a quadric cone in $\mathbb{P}^3$, with its vertex at p and passing through $p_1, \ldots, p_5$. Hence the five lines joining p to the p_i are on the Weddle surface $\mathcal{J}_\alpha$. The intersection

$$\mathcal{J}_\alpha \cap q^{-1}(\mathcal{H} \cap Q)$$

consists of the five lines and the cubic curve on the six points (as it is of degree eight), as stated. □

This identifies the binary sextic with the six points on the space cubic curve of Lemma 5.4.4. Note that the space of six points on a rational curve is just the configuration space $\mathbf{P}_6^3$, which is associated with (hence isomorphic to) $\mathbf{P}_6^1$. Note that the coordinates of $\mathbf{P}_6^3$ are just the coordinates of the point p, while the coordinates of $\mathbf{P}_6^1$ are the three double ratios of the six points. For this space (which is, via the quadrics $q_a, \ldots, q_f$, identified with the Segre cubic) there is a well developed theory of invariants. This is reviewed in the appendix, in section B.3.3. The result is as follows. Let $q_i(q_a)$ be the elementary symmetric functions in the quadrics q_a; the sextic is

$$\Sigma = \prod (X - q_a) = X^6 + 15q_2 X^4 + 15q_4 X^2 + 6q_5 X + q_6,$$

and this can be expressed in terms of the invariants I_2', I_4', I_6' of (B.31) and $I_5 = \sqrt{\Delta}$ as

$$\Sigma = (X^2 - I_2')^3 + 15 I_4' X^2 + \frac{1}{3^5} I_5 X + I_6'.$$

It remains to relate the invariants I_j' or also the $q_i(q_a)$ with the invariants of the Burkhardt group. This is very computational, although Coble makes it relatively geometric by using the symbolic method, which is, unfortunately, not always very enlightening. Let us give a single example, then quote the result. The discriminant Δ of the sextic vanishes when two of the roots coincide; if this occurs, then the curve is singular and the moduli point is in the boundary of $(\Gamma(3)\backslash \mathbb{S}_2)^*$, i.e., on one of the 40 lines of that Baily-Borel compactification. Recalling that under duality these 40 lines are the 40 j-planes on $\mathcal{B}$, we see that this, geometrically, is the discriminant. Now recall that the Hessian $\mathbf{Hess}(\mathcal{B})$ intersects $\mathcal{B}$ in exactly those 40 planes. Let J_{10} be the Hessian invariant of $\mathcal{B}$, so $\mathbf{Hess}(\mathcal{B}) = \{J_{10} = 0\}$. It is natural to take

$$I_5 = \lambda J_{10}^\mu,$$

for some undetermined constants λ and μ. To determine these constants one must consider the transformations of both I_5 and J_{10} under transformations of the sextic. The result is

$$I_5^2 = \Delta = 16^4 3^5 \lambda J_{10}^3.$$

The complete result is then as follows.

Theorem 5.4.5 ([C1] III (79)) *The resolvent sextic Σ of (B.36) is*

$$Q^6 - 2J_6 Q^4 + 16^2 \cdot 6 \cdot J_{12} Q^2 + 16^2 \sqrt{\lambda J_{10}^3} Q + 16^4 J_{18} = 0.$$

In terms of the invariants I_2, I_4, I_6, Δ of the sextic T this is

$$-6J_6 = 5I_2, \quad 16^2 \cdot 4 \cdot 3^3 \cdot J_{12} = 5(6I_2^2 - 5^2 I_4),$$

$$16^4 \cdot 2 \cdot 3^4 \cdot J_{18} = 5(-6I_2^3 + 3 \cdot 5^2 I_2 I_4 + 2 \cdot 5^3 I_6), \quad 16^4 \cdot 3^5 \cdot \lambda J_{10}^3 = \Delta = \prod (\eta_i - \eta_k).$$

This now completes 2°.

As we already mentioned, 3° and 4° are parts of the general solution of a sextic equation and use none of the special geometry of the situation. It remains, from the resolvent Σ, to display an explicit sextic T (whoose roots are denoted $\eta_1, \ldots, \eta_6$ in the Appendix, while the roots of Σ are the quadrics Q_i there or q_α here). The three linear covariants (B.28) fulfill a quadratic relation whose coefficients are rational in I_2, I_4, I_6 and Δ. Denoting this quadric by $K_2(\ell, m, n)$, the sextic is given by

$$T = K_3(\ell, m, n) = 0,$$

where K_3 is cubic in ℓ, m, n with coefficients rational in I_2, I_4, I_6 and Δ; in other words the six roots of the sextic are described as the intersection

$$(\eta_1, \ldots, \eta_6) = \{K_2 = K_3 = 0\}.$$

Viewing $\{K_2 = 0\}$ as a $\mathbb{P}^1$ by a parametrization introduces an accessory square root.

Once we have the sextic, the periods give us τ and the $Y_{\alpha\beta}(\tau)$ give us the point $\alpha \in \mathcal{B}$. Coble's solution requires a quartic and a square accessory irrationality. In the last paragraph of [C1], III, §7, Coble remarks that this should be compared with Klein's solution, for which two accessory *square* roots must be introduced. The main difference is that Klein's method required seperating a root of the sextic (the Weierstraß form). He also mentions that a quartic irrationality may be unavoidable in effecting the solution of the problem of the y's in terms of the problem of the z's.

5.5   Ball quotients

5.5.1   Logarithmic Chern classes

Recall that the logarithmic Chern classes of a pair $(\overline{X}, \Delta)$ consisting of a smooth projective variety $\overline{X}$ and a normal crossings divisor Δ on $\overline{X}$, are defined in terms of the sheaf $\Omega^1_{\overline{X}}(log(\Delta))$ of differential 1-forms with logarithmic singularities along Δ (i.e., at worst $\frac{dz}{z}$ if Δ is given locally by $z = 0$) by

$$\overline{c}_i(\overline{X}, \Delta) := (-1)^i c_i(\Omega^1_{\overline{X}}(log\Delta)), \tag{5.25}$$

where the right hand side is the usual c_i of a rank 1 sheaf. These cohomology classes *on* $\overline{X}$, are supposed to represent the corresponding class of the *open part* $X := \overline{X} - \Delta$. So, for example,

$$\overline{c}_1(\overline{X}, \Delta) = c_1(\overline{X}) - (\Delta), \tag{5.26}$$

where of course (Δ) denotes the cohomology class of Δ in this equation. From the exact sequence

$$0 \longrightarrow \Omega^1_{\overline{X}} \longrightarrow \Omega^1_{\overline{X}}(log\Delta) \longrightarrow j_*\mathcal{O}_\Delta \longrightarrow 0, \tag{5.27}$$

where $j : \Delta \hookrightarrow \overline{X}$ is the inclusion, one gets the following relation for the total Chern class:

$$c(\Omega^1_{\overline{X}}(\log \Delta)) = c(\Omega^1_{\overline{X}})c(j_*\mathcal{O}_\Delta).$$

Using the relation $c(j_*\mathcal{O}_\Delta) = 1 + \Delta + \Delta^2 + \cdots + \Delta^N$, $N = \dim(X)$, this can be written

$$
\begin{aligned}
\bar{c}_1(\overline{X},\Delta) &= c_1(\overline{X}) - (\Delta) \\
\bar{c}_2(\overline{X},\Delta) &= c_2(\overline{X}) - c_1(\overline{X})(\Delta) + (\Delta)^2 \\
&\vdots \\
\bar{c}_N(\overline{X},\Delta) &= c_N(\overline{X}) - c_{N-1}(\overline{X})(\Delta) + \cdots \pm (\Delta)^N.
\end{aligned}
\tag{5.28}
$$

Furthermore, assuming now $\Delta = \dot{\cup}\Delta_i$ is a *disjoint* union of *smooth* components Δ_i we may apply adjunction to each component Δ_i and then add up.

Lemma 5.5.1 *Let $(\overline{X},\Delta)$ be as above, with $\Delta = \dot{\cup}\Delta_i$ a disjoint union of smooth subvarieties. Then*

$$\bar{c}_j(\overline{X},\Delta) = c_j(\overline{X}) - c_{j-1}(\Delta).$$

Proof: We have

$$
\begin{aligned}
\bar{c}_j(\overline{X},\Delta) &= c_j(\overline{X}) - c_{j-1}(\overline{X})(\Delta) + \cdots + (-1)^{j+1}(\Delta)^j \\
&= c_j(\overline{X}) - (c_{j-1}(\Delta) + c_{j-2}(\Delta)(\Delta)) \\
&\quad + (c_{j-2}(\Delta)(\Delta) - c_{j-3}(\Delta)(\Delta)^2) \\
&\vdots \\
&\quad + (-1)^j(c_1(\Delta)(\Delta)^{j-1} - (\Delta)^j) \\
&\quad + (-1)^{j-1}(\Delta)^j \\
&= c_j(\overline{X}) - c_{j-1}(\Delta).
\end{aligned}
$$

□

Now suppose X is a ball quotient, $X = X_\Gamma$ (with Γ torsion free), and $\overline{X}$ is a toroidal compactification. Then $\Delta = \dot{\cup}\Delta_i$, and each Δ_i is an abelian variety (see Proposition 2.2.10 for the case of arithmetic groups; in the general case this holds also, but is much less elementary, as there is no *a priori* reason why the complex tori used to compactify should be abelian varieties), so in particular $c_j(\Delta_i) = 0$. From this it follows by adjunction $c_1(\overline{X}_\Gamma)_{|\Delta} = c_1(\Delta) + c_1(N_{\overline{X}_\Gamma}\Delta)$ that from $\bar{c}_1 = c_1 - \Delta$ we get

$$\bar{c}_1(\overline{X}_\Gamma,\Delta)_{|\Delta_i} = (c_1(\overline{X}_\Gamma) - \Delta)_{|\Delta_i} = c_1(N_{\overline{X}_\Gamma}\Delta) - \Delta_{|\Delta_i} = 0, \tag{5.29}$$

since $\Delta_{|\Delta_i} = \Delta_{i|\Delta_i} = c_1(N_{\overline{X}_\Gamma}\Delta_i)$. It follows that for ball quotients, or more generally for any $(\overline{X},\Delta)$ for which Δ consists of a finite disjoint union of abelian varieties (or varieties with vanishing first Chern class), $\bar{c}_1(\overline{X},\Delta)$ does not contain any component Δ_i of Δ, and $\bar{c}_1(\overline{X},\Delta)_{|\Delta_i} = 0$.

Applying Mumford's proportionality in the non-compact case, which states here that

$$\bar{c}_1^N(\overline{X}, \Delta) = \frac{2(N+1)}{N}\bar{c}_1^{N-2}(\overline{X}, \Delta)\bar{c}_2(\overline{X}, \Delta), \qquad (5.30)$$

to the Lemma 5.5.1 we get the following.

Proposition 5.5.2 *Let* $(\overline{X}, \Delta)$ *be a smooth compactification of an* N-*dimensional ball quotient* X_Γ *(*Γ *torsion free). Then*

$$N c_1^N(\overline{X}) - 2(N+1)c_1^{N-2}(\overline{X})c_2(\overline{X}) = (-1)^{N-1}N(\Delta)^N,$$

where $(\Delta)^N = \sum \Delta_i^N$ *is the sum of the self-intersection numbers.*

There is in fact a converse of this, due independently to R. Kobayashi for the case of surfaces and to S.-T. Yau for the general case. A proof can be found in [TY]. The statement is as follows.

Theorem 5.5.3 *Let* M *be a compact Kähler manifold of dimension* N, $D \subset M$ *a divisor, and assume:*

i) D *is a disjoint union of abelian varieties;*

ii) $K_M + D$ *is ample on* $M - D$.

Then

$$(-1)^N\bar{c}_1^N(\overline{M}, D) \le (-1)^N \frac{2(N+1)}{N}\bar{c}_1^{N-2}(\overline{M}, D)\bar{c}_2(\overline{M}, D),$$

with equality holding if and only if $M - D$ *is a quotient of the ball by a torsion free group.*

5.5.2 Relative proportionality

Next we recall the following relative proportionality property for subballs and subball quotients (modular subvarieties). They result from the fact that the subballs considered here are totally geodesic, which implies that the curvature tensor of the ambient space restricted to the subball *is* the curvature of the subball.

Lemma 5.5.4 *Let* $\overline{D}_\alpha \subset \overline{X}$ *be the compactification of a codimension 1 subball quotient* $D_\alpha \subset X = X_\Gamma$, *and set* $\Delta_\alpha = \overline{D}_\alpha \cap \Delta$. *Then the following proportionalities between the logarithmic Chern numbers hold (*$N = \dim(\overline{X})$*):*

i) $\bar{c}_1^{N-1}(\overline{X}, \Delta)_{|\overline{D}_\alpha} = \left(\frac{N+1}{N}\right)^{N-1} \bar{c}_1^{N-1}(\overline{D}_\alpha, \Delta_\alpha);$

ii) $\bar{c}_1^{N-3}(\overline{X}, \Delta)\bar{c}_2(\overline{X}, \Delta)_{|\overline{D}_\alpha} = \frac{N+1}{N-1}\left(\frac{N+1}{N}\right)^{N-3} \bar{c}_1^{N-3}(\overline{D}_\alpha, \Delta_\alpha)\bar{c}_2(\overline{D}_\alpha, \Delta_\alpha).$

Proof: Recall that the Kähler form on a bounded domain $\mathcal{D}$ is

$$\omega = \partial\bar{\partial}\log K_{\mathcal{D}}(z,\bar{z}),$$

where $K_{\mathcal{D}}(z,w)$ is the Bergmann kernel function, a (global) Kähler potential. For the complex N-ball, the Bergmann kernel function, denoted $K_N(z,w)$, is (see [KN] II, p. 163)

$$K_N(z,w) = (1 - {}^t z\bar{w})^{-(N+1)}.$$

From this it follows that on $\mathbb{B}_{N-1} = \{z_N = 0\}$ we have

$$K_N(z,w)_{|\mathbb{B}_{N-1}} = K_{N-1}(z,w)^{\frac{N+1}{N}}.$$

Consequently, we have

$$\omega_{N|\mathbb{B}_{N-1}} = \left(\frac{N+1}{N}\right)\omega_{N-1}.$$

Since the first logarithmic Chern class of the quotient X_Γ is just the cohomology class of $\frac{i}{2\pi}\omega$, we get the relation

$$\bar{c}_1(\overline{X},\Delta)_{|\overline{D}_\alpha} = \frac{N+1}{N}\bar{c}_1(\overline{D}_\alpha,\Delta_\alpha) \quad (\in H^2(\overline{D}_\alpha,\mathbb{Q})). \tag{5.31}$$

Similarly, calculating the second Chern class in terms of the curvature of the Bergmann metric

$$ds^2 = \frac{\partial^2}{\partial z_\alpha \partial \bar{z}_\beta}\log K_{\mathcal{D}}(z,\bar{z}),$$

we get the relation

$$\bar{c}_2(\overline{X},\Delta)_{|\overline{D}_\alpha} = \frac{N+1}{N-1}\bar{c}_2(\overline{D}_\alpha,\Delta_\alpha). \tag{5.32}$$

The proposition follows from (5.31) and (5.32). $\qquad\square$

Definition 5.5.5 Let $\overline{X}$ be projective and smooth of dimension N, $\Delta \in \overline{X}$ a normal crossings divisor. The *Yau proportionality factor* is:

$$\begin{aligned}
\mathcal{Y}(\overline{X},\Delta) &= N\bar{c}_1^N(\overline{X},\Delta) - 2(N+1)\bar{c}_1^{N-2}(\overline{X},\Delta)\bar{c}_2(\overline{X},\Delta)\\
&= Nc_1^N(\overline{X}) - 2(N+1)c_1^{N-2}(\overline{X})c_2(\overline{X}) + (-1)^N N(\Delta)^N.
\end{aligned}$$

Let $\overline{D}_\alpha \subset \overline{X}$ be the compactification of a codimension one subvariety with compactification divisor $\Delta_\alpha = \overline{D}_\alpha - D_\alpha = \overline{D}_\alpha \cap \Delta$. The *relative proportionality factors* are

$$\begin{aligned}
\mathcal{R}_1(\overline{D}_\alpha,\Delta_\alpha) &= \bar{c}_1^{N-1}(\overline{X},\Delta)_{|\overline{D}_\alpha} - \left(\frac{N+1}{N}\right)^{N-1}\bar{c}_1^{N-1}(\overline{D}_\alpha,\Delta_\alpha)\\
\mathcal{R}_2(\overline{D}_\alpha,\Delta_\alpha) &= \bar{c}_1^{N-3}(\overline{X},\Delta)\bar{c}_2(\overline{X},\Delta)_{|\overline{D}_\alpha} - \frac{N+1}{N-1}\left(\frac{N+1}{N}\right)^{N-3}.\\
&\quad\ \bar{c}_1^{N-3}(\overline{D}_\alpha,\Delta_\alpha)\bar{c}_2(\overline{D}_\alpha,\Delta_\alpha).
\end{aligned}$$

Note that, if $\overline{D}_\alpha$ is a surface, then $\mathcal{R}_2(\overline{D}_\alpha,\Delta_\alpha) = 0$ is a vacant condition.

Theorem 5.5.6 *Let* $(\overline{X}, \Delta)$ *be as in Lemma 5.5.1, with* Δ *a disjoint union of abelian varieties. If* $N \geq 3$ *and there exist divisors* $\overline{D}_\alpha \subset \overline{X}$ *and rational numbers* λ_α *such that, in* $H^2(\overline{X}, \mathbb{Q})$, $\overline{c}_1(\overline{X}, \Delta) = \sum \lambda_\alpha \overline{D}_\alpha$ *holds and the* $\overline{D}_\alpha$ *fulfill:* $\mathcal{R}_1(\overline{D}_\alpha, \Delta_\alpha) = \mathcal{R}_2(\overline{D}_\alpha, \Delta_\alpha) = \mathcal{Y}(\overline{D}_\alpha, \Delta_\alpha) = 0$, *then*

$$\mathcal{Y}(\overline{X}, \Delta) = 0.$$

Proof: From the assumption $\overline{c}_1(\overline{X}, \Delta) = \sum \lambda_\alpha \overline{D}_\alpha$ we can write

$$
\begin{aligned}
\mathcal{Y}(\overline{X}, \Delta) &= (\sum \lambda_\alpha \overline{D}_\alpha)(N\overline{c}_1^{N-1}(\overline{X}, \Delta) - 2(N+1)\overline{c}_1^{N-3}(\overline{X}, \Delta)\overline{c}_2(\overline{X}, \Delta)) \\
&= (\sum \lambda_\alpha \overline{D}_\alpha)\mathcal{Y}_1(\overline{X}, \Delta)
\end{aligned}
$$

and it suffices to show $\mathcal{Y}_1(\overline{X}, \Delta)_{|\overline{D}_\alpha} = 0$ for all $\overline{D}_\alpha$. Now calculating,

$$
\begin{aligned}
\mathcal{Y}_1(\overline{X}, \Delta)_{|\overline{D}_\alpha} &= N\overline{c}_1^{N-1}(\overline{X}, \Delta)_{|\overline{D}_\alpha} - 2(N+1)\overline{c}_1^{N-3}(\overline{X}, \Delta)\overline{c}_2(\overline{X}, \Delta)_{|\overline{D}_\alpha} \\
(\mathcal{R}_1 = \mathcal{R}_2 = 0) \quad &= N\left(\frac{N+1}{N}\right)^{N-1} \overline{c}_1^{N-1}(\overline{D}_\alpha, \Delta_\alpha) \\
&\quad -2(N+1)\left(\frac{N+1}{N-1}\right)\left(\frac{N+1}{N}\right)^{N-3} \overline{c}_1^{N-3}(\overline{D}_\alpha, \Delta_\alpha)\overline{c}_2(\overline{D}_\alpha, \Delta_\alpha) \\
(\mathcal{Y}(\overline{D}_\alpha, \Delta_\alpha) = 0) \quad &= \left[N\left(\frac{N+1}{N}\right)^{N-1}\frac{2N}{N-1}\right. \\
&\quad \left. -2(N+1)\left(\frac{N+1}{N-1}\right)\left(\frac{N+1}{N}\right)^{N-3}\right]\overline{c}_1^{N-3}(\overline{D}_\alpha, \Delta_\alpha)\overline{c}_2(\overline{D}_\alpha, \Delta_\alpha) \\
&= \left[\frac{2(N+1)^{N-1}}{N^{N-3}(N-1)} - \frac{2(N+1)^{N-1}}{N^{N-3}(N-1)}\right]. \\
&\quad \overline{c}_1^{N-3}(\overline{D}_\alpha, \Delta_\alpha)\overline{c}_2(\overline{D}_\alpha, \Delta_\alpha) \\
&= 0.
\end{aligned}
$$

This completes the proof of Theorem 5.5.6.   □

This result reduces the verification of $\mathcal{Y}(\overline{X}, \Delta) = 0$ to that of the $\mathcal{Y}$, $\mathcal{R}_i$ for some divisors fulfilling the assumptions of 5.5.6, something lower-dimensional.

Remark 5.5.7 It is clear that one may apply induction on the dimension of the subvarieties $\overline{D}_\alpha$, reducing essentially to the case of *surfaces* on $\overline{X}$ which are the two-dimensional intersections of the various $\overline{D}_\alpha$. One such formulation is given in the author's dissertation, [H0], Corollary 4.2.3. We leave this to the interested reader.

5.6 Uniformization

The purpose of this section is to prove the following result.

Theorem 5.6.1 $\mathcal{B}_4$ *is biregularly the Baily-Borel embedding of the arithmetic quotient* $X_{\Gamma_{\sqrt{-3}}(2)} = \Gamma_{\sqrt{-3}}(2)\backslash\mathbb{B}_3$, *where*

$$\Gamma_{\sqrt{-3}}(2) \subset U(3,1;\mathcal{O}_{\mathbb{Q}(\sqrt{-3})})$$

is the principal congruence subgroup of level 2.

Proof: Note that $\Gamma_{\sqrt{-3}}(2)$ is *not* a torsion free group, so one cannot apply Theorem 5.5.3 directly. Our proof will be performed in two steps:

I Show the existence of a finite Galois branched cover $Y^* \longrightarrow \mathcal{B}_4$, such that Y^* is the Satake compactification of X_{Γ_Y} for some torsion free subgroup Γ_Y.

II By I we know that the open part $\mathcal{B} - \{45 \text{ nodes}\}$ is a ball quotient, say $\Gamma_{\mathcal{B}}$, where $\Gamma_{\mathcal{B}}$ has torsion. We then use all the geometry of $\mathcal{B}$ to identify the group $\Gamma_{\mathcal{B}}$, showing it is $\Gamma_{\sqrt{-3}}(2)$.

5.6.1 Step I

We first consider the j−planes contained in $\mathcal{B}_4$, each of which contain 12 of the κ−lines which meet four at a time in the nine inflection points of the Hesse pencil.

Lemma 5.6.1.1 $\mathbb{P}^2$ *is the compactification of* $\Gamma_{\sqrt{-3}}(2)\backslash\mathbb{B}_2$, *where* $\Gamma_{\sqrt{-3}}(2) \subset U(2,1;\mathcal{O}_{\mathbb{Q}(\sqrt{-3})})$ *is the principal congruence subgroup of level 2.* $\mathbb{P}^2 - \{$the nine inflection points of the Hesse pencil$\} = \Gamma_{\sqrt{-3}}(2)\backslash\mathbb{B}_2$, *i.e., the inflection points are the cusps.*

Proof: It follows already from Höfer's thesis [Hö] that $\mathbb{P}^2 - \{$nine inflection points of the Hesse pencil$\}$ is a two-dimensional ball quotient. The group is identified in [J]; there it is essentially reduced to identifying each of the 12 lines, without the three inflection points lying on each, as a quotient $\Gamma_{\sqrt{-3}}(2)\backslash\mathbb{B}_1$, this time with $\Gamma_{\sqrt{-3}}(2) \subset U(1,1;\mathcal{O}_{\mathbb{Q}(\sqrt{-3})})$, as well as the blown up inflection points as quotients of $E_\varrho = \mathbb{C}/\mathbb{Z} \oplus \varrho\mathbb{Z}$, everything in an equivariant way. $\qquad\square$

Remark $\mathbb{P}^2$ blown up in those nine inflection points is the elliptic modular surface $S(3)$; the 12 lines give the four triangles of singular fibres of $S(3)$, the nine exceptional divisors are the nine sections.

Corollary 5.6.1.1 *There are 40 two-dimensional ball quotients on the Burkhardt quartic, which intersect along one-dimensional subball quotients. The cusps of these two-dimensional ball quotients are nodes of $\mathcal{B}_4$, and conversely, any node of $\mathcal{B}_4$ lies in some (in eight, in fact) of these surfaces.*

Proof: It follows from Lemma 5.6.1.1 that each j−plane is the said ball quotient, and that the nine cusps of that surface are the nine of the 45 nodes which lie in the j−plane. It remains to justify the statement on intersections; but the j−planes intersect each other along the κ−lines, which are the one-dimensional modular curves mentioned in the proof of Lemma 5.6.1.1. The fact that each node is contained in eight j–planes is mentioned in the last paragraph of 5.2.1 II.   □

Let $\varrho : \widetilde{B}_4 \longrightarrow B_4$ denote the blow up of B_4 at the 45 nodes (i.e., a desingularization), and let Δ denote the union of the 45 exceptional divisors. Note that because the j–planes contain all 45 nodes and B_4 is a hypersurface, $\bar{c}_1(\widetilde{B}_4, \Delta)$ *can* be written as total transforms of $\sum \lambda_\alpha \overline{D}_\alpha$, $\overline{D}_\alpha$ the j–planes. More precisely, we have

Lemma 5.6.1.2 *In* $H^2(\widetilde{B}_4, \mathbb{Q})$ *we have* $\bar{c}_1(\widetilde{B}_4, \Delta) = \sum \lambda_\alpha \varrho^*(\overline{D}_\alpha)$ *for suitable coefficients* λ_α, *and*

$$\bar{c}_1(\widetilde{B}_4, \Delta) = c_1(\widetilde{B}_4) - \Delta.$$

Proof: Since B_4 is a quartic, we have $c_1(\widetilde{B}_4) = \varrho^*(\Omega)$, for a hyperplane class Ω on B_4: by adjunction on the quartic we have

$$\begin{aligned} c_1(B_4) &= c_1(\mathbb{P}^4)_{|B_4} - c_1(N_{\mathbb{P}^4}B_4) \\ &= 5H_{|B_4} - 4H_{|B_4} = H_{|B_4} = \Omega. \end{aligned}$$

Since the $\overline{D}_\alpha$ contain all nodes, their total transforms contain all exceptional divisors, so over $\mathbb{Q}$ we can write $\bar{c}_1(\widetilde{B}_4, \Delta)$ as stated.   □

We now consider a special branched cover of B_4. Consider among the set of the 40 Steiner primes, ten of which intersect B_4 in the union of all 40 j–planes, say $\mathcal{S}_1, \ldots, \mathcal{S}_{10}$. We may apply the construction of Fermat covers (3.12) to the arrangement $\mathcal{A} = \bigcup_{i=1}^{10} \mathcal{S}_i$. Note that $\mathcal{A}$ is an arrangement in general position, i.e., no singular locus (see Definition 3.1.1). Hence $Y(\mathcal{A}, d) \longrightarrow \mathbb{P}^4$ is smooth, no resolution is necessary. We consider now $d = 2$. Let $Y \subset Y(\mathcal{A}, 2)$ be the induced cover over B_4, i.e., as in the following diagram:

$$\begin{array}{ccc} Y(\mathcal{A}, 2) & \longrightarrow & \mathbb{P}^4 \\ \cup & & \cup \\ Y & \longrightarrow & B_4. \end{array}$$

Y is of course singular at the nodes of B_4, but is smooth elsewhere. Let $\widetilde{Y}(\mathcal{A}, 2) \longrightarrow Y(\mathcal{A}, 2)$ be the blowup of $Y(\mathcal{A}, 2)$ at the inverse images of the nodes, and let $\widetilde{Y} \subset \widetilde{Y}(\mathcal{A}, 2)$ be the proper transform of Y. Then

$$\begin{array}{ccc} \widetilde{Y} & \xrightarrow{\ \tilde{\varrho}\ } & Y \\ \scriptstyle\tilde{\pi} \downarrow & & \downarrow \\ \widetilde{B}_4 & \xrightarrow{\ \varrho\ } & B_4 \end{array} \qquad (5.33)$$

is a resolution of singularities, so $\widetilde{Y}$ is smooth. Note in particular that the compactifying divisors on $\widetilde{Y}$ are $E_\varrho \times E_\varrho$. Let, for each proper transform of j–plane $\overline{D}_\alpha \subset \widetilde{B}_4$, $\overline{E}_\alpha$ denote any of the components of $\tilde{\pi}^{-1}(\overline{D}_\alpha)$ (they are all isomorphic, as one easily sees). We wish to apply Theorem 5.5.6 to $(\widetilde{Y}, \cup \overline{E}_\alpha)$, so we first check that the assumptions are fulfilled.

Lemma 5.6.1.3 *There are* $\lambda_\alpha \in \mathbb{Q}$ *such that* $\bar{c}_1(\tilde{Y}, \tilde{\Delta})$ *can be written*

$$\bar{c}_1(\tilde{Y}, \tilde{\Delta}) = \sum \lambda_\beta \overline{E}_\beta,$$

where the sum on the right hand side is over all components $\overline{E}_\beta \subset \tilde{Y}$ *which cover some* $\overline{D}_\alpha$.

Proof: First apply the standard formula for c_1 of a branched cover:

$$c_1(\tilde{Y}) = \tilde{\pi}^*(c_1(\tilde{B}_4) - \frac{1}{2}\mathcal{R}) = \tilde{\pi}^*(c_Y), \tag{5.34}$$

where $\mathcal{R} \in H^2(\tilde{B}_4, \mathbb{Z})$ is the ramification class, and the equation defines the class c_Y. Since the cover $\tilde{\pi} : \tilde{Y} \longrightarrow \tilde{B}_4$ is branched along the $\overline{E}_\beta$ and the exceptional locus $\tilde{\Delta}$, it follows that $\mathcal{R}$ can be rationally expressed in terms of the $\overline{E}_\beta$ and $\tilde{\Delta}_i$,

$$\mathcal{R} = \sum r_\beta \overline{E}_\beta + \sum s_i \tilde{\Delta}_i,$$

with coefficients r_β, $s_i \in \mathbb{Q}$. Similarly, since $H^2(\tilde{B}_4, \mathbb{Z})$ can be generated by a hyperplane class and the 45 exceptional divisors, we see that $\bar{c}_1(\tilde{B}_4)$ can also be expressed in terms of the $\overline{E}_\beta$ and $\tilde{\Delta}_i$,

$$c_1(\tilde{B}_4) = \sum l_\beta \overline{E}_\beta + \sum m_i \tilde{\Delta}_i.$$

Therefore we get for the usual $c_1(\tilde{Y})$ an expression (note that since Δ is in the branch locus, we have $\tilde{\pi}^*(\Delta) = 2\tilde{\Delta}$)

$$c_1(\tilde{Y}) = \tilde{\pi}^* \left(\sum \left(l_\beta - \frac{1}{2}r_\beta \right) \overline{E}_\beta + \sum \left(m_i - \frac{1}{2}s_i \right) \Delta_i \right).$$

Consequently,

$$\bar{c}_1(\tilde{Y}, \tilde{\Delta}) = c_1(\tilde{Y}) - \tilde{\Delta} = \tilde{\pi}^* \left(\sum \left(l_\beta - \frac{1}{2}r_\beta \right) \overline{E}_\beta + \sum \left(m_i - \frac{1}{2}s_i - \frac{1}{2} \right) \Delta_i \right),$$

and by (5.29), each coefficient $m_i - \frac{1}{2}s_i - \frac{1}{2} = 0$ as each $\tilde{\Delta}_i$ is an abelian variety, yielding (with $\lambda_\beta = l_\beta - \frac{1}{2}r_\beta$) the statement of the lemma.   □

More explicitly, we have coefficients $\sum r_\beta = 10$, since there are 10 branching hyperplanes in $\mathbb{P}^4$; since there are eight j-planes passing through each node, we have $s_i = 1 - 8 = -7$, and consequently $m_i = -3$. This latter coefficient can also be obtained as follows: $\tilde{B}_4$ is the proper transform of a quartic hypersurface in $\mathbb{P}^4$, so has normal bundle the class of $4H$, where H denotes the hyperplane class on $\tilde{\mathbb{P}}^4$. Then, applying adjunction we have $(c_1(\tilde{\mathbb{P}}^4) = 5H - 3\sum P_i$, $P_{i|\tilde{B}_4} = \Delta_i$, $H_{|\tilde{B}_4} = h)$

$$c_1(\tilde{\mathbb{P}}^4)_{|\tilde{B}_4} = c_1(\tilde{B}_4) + c_1(N_{\tilde{\mathbb{P}}^4}\tilde{B}_4) = c_1(\tilde{B}_4) + 4h,$$

which implies $c_1(\widetilde{B}_4) = h - 3\sum \Delta_i$. From this we get

$$c_Y - \frac{1}{2}\Delta = c_1(\widetilde{B}_4) - \frac{1}{2}\mathcal{R} - \frac{1}{2}\Delta = (1-5)h = -4h. \tag{5.35}$$

For each $\overline{E}_\beta$, let Δ_β denote the intersection $\overline{E}_\beta \cap \Delta$.

Lemma 5.6.1.4 *Each component* $(\overline{E}_\beta, \Delta_\beta)$ *fulfills* $\mathcal{R}_1(\overline{E}_\beta, \Delta_\beta) = 0$.

Proof: The calculation of $\bar{c}_1^2(\overline{E}_\beta, \Delta_\beta)$ is an easy task; if $\deg\pi$ denotes the degree of the cover $\overline{E}_\beta \longrightarrow \overline{D}_\alpha$, then

$$\bar{c}_1^2(\overline{E}_\beta, \Delta_\beta) = (\deg\pi)3^2. \tag{5.36}$$

Indeed, $c_1(\overline{E}_\beta) = \widetilde{\pi}^*(c_1(\overline{D}_\alpha) - \frac{1}{2}\mathcal{R}_\alpha) =: \widetilde{\pi}^*(c_{E_\beta})$ (with the notations c_{E_β} a class on $\overline{D}_\alpha$, and $\mathcal{R}_\alpha$ the ramification locus on $\overline{D}_\alpha$ of the cover $\overline{E}_\beta \longrightarrow \overline{D}_\alpha$), and $\left(c_1(\overline{D}_\alpha) - \frac{1}{2}\mathcal{R}_\alpha\right)^2$ is calculated as follows. $c_1(\overline{D}_\alpha) = 3\omega - \sum_1^9 E_i$ ($E_i \in \overline{D}_\alpha$ one of the nine exceptional divisors on $\overline{D}_\alpha$, ω the total transform of the hyperplane class), $\mathcal{R}_\alpha = 12\omega + \sum(1-4)E_i$ (since through each of the nodes, four of the 12 lines pass), so

$$c_{E_\beta} = 3\omega - \frac{1}{2}(12\omega) + \frac{1}{2}\sum E_i \tag{5.37}$$
$$= -3\omega + \frac{1}{2}\sum E_i$$

and $\bar{c}^2(\overline{E}_\beta, \Delta_\beta) = \widetilde{\pi}^*\left((c_{E_\beta} - \frac{1}{2}\sum E_i)^2\right) = (\deg\pi)3^2$, as stated.

To calculate $\bar{c}_1(\widetilde{Y}, \widetilde{\Delta})_{|\overline{E}_\beta}$, we utilize (5.35) and the relation

$$\bar{c}_1^2(\widetilde{Y}, \widetilde{\Delta})_{|\overline{E}_\beta} = \widetilde{\pi}^*\left((c_Y - \frac{1}{2}\Delta)^2 \cdot \overline{D}_\alpha\right) = (\deg\pi)\left((c_Y - \frac{1}{2}\Delta)^2 \cdot \overline{D}_\alpha\right).$$

This then yields, together with (5.35)

$$\bar{c}_1^2(\widetilde{Y}, \widetilde{\Delta})_{|\overline{E}_\beta} = (\deg\pi)\left((-4h)^2 \cdot \overline{D}_\alpha\right) = (\deg\pi)4^2.$$

This implies $\bar{c}_1^2(\widetilde{Y}, \widetilde{\Delta})_{|\overline{E}_\beta} = (\deg\pi)4^2 = \left(\frac{4}{3}\right)^2 \bar{c}_1^2(\overline{E}_\beta, \Delta_\beta)$, and with this the lemma (see 5.5.5). $\square$

By Lemmas 5.6.1.4 and 5.6.1.3 we may apply Theorem 5.5.6 to $(X, \Delta, \overline{D}_\alpha) = (\widetilde{Y}, \widetilde{\Delta}, \overline{E}_\beta)$ in the situation here, and this results in

Corollary 5.6.1.2 *Let* $\widetilde{\pi} : \widetilde{Y} \longrightarrow \widetilde{B}_4$ *be the cover introduced above. Let* $\Delta \subset \widetilde{B}_4$ *be the union of the exceptional divisors,* $\widetilde{\Delta} \subset \widetilde{Y}$ *the inverse image. Then*

$$\mathcal{Y}(\widetilde{Y}, \widetilde{\Delta}) = 0.$$

Hence $\widetilde{Y}$ *is the compactification of a ball quotient* $\widetilde{Y} - \widetilde{\Delta}$.

Proof: Since in each exceptional $\mathbb{P}^1 \times \mathbb{P}^1$ the intersections with the proper transforms of the j-planes are eight curves of the form $\{pt\} \times \mathbb{P}^1$ and $\mathbb{P}^1 \times \{pt\}$, the branched cover over them is a product. For $d = 2$ the Fermat cover of $\mathbb{P}^1$, branched at the four designated points, is just the curve E_ϱ, hence the cover on $\tilde{Y}$ of the exceptional $\mathbb{P}^1 \times \mathbb{P}^1$ is a product $E_\varrho \times E_\varrho$ of elliptic curves, so condition i) in Theorem 5.5.3 is satisfied. As for ii) it is sufficient to refer to (5.34) and (5.35). Hence, by Theorem 5.5.3, $\tilde{Y}$ is the compactification of a ball quotient. $\square$

This Corollary completes Step I.

Remark: Both Lemma 5.6.1.1 (with a suitable Γ) and Lemma 5.6.1.4 remain true for *any* Galois cover $Y \longrightarrow B_4$, which is branched to degree two along the j-planes. We have choosen the particular cover (5.33) in order to identify the group below, see Remark 5.6.14. But the proofs show that, in particular, the double cover of B_4 branched along the 40 j-planes is also a ball quotient.

5.6.2   Step II

From Step I we have $Y = \tilde{Y} - \tilde{\Delta} = \Gamma_Y \backslash \mathbb{B}_3$. The cover $\tilde{Y} \longrightarrow \tilde{B}$ is branched at the exceptional divisors and along the (proper transforms of the) 40 j-planes, and it follows immediately that $B - \{45 \text{ nodes}\}$ is an open ball quotient, say $\Gamma_B \backslash \mathbb{B}_3$, and that Γ_Y is a normal subgroup in Γ_B, with quotient group the Galois group of the cover, $(\mathbb{Z}/2\mathbb{Z})^9$. We now proceed to identify the group Γ_B. We wish to show that $\Gamma_B \cong \Gamma_{\sqrt{-3}}(2)$. We begin with some general material on compactifications of ball quotients. Though well-known, it is particularly of importance that this also applies to *non-arithmetic* groups, as we do not know whether Γ_B is arithmetic or not.

We fix the following unbounded realization of the n-ball:

$$\mathbb{B}_n = \left\{ (u, v_1, \ldots, v_{n-1}) \,\middle|\, \mathrm{Im}(u) - \frac{1}{2}\sum |v_i|^2 > 0 \right\},$$

and identify $U(3,1)$ with the automorphism group of this domain preserving the hermitian form given by the matrix

$$H = \begin{pmatrix} 0 & 0 & i \\ 0 & -1_{n-2} & 0 \\ -i & 0 & 0 \end{pmatrix}.$$

The cusp at ∞ is given by the limit of $\mathrm{Im}(u)$ going to infinity, or put differently, the limit $w \to 0$, where $w = \exp(2\pi i u)$. The stabilizer of the cusp ∞ will be denoted $P(\infty)$, and is a maximal $\mathbb{R}$-parabolic of $U(3,1)$. The group $P(\infty)$ splits $P(\infty) = A_\infty \cdot M_\infty \cdot U(\infty)$, where $U(\infty)$ is the unipotent radical. Under the identification mentioned above, $U(\infty)$ may be written as the set of matrices of the form:

$$U(\infty) = \left\{ [\alpha, x] = \begin{pmatrix} 1 & \alpha & x + \frac{i}{2}|\alpha|^2 \\ 0 & 1 & i^t\overline{\alpha} \\ 0 & 0 & 1 \end{pmatrix} \,\middle|\, \alpha \in \mathbb{C}^{n-1}, x \in \mathbb{R} \right\},$$

with multiplication $[\alpha, x][\alpha', x'] = [\alpha + \alpha', x + x' - \text{Im}(^t\overline{\alpha}\alpha')]$. The center of $U(\infty)$ is given by $Z(\infty) = \{[0, x] \in U(\infty)\}$, and we have an exact sequence

$$1 \longrightarrow Z(\infty) \longrightarrow U(\infty) \longrightarrow L(\infty) \longrightarrow 1,$$

where $L(\infty)$ can be identified with $\mathbb{C}^{n-1}$. Now consider any discrete subgroup $\Gamma \subset U(3,1)$, for which we assume that $U(\infty)/(U(\infty) \cap \Gamma)$ is *compact*; this is equivalent to the statement that the quotient $\Gamma \backslash \mathbb{B}_n$ is, locally near the cusp, of finite volume. Set $\Gamma(\infty) = U(\infty) \cap \Gamma$; we then have an exact sequence

$$1 \longrightarrow \Delta(\infty) \longrightarrow \Gamma(\infty) \longrightarrow \Lambda(\infty) \longrightarrow 1,$$

where $\Delta(\infty) = \Gamma(\infty) \cap Z(\infty) \cong \mathbb{Z}$ is generated by an element $[0, q]$, $q \in \mathbb{R}$. Moreover, one has (cf. [Yo], Lemma 1.3)

$$\frac{2}{q}\text{Im}(^t\overline{\alpha}\alpha') \in \mathbb{Z}.$$

It follows from this that the complex torus $L(\infty)/\Lambda(\infty)$ carries a canonical polarization, given by (cf. *loc. cit.*, Corollary 1.1)

$$(x, y) \mapsto < x, y >:= \frac{2}{q}\text{Im}(^t xy).$$

Then one has, also in the non-arithmetic case

Lemma 5.6.2 (Compactification of ball quotients) *Let* $X = \Gamma \backslash \mathbb{B}_n$, Γ *torsion free, a finite volume ball quotient. Then X can be smoothly compactified by adding an abelian variety of dimension $n - 1$ at each cusp.*

Proof: This was done in great detail in [Hem] §2 for $n = 2$ and in [Yo] §1 in arbitrary dimension. We therefore just explain the compactification. First we may assume the cusp is ∞, and the description above applies to $U(\infty) \cap \Gamma = \Gamma(\infty)$. A neighborhood of the cusp is given by

$$\mathcal{U} = \left\{ (u, v_1, \ldots, v_{n-1}) \middle| \text{Im}(u) - \frac{1}{2}\sum |v_i|^2 > N \right\}$$

for some $N > 0$. $\Gamma(\infty)$ acts on $\mathcal{U}$, and the quotient $\Gamma(\infty) \backslash \mathcal{U}$ is analytically isomorphic to a neighborhood of the cusp on the open ball quotient (for N sufficiently large). One glues in the abelian variety $L(\infty)/\Lambda(\infty)$ by noting that $\Gamma(\infty) \backslash \mathcal{U}$ is the total space of a line bundle over $L(\infty)/\Lambda(\infty)$ minus the zero section. The zero section is given by setting $w = 0$ for the w defined above, and v_i are viewed as coordinates on $L(\infty)$. Adding the zero section is the compactification near the given cusp. Further details are given in [Hem] and [Yo], both of which do not assume Γ is arithmetic.   □

Corollary 5.6.3 *Let $\Gamma \subset P(\infty)$ be a discrete, torsion free subgroup of finite volume in the parabolic stabilizing the cusp at ∞. Then Γ determines and is determined by the lattice $\Lambda_\Gamma := L(\infty) \cap \Gamma$ and the generator $[0, q]$ of the center $\Delta_\Gamma := Z(\infty) \cap \Gamma$. These two data in turn determine and are determined by the compactifying divisor Δ and the first Chern class of its normal bundle in the compactified ball quotient.*

Proof: The first statement is obvious from the above analysis. For the second, we refer for details to [Hem] §2, the idea being as follows. The assumption that Γ is torsion free implies that $\Gamma \subset U(\infty)$. Since $N_X(\Delta)$ is a line bundle on the abelian variety Δ, and since $N_X(\Delta)$ is negative (because Δ is exceptional), $N_X^*(\Delta)$ is positive, and the first Chern class is given by the zero locus of a theta function. As a local coordinate on the line bundle one takes $w = \exp(\frac{2\pi i u}{q})$, where q is the generator of Δ_Γ. In terms of the coordinates $(w, v_1, \ldots, v_{n-1})$, the element $[\alpha, r_\alpha + jq]$ (where r_α is the unique minimal $\mathrm{mod}(q)$ real number such that $[\alpha, r_\alpha] \in \Gamma(\infty) = U(\infty) \cap \Gamma$) acts as $(\alpha^* = {}^t\bar{\alpha})$

$$[\alpha, r_\alpha + jq](w, v_1, \ldots, v_{n-1}) = (w + \alpha \cdot v + \frac{i}{2}\alpha \cdot \alpha^* + r_\alpha + jq, v + i\alpha^*).$$

It follows that the theta function which is a section of $N_X^*(\Delta)$, $\theta_n(v)$, satisfies:

$$\theta_n(v + i\alpha^*)\exp(\frac{2\pi i n}{q}(\alpha \cdot v + \frac{i}{2}\alpha \cdot \alpha^* + r_\alpha)) = \theta_n(v), \quad \forall [\alpha, r_\alpha] \in \Gamma(\infty).$$

Clearly then q determines and is determined by this bundle (up to translations), hence also determines and is determined by the dual bundle $N_X(\Delta)$ (up to translations). But on an abelian variety a line bundle is determined up to translations by the first Chern class, yielding the second statement. $\qquad \square$

Now we return to subballs.

Lemma 5.6.4 *Let X be an n-dimensional ball quotient, $D \subset X$ an $(n-1)$-dimensional subball quotient, $D = \Gamma_{n-1}\backslash \mathbb{B}_{n-1}$. Fix an inclusion $\mathbb{B}_{n-1} \subset \mathbb{B}_n$ covering the inclusion $D \subset X$,*

$$
\begin{array}{ccc}
\mathbb{B}_{n-1} & \hookrightarrow & \mathbb{B}_n \\
\Gamma_{n-1} \downarrow & & \downarrow \Gamma \\
D & \hookrightarrow & X
\end{array}.
$$

Then $\Gamma_{n-1} = \mathrm{Aut}(\mathbb{B}_{n-1}) \cap \Gamma$, viewing $\mathrm{Aut}(\mathbb{B}_{n-1})$ as (up to a torus factor) the normalizer of $\mathbb{B}_{n-1}$ in $\mathbb{B}_n$.

Proof: The elements of $\mathrm{Aut}(\mathbb{B}_{n-1})$ normalize $\mathbb{B}_{n-1}$, while the elements of Γ act properly discontinuously on $\mathbb{B}_n$. The statement then follows from the definitions. $\qquad \square$

In particular, a subball quotient $D \subset X$ determines the discrete subgroup $\Gamma_{n-1} \subset \Gamma$ once $\mathbb{B}_{n-1}$ has been fixed.

Lemma 5.6.5 *Let $\mathcal{D}$ be a bounded symmetric domain, $\Gamma \subset \mathrm{Aut}(\mathcal{D})$ a properly discountinuous subgroup and $X = \Gamma_X\backslash\mathcal{D}$ some quotient of $\mathcal{D}$ by a discrete subgroup Γ_X. Suppose we have a commutative diagram*

$$
\begin{array}{ccc}
\phi: \mathcal{D} & \xrightarrow{\sim} & \mathcal{D} \\
p_X \downarrow & & \downarrow p \\
\phi: X & \xrightarrow{\sim} & \Gamma\backslash\mathcal{D}.
\end{array}
$$

Then Γ_X is conjugate to Γ.

Proof: The isomorphism $\mathcal{D} \xrightarrow{\sim} \mathcal{D}$ is given by some element $g \in \mathrm{Aut}(\mathcal{D})$; from the commutativity of the diagram we have for all $x \in \mathcal{D}$

$$\phi(p_X(x)) = p(\phi(x)) \iff g(p_X(x)) = p(g(x)),$$

from which it follows that $g\Gamma_X g^{-1} = \Gamma$. $\qquad \square$

We now apply these general results to our case at hand. Fix, for the rest of the proof, the subgroup $\Gamma_{\sqrt{-3}}(2) \subset U(3,1;\mathcal{O}_K)$, the projectivization of which is identified with a discrete subgroup of the automorphism group of the domain

$$\mathbf{D}_F = \left\{ (u, v_1, v_2) \in \mathbb{C}^3 \,\Big|\, \mathrm{Im}(u) - \frac{\sqrt{3}}{2} \sum |v_i|^2 > 0 \right\}.$$

This is a slightly different realization of the ball than that above, and the corresponding hermitian form is given by the matrix

$$\mathbf{J} = \begin{pmatrix} \sqrt{3} & 0 & 0 & 0 \\ 0 & \sqrt{3} & 0 & 0 \\ 0 & 0 & 0 & -i \\ 0 & 0 & i & 0 \end{pmatrix}. \qquad (5.38)$$

The reason for choosing this form is because it determines the skew-hermitian matrix $\mathbf{R} = i\mathbf{J}$, which is a matrix T for Shimura's construction (cf. sections 1.1.3 and 1.1.4). This unbounded realization of the three-ball is the one occuring in Shimura's original construction, see [Sh2], for the general construction; see also the first few pages of [Sh5] for this particular case. For the matrix $\mathbf{R}$, Theorem 1.1.8 gives a precise moduli interpretation of the arithmetic quotient. We fix the one- and two-dimensional subballs

$$\mathbf{D}_{F,1} = \{(u,0,0)\} \subset \mathbf{D}_{F,2} = \{(u,0,v_2)\} \subset \mathbf{D}_F;$$

let $G_{F,1}$ and $G_{F,2}$ denote the corresponding automorphism groups $G_{F,i} := \mathcal{N}_{G_F}(\mathbf{D}_{F,i})/\mathcal{Z}_{G_F}(\mathbf{D}_{F,i})$ in $G_F = U(\mathbf{J}, K^4)$, for which we make the identification $PG_F(\mathbb{C}) = PU(\mathbf{J}, \mathbb{C}^4) = \mathrm{Aut}(\mathbf{D}_F)$. Finally, let $P(\infty)$ be the stabilizer of the cusp at infinity in G_F, and $U(\infty)$ the unipotent radical. Then

$$U(\infty) = \left\{ [x,y] = \begin{pmatrix} 1 & 0 & 0 & x_1 \\ 0 & 1 & 0 & x_2 \\ \sqrt{-3}\bar{x}_1 & \sqrt{-3}\bar{x}_2 & 1 & y + \frac{\sqrt{-3}}{2}|x|^2 \\ 0 & 0 & 0 & 1 \end{pmatrix} \,\Big|\, x \in \mathbb{C}^2,\, y \in \mathbb{R} \right\}.$$

We use the notations as above for $Z(\infty)$, $L(\infty)$, and consider the subgroup $\Gamma = \Gamma_{\sqrt{-3}}(1) = U(\mathbf{J}, \mathcal{O}_K^4)$ and denote its intersections as above by $\Delta(\infty), \Lambda(\infty)$ and $\Gamma(\infty)$. We then have for some $q \in \mathbb{R}$

$$\Delta(\infty) = \{[0, qj] | j \in \mathbb{Z}\} \subset \Gamma(\infty), \quad \Lambda(\infty) \cong \mathcal{O}_K^2.$$

Clearly, $L(\infty)/\Lambda(\infty) = E_\varrho \times E_\varrho$, with a product polarization. Hence, for any torsion-free subgroup of finite index $\Gamma \subset U(\mathbf{J}; \mathcal{O}_K^4)$, the compactifying divisors on the quotient will be copies of $E_\varrho \times E_\varrho$, but with different polarizations.

We know from Step I that both $\Gamma_{\tilde{\gamma}}$ and Γ_B are discrete subgroups of $U(\mathbf{J}, \mathbb{C}^4)$; we pick one of the parabolics of each, say $P' \subset \Gamma_{\tilde{\gamma}}$, $P \subset \Gamma_B$. These are contained in uniquely determined $\mathbb{R}$-subgroups $P' \subset \mathbf{P}'$, $P \subset \mathbf{P}$, and we conjugate

P', **P** in $U(\mathbf{J}, \mathbb{C})$ onto $P(\infty)$. In this way, if $\Gamma_B(\infty)$ and $\Gamma_{\widetilde{Y}}(\infty)$ denote these parabolics, then we have $\Gamma_B(\infty)$, $\Gamma_{\widetilde{Y}}(\infty) \subset P(\infty)$. Since $\Gamma_{\widetilde{Y}}$ is torsion free, with compactification divisor $E_\varrho \times E_\varrho$, we may apply Corollary 5.6.3 and get that $\Lambda_{\Gamma_{\widetilde{Y}}} = \mathcal{O}_K^2$. As to the number q we claim that the normal bundle of $E_\varrho \times E_\varrho$ in $\widetilde{Y}$ is equal to that of $E_\varrho \times E_\varrho$ in the compactification of $\Gamma_{\sqrt{-3}}(4) \backslash \mathbb{B}_3$. Note first that it is sufficient to show the corresponding fact for the $\mathbb{P}^1 \times \mathbb{P}^1$'s on $\widetilde{B}$ and on the compactification of $\Gamma_{\sqrt{-3}}(2) \backslash \mathbb{B}_3$, once we know the ramification of the cover $E_\varrho \times E_\varrho \longrightarrow \mathbb{P}^1 \times \mathbb{P}^1$ is the same for both $\widetilde{Y} \longrightarrow \widetilde{B}$ as for $\Gamma_{\sqrt{-3}}(4) \backslash \mathbb{B}_3 \longrightarrow \Gamma_{\sqrt{-3}}(2) \backslash \mathbb{B}_3$, as these data determine the Chern class of the line bundle. On $\widetilde{B}$ we know the Chern class is $(-1, -1)$, and an easy calculation shows the same for the compactification of $\Gamma_{\sqrt{-3}}(2) \backslash \mathbb{B}_3$ (use the fact ([J], p. 542) that for the cover $\Gamma_{\sqrt{-3}}(2\sqrt{-3}) \backslash \mathbb{B}_3$ the class is $(-6, -6)$, and the cover induced on the boundary components $E_\varrho \times E_\varrho \longrightarrow \mathbb{P}^1 \times \mathbb{P}^1$ is $p(3) \times p(3)$, where $p(3)$ is the map of *loc. cit.*, Figure 2, p. 534). Furthermore, we know the ramification of $E_\varrho \times E_\varrho \longrightarrow \mathbb{P}^1 \times \mathbb{P}^1$ on $\widetilde{Y} \longrightarrow \widetilde{B}$ is the Fermat cover, branched at the four points; it is easily seen that this is precisely the same cover induced on the compactification of $\Gamma_{\sqrt{-3}}(4) \backslash \mathbb{B}_3 \longrightarrow \Gamma_{\sqrt{-3}}(2) \backslash \mathbb{B}_3$. Now we may apply Corollary 5.6.3 and get the following.

Proposition 5.6.6 *The subgroups* $\Gamma_B(\infty)$, $\Gamma_{\widetilde{Y}}(\infty) \subset P(\infty)$ *are subgroups of finite index in the corresponding parabolic lattice* $P(\infty) \cap \Gamma_{\sqrt{-3}}(1)$, *namely* $\Gamma_B(\infty) = P(\infty) \cap \Gamma_{\sqrt{-3}}(2)$, $\Gamma_{\widetilde{Y}}(\infty) = P(\infty) \cap \Gamma_{\sqrt{-3}}(4)$.

Proof: For $\Gamma_{\widetilde{Y}}(\infty)$ this follows from 5.6.3. For it to apply to $\Gamma_B(\infty)$, we must know that the morphism $E_\varrho \times E_\varrho \longrightarrow \mathbb{P}^1 \times \mathbb{P}^1$ on the boundary components induced by the cover $\widetilde{Y} \longrightarrow \widetilde{B}$ is the same morphism as is determined by the finite group $\Gamma_B(\infty)/\Gamma_{\widetilde{Y}}(\infty)$ acting naturally on $E_\varrho \times E_\varrho$, which we just have shown. $\quad\square$

Next consider one of the j-planes, and the corresponding stabilizer N_1 in Γ_B; we embedd Γ_B in $U(\mathbf{J}, \mathbb{C}^4)$ such that N_1 is contained in the stabilizer $G_{F,2}$. Then $G_{F,2} \cap N_1$ is a lattice in $G_{F,2}$. It follows from Lemma 5.6.1.1 and Corollary 5.6.1.1 that the assumptions of Lemma 5.6.5 are satisfied. Indeed, for torsion free subgroups Γ and Γ_X, the quotients, being $K(\pi, 1)$ spaces, always possess such an extension of an isomorphism $\Gamma \backslash \mathcal{D} \longrightarrow \Gamma_X \backslash \mathcal{D}$ to an isomorphism $\mathcal{D} \longrightarrow \mathcal{D}$ as in Lemma 5.6.5. For Γ, Γ_X with torsion, this is a question of finite commutative diagrams. Hence consider

$$
\begin{array}{ccc}
M_1 \backslash \mathbb{B}_2 & \longrightarrow & \Gamma_{2,\sqrt{-3}}(4) \backslash \mathbb{B}_2 \\
\downarrow & & \downarrow \\
N_1 \backslash \mathbb{B}_2 & \xrightarrow{\sim} & \Gamma_{2,\sqrt{-3}}(2) \backslash \mathbb{B}_2.
\end{array}
$$

The fact that this diagram commutes amounts to identifying the cover

$$\Gamma_{2,\sqrt{-3}}(4) \backslash \mathbb{B}_2 \longrightarrow \Gamma_{2,\sqrt{-3}}(2) \backslash \mathbb{B}_2$$

as the corresponding (open part of) the Fermat cover $Y(\mathcal{A}, 2)$, where $\mathcal{A}$ is the arrangement of the 12 κ-lines. This is straightforward and is left to the reader. Hence by 5.6.5 $N_1 \subset G_{F,2}$ is conjugate to $G_{F,2} \cap \Gamma_{\sqrt{-3}}(2) = \Gamma_{2,\sqrt{-3}}(2)$. It follows that, for a given inclusion such that $N_1 \subset G_{F,2}$, we may conjugate it to get an inclusion $i_B : \Gamma_B \subset U(\mathbf{J}, \mathbb{C}^4)$ which satisfies

$$i_B(N_1) = \Gamma_{2,\sqrt{-3}}(2). \tag{5.39}$$

We now fix the embedding i_B fulfilling (5.39). Then ∞ is a cusp of both $\mathbf{D}_F$ and $\mathbf{D}_{F,2}$, with corresponding parabolics $P(\infty)$, $P_2(\infty)$. The parabolic lattices $P_1 \subset \Gamma_B$, $P_{N_1} \subset N_1$ clearly satisfy $i_B(P_{N_1}) \subset i_B(P_1) \subset P(\infty)$. It now follows from Proposition 5.6.6 that, in fact,

$$i_B(P_1) = P(\infty) \cap \Gamma_{\sqrt{-3}}(2). \tag{5.40}$$

The next step is to study the eight groups $N_1, \ldots, N_8$ which are normalizers of subballs, all of which contain the given cusp ∞ in their closures (that there are eight such is the fact that on $\mathcal{B}$, there are eight j-planes which contain a given node). In this case the situation is quite easy, as the abelian surface (for $\Gamma_{\tilde{Y}}$) or quotient thereof (for Γ_B) which is the compactification divisor (they are all isomorphic) is a product. The eight N_κ form two sets of four parallel ones, i.e., the intersections of the (proper transforms of the) subball quotients with the compactification divisors are of the form $\{pt\} \times E$, $E \times \{pt\}$, where E is the elliptic curve E_ϱ (for $\Gamma_{\tilde{Y}}$) or a quotient of E_ϱ, a $\mathbb{P}^1$ (for Γ_B). If we consider the action of $P(\infty) \cap \Gamma_{\sqrt{-3}}(1) / P(\infty) \cap \Gamma_{\sqrt{-3}}(2)$ on the factors, all eight curves are identified. In other words, if we now number these curves $C_1, \ldots, C_8$, then $C_1, \ldots, C_8$ are $\Gamma_{\sqrt{-3}}(1)$-equivalent. The same is true for Γ_B: consider the action of $G_{25,920}$ on the subgroups N_κ, $\kappa = 1, \ldots, 8$; then these subgroups are all $G_{25,920}$-equivalent. Next we require a lemma which is similar to [J], Lemma 3.14. For this we fix some fundamental domain for $\Gamma_{\sqrt{-3}}(2)$ (resp. for Γ_B) and consider the action of $\Gamma_{\sqrt{-3}}(1)$ (resp. of Γ'_B) on it, where Γ'_B is a discrete subgroup of $U(\mathbf{J}, \mathbb{C}^4)$ such that $\Gamma'_B / \Gamma_B = G_{25,920}$. Note that Γ'_B exists, since upon forming the quotient of $\mathcal{B}$ by $G_{25,920}$, all branch loci are either the subballs or the compactification divisors, hence this quotient is also a ball quotient.

Lemma 5.6.7 *(i) Two cusps c_1, c_2 of $\mathbb{B}_3$ are equivalent under $\Gamma_{\sqrt{-3}}(1)$ (resp. under Γ'_B) if and only if they are equivalent under the normalizer of B in $\Gamma_{\sqrt{-3}}(1)$ (resp. in Γ'_B), where B is the unique $\mathbb{B}_1$ with c_1 and c_2 in its closure, and if and only if they are equivalent under the normalizer of any $\mathbb{B}_2$ with both c_1 and c_2 in its closure, for which the group induced is $\Gamma_{2,\sqrt{-3}}(1)$.*

(ii) Two subballs B_1, B_2 ($\cong \mathbb{B}_2$) of $\mathbb{B}_3$ with the same cusp c in their closures are equivalent under $\Gamma_{\sqrt{-3}}(1)$ (resp. under Γ'_B) if and only if they are equivalent under the normalizer of c in $\Gamma_{\sqrt{-3}}(1)$ (resp. in Γ'_B).

Proof: (i) follows from the fact that for a given $\mathbb{B}_1$, the set of (rational) cusps of the $\mathbb{B}_1$ is the restriction of the set of (rational) cusps of $\mathbb{B}_3$ to the closure of the $\mathbb{B}_1$ in the closure of the $\mathbb{B}_3$. All rational cusps are equivalent under $\Gamma_{\sqrt{-3}}(1)$

(resp. under Γ'_B), and the same is true for the action of $\Gamma_{\sqrt{-3}}(1)$ (resp. Γ'_B) on the $\mathbb{B}_1$ or a $\mathbb{B}_2$ as specified in the Lemma (all groups $\Gamma_{\sqrt{-3}}(1)$, $\Gamma_{2,\sqrt{-3}}(1)$ and $\Gamma_{1,\sqrt{-3}}(1)$ have class number one, i.e., a single cusp).

(ii) Let B_1, B_2 be equivalent under $\Gamma_{\sqrt{-3}}(1)$ (resp. under Γ'_B), $\gamma(B_1) = B_2$, and let c denote the common cusp. Then $\gamma(c) \in B_2^*$ and it is equivalent to c in B_2^* (as all cusps are equivalent). Then by (i), there is a δ in the normalizer of the B_2 such that $\delta\gamma(c) = c$, while $\delta\gamma(B_1) = B_2$. It follows that $\delta\gamma$ is an element of the normalizer of c in $\Gamma_{\sqrt{-3}}(1)$ (resp. in Γ'_B). □

We may apply this result to the $i_B(N_\kappa)$, $\kappa = 1,\ldots,8$; it follows that $i_B(N_\kappa)$ are equivalent by an element of the normalizer of the cusp ∞ in $\Gamma_{\sqrt{-3}}(1)$ (resp. in Γ'_B). On the other hand, this normalizer is $P(\infty) \cap \Gamma_{\sqrt{-3}}(1)$ in both cases as follows from (5.40), and the element conjugating N_1 into the other N_κ, $\kappa = 2,\ldots,8$ may be taken as an element of $P(\infty) \cap \Gamma_{\sqrt{-3}}(1)$. In particular,

Lemma 5.6.8 *The subgroups $i_B(N_\kappa)$, $\kappa = 1,\ldots,8$ are all mutually conjugate by elements in $\Gamma_{\sqrt{-3}}(1)$, hence in particular, $i_B(N_\kappa) \subset \Gamma_{\sqrt{-3}}(1)$, $\kappa = 1,\ldots,8$.*
□

Up to now we have been working with a local situation near the cusp ∞. But we can get global results from this, utilizing the following description of subgroups of Γ_B and $\Gamma_{\sqrt{-3}}(2)$. In the following statements all unitary groups are with respect to the form $\mathbf{J}$ and its restrictions to the subspaces in K^4 which give rise to the subdomains $\mathbf{D}_{F,i}$, $i = 1,2$.

Proposition 5.6.9 *We know the following facts about the two groups Γ_B and $\Gamma_{\sqrt{-3}}(2)$:*

1) *There are 45 subgroups $P_1,\ldots,P_{45} \subset \Gamma_B$, each a lattice in a parabolic subgroup of $U(3,1)$. There are 45 subgroups $Q_1,\ldots,Q_{45} \subset \Gamma_{\sqrt{-3}}(2)$, each a lattice in a parabolic of $U(3,1;K)$. All groups P_i, Q_i are isomorphic; as a subgroup of $P(\infty)$ (conjugating the cusp normalized by P_i to the cusp ∞), each is the subgroup $P(\infty) \cap \Gamma_{\sqrt{-3}}(2)$.*

2) *There are 40 subgroups $N_1,\ldots,N_{40} \subset \Gamma_B$, each a lattice in a group isomorphic to $U(2,1)$. There are 40 subgroups $M_1,\ldots,M_{40} \subset \Gamma_{\sqrt{-3}}(2)$, each a lattice in a group isomorphic to $U(2,1;K)$. All groups N_i, M_i are isomorphic; each is isomorphic to $\Gamma_{2,\sqrt{-3}}(2)$, the principal congruence subgroup of level 2 in $U(2,1;\mathcal{O}_K)$.*

3) *There are 360 subgroups $P_{i\kappa} \subset N_i$, $\kappa = 1,\ldots,9$, $i = 1,\ldots,40$, each a lattice in a group isomorphic to $U(1,1)$. There are 360 subgroups $Q_{i\kappa} \subset M_i$, $\kappa = 1,\ldots,9$, $i = 1,\ldots,40$, each a lattice in a group isomorphic to $U(1,1;K)$. All groups $P_{i\kappa}$, $Q_{i\kappa}$ are isomorphic; as a subgroup of the normalizer of the cusp ∞ in $G_{F,2}$, each is the subgroup $P(\infty) \cap \Gamma_{2,\sqrt{-3}}(2)$.*

4) *There are 240 subgroups $N_{ij} = N_i \cap N_j$, determined by j-planes meeting in a line (as opposed to a point). There are 240 subgroups $M_{ij} = M_i \cap M_j$, determined by two-dimensional subball quotients meeting in one-dimensional*

subball quotients. All groups N_{ij} and M_{ij} are isomorphic to $\Gamma_{1,\sqrt{-3}}(2)$, the principal congruence subgroup of level 2 in $U(1,1;\mathcal{O}_K)$.

5) *The following incidences hold for the above objects.*

    i) *Fix a parabolic P_i (resp. Q_i). There are eight N_κ (resp. M_κ) for which $P_i \cap N_\kappa = P_{i\kappa}$ (resp. $Q_i \cap M_\kappa = Q_{i\kappa}$) is one of the groups of 3).*

    ii) *Fix one of the N_κ (resp. M_κ). There are nine parabolics P_i (resp. Q_i) for which $P_i \cap N_\kappa = P_{i\kappa}$ (resp. $Q_i \cap M_\kappa = Q_{i\kappa}$) is one of the groups of 3).*

6) *The sets $\{P_i\}$ and $\{Q_i\}$ are $G_{25,920}$-isomorphic, as are the sets $\{N_\kappa\}$ and $\{M_\kappa\}$.*

Proof: 1) Follows from Corollary 5.6.3, (5.40) and the fact that all groups mentioned are isomorphic.

2) follows from Lemma 5.6.1.1, 3) follows from the same Lemma, together with *loc. cit.*, Lemma 3.13, together with the fact that each j-plane contains nine of the nodes, and 4) follows from these results together with *loc. cit.*, Lemma 3.3.

5) i) For the P_i this is explained in the last paragraph of 5.2.1, II; for the Q_i it is contained in *loc. cit.*, Theorem 2.11: the number $e(c)$ of cross-simplices divided by the number of parabolics, 45, $360/45 = 8$. ii) For the N_κ this is the fact that each j-plane contains nine of the nodes of $\mathcal{B}$; for the M_κ it is given by the same number 360, divided by the number of M_κ's: $360/40 = 9$.

6) The set $\{P_i\}$ may be $G_{25,920}$-equivariantly identified with the set of 45 tritangents of a cubic surface, see (5.8); the set $\{Q_i\}$ may be $G_{25,920}$-equivariantly identified with the set of isotropic lines in the finite geometry $\mathbb{F}_4^4$ with the hermitian form as in Corollary 2.2.9, and this set is identified in the proof of Lemma 2.4.3. The identification of these two $G_{25,920}$-sets is the last line of the table of [C1], p. 26, under the columns "Unitary" and "Schläfli", respectively. Recall that in $G_{25,920}$, there are two conjugacy classes of subgroups of index 40; the first is the class of normalizers of the Steiner primes on $\mathcal{B}$, the second the normalizers of the j-planes. The former can be $G_{25,920}$-equivariantly identified with the set of triples of trihedral pairs, as each Steiner prime contains 18 of the nodes, corresponding to the 18 tritangent planes of such a triple. It follows that the normalizers of the j-planes, hence the $\{N_\kappa\}$, can be $G_{25,920}$-equivariantly identified with groups from the other conjugacy class. The set $\{M_\kappa\}$ can be $G_{25,920}$-equivariantly identified with the set of non-degenerate subspaces of dimension three in the finite geometry $\mathbb{F}_4^4$, which in turn are determined by their orthogonal complements, an anisotropic vector. The identification of the two sets $\{N_\kappa\}$ and $\{M_\kappa\}$ is then the third row of the table just mentioned: in the column "Unitary", we see "non-isotropic line" (in [C1], the authors work projectively, hence they have a point instead of a line), while in the column "Schläfli" there is no entry; the entry "trisection" one row down is the same as triples

of trihedral pairs, which therefore correspond to the normalizers of the Steiner primes.   □

Lemma 5.6.10 $i_B(N_\kappa) = M_\kappa$ *for* $\kappa = 1, \ldots, 8$.

Proof: By Lemma 5.6.7 M_κ (resp. $i_B(N_\kappa)$) is conjugate to $M_1 = i_B(N_1)$ by an element of $P(\infty) \cap \Gamma_{\sqrt{-3}}(1)$. The element conjugating M_1 to M_κ is just a translation in the lattice $\Lambda(\infty) \subset P(\infty) \cap \Gamma_{\sqrt{-3}}(1)$, and this is the same element conjugating $i_B(N_1)$ to $i_B(N_\kappa)$. This is because the geometry is the same, i.e., in the compactifying divisor Δ ($\cong \mathbb{P}^1 \times \mathbb{P}^1$), the intersection curves of Δ with the eight subballs normalized by $i_B(N_1), \ldots, i_B(N_8)$ are the same as those of the eight subballs normalized by $M_1, \ldots, M_8$; such a subball is determined by this intersection (which gives the tangent directions of the subball at the cusp).   □

Proposition 5.6.11 Γ_B *(resp.* $\Gamma_{\sqrt{-3}}(2)$*) is generated by the 45 parabolics* P_i *(resp.* Q_i*) and the 40 subgroups* N_κ *(resp.* M_κ*).*

Proof: Consider for Γ_B and $\Gamma_{\sqrt{-3}}(2)$ the projections

$$\pi_1 : \mathbb{B}_3^* \longrightarrow (\Gamma_B \backslash \mathbb{B}_3)^*; \quad \pi_2 : \mathbb{B}_3^* \longrightarrow (\Gamma_{\sqrt{-3}}(2) \backslash \mathbb{B}_3)^*,$$

and define the complex $\mathcal{C}_B$ (resp. $\mathcal{C}_2$) as follows. The set of vertices of $\mathcal{C}_B$ (resp. $\mathcal{C}_2$) consists of i) the elements $x \in \pi_1^{-1}(p)$, p one of the 45 nodes (resp. $w \in \pi_2^{-1}(q)$, q one of the 45 cusps) and ii) the irreducible components $S \subset \pi_1^{-1}(T)$, T one of the 40 j-planes (resp. $Y \subset \pi_1^{-1}(Z)$, Z one of the 40 two-dimensional subball quotients). Two vertices x_1, x_2 are connected by an edge if and only if one of x_i, $i = 1, 2$ is a cusp and the other is a surface whose image in $(\Gamma_B \backslash \mathbb{B}_3)^*$ (resp. in $(\Gamma_{\sqrt{-3}}(2) \backslash \mathbb{B}_3)^*$) contains the image of the cusp.

Lemma 5.6.11.1 $\mathcal{C}_B$ *(resp.* $\mathcal{C}_2$*) is a connected graph on which* Γ_B *(resp.* $\Gamma_{\sqrt{-3}}(2)$*) acts by conjugation of subgroups. The quotient* $\Gamma_B \backslash \mathcal{C}_B$ *(resp.* $\Gamma_{\sqrt{-3}}(2) \backslash \mathcal{C}_2$*) has* 45+40 *vertices and 360 edges.*

Proof: By definition $\mathcal{C}_B$ and $\mathcal{C}_2$ are one-dimensional simplicial complexes. To see they are connected, view the vertices as being the cusps and two-dimensional subballs, and note that the condition of vertices being connected in $\mathcal{C}_B$ and $\mathcal{C}_2$ is that the suball contains the cusp in its own closure. Given a cusp, c, there is a subball in $\mathcal{C}_B$ (resp. in $\mathcal{C}_2$) connected to c which contains any other cusp c', hence the distance of any two different cusps is 2. Given a cusp c and a surface S which are not connected, let c' be any cusp connected to S; then by the above, the distance from c to c' is 2, hence the distance from c to S is at most 3. Finally, given two surfaces S and S' with no common cusp, let c and c' be any cusps of S resp. S', i.e., connected to S resp. S'. The distance from c to c' is 2, hence from S to S' it is 4. It follows that these complexes are connected. The number of vertices and edges of the quotients is by definition.   □

Recall the following.

Lemma 5.6.11.2 *([BHC], 6.6 & [Se1], I.3, Appendix) Let X be a connected space, G a group acting on X, and let U be an open fundamental domain of the action (i.e. $U \cdot G = X$). Then G is generated by the following set:*

$$\Sigma := \{g \in G | gU \cap U \neq \varnothing\}.$$

We want to apply this to simplicial complexes X. For these we have:

Corollary 5.6.11.1 *In the situation of the last lemma, assume that X is a simplicial complex and that G acts simplicially on X. Then we may take a closed fundamental domain for the action of G. Moreover, G is generated by the normalizers of the sub-simplices of the closed fundamental domain.*

Proof: The point is quite simply that if an element of G normalizes an open subset of a simplex, then it normalizes the entire simplex. The second statement follows directly from this.   □

We note that the normalizers of the vertices of $\mathcal{C}_B$ (resp. of $\mathcal{C}_2$) are the groups P_i and N_κ (resp. Q_i and M_κ), and that the normalizers of the edges are the intersections $P_i \cap N_\kappa$, such that the cusp c_i is a cusp of $\mathcal{D}_\kappa$, the subball normalized by N_κ (resp. the intersections $Q_i \cap M_\kappa$, such that the cusp which Q_i normalizes is a cusp of the subball normalized by M_κ). Then, applying the above corollary to the action of Γ_B on $\mathcal{C}_B$ (resp. $\Gamma_{\sqrt{-3}}(2)$ on $\mathcal{C}_2$), we get the desired result, completing the proof of the proposition.   □

Proposition 5.6.12 $i_B(\Gamma_B) = \Gamma_{\sqrt{-3}}(2)$, *where $i_B : \Gamma_B \hookrightarrow U(\mathbf{J}, \mathbb{C}^4)$ is the embedding satisfying (5.39) and (5.40).*

Proof: By Lemma 5.6.10, $i_B(N_\kappa) = M_\kappa$ for $\kappa = 1, \ldots, 8$. By Proposition 5.6.9, 6) we know that the sets $\{N_\kappa\}$ and $\{M_\kappa\}$ are isomorphic as $G_{25,920}$-sets. For each parabolic incident with any of the N_κ, $\kappa = 1, \ldots, 8$, the Proposition 5.6.6 and (5.40) implies that $i_B(P_i) = Q_i$. Again, for any of these, a result analogous to Lemma 5.6.10 holds, hence we have

$$i_B(N_\kappa) = i(M_\kappa), \ i = 1, \ldots, 40, \ i_B(P_i) = i(Q_i), \ i = 1, \ldots, 45.$$

Hence the proposition follows from Proposition 5.6.11: we know that Γ_B is generated by the N_κ and P_i, while $\Gamma_{\sqrt{-3}}(2)$ is generated by the M_κ and Q_i. It now follows clearly that the $i_B(N_\kappa)$, $i_B(P_i)$ and the M_κ, Q_i generate the same subgroup, namely $\Gamma_{\sqrt{-3}}(2)$.   □

Corollary 5.6.13 $\mathcal{B}_4$ *is the Satake compactification of $X_{\Gamma_{\sqrt{-3}}(2)}$. In fact, $\mathcal{B}_4$ is a Baily-Borel embedding of the Satake compactification.*

Proof: By the above, the open part $\mathcal{B}_4^0 = \mathcal{B}_4 - \{45 \text{ nodes}\}$ is the quotient $\Gamma_{\sqrt{-3}}(2) \backslash \mathbb{B}_3$. By Satake compactification of this quotient we mean the quotient space of [BB], Corollary 4.11. This space is determined as the quotient of the union of $\mathbb{B}_3$ and the *rational* boundary components (here cusps), endowed with the unique topology (Satake topology) of *loc. cit.*, Theorem 4.9. Consider the complex C_2 defined above. The set of cusps determined by the cusp vertices of that complex are precisely the rational cusps, as it is known that $U(3,1;\mathcal{O}_K)$ acts transitively on these (by class number equal to 1). It is then clear that the compactification $\mathcal{B}$ of $\mathcal{B}^0$ is the Satake compactification, and the resolution $\widetilde{\mathcal{B}}$ is a toroidal compactification. The term Baily-Borel embedding refers to the result *loc. cit.*, Theorem 10.11, giving a normal embedding of the Satake compactification. Since $\mathcal{B}_4$ is normal, it is a Baily-Borel embedding.   □

This result completes the proof of Theorem 5.6.1.   □

Remark 5.6.14 The same methods can be applied to show that the cover $\widetilde{Y}$ is the ball quotient by the group $\Gamma_{\sqrt{-3}}(4)$, and that $(\mathbb{Z}/2\mathbb{Z})^9 \cong \Gamma_{\sqrt{-3}}(2)/\Gamma_{\sqrt{-3}}(4)$ is the Galois group of the cover $\widetilde{Y} \longrightarrow \widetilde{\mathcal{B}}$. Then Theorem 5.6.1 can be deduced from this by identifying the action of the group of $\widetilde{Y}$, as is done in Lemma 3.2.5.7.

5.7   Moduli interpretation

5.7.1   Abelian surfaces with a level 3 structure

Recall from Theorem 5.3.4 that a Zariski open subset of $\mathcal{B}$ is biregular to a Zariski open subset of the moduli space of principally polarized complex abelian surfaces with a level 3 structure (see Definition 1.2.7), i.e., there are Zariski open subsets ${}^0\mathcal{B} \subset \mathcal{B}$ and $X^0_{\Gamma(3)} \subset \Gamma(3)\backslash \mathbb{S}_2 = X_{\Gamma(3)}$ such that

$$ {}^0\mathcal{B} \cong X^0_{\Gamma(3)} $$

is a biregular isomorphism. We can easily determine ${}^0\mathcal{B}$ and $X^0_{\Gamma(3)}$.

Lemma 5.7.1 ${}^0\mathcal{B} = \mathcal{B}\backslash\{40\ j-planes\},$

$$ X^0_{\Gamma(3)} = X_{\Gamma(3)}\backslash\{45\ translates\ of\ the\ 'diagonal'\}. $$

Proof: Recall that the principally polarized abelian surface A_α defined by $\alpha \in \mathcal{B}$ had its Weddle surface mapping to the hyperplane tangent section of $\mathcal{B}$ at α; since outside of the 40 $j-$planes the tangent hyperplane section has an isolated node at α, the Weddle is a genuine Weddle determining a Kummer, and A_α is a simple abelian surface. On the other hand, the embedding of the abelian surface given by the quadrics above works as long as A_τ *does not split*; but A_τ splits exactly for τ in one of the 45 translates of the diagonal (i.e., the image of

$$ \mathbb{S}_1 \times \mathbb{S}_1 \subset \mathbb{S}_2 : \begin{pmatrix} \tau_1 & 0 \\ 0 & \tau_2 \end{pmatrix} \text{ under } \Gamma(3)). $$ □

As a corollary we get the following

Theorem 5.7.2 *Let $\widetilde{B}$ be the desingularization; then there are maps*

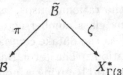

$$\widetilde{B}$$
$$\pi \swarrow \qquad \searrow \zeta$$
$$B \qquad\qquad X^*_{\Gamma(3)}$$

where $X^*_{\Gamma(3)}$ denotes the Satake compactification. π blows up the 45 nodes; the exceptional divisors on $\widetilde{B}$ map under ζ to the 45 translates of the diagonal. The inverse images of the 40 j-planes under π are blown down under ζ to the 40 singular boundary varieties of $X^*_{\Gamma(3)}$. $\widetilde{B}$ is isomorphic to the Igusa desingularization $\overline{X}_{\Gamma(3)}$ of $X^*_{\Gamma(3)}$.

Compare this result with Theorem 3.3.11.

Remark 5.7.3 It was noted in [Bu] that for three subgroups, of indices 45, 40 and 40, respectively, there is an immediate interpretation in terms of hyperelliptic functions. These are as follows (cf. [Bu], §40-42), which are expressed in terms of *non-normalized* periods $\omega_{i,1}, \ldots, \omega_{i,4}$ for $i = 1, 2$.

- Index 45: the images of $\omega_{i,1}$ and $\omega_{i,3}$ (under the action of $G_{25,920}$) are linearly expressed in terms of themselves, and the same holds for $\omega_{i,2}$ and $\omega_{i,4}$. In addition there is a transformation switching $(\omega_{i,1}, \omega_{i,3})$ and $(\omega_{i,2}, \omega_{i,4})$.

- Index 40: the images of $\omega_{i,3}$ and $\omega_{i,4}$ are linearly expressed in terms of themselves.

- Index 40: the images of $\omega_{i,1}$, $\omega_{i,2}$ and $\omega_{i,3}$ are linearly expressed in terms of themselves.

For the first locus we clearly see the splitting of the abelian surface. For the other loci this description implies the abelian surface degenerates, but this is not so immediate. Burkhardt's condition implies the monodromy is unipotent, and by Hodge theory this corresponds to degenerations.

5.7.2   Abelian fourfolds with complex multiplication

This section is devoted to another moduli interpretation. We already know by Theorem 5.6.1 that B_4 is the Satake compactification of a ball quotient. Then we just invoke Propositions 2.2.3 and 2.2.6 and conclude:

Theorem 5.7.4 *Let $B_4^0 = B_4 - \{45 \text{ nodes}\}$ as above. Then B_4^0 is the moduli space of principally polarized abelian fourfolds given by the following PEL-data: $\{(K, \Phi, *), (Q, T, \mathcal{M}), y_1, \ldots, y_s\}$,, where $Q = K^4$, T is the skew-hermitian form iJ, where J is the form of (5.38), $\mathcal{M} = \mathcal{O}_K^4$, and the points $y_1, \ldots, y_s$ are the points of a level 2 structure in the sense of Definition 2.2.4, i.e., the points of the kernel of multiplication by 2.*

Proof: By Theorem 5.6.1, $B_4^0 = X_{\Gamma(2)}$, where $\Gamma(2)$ is the principal congruence subgroup of level 2 in $U(\mathbf{J}, K^4)$. The Theorem then follows from Proposition 2.2.6. □

Furthermore it is known (see [J]) that this family is the family of *Jacobians* of the following, so-called *Picard curves*:

$$C_\lambda = \{y^3 = x(x-1)(x-\lambda_1)(x-\lambda_2)(x-\lambda_3)\}, \tag{5.41}$$

that is, trigonal curves of genus 4. In fact, since C_λ has an automorphism of order 3, so does its Jacobian, hence this Jacobian has complex multiplication by $\mathbb{Q}(\varrho)$. But $\mathbb{Q}(\varrho) = \mathbb{Q}(\sqrt{-3})$, so this is exactly the set of abelian fourfolds of Theorem 5.7.4, without the level structure. This is the same set of abelian fourfolds as occurs in Theorem 3.2.6. B. v. Geemen has identified the level structure 2 in terms of theta characteristics on the curve C_λ, see [Ge].

We now consider the moduli interpretation of the modular subvarieties on B_4. First we note that Lemma 5.6.1.1, applied to the j-planes, can be interpreted by Proposition 2.2.11, and we readily conclude:

Theorem 5.7.5 *The 40 j-planes on B_4 are the sublocus where the abelian fourfolds of 5.7.4 split*

$$A_\tau = A^1 \times A_\tau^3.$$

A^1 is an elliptic curve with complex multiplication by K (and hence rigid), and A_τ^3 is the Jacobian of a genus 3 Picard curve.

The 12 κ-lines on each j-plane are the sublocus of the above where the abelian threefold A_τ^3 splits,

$$A_\tau^3 = B^1 \times A_\tau^2,$$

where B^1 is an elliptic curve and A_τ^2 is the Jacobian of a Picard curve of genus 2.

Perhaps much more interesting is the situation for the desmic surfaces S_{12} discussed above. First we give an independent proof of Proposition 5.2.3.

Lemma 5.7.6 *The desmic suface $S_{12} = J \cap B_4$, where J is one of the Jordan primes, is the Kummer surface of $E_\varrho \times E_\varrho$.*

Proof: Consider the double cover $T \longrightarrow S_{12}$ branched along the 16 κ-lines contained in S_{12}. It is desingularized by blowing up the 12 nodes on the surface S_{12}; denote this by $\widetilde{T} \longrightarrow T$. Since each κ-line contains three of the nodes, after blowing up it is exceptional of the first kind and being in the branch locus, yields such a curve on $\widetilde{T}$. In other words, the 16 proper transforms of the κ-lines are exceptional on $\widetilde{T}$ and may be blown down (after blowing up the nodes, the 16 curves are disjoint); we denote this by $\widetilde{T} \longrightarrow \widehat{T}$. Consider the images in $\widetilde{T}$ of the 12 exceptional divisors of $\widetilde{T} \longrightarrow T$. Since there are four κ-lines through each node, this will be a double cover of $\mathbb{P}^1$ branched at four points, i.e., an elliptic curve, and since there is a tetrahedral group acting on the four κ-lines through the node, the four branch points are an orbit of the tetrahedral group (see Figure 5.4). It follows that the elliptic curve is E_ϱ. Hence on $\widetilde{T}$ we have 12 copies of E_ϱ. Furthermore, two such intersect on $\widetilde{T}$ $\Longleftrightarrow$ the corresponding nodes of S_{12} lie on a common κ-line. Hence there are two sets of copies of E_ϱ with self intersection 0 on $\widehat{T}$, so clearly $\widehat{T} = E_\varrho \times E_\varrho$. $\square$

Note that $E_\varrho \times E_\varrho$ is a very special abelian *surface* with complex multiplication by K. In fact, the surface S_{12} is a moduli space of some kind of abelian varieties with complex multiplication, apparently surfaces. First we have

Theorem 5.7.7 S_{12} *is the Satake compactification of a two-dimensional ball quotient* $\Gamma_{S_{12}} \backslash \mathbb{B}_2$, *where* $\Gamma_{S_{12}}$ *is an arithmetic subgroup of* $\Gamma_{\sqrt{-3}}(2)$. *In other words, the 45 (open) desmic surfaces are modular subvarieties of the ball quotient* $\mathcal{B}_4$.

Proof: Since the centralizer in $G_{25,920}$ of S_{12} has order 6, the set of abelian fourfolds parameterized by points in S_{12} have extra automorphisms. Any automorphism of $\mathcal{B}_4$ which centralizes S_{12} (in particular, the projection from the node which is opposite to the given Jordan prime) is an element of $\Gamma_{\sqrt{-3}}/\Gamma_{\sqrt{-3}}(2)$, hence can be lifted to an automorphism of $\mathbb{B}_3$ contained in $\Gamma_{\sqrt{-3}}$. This establishes that the desmic surfaces are subball quotients, and the statement that $\Gamma_{S_{12}} \subset \Gamma_{\sqrt{-3}}(2)$ is then quite obvious. $\square$

We note that the desmic surfaces parameterize the Jacobians of the curves (5.41) which take the special form:

$$C_{a,b} = \{ y^3 = (x^2 - 1)(x^2 - a)(x^2 - b) \}, \tag{5.42}$$

and an extra automorphism is given by $x \mapsto -x$. In terms of Jacobians, this seems to correspond to those abelian fourfolds which are isogenous to the product of two abelian surfaces.

Remark 5.7.8 These results show that S_{12} is a new kind of Janus-like variety, on the one hand the quotient of an abelian variety, on the other a ball quotient. In fact, letting $\widetilde{S}_{12}$ denote the blow up of S_{12} at the 12 nodes with exceptional divisors $s_1, \dots, s_{12}$, we have

i) $\widetilde{S}_{12} - \{s_1, \dots, s_{12}\} \cong \Gamma_{S_{12}} \backslash \mathbb{B}_2$;

ii) $\widetilde{S}_{12} - \{l_1, \dots, l_{16}\} \cong (\mathbb{C}^2/\Lambda - \text{two-torsion points})/(\mathbb{Z}/2\mathbb{Z})$.

In i) the 16 curves l_i are subball quotients; in ii) the 12 s_i are fibres which are fixed by the involution. Furthermore, looking at the fibre structure of S_{12} as elliptic surface (see Figure 5.6), we see the similarity with Shioda's $S(6)$. Indeed, the latter has 12 fibres of type I_6, and 36 sections. S_{12} has three fibres of type I_6, one of type I_2 (which is a quotient of I_6 by $\mathbb{Z}/3\mathbb{Z}$), and four of type I_1 (a quotient of I_6 by a $\mathbb{Z}/6\mathbb{Z}$). Furthermore, we know from [J] the following.

Theorem 5.7.9 *Let* $S(6)^0$ *denote the elliptic modular surface of level 6 minus the 36 sections. Then*

$$S(6)^0 \cong \Gamma_{\sqrt{-3}}(2\sqrt{-3}) \backslash \mathbb{B}_2.$$

It is then natural to believe that S_{12} is a *quotient* of this. Consider the symmetry groups. $\text{Aut}(S_{12})$ is a quotient of G_{576} by a subgroup of order 6, so has order 96. $\text{Aut}(S(6)) \simeq PSp(2, \mathbb{Z}/6\mathbb{Z}) \ltimes (\mathbb{Z}/6\mathbb{Z})^2 \cong (\Sigma_4 \times G_{216}) \ltimes (\mathbb{Z}/6\mathbb{Z})^2$. It is easy to find (normal) subgroups of $\text{Aut}(S(6))$ of the right order. This seems to be in some sense a non-compact version of what happens for the elliptic modular surface $S(7)$. Livné has constructed a 7-fold cover of $S(7)$, $S_7(7)$ which is a compact ball quotient, and Naruki constructed a K3-surface S' which is a quotient of a ball quotient S_7 such that there is a commutative diagram of ball quotients:

$$
\begin{array}{ccc}
S_7 & \xrightarrow{\pi} & S' \\
\downarrow & & \uparrow \\
S_7(7) & \longrightarrow & S(7).
\end{array}
$$

The map $\pi : S_7 \longrightarrow S'$ has degree 7^5 (and group $(\mathbb{Z}/7\mathbb{Z})^5$), while both vertical arrows have degree 7^2. The elliptic surface S' has three singular fibres of type I_7 and three of type I_1, and the similarity to the elliptic fibre structure of S_{12} (see Figure 5.6) is obvious. See [H1], §4.4 for more details on this example.

5.7.3   The dual variety

Unfortunately we do not know much about the dual variety. By the formula in section B.1.1.6, the degree of the dual variety is 18. It is of course also an invariant of $G_{25,920}$, and could be taken as the degree 18 invariant for $G_{25,920}$. However it is exceedingly difficult to get an explicit equation. What we can say about its geometry is what follows by general theory from the fact that it is dual to B.

- There are 45 double tangent planes (tropes).

- There is a singular locus of 40 lines meeting in 40 points.

Recall Theorem 5.7.2, which implies that $\widetilde{B} = \varphi(B_4)$, where φ is the map defined by the ideal of the 45 nodes, and the fact that the map of B_4 onto the dual is given by the Jacobian ideal, i.e., by cubics on the 45 nodes. It follows from $\mathfrak{Jac}(B_4) \subset \mathfrak{J}(45)$, $\mathfrak{J}(45) = $ ideal of the 45 nodes, that the projection of $\varphi(B_4)$ onto a $\mathbb{P}^4$ displays the dual of B_4 as the projection of the Satake compactification of the Siegel modular threefold of level 3. We do not know whether the dual of the Burkhardt quartic is normal or not. We have calculated the ideal $\mathfrak{J}(45)$ of the 45 lines, and found it is generated by five cubics (this is the Jacobian ideal of B_4), and by five quintics.

Chapter 6

A Gem of the modular universe

In this chapter we will introduce a final variety whose existence derives from the set of 27 lines on a cubic surface – the invariant quintic fourfold $\mathcal{I}_5 \subset \mathbb{P}^5$, the unique invariant hypersurface of degree 5 under the natural action of $\mathrm{Aut}(\mathcal{L}) \cong W(E_6)$ on $\mathbb{P}^5$. It is also probably a modular variety, and in fact a (very) Janus-like variety, but the proof of this fact remains to be completed. At any rate it is full of modular subvarieties, i.e., subvarieties which *are* modular varieties, and as an algebraic variety it has the following very special properties.

- $\mathcal{I}_5$ is the unique invariant of degree $d = 5$ for the reflection group $W(E_6)$ acting on $\mathbb{P}^5$. Hence the symmetry group of $\mathcal{I}_5$ is $W(E_6)$.

- The singular locus of $\mathcal{I}_5$ consists of a set of 120 double lines which meet ten at a time in 36 triple points.

- The 36 triple points correspond in a $W(E_6)$-equivariant manner with the 36 double sixes, the 120 double lines correspond in the same equivariant manner with the 120 trihedral pairs, of the set of 27 lines on a smooth cubic surface.

- There are 27 special hyperplane sections of $\mathcal{I}_5$ which split into a pentahedron (five $\mathbb{P}^3$'s) in that hyperplane (a $\mathbb{P}^4$). These correspond in an equivariant manner with the 27 lines.

- There are 45 $\mathbb{P}^3$'s which are contained in $\mathcal{I}_5$; these correspond in an equivariant manner with the 45 tritangents.

- Each of the 27 hyperplane sections contains five of the 45 $\mathbb{P}^3$'s, corresponding to the five tritangents containing a given line; each of the 45 $\mathbb{P}^3$'s is contained in three of the 27 hyperplanes, corresponding to the three lines which lie in a given tritangent.

- The Hessian variety of $\mathcal{I}_5$, a degree $d = 18$ hypersurface in $\mathbb{P}^5$, intersects $\mathcal{I}_5$ in the union of the 45 $\mathbb{P}^3$'s – each with multiplicity 2.

- $\mathcal{I}_5$ is a self-Steinerian variety.

- There are 36 hyperplane sections which are a quintic threefold which is isomorphic to the Hessian variety $\text{Hess}(\mathcal{S}_3)$ of the Segre cubic.

- The 36 singular points are resolved by cubic hypersurfaces in $\mathbb{P}^4$ which are all copies of the Segre cubic $\mathcal{S}_3$.

As in the case of the Burkhardt quartic one may consider the configuration of the 36 points, which contains a lot of geometry, but generally speaking the *intersections* with $\mathcal{I}_5$ are not as interesting as was the case for $\mathcal{B}_4$. The most special such space section which is an irreducible quintic is depicted in the frontispiece.

Since the dimension of the moduli space of cubic surfaces is four and $\mathcal{I}_5$ is so intricately related with the 27 lines, it is natural to inquire about a more immediate relation. For example, *is* the quintic $\mathcal{I}_5$ *the* moduli space of marked cubic surfaces? This turns out to not quite be true, and the relationship will be explicitly described below. This involves in particular the Coble variety $\mathcal{Y}$, giving a surprising relation of $\mathcal{I}_5$ with moduli spaces of certain K3 surfaces.

6.1 The Weyl group $W(E_6)$

In this section we recall some well-known facts on the root system of type E_6, and describe also the arrangement and the configuration defined by the 36 reflection hyperplanes and 36 dual points, respectively. For the roots, etc., we adhere to the notations of Bourbaki.

6.1.1 Notations

We use the same notation as above for the 27 lines on a cubic surface in $\mathbb{P}^3$: $a_1, ... a_6$, $b_1, ..., b_6$, $c_{12}, ..., c_{56}$. The 36 double sixes are:

$$N = \begin{bmatrix} a_1 & a_2 & a_3 & a_4 & a_5 & a_6 \\ b_1 & b_2 & b_3 & b_4 & b_5 & b_6 \end{bmatrix}, \qquad (1)$$

$$N_{ij} = \begin{bmatrix} a_i & b_i & c_{jk} & c_{jl} & c_{jm} & c_{jn} \\ a_j & b_j & c_{ik} & c_{il} & c_{im} & c_{in} \end{bmatrix}, \qquad (15)$$

$$N_{lmn}{}^1 = \begin{bmatrix} a_i & a_j & a_k & c_{mn} & c_{ln} & c_{lm} \\ c_{jk} & c_{ik} & c_{ij} & b_l & b_m & b_n \end{bmatrix} \qquad (20).$$

The 45 tritangents are:

$$\begin{aligned} (ij) &= < a_i\, b_j\, c_{ij} >, \quad i \neq j \quad (30) \\ (ij.kl.mn) &= < c_{ij}\, c_{kl}\, c_{mn} > \quad (15). \end{aligned} \qquad (6.1)$$

---

[1]here we switch notations from N_{ijk} in equation (4.1) to N_{lmn} for convenience

Two double sixes are *syzygetic* if they contain four lines in common, for example:

$$N \text{ and } N_{12} \text{ have } a_1, a_2, b_1, b_2 \text{ in common,}$$

and *azygetic* if they have six lines in common, for example:

$$N \text{ and } N_{456} \text{ have } a_1, a_2, a_3, b_4, b_5, b_6 \text{ in common.}$$

Two azygetic double sixes have six lines in common and contain 12 other lines; these 12 lines form another double six, azygetic with respect to both, for example N, N_{123}, N_{456}. Such triples are referred to as triples of azygetic double sixes or, because of the interpretation in terms of tritangents, a trihedral pair . Each double six is syzygetic to 15 others, forming 270 such pairs, and azygetic to 20 others, forming 120 triples. Our notation for the 120 triples are:

$$
\begin{aligned}
\{ijk\} &= \ <N, N_{ijk}, N_{lmn}>, \quad (10) \\
\{ij.jk\} &= \ <N_{ij}, N_{ik}, N_{jk}>, \quad (20) \\
\{ij.kl\} &= \ <N_{ij}, N_{ikl}, N_{jkl}> \quad (90).
\end{aligned}
\qquad (6.2)
$$

We recognize these as the trihedral pairs of (4.2) under the correspondence

$$
\begin{bmatrix}
a_i & b_j & c_{ij} \\
b_k & c_{jk} & a_j \\
c_{ik} & a_k & b_i
\end{bmatrix}
\longleftrightarrow <N_{ij}, N_{ik}, N_{jk}>,
$$

$$
\begin{bmatrix}
c_{il} & c_{jm} & c_{kn} \\
c_{mn} & c_{ik} & c_{jl} \\
c_{jk} & c_{ln} & c_{im}
\end{bmatrix}
\longleftrightarrow <N, N_{ijk}, N_{lmn}>,
$$

$$
\begin{bmatrix}
a_i & b_j & c_{ij} \\
b_l & a_k & c_{kl} \\
c_{il} & c_{jk} & c_{mn}
\end{bmatrix}
\longleftrightarrow <N_{ij}, N_{ikl}, N_{jkl}> .
$$

Hence the triads of trihedral pairs discussed there are expressed in condensed form as follows:

$$
[ijk.lmn] =
\begin{bmatrix}
N_{ij} & N_{jk} & N_{ik} \\
N_{lm} & N_{mn} & N_{ln} \\
N & N_{ijk} & N_{lmn}
\end{bmatrix},
\qquad (10)
$$

$$
[ij.kl.mn] =
\begin{bmatrix}
N_{ij} & N_{ikl} & N_{jkl} \\
N_{kl} & N_{kmn} & N_{lmn} \\
N_{mn} & N_{nij} & N_{mij}
\end{bmatrix},
\qquad (30).
$$

$$(6.3)$$

The group of incidence preserving permutations of the 27 lines, a group of order 51840, can be generated by the following six operations:

$$(i, i+1), \ i = 1, ..., 5 : \text{transposition of the indices,}$$

and

$$(123) : \text{map } N \mapsto N_{123},$$

and the graph of this presentation is shown in Figure 6.1. This is the graph whose vertices correspond to generators, two vertices A, B corresponding to generators α, β being connected if $\alpha\beta\alpha = \beta\alpha\beta$ and not connected if $\alpha\beta = \beta\alpha$.

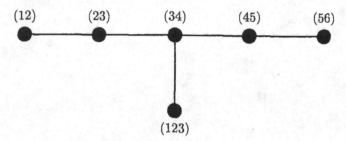

Figure 6.1: The graph of the group of the permutations of the 27 lines

6.1.2 Roots

Let $\mathfrak{t}$ be a maximal abelian subalgebra of the compact Lie algebra $\mathfrak{e}_{6,u}$ over $\mathbb{R}$, i.e., $\mathfrak{t} \cong \mathbb{R}^6$. Let $x_1, ..., x_6$ be coordinates such that the root forms of E_6 are:

$$(40) \qquad \pm(x_i \pm x_j), \qquad 1 \le i < j \le 5$$

$$(32) \qquad \pm\frac{1}{2}(\pm x_1 \pm x_2 \pm x_3 \pm x_4 \pm x_5 + x_6), \qquad \begin{array}{l}\text{even number of ``$-$'' signs inside}\\ \text{the parenthesis.}\end{array}$$

(Note that in Bourbaki notation, our variables $x_i = \varepsilon_i$, $i = 1, .., 5$, while our coordinate x_6 is denoted $\varepsilon_8 - \varepsilon_7 - \varepsilon_6$ there). The 36 positive root forms are given by $\pm x_i + x_j$ and $\frac{1}{2}(\pm x_1 \pm x_2 \pm x_3 \pm x_4 \pm x_5 + x_6)$, and they correspond to the 36 double sixes of the 27 lines on a cubic surface. We use the following notations for these forms

$$
\begin{aligned}
h &= \tfrac{1}{2}(x_1 + ... + x_6), \\
h_{1j} &= x_{j-1} - \tfrac{1}{2}(x_1 + ... + x_5 - x_6), \quad j = 2, ..., 6 \\
h_{jk} &= -x_{j-1} + x_{k-1}, \quad 1 \ne j < k \\
h_{1jk} &= x_{j-1} + x_{k-1}, \quad j, k = 2, ..., 6 \\
h_{jkl} &= +x_{j-1} + x_{k-1} + x_{l-1} - \tfrac{1}{2}(x_1 + ... + x_5 - x_6), \quad j, k, l \ne 1.
\end{aligned}
\qquad (6.4)
$$

The Weyl group of E_6 is generated by the reflections on these 36 hyperplanes; we denote these reflections by s, s_{ij}, and s_{ijk}. As a system of simple roots we take :

$$
\begin{aligned}
\alpha_1 &= -\tfrac{1}{2}(-x_1 + ... + x_5 - x_6) &= h_{12} \\
\alpha_2 &= x_1 + x_2 &= h_{123} \\
\alpha_3 &= -x_1 + x_2 &= h_{23} \\
\alpha_4 &= -x_2 + x_3 &= h_{34} \\
\alpha_5 &= -x_3 + x_4 &= h_{45} \\
\alpha_6 &= -x_4 + x_5 &= h_{56}.
\end{aligned}
\qquad (6.5)
$$

Then the Dynkin diagram is as shown in Figure 6.2; we recover Figure 6.1 by replacing α_i by the corresponding *reflection* s, s_{ij}, s_{ijk} on the hyperplanes where h, h_{ij}, h_{ijk}, respectively, vanish. This shows clearly the isomorphism of $W(E_6)$ and the group of the permutations of the 27 lines,

$$\mathrm{Aut}(\mathcal{L}) \cong W(E_6).$$

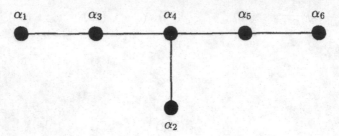

Figure 6.2: The Dynkin diagram of the Weyl group of E_6

The action of the reflections on the root forms can be described as follows:

$s(h_{ij}) = h_{ij}$	$s(h_{ijk}) = h_{lmn}$	$s_{ijk}(h) = h_{lmn}$	$s_{ij}(h) = h$
$s_{ijk}(h_{lmn}) = h$	$s_{ijk}(h_{kmn}) = h_{kmn}$	$s_{ijk}(h_{jkn}) = h_{in}$	$s_{ijk}(h_{ln}) = h_{ln}$
$s_{ijk}(h_{kn}) = h_{ijn}$	$s_{ijk}(h_{jk}) = h_{jk}$	$s_{ij}(h_{klm}) = h_{klm}$	$s_{ij}(h_{jlm}) = h_{ilm}$
$s_{ij}(h_{ijm}) = h_{ijm}$	$s_{ij}(h_{ij}) = h_{ij}$	$s_{ij}(h_{jk}) = h_{ik}$	$s_{ij}(h_{lk}) = h_{lk}$

6.1.3   Vectors

The Killing form of E_6, a quadratic invariant, can be calculated as the sum of the squares of all roots, and evaluates to (a constant times):

$$I_2 = x_1^2 + x_2^2 + x_3^2 + x_4^2 + x_5^2 + \frac{1}{3}x_6^2.$$

With respect to the Killing form we have the vectors dual to the root forms:

$$
\begin{aligned}
H &= \tfrac{1}{2}(1,1,1,1,1,3); \\
H_{1j} &= -\tfrac{1}{2}(1,...,-1,...,-3), && \text{1 in the } (j-1)^{st} \text{ spot, } j = 2,\ldots,6; \\
H_{jk} &= -\tfrac{1}{2}(0,..1,..,-1,..,0), && \pm 1 \text{ in the } (j-1)^{st}, (k-1)^{st} \text{ spot,} \\
& && 1 < j < k \le 6; \\
H_{1jk} &= \tfrac{1}{2}(0,..1,..,1,..,0), && \text{1 in the } (j-1)^{st}, (k-1)^{st} \text{ spot,} \\
& && 1 < j < k \le 6; \\
H_{jkl} &= -\tfrac{1}{2}(1,-1..,-1,..,-1,..,-3), && \text{1's in the } (j-1), (k-1), \\
& && (l-1)\text{spots}, 1 < j < k < l \le 6;
\end{aligned}
$$

$$(6.6)$$

which may be thought of as the root vectors (of the positive roots; the negative roots have a "$-$" sign in front).

As is well-known, there is also a set of 27 fundamental weights which form

Table 6.1: The arrangement in $\mathbb{P}^5$

$N(\mathcal{O})$	A_1	A_1^2	A_2	A_1^3	$A_{1,2}$	A_3	$A_{1^2,2}$	A_2^2	$A_{1,3}$	A_4	D_4	$A_{1,2^2}$	$A_{1,4}$	A_5	D_5
$\mathcal{O}$	k	$t_2(3)$	$t_3(3)$	$t_3(2)$	$t_4(2)$	$t_6(2)$	$t_5(1)$	$t_6(1)$	$t_7(1)$	$t_{10}(1)$	$t_{12}(1)$		t_7	t_{11}	t_{15}	t_{20}
#	36	270	120	540	720	270	1080	120	540	216	45		360	216	36	27
$t(4)$	1	15	10	45	80	45	150	20	105	60	15		70	66	15	15
$t_2(3)$		1	0	6	8	3	28	4	18	12	3		20	20	6	7
$t_3(3)$			1	0	6	9	9	2	18	18	6		6	18	6	9
$t_3(2)$				1	0	0	6	0	3	0	1		6	6	1	3
$t_4(2)$					1	0	3	1	3	3	0		4	6	3	3
$t_6(2)$						1	0	0	2	4	2		0	4	2	5
$t_5(1)$							1	0	0	0	0		2	2	0	1
$t_6(1)$								1	0	0	0		3	0	3	0
$t_7(1)$									1	0	0		0	2	1	1
$t_{10}(1)$										1	0		0	1	1	2
$t_{12}(1)$											1		0	0	0	3

an orbit of $W(E_6)$, namely:

$$
\begin{aligned}
a_1 &= -\tfrac{2}{3}x_6; & a_j &= x_{j-1} - \tfrac{1}{2}(x_1 + ... + x_5 + \tfrac{1}{3}x_6); \\
b_1 &= \tfrac{1}{2}(x_1 + ... + x_5 - \tfrac{1}{3}x_6); & b_j &= x_{j-1} + \tfrac{1}{3}x_6; \\
c_{1j} &= -x_{j-1} + \tfrac{1}{3}x_6; & c_{ij} &= -x_{j-1} - x_{i-1} \\
& & & \quad + \tfrac{1}{2}(x_1 + ... + x_5 - \tfrac{1}{3}x_6).
\end{aligned}
$$

(6.7)

These form the $W(E_6)$-orbit of the fundamental weights denoted $\overline{\omega}_1$ and $\overline{\omega}_6$ in Bourbaki, which are just our $-a_1$ and b_6, respectively. Note that the following relation holds:

$$
\sum_{i=1}^6 a_i = -3h = -3\left(\sum_{i=1}^6 x_i\right) = -\sum_{i=1}^6 b_i.
$$

(6.8)

Also note that the a_i and b_i are related by

$$
b_i = a_i - \frac{1}{3}(a_1 + \cdots + a_6).
$$

(6.9)

Finally note also the 45 relations given by the tritangents:

$$
a_i + b_j + c_{ij} = 0, \quad c_{ij} + c_{kl} + c_{mn} = 0.
$$

(6.10)

The vectors which are dual to the a_i, b_i, c_{ij} with respect to the Killing form are:

$$
\begin{aligned}
A_1 &= (0,...,0,-2); & A_j &= \tfrac{1}{2}(-1,...,+1,..,-1) + 1 \text{ in the } j-1 \text{ spot}; \\
B_1 &= \tfrac{1}{2}(1,...,1,-1); & B_j &= (0,...,1,..,1) + 1 \text{ in the } j-1 \text{ spot}; \\
C_{1j} &= (0,..,-1,...,1); & C_{ij} &= \tfrac{1}{2}(1,...,-1,...,-1,...,-1) \\
& & & \quad -1 \text{ in the } j-1, i-1 \text{ spots}.
\end{aligned}
$$

(6.11)

6.1.4   The arrangement defined by $W(E_6)$

The 36 hyperplanes in $\mathbb{P}^5$ defined by the vanishing of the 36 root forms form the arrangement $\mathcal{A}(E_6)$ of (3.3). For later reference we give the combinatorial data of the arrangement here. We denote as in (3.4) a $\mathbb{P}^m$ through which k of the hyperplanes pass by $t_k(m)$. For the normalisers we use the notation A_{i^k,j^l} for $A_i^k \times A_j^l$. The data of the arrangement is given in Table 6.1.

6.1.5   Special Loci

In Table 6.2 we give a list of special loci which will be particularly important in what follows, so we give a brief description of each.

Table 6.2: Special loci in $\mathbb{P}^5$

#	space	Symmetry	$N(\mathcal{O})$	notation in Table 6.1
36	$\mathbb{P}^4$	A_5	A_1	–
120	$\mathbb{P}^3$	D_4	A_2	$t_3(3)$
120	$\mathbb{P}^1$	A_2	$A_2 \times A_2$	$t_6(1)$
216	$\mathbb{P}^1$	A_2	A_4	$t_{10}(1)$
45	$\mathbb{P}^1$	A_1	D_4	$t_{12}(1)$
36	point	–	A_5	t_{15}
27	point	–	D_5	t_{20}

6.1.5.1   36 $\mathbb{P}^4$'s

In each of the 36 hyperplanes given by the vanishing of one of the 36 forms (6.4), h say, the induced group is Σ_6, and as a reflection group on $\mathbb{P}^4$ it defines a projective arrangement of 15 planes; since each double six is syzygetic to 15 and azygetic to 20 others, there are 15 hyperplanes through which one of the other 35 intersect h, and ten planes through which two others of the 35 meet h. We immediately recognize this geometry as that in $\mathbb{P}^4$ discussed in Chapter 3. The 15 hyperplanes are the 15 $\mathcal{H}_{ij}$ of (3.34), each of which cuts out three planes on $\mathcal{S}_3$, and the ten are the hyperplanes mentioned in (3.46) and Proposition 3.3.9. These in turn are the dual hyperplanes to the ten nodes on $\mathcal{S}_3$.

6.1.5.2   120 $\mathbb{P}^3$'s

These $\mathbb{P}^3$'s correspond to the 120 triples of azygetic double sixes, i.e., each is cut out by three of the 36 hyperplanes of 6.1.5.1. In each such hyperplane, these $\mathbb{P}^3$'s correspond to the ten hyperplanes in h just mentioned, given by the K_{ijk} of (3.46). Each of these contains 15 planes, and one can check that these are just the faces and symmetry planes of a cube. The six lines in K_{ijk} which are the singular locus $\mathcal{I}_4 \cap K_{ijk}$, are easily identified with the six 12-fold lines $t_{12}(1)$

which are contained in $K_{ijk}{}^2$, and the nine points t_{20} contained in K_{ijk} are the intersection points of those six lines[3]. Equations of the 120 $\mathbb{P}^3$'s are given by a triple of azygetic double sixes, e.g., by $< h, h_{ijk}, h_{lmn} >$.

6.1.5.3   120 $\mathbb{P}^1$'s

The 120 lines correspond exactly to A_2 subroot systems, each containing three (positive) roots, so that each line contains three of the 36 points. The 120 lines are determined as follows. Consider a triad of triples of azygetic double sixes and the corresponding matrix of linear forms (see (6.3)), say

$$[ijk.lmn] = \begin{bmatrix} h_{ij} & h_{jk} & h_{ik} \\ h_{lm} & h_{mn} & h_{ln} \\ h & h_{ijk} & h_{lmn} \end{bmatrix}.$$

Taking the ideal defined by the vanishing of two rows defines the corresponding line, i.e.,

$$\begin{aligned} L_{\{ij.jk\}} &= < h_{lm}, h_{mn}, h_{ln}, h, h_{ijk}, h_{lmn} > \\ L_{\{lm.mn\}} &= < h_{ij}, h_{jk}, h_{ik}, h, h_{ijk}, h_{lmn} > \\ L_{\{ijk\}} &= < h_{ij}, h_{jk}, h_{ik}, h_{lm}, h_{mn}, h_{ln} > . \end{aligned} \qquad (6.12)$$

Each of the 120 lines contains three of the root vectors, so for example,

$$H_{ij}, H_{jk}, H_{ik} \in L_{\{ij.jk\}}. \qquad (6.13)$$

There are 40 such triples of the 120 lines, which have the characterising property that they span $\mathbb{P}^5$. These correspond to subroot systems of the type $A_2 \times A_2 \times A_2$, where all three copies are orthogonal to one another. Note that given an A_2 subroot system, there is a unique $A_2 \times A_2$ subroot system orthogonal to it. Thus the A_2 subroot system is defined by the vanishing of the six root forms of the complementary $A_2 \times A_2$. There are 120 of each of both types of subroot systems. Summing up, there are six of the 36 hyperplanes passing through each of these 120 lines while each such line contains three of the 36 nodes.

The *induced arrangement* is as follows. Blowing up along the line introduces an exceptional $\mathbb{P}^3$ over each point of the line; the intersection of it with the proper transforms of the six planes passing through it is the induced arrangement. It is of type $A_2 \times A_2$, i.e., is given by two skew lines in $\mathbb{P}^3$ and two sets of three hyperplanes through each line.

6.1.5.4   216 $\mathbb{P}^1$'s

Consider a pair of *skew* lines, say a_1, a_2. There is a unique double six containing the given pair as a column, e.g.,

$$N_{12} = \begin{bmatrix} a_1 & b_1 & c_{23} & c_{24} & c_{25} & c_{26} \\ a_2 & b_2 & c_{13} & c_{14} & c_{15} & c_{16} \end{bmatrix}.$$

---

[2]The arrangement is $\mathcal{A}(\mathbf{D_4})$, minus the plane at infinity. Of the 16 $t_3(1)$ of $\mathcal{A}(\mathbf{D_4})$ (see (3.5)), four lie in the plane at infinity.

[3]Likewise, nine of the 12 t_6 of $\mathcal{A}(\mathbf{D_4})$ lie in the plane at infinity

There are 216 lines in $\mathbb{P}^5$ which join points such as A_1, A_2, H_{12} (see (6.6) and (6.11)). The ideal of these 216 lines is generated by 24 sextics, forming the irreducible representation denoted 24_p in [BL]. We can exhibit these sextics explicitly, as follows. The 216 lines are given by the equations:

$$
\begin{aligned}
< A_i, A_j, H_{ij} > &= < h_{kl}|k, l \neq i, j; h_{ijk} > & (15)\\
< B_i, B_j, H_{ij} > &= < h_{kl}|k, l \neq i, j; h_{klm}|k, l, m \neq i, j > & (15)\\
< A_i, B_i, H > &= < h_{kl}|k, l \neq i > & (6)\\
< A_i, C_{jk}, H_{lmn} > &= < h_{jk}, h_{\lambda\mu}|\lambda, \mu \neq i, j, k; h_{ij\lambda}, h_{ik\lambda}|\lambda \neq i, j, k > & (60)\\
< B_i, C_{jk}, H_{ijk} > &= < h_{jk}, h_{\lambda\mu}|\lambda, \mu \neq i, j, k; h_{\lambda\mu\nu}| & \\
& \qquad \lambda \neq i, j, k, \mu \neq i, k, \nu \neq i > & (60)\\
< C_{ik}, C_{jk}, H_{ij} > &= < h_{ijm}, h_{mn}, h, h_{kmn}|m, n \neq i, j, k > & (60),
\end{aligned}
$$
$$
(6.14)
$$

i.e., each is defined by the vanishing of ten of the h's; these lines are the $t_{10}(1)$ listed in the table of the arrangement. We claim the sextics are the products of the six root forms of an $A_2 \times A_2$ subroot system. To see this, pick one, say $\Phi = h_{12} \cdot h_{13} \cdot h_{23} \cdot h_{45} \cdot h_{46} \cdot h_{56}$. It will suffice to check that for any of the 216 lines listed in (6.14), at least one of the hyperplanes on the right hand side is among the set $h_{12}, h_{13}, h_{23}, h_{45}, h_{46}, h_{56}$. This is at most a tedious, but straightforward task.

The dual $\mathbb{P}^3$'s, which are defined by the vanishing of the forms which are dual to the points of the left-hand sides, each *contain* the ten points which are dual to the forms on the right, for example

$$
P_{<A_i, A_j>} = \{a_i = a_j = h_{ij} = 0\} \ni (H_{kl}, H_{ijk}).
$$

The induced arrangement over each line is the arrangement $\mathcal{A}(W(\mathbf{A_4}))$ of (3.5). Among the ten hyperplanes defining the line, say $< A_1, B_1, H >$, there are ten triples of azygetic double sixes, $\{ij.jk\}$ in (6.2), with $i \neq 1$: $\{23.34\}$, $\{23.35\}$, $\{23.36\}$, $\{24.45\}$, $\{24.46\}$, $\{25.56\}$, $\{34.45\}$, $\{34.46\}$, $\{35.56\}$, and $\{45.56\}$, and these determine the ten $t_3(1)$ of (3.5).

6.1.5.5   45 $\mathbb{P}^1$'s

The 45 lines are the lines joining the 27 points of (6.11) in threes, for example,

$$
L_{(ij)} =< A_i, B_j, C_{ij} > .
$$

These lines are defined by the vanishing of 12 of the h's, so for example

$$
L_{(12)} =< h_{34}, h_{35}, h_{36}, h_{45}, h_{46}, h_{56}, h_{234}, h_{235}, h_{236}, h_{245}, h_{246}, h_{256} >; \quad (6.15)
$$

these are the hyperplanes corresponding to the 12 double sixes *not* containing any of a_1, b_2, c_{12}.

The induced arrangement is $\mathcal{A}(W(\mathbf{D_4}))$, with 12 planes corresponding to the 12 hyperplanes through the line. Again there will be $t_2(1)$'s and $t_3(1)$'s, corresponding to triples of azygetic double sixes (respectively to syzygetic double sixes).

The ideal of the 45 lines is generated by 15 quartics which form the irreducible representation denoted 15_q in [BL]. It is easy to see that this space of quartics is generated by a product of four pairwise azygetic h's, for example by $h_{24} \cdot h_{124} \cdot h_{35} \cdot h_{135}$. In fact, each hyperplane of type h_{ij} contains the 15 lines numbered (like the tritangents) by:

$$
\begin{array}{llll}
(kl) & \text{for} & k,l \neq i,j & \text{(12 of these)} \\
(ij.kl.mn) & \text{for} & k,l,m,n \neq i,j & \text{(3 of these)}
\end{array}
$$

while the hyperplanes of type h_{ijk} contain the 15 lines numbered by:

$$
\begin{array}{llll}
(mn) & \text{for} & n = i \text{ or } j, m \neq i,j & \text{(9 of these)} \\
(il.jm.kn) & & & \text{(6 of these).}
\end{array}
$$

It is now easy to check that every line is contained in at least one of the four hyperplanes. Alternatively we can argue as follows: each h contains 15 of the lines; there are six $\mathbb{P}^3$'s which are the intersection of two of the four, three of which are contained in each h. These three meet in a common line in the h, so the number of lines contained in the union is: $4 \cdot (15 - 7) + 2 \cdot 6 + 1 = 45$, where the 7=number of lines in the union of the three $\mathbb{P}^2$'s in each h, 2=3-1 is the number of lines in each such $\mathbb{P}^2$, not in the others, and one is the common line. Note that this is the Macdonald representation corresponding to the four roots of an $A_1 \times A_1 \times A_1 \times A_1 \subset D_4$ subroot system. Five of these lines meet at each of the 27 points, corresponding to the five tritangents through each of the 27 lines.

6.1.5.6 36 points

These are the 36 points (6.6) dual to the 36 hyperplanes of 6.1.5.1. The induced arrangement is of course just the arrangement in $\mathbb{P}^4$ above. There are 15 hyperplanes passing through each of the 36 points, and these are just the hyperplanes which are coded by the double sixes which are syzygetic to the one with the notation of the point as in (6.6). So, for example, the 15 $\mathbb{P}^4$'s through the point H are the 15 h_{ij}.

These points correspond to $(\pm)$ the roots of E_6. The orthogonal complement in $\mathbb{R}^6$ of the root α is projectively equivalent to the *dual* hyperplane to the point. For example, H is dual to h, and one of the hyperplanes P will contain $H \iff$ the dual point p is contained in h. The ideal of the 36 points is generated by 20 cubics, forming the irreducible representation of $W(E_6)$ denoted 20_p in [BL]. We can find these cubics explicitly as follows. Consider the hyperplanes a_1, b_2, c_{12} corresponding to a tritangent. From Table 6.1 above we see that each of these hyperplanes contains 20 of the 36 points (actually, the table contains the dual information: there are 20 of the h_{ij}, etc, passing through each of the 27 points), and the $\mathbb{P}^3$ which is the common intersection of these three contains 12 of the 36 (the dual information is contained in the table: the 45 lines are 12-fold lines). Hence the product $a_1 \cdot b_2 \cdot c_{12}$ contains $3.(20-12)+12 = 36$ of the 36 points.

Through each of the 36 points, also 15 of the 27 hyperplanes of (6.7) pass, corresponding to the 15 lines *not* contained in the double six whose notation the point has. For example, the point $H = \frac{1}{2}(1, \ldots, 1, 3)$ is contained in all the c_{ij}. In the exceptional $\mathbb{P}^4$ at the point, both sets of 15 hyperplanes (coming from the 36, respectively 27 hyperplanes) *coincide*.

6.1.5.7   27 points

These are the points A_i, B_i, C_{ij} of (6.11). There are 20 of the 36 hyperplanes meeting at each, so the induced arrangement is one of 20 $\mathbb{P}^3$'s in $\mathbb{P}^4$, and one sees easily that it is $\mathcal{A}(W(\mathbf{D_5}))$. This arrangement is also induced in any of the hyperplanes a_i, b_i, c_{ij} of (6.7); we note that there are two kinds of $\mathbb{P}^2$, namely $t_2(2)$'s, corresponding to pairs of skew lines, and $t_3(2)$'s, corresponding to tritangents. Since each line is contained in five tritangents, there are five of the latter and 15 of the former (in each a_i, etc.). These 15 form an arrangement of type $\mathcal{A}(W(\mathbf{A_5}))$ as discussed above. The ideal of these 27 points is generated by 30 cubics, forming the irreducible representation denoted 30_p in [BL]. It is easy to see that this space of cubics is generated by a product of three members of a syzgetic triple as $h_{12} \cdot h_{13} \cdot h_{23}$, for example: Each of the hyperplanes contains 15 of the 27, the $\mathbb{P}^3$ which is thier common intersection contains nine, so the union contains 3.(15-9)+9=27, or all of the points. Note that this is just the Macdonald representation corresponding to the (3) roots of an A_2 subroot system.

We need, in addition to the above, certain information on the dual spaces.

6.1.5.8   45 $\mathbb{P}^3$'s

Consider one of the 45 $\mathbb{P}^3$'s which is dual to one of the 45 lines of 6.1.5.5; it is cut out by three of the forms (6.7), and can be denoted as one of the 45 tritangents, for example, if $L_{(ij)}$ denotes the line $< A_i, B_j, C_{ij} >$ as in (6.15), the dual $\mathbb{P}^3$ may be denoted by $l_{(ij)}$, and

$$l_{(ij)} = a_i \cap b_j \cap c_{ij} \qquad (6.16)$$
$$l_{(ij.kl.mn)} = c_{ij} \cap c_{kl} \cap c_{mn}.$$

Consider the $\mathbb{P}^3$ $l_{(12)}$, given by $a_1 = b_2 = c_{12} = 0$, or $x_1 = x_6 = 0$. Then one checks easily that the hyperplanes (6.4) reduce in $l_{(12)}$ to the arrangement $\mathcal{A}(\mathbf{F_4})$ of (3.3). Considering how the 27 hyperplanes (6.7) intersect $l_{(12)}$, we find that these restrict to the set of short roots, that is, give a subarrangement of type $\mathcal{A}(\mathbf{D_4})$. See also Proposition 6.3.1 below.

6.1.6   The configuration defined by the 36 points

We now add to the above the description of the configuration of the 36 points. We will use notations which hopefully convey a similarity with the configuration of the 45 points in $\mathbb{P}^4$ (section 5.2.1). We again list these loci by dimension.

0. **Points:** There are 36 points which correspond to the 36 positive roots of E_6 and form an orbit of $W(E_6)$. They correspond also in a $W(E_6)$-equivariant manner to the 36 double sixes. The stabilizer of a point has order $|W(E_6)|/36 = 720$ and is isomorphic to Σ_6.

1. **Lines:** There are again lines joining the 36 nodes in twos and in threes. There are 270 $\varepsilon-lines$, each containing two nodes, and 120 $\kappa-lines$, each containing three nodes. For any two nodes n_1, n_2 let N_1, N_2 denote the corresponding double sixes. Then N_1 and N_2 are syzegetic if and only if n_1 and n_2 lie on a common $\varepsilon-$line. N_1 and N_2 are azegetic if and only if n_1 and n_2 lie on a common $\kappa-$line. The $\kappa-$lines are the lines of 6.1.5.3.

2. **Planes:** There are 540 $f-planes$, each containing three nodes and the three $\varepsilon-$lines joining them in pairs.

   There are 720 $d-planes$, each containing four nodes, joining a $\kappa-$line with a node, containing 4 nodes, 3 $\varepsilon-$lines and 1 $\kappa-$line.

   There are 270 $c-planes$, each containing six nodes, whcih are the intersection points of four $\kappa-$lines contained in the $c-$plane. Each $c-$plane contains 6 nodes, 3 $\varepsilon-$lines and 4 $\kappa-$lines.

3. **Solids:** There are 1080 $X-$solids, each containing five nodes, joining an $f-$plane with two "other" nodes.

   There are 120 $\ell-solids$, each containing six nodes which lie in threes on two skew $\kappa-$lines, which span the solid. The nine lines joining nodes N_1 and N_2 on different $\kappa-$lines are $\varepsilon-$lines. The six planes each containing one of the $\kappa-$lines and one of the nodes on the other $\kappa-$line are $d-$planes. Hence an $\ell-$solid contains 6 nodes, 9 $\varepsilon-$lines, 2 $\kappa-$lines and 6 $d-$planes.

   There are 540 $x-solids$, each containing seven nodes. Such a solid joins a $c-$plane with a node. The six lines joining the nodes of the $c-$planes with the other node are $\varepsilon-$lines; each join of one of the $\kappa-$lines in the $c-$plane with the other node is a $d-$plane; each join of an $\varepsilon-$line in the $c-$plane (of which there are three) with the other node is an $f-$plane. Each $x-$solid contains 7 nodes, 9 $\varepsilon-$lines, 4 $\kappa-$lines, 3 $f-$planes, 4 $d-$planes and 1 $c-$plane.

   There are 216 $n-solids$, each containing ten nodes, forming the same configuration as for the Burkhardt group: 10 nodes, 15 $\varepsilon-$lines, 10 $\kappa-$lines, 10 $d-$planes and 5 $c-$planes.

   There are 45 $J-solids$, each containing 12 nodes which are the vertices of desmic tetrahedra just as the Jordan primes for the Burkhardt group. Hence a $J-$solid contains 12 nodes, 18 $\varepsilon-$lines, 16 $\kappa-$lines, 12 $f-$planes, and 12 $c-$planes. These 45 $\mathbb{P}^3$'s are of course the same mentioned in section 6.1.5.8, and the $f-$planes and $c-$planes are identified there.

4. **Primes:** There are 360 $x\ell-primes$, each containing 7 nodes, joining an $\ell-$solid with a node not in that solid.

There are 216 $xn-primes$, each containing 11 nodes, joining an n-solid with a node not in the solid.

There are 36 $W-primes$, each containing 15 nodes. These primes are just the reflection hyperplanes, an analog of the Jordan primes, here each being opposite one of the 36 nodes. See section 6.1.5.1 for details.

There are 27 $S-primes$ corresponding to the 27 lines, each containing 20 nodes, these corresponding to the 20 double sixes not containing the given line. In each S-prime there is a pentahedron of reference consisting of j-solids, corresponding to the five tritangents containing a given line. There are 15 n-solids contained in an S-prime, corresponding to the 15 pairs of skew lines formed with the given line, i.e., the 15 lines of the 27 with which the given line is skew. Recall that the stabilizer of a line, of order 920, is isomorphic to $W(D_5) \subset W(E_6)$, hence the configuration of four pentahedra in the S-prime is a higher-dimensional analog of the configuration of the triad of desmic tetrahedra in $\mathbb{P}^4$.

We refrain from further details, as this will suffice for our purposes. It is at any rate a truely remarkable configuration.

6.1.7 Invariants

As in section B.1.2.2 there is a general method of deriving the degrees of the invariants of a unitary reflection group. Since the 27 forms (6.7) are (as a set) invariant under the Weyl group the expression

$$I_k := \sum_{i,j} \left\{ a_i^k + b_i^k + c_{ij}^k \right\}, \qquad (6.17)$$

if non-vanishing, is an invariant of degree k. The ring of invariants of $W(E_6)$ is generated by elements in degrees 2, 5, 6, 8, 9 and 12, which can be taken to be $I_2, \ldots, I_{12}$. We note that while I_2 and I_5 are *unique*, the other invariants are only defined up to addition of terms coming from lower degrees.

6.2 The invariant quintic

6.2.1 Equation

There is a unique (up to scalars) $W(E_6)$-invariant polynomial of degree 5. Written with integer coefficients in the variables x_i it is

$$f(x_1, \ldots, x_6) = x_6^5 - 6x_6^3\sigma_1(x) - 27x_6 \left(\sigma_1^2(x) - 4\sigma_2(x)\right) - 648\sqrt{\sigma_5(x)}, \quad (6.18)$$

where $\sigma_i(x)$ is the ith elementary symmetric polynomial of the $x_1^2, \ldots, x_5^2$, so in particular $\sqrt{\sigma_5(x)} = x_1 x_2 x_3 x_4 x_5$. The polynomial $f(x)$ displays manifestly the $W(D_5)$-invariance of the quintic. Under the change of variables from the x_i to the a_i of (6.7), the equation $g(a)$ can be derived as follows. By (6.9), we have

$b_i = a_i - \frac{1}{3}(a_1 + \cdots + a_6)$, which by equation (6.8) can be written $b_i = a_i - h$. The following trick was shown to me by I. Naruki. Consider $\prod_{i=1}^{6} a_i - \prod_{i=1}^{6} b_i$. This sextic divides the root h, and the quotient is $W(E_6)$-invariant. To see this, calculate

$$a_1 \cdots \cdots a_6 - (b_1 \cdots \cdots b_6) \quad = \quad \prod a_i - \prod (a_i - h) \qquad (6.19)$$
$$= \sigma_6(a) - [\sigma_6(a) - h\sigma_5(a) + h^2\sigma_4(a) - h^3\sigma_3(a) + h^4\sigma_2(a) - h^5\sigma_1(a)] ,$$

where here $\sigma_i(a)$ are the elementary symmetric functions of the a_i. Consequently,

$$a_1 \cdots \cdots a_6 - (b_1 \cdots \cdots b_6) \quad = \quad h \left(\sigma_5(a) - h\sigma_4(a) + h^2\sigma_3(a) - h^3\sigma_2(a) + h^4\sigma_1(a) \right) ,$$

and since by (6.8) $h = -\frac{1}{3}\sigma_1(a)$, this yields

$$g(a) = 81\sigma_5(a) + 27\sigma_4(a)\sigma_1(a) + 9\sigma_3(a)\sigma_1^2(a) + 3\sigma_2(a)\sigma_1^3(a) + \sigma_1^5(a), \quad (6.20)$$

giving the expression of the invariant quintic expressing manifestly the $W(A_5) = \Sigma_6$-invariance.

Definition 6.2.1 The *invariant quintic* $\mathcal{I}_5$ is the hypersurface of degree 5

$$\mathcal{I}_5 := \{x \in \mathbb{P}^5 \big| f(x) = 0\} \cong \{a \in \mathbb{P}^5 \big| g(a) = 0\},$$

where the isomorphism is given by the change of coordinates from the x_i to the a_i.

6.2.2 Singular locus

Because of the equivalence of the coordinates x_i, $i = 1, \ldots, 5$, there are essentially two different partial derivatives of f, namely

$$j_1 : \quad = \quad \frac{\partial f}{\partial x_1} \cong \cdots \cong \frac{\partial f}{\partial x_5} \qquad (6.21)$$
$$j_2 : \quad = \quad \frac{\partial f}{\partial x_6} .$$

Calculating these forms gives

$$-\frac{\partial f}{\partial x_i} \quad = \quad 12x_6^3 x_i + 54x_6 x_i(x_j^2 + x_k^2 + x_l^2 + x_m^2) + 648x_j x_k x_l x_m, \qquad (6.22)$$
$$\frac{\partial f}{\partial x_6} \quad = \quad 5x_6^4 - 18x_6^2\sigma_1(x) + 27(\sigma_1^2(x) - 4\sigma_4(x)).$$

These are quartics with manifest $W(D_4)$ and $W(D_5)$ symmetry, respectively.

Theorem 6.2.2 *The singular locus of $\mathcal{I}_5$ consists of the 120 lines of 6.1.5.3, which meet ten at a time in the 36 points of 6.1.5.6.*

Proof: Consider first the hyperplane section $x_6 = 0$. Then the equations to be solved are

$$x_i x_j x_k x_l = 0, \quad (i, j, k, l < 6); \qquad (6.23)$$
$$\sigma_1^2(x) - 4\sigma_2(x) = 0. \qquad (6.24)$$

From (6.23) we get: two of the x_i must vanish, say x_4, x_5, and then (6.24) takes the form

$$\left(x_1^2 + x_2^2 + x_3^2\right)^2 - 4\left(x_1^2 x_2^2 + x_1^2 x_3^2 + x_2^2 x_3^2\right) = 0,$$
$$(x_1 + x_2 + x_3)(x_1 - x_2 - x_3)(x_1 + x_2 - x_3)(x_1 - x_2 + x_3) = 0 \quad (6.25)$$

which splits into a product of four lines. Since the product $x_1 \cdots \cdots x_5 = 0$ is a coordinate simplex in $\mathbb{P}^4 = \{x_6 = 0\}$, it follows that the 2-simplices of this simplex correspond to planes where two of the coordinates vanish, hence there are $\binom{5}{2} = 10$ such 2-simplices; in each we have the four lines given by (6.25). This gives the 40 of the 120 lines contained in $x_6 = 0$. This implies that, in the union of the 27 hyperplanes (6.7), the singular locus of $\mathcal{I}_5$ consists of 120 lines.

Suppose that $x_6 \neq 0$. Then the simultaneous vanishing of the partials $\frac{\partial f}{\partial x_i}$, $i = 1, \ldots, 5$ imply that four of the x_i must vanish, say $x_2 = x_3 = x_4 = x_5 = 0$. But the intersection of $\mathcal{I}_5$ with the line $\{x_2 = x_3 = x_4 = x_5 = 0\}$ is given by

$$x_6^5 - 6x_6^3 x_1^2 - 27x_6 x_1^4 = x_6(x_6 + i\sqrt{3}x_1)(x_6 - i\sqrt{3}x_1)(x_6 + 3x_1)(x_6 - 3x_1), \quad (6.26)$$

and the last two terms are the equations of b_2 and c_{12}, two other of the 27 hyperplanes of (6.7). From this we conclude that any singular point is contained in one of the 27 hyperplanes, and by the above, that the singular locus of $\mathcal{I}_5$ consists of the 120 lines, as stated.   □

The types of singularities are given by the following.

Proposition 6.2.3 *The singularities of $\mathcal{I}_5$ can be described as follows.*

i) *At a generic point $x \in$ one of the 120 lines, a transversal hyperplane section has an ordinary A_1-singularity, so the singularity is of type $disc \times A_1$.*

ii) *At one of the 36 intersection points p, the singularity has multiplicity 3, and the tangent cone is of the form*

$$s_5 + s_4 t + s_3 t^2,$$

where s_3 is the cone over the Segre cubic, $s_4 = s_3 \cdot h_p$, where h_p is the hyperplane of 6.1.5.1 dual to p, and s_5 is the cone over the intersection $\mathcal{I}_5 \cap h_p$.

Proof: i) follows from consideration of generic hyperplane sections of $\mathcal{I}_5$, which are quintic threefolds with 120 isolated singularities. Recall the Varchenko bound for such quintics, which imply that a quintic threefold can have at most 136 ordinary double points. A singularity with more than one vanishing cycle contributes more than one to the number, hence each singularity has at most one vanishing cycle. Since furthermore all singularities are rational, it follows they must be of the stated form. This can also be seen from ii), as the singularities in the infinitesimal S_3 are ordinary double points. ii) is just a computation, done as follows. Suppose the point is $p = H_{23} = (1, -1, 0, 0, 0, 0)$. Then inhomogenizing by setting $t_i = x_i/x_1 - tp_i$ (where p_i denotes the ith coordinate of p), inserting into the equation of $\mathcal{I}_5$ gives the stated result. The fact that s_3 is the cone over S_3 can be seen as follows. We can write

$$f = s_5 + s_3(th_p + t^2) \tag{6.27}$$

and it follows that blowing up $\mathcal{I}_5$ at p is given by setting $t = \infty$ and that the proper transform of $\mathcal{I}_5$ in the exceptional $\mathbb{P}^4$ (of the blow up of $\mathbb{P}^5$ at the point p) is a cubic $S_3 = 0$, where $s_3 = 0$ is the cone over $S_3 = 0$. Since there are ten of the 120 lines meeting at p, the resolving divisor of the blow up, which is a cubic threefold, will have ten isolated singularities. As mentioned already above, this implies the cubic threefold is isomorphic to S_3. One can also see the 15 special hyperplane sections of S_3: these are the proper transforms, under the blowing up of p, of the 15 of the 36 hyperplanes 6.1.5.1 passing through the point. The rest is calculation.   □

We have (using Macaulay) calculated the ideal $\mathcal{J}(120)$ of the 120 lines, and it turns out to be just the Jacobian ideal of $\mathcal{I}_5$. I know of no simple proof of this fact.

6.2.3 Resolution of singularities

It turns out to be very easy to desingularize $\mathcal{I}_5$. By the proof of Proposition 6.2.3, we know the 36 triple points can be resolved by blowing up each such point p. Let $\varrho^{(1)} : \mathcal{I}_5^{(1)} \longrightarrow \mathcal{I}_5$ denote this blow up of $\mathcal{I}_5$. This has the effect of seperating all 120 lines of $\mathcal{I}_5$, and the singularities along the lines are just A_1, by Proposition 6.2.3, i). Hence a desingularisation is achieved by resolving each of the 120 lines. There are two possible ways to do this. First, one can blow up the lines in $\mathbb{P}^{5(1)} = \mathbb{P}^5$ blown up in the 36 points, and take the proper transform of $\mathcal{I}_5^{(1)}$; this has the effect of replacing each singular line by a quadric surface bundle, a $\mathbb{P}^1 \times \mathbb{P}^1$-bundle, over the line. Hence there are 120 exceptional divisors, each isomorphic to $\mathbb{P}^1 \times \mathbb{P}^1 \times \mathbb{P}^1$. We call this resolution of $\mathcal{I}_5$ the *big resolution* and denote it by $\widetilde{\mathcal{I}}_5$. Secondly, we can take a small resolution by blowing down one of the fiberings in each of the exceptional $\mathbb{P}^1 \times \mathbb{P}^1$ over a point of the line. In this way, each of the singular lines is replaced by a $\mathbb{P}^1$-bundle over the line, in other words by a $\mathbb{P}^1 \times \mathbb{P}^1$. We call any such resolution a *small resolution* and denote it by $\mathcal{I}_5^{(s)}$; there are 2^{120} possible such resolutions. Here no further (beyond the 36 on $\mathcal{I}_5^{(1)}$) exceptional divisors are introduced.

Lemma 6.2.4 *The quintic $\mathcal{I}_5$ has two types of resolutions, which we denote by $\widetilde{\mathcal{I}}_5$ and $\mathcal{I}_5^{(s)}$. On $\widetilde{\mathcal{I}}_5$ there are 36+120 exceptional divisors, 36 copies of the resolution of the Segre cubic, and 120 copies of $\mathbb{P}^1 \times \mathbb{P}^1 \times \mathbb{P}^1$. On $\mathcal{I}_5^{(s)}$ there are only 36 exceptional divisors, each a small resolution of the Segre cubic.*

6.2.4 $\mathcal{I}_5$ is rational

Quite generally, in $\mathbb{P}^5$, taking four $\mathbb{P}^3$'s which meet only in lines, there is a unique line which meets each of them and passes through a given point $P \in \mathbb{P}^5$, namely the line $< \alpha, P > \cap < \beta, P > \cap < \gamma, P > \cap < \delta, P >$, if $\alpha, \beta, \gamma, \delta$ denote the $\mathbb{P}^3$'s and $< \alpha, P >$ is the hyperplane spanned by α and P. Now let $P \in \mathcal{I}_5$, and choose four of the 15 of the 45 $\mathbb{P}^3$'s through one of the triple points p, such that the four $\mathbb{P}^2$'s on $(S_3)_p$ meet each other only in points; then the four $\mathbb{P}^3$'s meet only in lines, and we may apply this reasoning to conclude:

for each $P \in \mathcal{I}_5 - \{4\ \mathbb{P}^3$'s $\}$, there is a unique line L_p, which joins P and $\alpha, \beta, \gamma, \delta$.

Then, fixing a generic hyperplane $F \subset \mathbb{P}^5$. the line L_p intersects F in a single point; we get a rational map:

$$\mathcal{I}_5 - - - \to F$$

$$P \mapsto L_p \cap F.$$

We now carry out this argument to derive an explicit rationalisation. I am indebted to B. v. Geemen for help in performing this. We *choose* four convenient $\mathbb{P}^3$'s which only meet in lines (although these do not all pass through a point). The four $\mathbb{P}^3$'s will be defined as follows:

$$
\begin{aligned}
P_1 &= \{a_1 = b_4 = 0\} &=& \{l_1 = m_1 = 0\} \\
P_2 &= \{a_4 = b_5 = 0\} &=& \{l_2 = m_2 = 0\} \\
P_3 &= \{a_5 = b_6 = 0\} &=& \{l_3 = m_3 = 0\} \\
P_4 &= \{c_{35} = c_{12} = 0\} &=& \{l_4 = m_4 = 0\}
\end{aligned}
\tag{6.28}
$$

Letting F be an auxilliary $\mathbb{P}^4$ with homogenous coordinates $(y_0 : \ldots : y_4)$, the intersection of the line $< P_1, \alpha > \cap \cdots \cap < P_4, \alpha >$ with F is given by

$$y_0 l_i - y_i m_i = 0,$$

which leads to

$$
\begin{aligned}
y_0 &= m_1 \cdots m_4 \\
y_i &= l_i \cdot m_1 \cdots \widehat{m}_i \cdots m_4,
\end{aligned}
\tag{6.29}
$$

a system of quartics in $\mathbb{P}^5$, which, when restricted to $\mathcal{I}_5$, give the rational map $\varphi : \mathcal{I}_5 - - \to \mathbb{P}^4(= F)$. Inverting the equations for $x_1, \ldots, x_6$ we get the expressions listed in Table 6.3. This is a system of octics in $\mathbb{P}^4$, yielding the rational map

$$\psi : \mathbb{P}^4 \longrightarrow \mathcal{I}_5.$$

These rational maps are morphisms outside of the base locus.

Lemma 6.2.5 *The base locus of φ consists of the four $\mathbb{P}^3$'s P_1, P_2, P_3, P_4. The base locus of ψ is a surface of degree 32.* □

The first statement is clear from construction, while the second is a computation. We performed this with the help of Macaulay to calculate a standard basis of the ideal; the base locus is the intersection of the six octics.

6.3 Hyperplane sections

6.3.1 Lower-dimensional intersections

Before discussing the hyperplane sections, we describe first the lower-dimensional intersections. As mentioned at the beginning of the chapter, things are not quite as interesting for $\mathcal{I}_5$ as they are for $\mathcal{B}_4$. We use only the following fact, to be proved presently: all J–solids are contained in $\mathcal{I}_5$. Hence all lower-dimensional loci contained in a J–solid are likewise contained in $\mathcal{I}_5$, so we have

- $\mathcal{I}_5$ contains all nodes, all ε–lines, all κ–lines, all f–planes, all c–planes and all J–solids.

From this it follows that the n–solids, which contain five c–planes each, intersect $\mathcal{I}_5$ in five planes (a degenerate quintic). Consider an x–solid P. Since it contains a c–plane, the intersection $P \cap \mathcal{I}_5$ consists of a plane and a residual quartic surface. Hence the 120 ℓ–solids are the most special $\mathbb{P}^3$'s for which the intersection with $\mathcal{I}_5$ is an irreducible quintic. This quintic surface is depicted in the frontispiece.

6.3.2 Reducible hyperplane sections

Consider the hyperplane section $H_5 := \mathcal{I}_5 \cap \{x_6 = 0\}$; it is the union of five $\mathbb{P}^3$'s which form a coordinate simplex $x_1 \cdot x_2 \cdot x_3 \cdot x_4 \cdot x_5$ in the $\mathbb{P}^4$ given by $\{x_6 = 0\}$. Now $a_1 = -2/3x_6$ and invariance implies that the 27 hyperplane sections $a_i = 0$, $b_i = 0$, $c_{ij} = 0$ all have the same property. Each such hyperplane contains 40 of the 120 $\mathbb{P}^1$'s, which meet six at a time in 20 of the 36 points. Consider three lines in a tritangent, say (a_1, b_2, c_{12}). These three hyperplanes pass through a common $\mathbb{P}^3$, namely

$$l_{(12)} := \{x_6 = 0, \; x_1 = 0\}.$$

Such $\mathbb{P}^3$'s therefore correspond to the tritangents and there are 45 such on $\mathcal{I}_5$; these are just the 45 $\mathbb{P}^3$'s of 6.1.5.8. Hence we have

Proposition 6.3.1 *The quintic $\mathcal{I}_5$ contains 45 $\mathbb{P}^3$'s, which are cut out by the 27 hyperplane sections (6.7), and each such hyperplane section meets $\mathcal{I}_5$ in the union of five of the 45 $\mathbb{P}^3$'s. These can be numbered in terms of the tritangents of a cubic surface, i.e., for any 3 lines in a tritangent plane of a cubic surface, the corresponding hyperplanes of (6.7) intersect in a common $\mathbb{P}^3$, and this $\mathbb{P}^3$ lies on $\mathcal{I}_5$.*

Table 6.3: The octics giving the rationalization of $\mathcal{I}_5$

$$
\begin{aligned}
x_1 =\ & y_0^6 y_1 y_3 + y_0^5 y_1^2 y_3 - 2y_0^6 y_2 y_3 - y_0^5 y_1 y_2 y_3 + y_0^4 y_1^2 y_2 y_3 + 2y_0^4 y_1 y_2 y_3^2 \\
& - y_0^6 y_1 y_4 - y_0^5 y_1^2 y_4 - y_0^5 y_1 y_2 y_4 - 2y_0^4 y_1^2 y_2 y_4 - y_0^3 y_1^2 y_2^2 y_4 - y_0^5 y_1 y_3 y_4 \\
& - 2y_0^4 y_1 y_2 y_3 y_4 - 4y_0^3 y_1^2 y_2 y_3 y_4 - 2y_0^3 y_1 y_2^2 y_3 y_4 - 3y_0^2 y_1^2 y_2^2 y_3 y_4 \\
& - 3y_0^2 y_1^2 y_2 y_3^2 y_4 - 2y_0^2 y_1 y_2^2 y_3^2 y_4 - 2y_0 y_1^2 y_2^2 y_3^2 y_4 + y_0^3 y_1^2 y_2 y_3 y_4^2 \\
& + y_0 y_1^2 y_2^2 y_3 y_4^2 + y_0 y_1^2 y_2 y_3^2 y_4^2 + y_1^2 y_2^2 y_3^2 y_4^2 + 2y_0^3 y_1^2 y_2 y_3^2 \\
& - y_0^4 y_1^2 y_3 y_4 - 2y_0^3 y_1 y_2 y_3^2 y_4
\end{aligned}
$$

$$
\begin{aligned}
x_2 =\ & y_0^6 y_1 y_3 + y_0^5 y_1^2 y_3 + y_0^5 y_1 y_2 y_3 + y_0^4 y_1^2 y_2 y_3 + 2y_0^5 y_1 y_3^2 + 2y_0^4 y_1^2 y_3^2 \\
& + 2y_0^3 y_1^2 y_2 y_3^2 - y_0^6 y_1 y_4 - y_0^5 y_1^2 y_4 - y_0^5 y_1 y_2 y_4 - 2y_0^4 y_1^2 y_2 y_4 - y_0^3 y_1^2 y_2^2 y_4 \\
& - 5y_0^4 y_1^2 y_3 y_4 - 6y_0^4 y_1 y_2 y_3 y_4 - 8y_0^3 y_1^2 y_2 y_3 y_4 - 2y_0^3 y_1 y_2^2 y_3 y_4 \\
& - 4y_0^3 y_1^2 y_3^2 y_4 - 6y_0^3 y_1 y_2 y_3^2 y_4 - 7y_0^2 y_1^2 y_2 y_3^2 y_4 - 2y_0^2 y_1 y_2^2 y_3^2 y_4 \\
& + 2y_0^4 y_1^2 y_4^2 + 2y_0^4 y_1 y_2 y_4^2 + 4y_0^3 y_1^2 y_2 y_4^2 + 2y_0^2 y_1^2 y_2^2 y_4^2 + 4y_0^4 y_1 y_3 y_4^2 \\
& + 6y_0^3 y_1 y_2 y_3 y_4^2 + 9y_0^2 y_1^2 y_2 y_3 y_4^2 + 2y_0^2 y_1 y_2^2 y_3 y_4^2 + 5y_0 y_1^2 y_2^2 y_3 y_4^2 \\
& + 4y_0^3 y_1 y_2 y_3^2 y_4^2 + 5y_0 y_1^2 y_2 y_3^2 y_4^2 + 2y_0 y_1 y_2^2 y_3^2 y_4^2 + 3y_1^2 y_2^2 y_3^2 y_4^2 \\
& + 2y_0^4 y_1 y_2 y_3^2 - 5y_0^5 y_1 y_3 y_4 - 3y_0^4 y_1^2 y_2 y_3 y_4 - 4y_0^5 y_1 y_3^2 y_4 \\
& - 2y_0 y_1^2 y_2^2 y_3^2 y_4 + 2y_0^5 y_1 y_4^2 + 4y_0^3 y_1^2 y_3 y_4^2 \\
& + 2y_0^3 y_1 y_2 y_3^2 y_4^2 + 2y_0^2 y_1^2 y_3^2 y_4^2
\end{aligned}
$$

$$
\begin{aligned}
x_3 =\ & 2y_0^7 y_3 + 3y_0^6 y_1 y_3 + y_0^5 y_1^2 y_3 + 2y_0^6 y_2 y_3 + 3y_0^5 y_1 y_2 y_3 + y_0^4 y_1^2 y_2 y_3 \\
& - y_0^5 y_1^2 y_4 - 2y_0^6 y_2 y_4 - 5y_0^5 y_1 y_2 y_4 - 2y_0^4 y_1^2 y_2 y_4 - 2y_0^4 y_1 y_2^2 y_4 \\
& - 3y_0^5 y_1 y_3 y_4 - y_0^4 y_1^2 y_3 y_4 - 4y_0^5 y_2 y_3 y_4 - 6y_0^4 y_1 y_2 y_3 y_4 - 2y_0^3 y_1^2 y_2 y_3 y_4 \\
& - y_0^2 y_1^2 y_2^2 y_3 y_4 + 2y_0^3 y_1 y_2 y_3^2 y_4 + y_0^3 y_1^2 y_2 y_3 y_4^2 - 2y_0^3 y_1 y_2^2 y_3 y_4^2 \\
& - y_0^2 y_1^2 y_2 y_3 y_4^2 - 2y_0^2 y_1 y_2^2 y_3 y_4^2 - y_0 y_1^2 y_2^2 y_3 y_4^2 - 2y_0^2 y_1 y_2 y_3^2 y_4^2 \\
& - 2y_0 y_1 y_2^2 y_3^2 y_4^2 - y_1^2 y_2^2 y_3^2 y_4^2 - 2y_0^7 y_4 - 3y_0^6 y_1 y_4 \\
& - y_0^3 y_1^2 y_2^2 y_4 - 2y_0^6 y_3 y_4 - 2y_0^3 y_1 y_2^2 y_3 y_4 - y_0 y_1^2 y_2^2 y_3 y_4^2
\end{aligned}
$$

$$
\begin{aligned}
x_4 =\ & 2y_0^7 y_3 + 3y_0^6 y_1 y_3 + y_0^5 y_1^2 y_3 + y_0^5 y_1 y_2 y_3 + y_0^4 y_1^2 y_2 y_3 - 2y_0^5 y_1 y_3^2 \\
& - 3y_0^6 y_1 y_4 - y_0^5 y_1^2 y_4 - 3y_0^5 y_1 y_2 y_4 - 2y_0^4 y_1^2 y_2 y_4 - y_0^3 y_1^2 y_2^2 y_4 - 2y_0^6 y_3 y_4 \\
& + y_0^4 y_1^2 y_3 y_4 - 2y_0^4 y_1 y_2 y_3 y_4 - y_0^2 y_1^2 y_2^2 y_3 y_4 + 4y_0^4 y_1 y_3^2 y_4 + 4y_0^3 y_1^2 y_3^2 y_4 \\
& + 3y_0^2 y_1^2 y_2 y_3^2 y_4 - 2y_0^4 y_1 y_3 y_4^2 - 2y_0^3 y_1^2 y_3 y_4^2 - 2y_0^3 y_1 y_2 y_3 y_4^2 \\
& - y_0 y_1^2 y_2^2 y_3 y_4^2 - 2y_0^3 y_1 y_3^2 y_4^2 - 2y_0^2 y_1^2 y_3^2 y_4^2 - 2y_0^2 y_1 y_2 y_3^2 y_4^2 \\
& - 2y_0^4 y_1 y_3^2 - 2y_0^7 y_4 - y_0^5 y_1 y_3 y_4 + 2y_0^3 y_1 y_2 y_3^2 y_4 - 3y_0^2 y_1^2 y_2 y_3 y_4^2 \\
& - 3y_0 y_1^2 y_2 y_3^2 y_4^2 - y_1^2 y_2^2 y_3^2 y_4^2
\end{aligned}
$$

$$
\begin{aligned}
x_5 =\ & -y_0^6 y_1 y_3 - y_0^5 y_1^2 y_3 - y_0^5 y_1 y_2 y_3 - y_0^4 y_1^2 y_2 y_3 + y_0^6 y_1 y_4 + y_0^5 y_1^2 y_4 \\
& + 2y_0^4 y_1^2 y_2 y_4 + 2y_0^4 y_1 y_2^2 y_4 + y_0^3 y_1^2 y_2^2 y_4 + 3y_0^5 y_1 y_3 y_4 + 3y_0^4 y_1^2 y_3 y_4 \\
& + 4y_0^3 y_1^2 y_2 y_3 y_4 + 2y_0^3 y_1 y_2^2 y_3 y_4 + y_0^2 y_1^2 y_2^2 y_3 y_4 + y_0^2 y_1^2 y_2 y_3^2 y_4 - 2y_0^5 y_1 y_4^2 \\
& - 2y_0^4 y_1 y_2 y_4^2 - 4y_0^3 y_1^2 y_2 y_4^2 - 2y_0^2 y_1^2 y_2^2 y_4^2 - 2y_0^4 y_1 y_3 y_4^2 - 2y_0^3 y_1^2 y_3 y_4^2 \\
& - 5y_0^2 y_1^2 y_2 y_3 y_4^2 - 3y_0 y_1^2 y_2^2 y_3 y_4^2 - y_0 y_1^2 y_2 y_3^2 y_4^2 - y_1^2 y_2^2 y_3^2 y_4^2 \\
& + 2y_0^6 y_2 y_4 + 3y_0^5 y_1 y_2 y_4 + 2y_0^4 y_1 y_2 y_3 y_4 - 2y_0^3 y_1^2 y_4^2 - 2y_0^3 y_1 y_2 y_3 y_4^2
\end{aligned}
$$

$$
\begin{aligned}
x_6 =\ & -3y_0^6 y_1 y_3 - 3y_0^5 y_1^2 y_3 - 3y_0^5 y_1 y_2 y_3 - 3y_0^4 y_1^2 y_2 y_3 + 3y_0^6 y_1 y_4 + 3y_0^5 y_1^2 y_4 \\
& + 6y_0^4 y_1^2 y_2 y_4 + 3y_0^3 y_1^2 y_2^2 y_4 + 3y_0^5 y_1 y_3 y_4 + 3y_0^4 y_1^2 y_3 y_4 + 6y_0^4 y_1 y_2 y_3 y_4 \\
& + 3y_0^2 y_1^2 y_2^2 y_3 y_4 - 3y_0^2 y_1^2 y_2 y_3^2 y_4 + 3y_0^2 y_1^2 y_2 y_3 y_4^2 + 3y_0 y_1^2 y_2^2 y_3 y_4^2 \\
& + 3y_1^2 y_2^2 y_3^2 y_4^2 + 3y_0^5 y_1 y_2 y_4 + 6y_0^3 y_1^2 y_2 y_3 y_4 + 3y_0 y_1^2 y_2 y_3^2 y_4^2
\end{aligned}
$$

Also the intersections of the 45 $\mathbb{P}^3$'s can be described. Each such $\mathbb{P}^3$ contains 16 of the 120 lines which meet in 12 of the 36 points; these 12 points are the vertices of a triad of desmic tetrahedra. Consider the $\mathbb{P}^3$ $l_{(12)}$; the corresponding tritangent meets 12 others, namely (13), (14), (15), (16), (32), (42), (52), (62), (12.34.56), (12.35.46), (12.36.45) and (21), and the 12 $\mathbb{P}^3$'s corresponding to them meet $l_{(12)}$ in a $\mathbb{P}^2$ (the generic intersection has dimension 1). These 12 planes in $l_{(12)}$ form the arrangement $\mathcal{A}(\mathbf{D_4})$ in $\mathbb{P}^3$.

6.3.3 Special hyperplane sections

We now consider the intersections of $\mathcal{I}_5$ with the 36 reflection hyperplanes 6.1.5.1. Take for example the reflection hyperplane $\{h = 0\}$; since h is just a multiple of $\sigma_1(a)$ (see (6.8)), it follows from the equation (6.20) that the intersection $\{h = 0\} \cap \mathcal{I}_5$ is a quintic hypersurface in $\mathbb{P}^4$ with the equation:

$$Q_1 := \{h = 0\} \cap \mathcal{I}_5 = \left\{ \begin{array}{l} \sigma_1(a) = 0 \\ \sigma_5(a) = 0 \end{array} \right. \qquad (6.30)$$

Comparing with the equation (3.64), we see that this is a copy of the Nieto quintic! By symmetry, each of the 36 hyperplane sections is isomorphic to this one, and we denote them by

$$T = \{h = 0\} \cap \mathcal{I}_5, \; T_{ij} = \{h_{ij} = 0\} \cap \mathcal{I}_5, \; T_{ijk} = \{h_{ijk} = 0\} \cap \mathcal{I}_5. \qquad (6.31)$$

So we have:

Proposition 6.3.2 *There are 36 copies of the Nieto quintic (3.64) on $\mathcal{I}_5$.*

We can determine the singular locus of these hyperplane sections, independently of the discussion given in section 3.4.1. The reflection hyperplane contains 20 of the 120 lines, which meet in 15 of the 36 points (corresponding to the 15 roots of an $\mathbf{A_5}$ subsystem), so the quintic threefold has 20 singular lines, with 15 singular points of multiplicity 3. In fact, the resolving divisor of each of these 15 points is a four-nodal cubic surface, which is a hyperplane section of the Segre cubic S_3 (see the discussion following Problem 3.4.19, ii)). Furthermore, recalling that there are ten of the 120 lines which pass through the triple point which is *dual* to the given reflection hyperplane, each such intersects the reflection hyperplane transversally, giving the ten isolated ordinary double points on that quintic (see Proposition 3.4.1), and in some sense "explains" these isolated singularities.

6.3.4 Generic hyperplane sections

A generic hyperplane section is a quintic threefold in $\mathbb{P}^4$ with 120 nodes. This is a fascinating family of Calabi-Yau threefolds, which has a beautiful geometric configuration associated with it, in some sense "dual" to the configuration of the 27 lines on a cubic surface.

Proposition 6.3.3 *Let $H \in \mathbb{P}^5$ be a generic hyperplane and let $Q_H = \mathcal{I}_5 \cap H$ be the hyperplane section. Then we have*

1) *There are 45 $\mathbb{P}^2$'s on Q_H, which are cut out by 27 hyperplanes; these could appropriately be called* quintangent *planes.*

2) *The group of incidence preserving permutations of the 45 $\mathbb{P}^2$'s is $W(E_6)$; this is also the group of incidence preserving permutations of the 27 hyperplanes.*

3) *There are 36 hyperplane sections of Q_H, each of which is a 20-nodal quintic surface.*

4) *The 120 nodes of Q_H form an orbit under $W(E_6)$.*

Proof: For any of the 45 $\mathbb{P}^3$'s in $\mathcal{I}_5$ and hyperplane section H, it holds that $\mathbb{P}^3 \cap H = \mathbb{P}^2 \subset H \cap \mathcal{I}_5 = Q_H$, showing 1). The second point is evident, and in a sense "dual" to the situation with cubic surfaces. We have seen that a special hyperplane section as in (6.31) is isomorphic to the Nieto quintic and has 20 singular lines in its singular locus; therefore any generic hyperplane section has exactly 20 nodes. 4) follows since the $W(E_6)$-orbit consisting of the 120 lines, restricted to the hyperplane section, is still an orbit. $\square$

We now consider some of the invariants of the nodal quintic threefolds. Let V denote a nodal quintic, $\widehat{V} \longrightarrow V$ a small resolution and $\widetilde{V} \longrightarrow V$ a big resolution. Letting s denote the number of nodes, the betti numbers are

$$
\begin{aligned}
b_1(V) &= 1 = b_1(\widehat{V}), & b_2(\widetilde{V}) &= 1 + d + s; \\
b_2(V) &= 1 + d = b_2(\widehat{V}), & b_4(\widetilde{V}) &= 1 + d + s; \\
b_3(V) &= b_3(V_t) - s + d, & b_3(\widehat{V}) &= b_3(V) - s + d = b_3(\widetilde{V}),
\end{aligned}
\tag{6.32}
$$

where V_t is a smooth hypersurface of same degree as V and d is the *defect*. The defect may be calculated by the following result.

Theorem 6.3.4 ([We], p. 27) *Let $V \in \mathbb{P}^4$ be a nodal hypersurface of degree $n \geq 3$. Then*

$$
dim(\mathcal{P}_{2n-5}(V)) = dim \left\{ \begin{matrix} \text{homogenous} \quad \text{polynomi-} \\ \text{als of degree } 2n-5 \text{ in } \mathbb{P}^4, \\ \text{containing all nodes of } V \end{matrix} \right\} = \binom{2n-1}{4} - s + d.
$$

Applied to the case at hand, we need the dimension of the space of *quintics* vanishing at all the nodes. Clearly this is the degree five component in the ideal of the 120 points. As we mentioned above, we *know* the ideal of the 120 lines (it is the Jacobian ideal of $\mathcal{I}_5$, $\mathcal{J}ac(\mathcal{I}_5)$), so we know also the ideal of the 120 points; it is the restriction of $\mathcal{J}ac(\mathcal{I}_5)$ to the hyperplane, generated by six quartics.

Proposition 6.3.5 *The dimension of the space $\mathcal{P}_5(Q_H)$ is 30.*

Proof: Each of the quartics (which are clearly independent for a generic hyperplane H) of $\mathcal{J}ac(\mathcal{I}_5)$ can be multiplied by any hyperplane, giving a quintic which contains the 120 nodes. The set of hyperplanes is $(\mathbb{P}^5)^\vee$, so the dimension of $\mathcal{P}_5(Q_H)$ is $6 \cdot 5 = 30$. □

We can now apply Theorem 6.3.4 to calculate the defect d for Q_H. The formula is $126 - 120 + d = 30$, from which is follows that $d = 24$.

Corollary 6.3.6 *The small resolutions* $\widehat{Q}_H$ *of the quintic threefolds* Q_H *have the following betti and Hodge numbers:*

$$b_2(\widehat{Q}_H) = 25, \qquad\qquad h^{1,1} = 25,$$
$$b_3(\widehat{Q}_H) = 12 = 2 + 2h^{2,1}, \quad h^{2,1} = 5, \quad e = 2h^{1,1} - 2h^{2,1} = 40.$$

Proof: Insertion of $d = 24$ in (6.32). □

In the well-known manner for Calabi-Yau threefolds the isomorphism

$$H^2(V, \Omega^1) \cong H^1(V, \Theta)$$

identifies the Hodge space $H^{2,1}$ with the space of infinitesimal deformations of V, $H^1(V, \Theta)$. This is by the above five-dimensional, hence the moduli space of these 120 nodal quintics (a Zariski open subset of $(\mathbb{P}^5)^\vee$) is also a global space of complex deformations of the small resolution. We can describe the space $H^{2,1}$ more concretely as follows. Consider the space $\mathcal{P}_5(Q_H)$; let $\mathcal{J} \subset \mathcal{P}_5(Q_H)$ be the subspace generated by the Jacobi ideal of Q_H; since Q_H has five partial derivatives, $\mathcal{J}$ is $5 \cdot 5 = 25$ dimensional, and $\mathcal{J}$ cannot contribute to infinitesimal deformations, so we have

$$H^{2,1}(\widehat{Q}_H) \cong \mathcal{P}_5(Q_H)/\mathcal{J}.$$

As a final remark consider the Picard group $\mathrm{Pic}(\widehat{Q}_H)$ and the orthocomplement of the hyperplane section $\mathrm{Pic}^0(\widehat{Q}_H)$. Then $rk_\mathbb{Z} \mathrm{Pic}^0(\widehat{Q}_H) = 24$, and the 45 $\mathbb{P}^2$'s give us privledged representatives in Pic^0; the 27 hyperplanes represent relations, so we have an exact sequence

$$\mathbb{Z}^{27} \longrightarrow \mathbb{Z}^{45} \longrightarrow \mathrm{Pic}^0(\widehat{Q}_H) \longrightarrow 1,$$

and the kernel is six-dimensional. The sum sequence is then

$$1 \longrightarrow \mathbb{Z}^6 \longrightarrow \mathbb{Z}^{27} \longrightarrow \mathbb{Z}^{45} \longrightarrow \mathbb{Z}^{24} \longrightarrow 1, \qquad (6.33)$$

and this is really dual to the sequence (4.12) for cubic surfaces.

Remark 6.3.7 The period map for this five-dimensional family of Calabi-Yau threefolds maps to the domain $\mathcal{D} = Sp(6, \mathbb{R})/U(1) \times U(5)$. Note that any hyperplane passing through one of the 45 $\mathbb{P}^3$'s will intersect $\mathcal{I}_5$ in the union of that $\mathbb{P}^3$ and a residual quartic; clearly these constitute the set of cusps for the period map, i.e., the 45 lines in $(\mathbb{P}^5)^\vee$ (the dual $\mathbb{P}^5$) which parameterize the set of hyperplanes passing through one of the 45 $\mathbb{P}^3$'s, have image under the period map in the boundary of the domain $\mathcal{D}$ above. These 45 one-dimensional cusps meet in 27 points, i.e., zero-dimensional cusps, which correspond to the 27 hyperplane sections which split into the union of five $\mathbb{P}^3$'s. But we can say more. Noting that, excepting the hyperplanes above, all hyperplane sections of $\mathcal{I}_5$ are irreducible quintics, the worst that can happen is that the hyperplane passes through one of the 36 triple points of $\mathcal{I}_5$. We will see below that these are still Calabi-Yau (Proposition 6.4.8), hence *not* contained in the boundary.

6.3.5   Tangent hyperplane sections

We now consider the case of a hyperplane tangent to $\mathcal{I}_5$ at a point $p \in \mathcal{I}_5$. In this case the section Q_p aquires an additional node. Note that the 121 nodes fall into two "orbits", one set of 120 on which $W(E_6)$ acts as a permutation group, and the additional point p. For a 121-nodal quintic the same calculation as above gives $e(Q_p) = 42$, $h^{2,1} = 4$, $h^{1,1} = 25$. It follows that the $H_4(Q_p, \mathbb{Q})$ is the same as for Q_x, $x \in \mathbb{P}^5$ generic. The difference to the generic case is in H_3, more precisely in $H^{2,1}$. Indeed, we now require $\mathcal{P}_5(Q_p)$, that is, quintics through all 121 nodes, As above, for each of the five quartics in the Jacobi ideal of Q_p, since each contains p, we get a five-dimensional family of quintics. But for the quartics through the 120 nodes which are *not* in the Jacobi ideal, we must take a hyperplane *through the point p*, so

Proposition 6.3.8 *For a 121-nodal quintic Q_p, $p \in \mathcal{I}_5$, we have $\dim \mathcal{P}_5(Q_p) = 5 \cdot 5 + 1 \cdot 4 = 29$.*   □

We can now apply Theorem 6.3.4 to calculate the defect:

$$d = 29 - 126 + 121 = 24.$$

Corollary 6.3.9 *The betti numbers for the small resolution $\widehat{Q}_p$ are*

$$b_2(\widehat{Q}_p) = 25, \quad h^{1,1} = 25,$$
$$b_3(\widehat{Q}_p) = 10, \quad h^{2,1} = 4, \quad e = 42.$$

We remark that since $h^{1,1}$ is still 25, the sequence (6.33) still holds for Q_p.

6.4   Birational maps and the projection from a triple point

6.4.1   The cuspidal model

First we recall the notations $\mathcal{I}_5^{(1)}$ for the blow up of $\mathcal{I}_5$ at the 36 triple points, $\widetilde{\mathcal{I}}_5$ for the big resolution of $\mathcal{I}_5$, and $\mathcal{I}_5^{(s)}$ for a small resolution. Note that on $\mathcal{I}_5^{(1)}$, each of the 120 lines has normal bundle $\mathcal{O}(-2)^{\oplus 3}$, hence each line can be blown down to an isolated singular point.

Definition 6.4.1 Consider the following birational transformation of $\mathcal{I}_5$:

i) Blow up the 36 triple points, $\varrho^{(1)} : \mathcal{I}_5^{(1)} \longrightarrow \mathcal{I}_5$;

ii) Blow down the proper transforms of the 120 lines to 120 isolated singularities, $\varrho^{(2)} : \mathcal{I}_5^{(1)} \longrightarrow \widehat{\mathcal{I}}_5$.

Step ii) defines the *cuspidal model* $\widehat{\mathcal{I}}_5$.

This is a four-dimensional analogue of $\widehat{\mathcal{N}}_5$ of (3.68). Indeed, for each of the 36 hyperplane sections of Proposition 6.3.2, the proper transform on $\widehat{\mathcal{I}}_5$ is isomorphic to $\widehat{\mathcal{N}}_5$:

Lemma 6.4.2 *Let* $T \cong \mathcal{N}_5$ *be one of the 36 special hyperplane sections of (6.31), and let* $\widehat{T}$ *denote its proper transform on* $\widehat{\mathcal{I}}_5$. *Then* $\widehat{T} \cong \widehat{\mathcal{N}}_5$.

Proof: Just check that the steps i) and ii) of Definition 6.4.1, when restricted to T, coincide with those of (3.68). $\qquad\square$

Let us mention that $\widehat{\mathcal{I}}_5$ "looks like" a ball quotient too, at least assuming a positive answer to Problem 3.4.19. We explain what "looks like" means in the following items.

I1 Each isolated singularity is resolved by a $\mathbb{P}^1 \times \mathbb{P}^1 \times \mathbb{P}^1$; the arrangement induced in each by the proper transforms of the 36 hyperplanes and 36 exceptional divisors is a *product*, consisting of three fibres in each fibering (i.e., {3 points} $\times \mathbb{P}^1 \times \mathbb{P}^1$, $\mathbb{P}^1 \times$ {3 points} $\times \mathbb{P}^1$, $\mathbb{P}^1 \times \mathbb{P}^1 \times$ {3 points}, see 6.1.5.3). Hence this can be covered in an equivariant way by $E_\varrho \times E_\varrho \times E_\varrho$ (see Lemma 3.2.5.2).

I2 The proper transforms of the 36 hyperplane sections of Proposition 6.3.2 are by Lemma 6.4.2 isomorphic to $\widehat{\mathcal{N}}_5$, so, if the Problem 3.4.19 has an affirmative solution, these are ball quotients, with cusps being those isolated singularities of $\widehat{\mathcal{I}}_5$ which are contained in the given $\widehat{T}$.

I3 Consider the 45 $\mathbb{P}^3$'s of Proposition 6.3.1. These are (the proper transforms of) the 45 $\mathbb{P}^3$'s of 6.1.5.8. These $\mathbb{P}^3$'s are also ball quotients, in fact in two different ways.

1) There is a cover $Y \longrightarrow \mathbb{P}^3$, branched over the arrangement $\mathcal{A}(W(\mathbf{D_4}))$ in $\mathbb{P}^3$, which is a ball quotient. This example can be derived from the solution 4) of (3.22) by means of the natural squaring map m_2 : $\mathbb{P}^3 \longrightarrow \mathbb{P}^3$, $(x_0 : \ldots : x_3) \mapsto (x_0^2 : \ldots : x_3^2)$. Then the arrangement $\mathcal{A}(W(\mathbf{D_4}))$ is the pullback under m_2 of the six symmetry planes of the tetrahedron in the arrangement $\mathcal{A}(W(\mathbf{A_4}))$, and pulling back the solution 4), we get the cover $Y \longrightarrow \mathbb{P}^3$, branched along $\mathcal{A}(W(\mathbf{D_4}))$ (with branching degree 3 at each hyperplane), which is a ball quotient by a fix point free group.

2) There is a cover $Z \longrightarrow \mathbb{P}^3$, branched along the arrangement $\mathcal{A}(W(\mathbf{F_4}))$ in $\mathbb{P}^3$ (but not a Fermat cover), which is a ball quotient; this example is explained in [H1], Thm. 7.6.5, and is the *only* known ball quotient

related to a plane arrangement in $\mathbb{P}^4$ which does *not* derive from those given by solutions of the hypergeometric differential equation.

Both of the arrangements mentioned, $\mathcal{A}(W(\mathbf{D_4}))$ and $\mathcal{A}(W(\mathbf{F_4}))$, arise naturally on the 45 $\mathbb{P}^3$'s: the first is the intersection with the 27 hyperplanes, the second is the intersection with the 36 hyperplanes.

6.4.2 Projection from a triple point

Let $p \in \mathcal{I}_5$ be one of the 36 triple points, and let h_p be the dual hyperplane (one of the 36 $\mathbb{P}^4$'s of 6.1.5.1). The projection of $\mathbb{P}^5$ from p is defined as follows. Consider the $\mathbb{P}^4$ of all lines through p; this is just the dual h_p, and each line l_p through p corresponds to a unique point of h_p (its intersection with h_p). Since any point x of $\mathbb{P}^5$ is on a unique line $(l_x)_p$ through p, the map

$$\pi_p : \mathbb{P}^5 \longrightarrow h_p \qquad\qquad (6.34)$$
$$x \mapsto (l_x)_p \cap h_p$$

gives the *projection from p*. Restricting to $\mathcal{I}_5$ this gives a generically finite (rational) map, which we also denote by π_p, $\pi_p : \mathcal{I}_5 \; - - \to h_p$.

Lemma 6.4.3 $\pi_p : \mathcal{I}_5 \; - - \to h_p$ *is generically a double cover.*

Proof: Since the triple point has multiplicity 3, a generic line will meet $\mathcal{I}_5$ in $(5 - 3) = 2$ further points.   □

Lemma 6.4.4 $\pi_p : \mathcal{I}_5 \; - - \to h_p$ *is a quotient map by the group $G_p \cong \mathbb{Z}/2\mathbb{Z}$ generated by the reflection σ_p on the root p.*

Proof: The reflection σ_p fixes h_p; it is the inversion $((z_0 : z_1) \mapsto (z_1 : z_0))$ on any line l_p through p, where the homogenous coordinates are choosen such that $l_p \cap h_p = (1 : 1)$. Since $\mathcal{I}_5$ is mapped by σ_p onto itself, it follows that two points of $\mathcal{I}_5 \cap l_p$ are related by inversion on l_p. So the group action is manifest.   □

We now describe how to make π_p into a *morphism*. First of all, one must blow up p; let $\varrho_p : \mathcal{I}_{5,p} \longrightarrow \mathcal{I}_5$ denote this blow up. Let $(\mathcal{S}_3)_p$ be the copy of $\mathcal{S}_3$ which is the exceptional divisor at p. For any $x \in (\mathcal{S}_3)_p$, the line $(l_x)_p$ through p and intersecting h_p in the Segre cubic there, is tangent to $\mathcal{I}_5$ *at the triple point p*. Secondly, certain subvarieties get *blown down*. Indeed, suppose $(l_x)_p$ is *contained in $\mathcal{I}_5$* for some $x \in \mathcal{I}_5$. Then, clearly, $(l_x)_p \mapsto (l_x)_p \cap h_p$, the whole line maps to a point, or in other words, gets blown down.

Lemma 6.4.5 *The projection $\pi_p : \mathcal{I}_5 \; - - \to h_p$, which is well-defined on $\mathcal{I}_{5,p}$, blows down all linear subspaces on $\mathcal{I}_5$ which pass through p, and is a double cover outside the union $\mathcal{L}_p$ of all such linear subspaces on $\mathcal{I}_5$ passing through p.*   □

We now describe $\mathcal{L}_p$. Recall that the linear subspaces on $\mathcal{I}_5$ are the 45 $\mathbb{P}^3$'s and their intersections. Hence $\mathcal{L}_p$ consists of all the $\mathbb{P}^3$'s and their intersections, which pass through p. Recall from 6.1.5.6 that this is the set of 15 of the 45 $\mathbb{P}^3$'s of Proposition 6.3.1. Therefore, we get

Lemma 6.4.6 *The projection* $\pi_p, p : \mathcal{I}_{5,p} \longrightarrow h_p$ *blows down the union of 15* $\mathbb{P}^3$ *'s to the 15 planes in* h_p *which are the intersection of* $\mathcal{S}_3$ *and* $\mathcal{N}_5$.

Now let $X = \mathcal{I}_{5,p}^{\%}$, the double cover of h_p branched along the union of $\mathcal{S}_3$ and $\mathcal{N}_5$ (which is of degree 8, so a double cover exists). X is clearly *singular along* the 15 planes. Indeed:

Lemma 6.4.7 $\pi_p, p : \mathcal{I}_{5,p} \longrightarrow h_p$ *factors over* $\mathcal{I}_{5,p}^{\%}$, *and* $\Pi : X \longrightarrow h_p$ *is the double cover of* $\mathbb{P}^4$ *branched along the union* $\mathcal{S}_3 \cup \mathcal{N}_5$.

Proof: This follows from the discussion above; the branch locus $\mathcal{R}$ is the set:

$$\mathcal{R} = \{x \in \mathcal{I}_{5,p} \big| (l_x)_p \text{ is tangent to } \mathcal{I}_5 \text{ at } x\}.$$

This happens if either

   i) $x \in h_p$, since then x is fixed by σ_p;

   ii) $x \in (\mathcal{S}_3)_p$, the exceptional divisor over p.

Therefore $\mathcal{R} = \mathcal{S}_3 \cup \mathcal{N}_5$. By Lemmas 6.4.5 and 6.4.6, 15 $\mathbb{P}^3$'s are blown down to $\mathbb{P}^2$'s, and outside of this locus, Π is 2:1. $\qquad\square$

6.4.3   Double octics and quintic hypersurfaces

With the result of Lemma 6.4.7 at hand, we can get a new slant on the quintic threefolds which are hyperplane sections of $\mathcal{I}_5$. For this, consider a hyperplane section of the cover $\Pi : X \longrightarrow h_p$, that is, let $H \subset h_p$ be a hyperplane, and let X_H be its inverse image in X:

$$\Pi_H : X_H \longrightarrow H,$$

a double cover of $\mathbb{P}^3$. The branch locus is $H \cap (\mathcal{S}_3 \cup \mathcal{N}_5)$, which is the union of a cubic and a quintic surface in $\mathbb{P}^3$. Note the $H \cap \{$ one of the 15 $\mathbb{P}^2$'s $\subset \mathcal{S}_3 \cap \mathcal{N}_5\}$ is a *line*, contained in both $H \cap \mathcal{S}_3$ and in $H \cap \mathcal{N}_5$. In other words, $H \cap (\mathcal{S}_3 \cup \mathcal{N}_5) = \mathcal{S}_H \cup \mathcal{Q}_H$, where $\mathcal{S}_H$ is the cubic surface, $\mathcal{Q}_H$ is the quintic surface, and

$$\mathcal{S}_H \cap \mathcal{Q}_H = \{15 \text{ lines}\}.$$

Proposition 6.4.8 *Let* $X_H = \Pi^{-1}(H)$, *the double cover of* $\mathbb{P}^3$ *branched along* $\mathcal{S}_H \cup \mathcal{Q}_H$. *Then there is a canonical model* $\overline{X}_H$ *of* X_H *which is Calabi-Yau.*

Proof: We know the resolution of X; it is given by "inverting" the projection from the node, by blowing up along the 15 planes $\mathcal{S}_3 \cap \mathcal{N}_5$, yielding $\mathcal{I}_{5,p}$. Let $\overline{X}_H$ be the proper transform of X_H in $\mathcal{I}_{5,p}$. Assuming H to be sufficiently general, $\overline{X}_H$ clearly has canonical singularities (as $\mathcal{I}_{5,p}$ does), so we must only show that it is Calabi-Yau. We note, however, that $\overline{X}_H$ is (the proper transform on $\mathcal{I}_{5,p}$ of) a hyperplane section of $\mathcal{I}_5$! This is because the degree is invariant under projection, hence under Π. But this is a hyperplane section of $\mathcal{I}_5$ through the

Table 6.4: Degenerations of double octics and quintic hypersurfaces

Space of all quintic hypersurfaces
101-dimensional

∪

120-nodal quintics
5-dimensional

∪

quintic hypersurfaces with 111
nodes and one multiplicity 3
singular point
4-dimensional

‖

double cover $Y \longrightarrow \mathbb{P}^3$,
branched over $S \cup Q$
$S \cap Q = \{\ 15 \text{ lines}\}$

∩

double cover branched over cubic and
quintic, such that $S \cup Q$ is stable

∩

Space of all double octics
149-dimensional

triple point p. Hence the proper transform on $\overline{X}_H$ of the exceptional divisor $(S_3)_p$ is a hyperplane section of S_3, i.e., a (generically smooth) cubic surface. This singularity is known to be canonical, and $\overline{X}_{II}$ is, just as a nodal quintic, Calabi-Yau. □

Corollary 6.4.9 *The family of hyperplane sections of $\mathcal{I}_5$ passing through one of the 36 triple points p is, via projection, a family of Calabi-Yau threefolds which are degenerations of double octics.*

It is natural to ask the meaning of this in terms of variations of Hodge structures. Recalling that a Type III degeneration of a K3 surface, corresponding to a zero-dimensional boundary component of the period domain, is one like a quartic degenerating into four planes, it is natural to ask

Question 6.4.10 *Is a double cover of $\mathbb{P}^3$ branched over the union of a cubic and a quintic a semistable degeneration of a double octic?*

Remark 6.4.11 There is a notion of "connecting" moduli spaces of Calabi-Yau threefolds by degenerations, and the Corollary 6.4.9 shows that the moduli space of quintic hypersurfaces in $\mathbb{P}^4$ and the moduli space of double octics are connected; the birational transformations which are required for such "connections" are given here by projection in projective space, something very geometric.

In Table 6.4 we give a rough description of these relations.

6.4.4 The dual picture

In some cases we have seen that the Hessian variety of a modular variety is also modular. In the case of S_3, the Hessian is at least (a compactification of) a Siegel variety, and we have conjectured that a different birational model, $\widehat{\mathcal{N}_5}$, is the Satake compactification of a ball quotient (see section 3.4.4), making it Σ_6-equivariantly a birational Janus-like variety. In the case of B_4, from the self-Steinerian property we know that the Hessian is birational to B_4. It is natural to consider whether a similar fact is also true for $\mathcal{I}_4$. The Hessian $\mathcal{W}_{10}$ is singular along its intersection with $\mathcal{I}_4$, the union of 10 quadric surfaces. Hence the normalization will have a similar singularity structure as $\mathcal{I}_4$, so it is natural to consider the possibility of $\mathcal{W}_{10}$ being also a Siegel space quotient. Then, letting $\mathcal{W} \longrightarrow \mathbb{P}^4$ denote the double cover branched along $\mathcal{W}_{10}$, we may consider the fiber square:

$$
\begin{array}{ccc}
\mathcal{Z} & \longrightarrow & \mathcal{Y} \\
\downarrow & & \downarrow \pi \\
\tau : \mathcal{W} & \longrightarrow & \mathbb{P}^4
\end{array}
\tag{6.35}
$$

where $\pi : \mathcal{Y} \longrightarrow \mathbb{P}^4$ is defined in Definition 3.5.2. Then $\pi_{\mathcal{Z}} : \mathcal{Z} \longrightarrow \mathbb{P}^4$ is a Galois cover with Galois group $G_{\mathcal{Z}} \cong \mathbb{Z}/2\mathbb{Z} \times \mathbb{Z}/2\mathbb{Z}$. Let $H \cong \mathbb{Z}/2\mathbb{Z} \subset G_{\mathcal{Z}}$ be the diagonal subgroup; it is a normal subgroup, and we may form the quotient

$$
\eta : \mathcal{Z} \longrightarrow \mathcal{Z}', \quad \mathcal{Z}' = \mathcal{Z}/H.
$$

Lemma 6.4.12 $\pi_Z : Z \longrightarrow \mathbb{P}^4$ *factors over* η, *and* $\eta' : Z' \longrightarrow \mathbb{P}^4$ *is a double cover, hence Galois.*

Proof: This is a general fact about fibre squares of double covers like (6.35). □

By results of M. Yoshida and his coworkers discussed in section 3.5, one knows that $\mathcal{Y}$ is *precisely* the Satake compactification of an arithmetic quotient of the four-dimensional domain of type **IV** (or of type $\mathbf{I_{2,2}}$, they are the same), by a subgroup which was denoted $\Gamma(2)$ in Theorem 3.5.6. On the third page of the article [Mat] a prism of subgroups commensurable to this $\Gamma(2)$ is given (although the notations there differ; the group $\Gamma(2)$ ($\subset SO(4,2)$) is the group $\Gamma_M(i+1)$ ($\subset SU(2,2)$), corresponding to a domain of type $\mathbf{I_{2,2}}$; the corresponding covers of $\mathbb{P}^4$ fit into a diagram just like that of $\mathcal{W}$, Z', Z and $\mathcal{Y}$.

On the other hand, we can easily derive the following relation between our variety $\mathcal{I}_5$ and the 6:1 cover $Z \longrightarrow \mathcal{Y}$.

Theorem 6.4.13 $\mathcal{I}_5$ *is* Σ_6*-equivariantly birational to a* Σ_6*-equivariant deformation of the 6:1 cover of* $\mathcal{Y}$, Z.

Proof: We have already seen that $\mathcal{I}_5$ is Σ_6-equivariantly birational to $\mathcal{I}_{5,p}^{\%}$, which is a 2:1 cover of $\mathbb{P}^4$ branched over $\mathcal{S}_3 \cup \mathcal{N}_5$, while Z is a double cover of $\mathbb{P}^4$ branched over $\mathcal{S}_3 \cup \mathcal{P}_5$. We now just connect $\mathcal{N}_5$ and $\mathcal{P}_5$ in the space of Σ_6-symmetric quintic hypersurfaces of $\mathbb{P}^4$. □

It is natural to expect that a little more is true: that the two Calabi-Yau threefolds $\mathcal{N}_5$ and $\mathcal{P}_5$ are birational and that, moreover, $\mathcal{I}_5$ is actually $W(E_6)$-equivariantly birational to the variety $\widetilde{Z}$ mentioned in the introduction, which has a biregular $W(E_6)$ action. More evidence for this is the following:

i) $\widetilde{Z}$ has 36 divisors (each a Σ_6-equivariant modification of $\mathbb{P}^3$) isomorphic to the inverse image of $\mathcal{S}_3$ and 36 divisors isomorphic to the inverse image of $\mathcal{P}_5$.

ii) The Naruki cross ratio variety $\mathcal{C}$ has 36+40 divisors belonging to the compactification locus; the inverse images on $\widetilde{Z}$ of the first 36 are the 36+36 divisors mentioned in i). The 40 divisors are isomorphic to $\mathbb{P}^1 \times \mathbb{P}^1 \times \mathbb{P}^1$.

iii) The resolution $\widetilde{\mathcal{I}}_5$ has $36 + 36$ divisors similar to those on $\widetilde{Z}$ of i), copies of resolutions of $\mathcal{S}_3$ and $\mathcal{N}_5$.

iv) $\widetilde{\mathcal{I}}_5$ has 120 divisors isomorphic to $(\mathbb{P}^1)^3$, the resolving divisors of the 120 singular lines of $\mathcal{I}_5$.

So it seems natural that $\widetilde{\mathcal{I}}_5$ is birational to $\widetilde{Z}$, a resolution of a 6:1 cover of $\mathcal{Y}$, such that the divisors of type iii) and iv) map to those of ii). This would correspond to, in the description above, to a non-normal subgroup $\Gamma \subset \Gamma(2)$ of index 6. Whether this holds remains to be seen.

6.5 $\mathcal{I}_5$ and cubic surfaces

6.5.1 The Picard group

Let $A_1(\mathcal{I}_5)$ be the Chow group of Weil divisors modulo algebraic equivalence. Clearly a generic hyperplane section yields an element in $A_1(\mathcal{I}_5)$, which we denote by n. Recall the reducible hyperplane sections of Proposition 6.3.1 which split each into the union of five copies of $\mathbb{P}^3$. These subvarieties are divisors on $\mathcal{I}_5$, hence also yield classes in the Chow group. These 45 divisors are related by 27 relations, the sum of the five classes in the Chow group being equivalent to n. Since $\mathcal{I}_5$ is normal, we have an injection $\mathrm{Pic}(\mathcal{I}_5) \hookrightarrow A_1(\mathcal{I}_5)$. Let $\mathrm{Pic}^0(\mathcal{I}_5)$ denote the orthogonal complement of the class n in $A_1(\mathcal{I}_5)$ with respect to this injection. Then we have

Lemma 6.5.1 *We have an exact sequence of $\mathbb{Z}$-modules,*

$$0 \longrightarrow \mathbb{Z}^6 \longrightarrow \mathbb{Z}^{27} \longrightarrow \mathbb{Z}^{45} \longrightarrow \mathrm{Pic}^0(\mathcal{I}_5) \longrightarrow 0.$$

Proof: The 45 $\mathbb{P}^3$'s are classes in $A_1(\mathcal{I}_5)$ which generate $\mathrm{Pic}^0(\mathcal{I}_5)$ (as they contain all singularities), and the 27 relations are those just mentioned, given by the 27 hyperplane sections. So the sequence is clear as soon as we have shown that the rank of $\mathrm{Pic}^0(\mathcal{I}_5)$ is 24 (see the sequence (6.33)). This now follows from the Lefschetz hyperplane theorem, as the dimension of $\mathcal{I}_5$ is four, so there is an isomorphism between the H^2's of $\mathcal{I}_5$ and a hyperplane section. We may apply the Lefschetz theorem because the singularities of $\mathcal{I}_5$ and of a hyperplane section are local complete intersections (see [GM] for details). □

Note that this sequence displays $\mathrm{Pic}^0(\mathcal{I}_5)$ as an *irreducible* $W(E_6)$-module. Furthermore, we see that just as in (6.33), this sequence is dual to the corresponding sequence for cubic surfaces.

6.5.2 $\mathcal{I}_5$ and cubic surfaces: combinatorics

We collect the facts relating the combinatorics of the 27 lines with those of $\mathcal{I}_5$ in Table 6.5.

6.6 The dual variety

Let $\mathcal{I}_5^\vee$ be the projective dual variety to $\mathcal{I}_5$; since $\mathcal{I}_5$ is invariant under $W(E_6)$, so is $\mathcal{I}_5^\vee$. Although we do not have explicit equations for $\mathcal{I}_5^\vee$, we can say quite a bit about its geometry, just from the fact that it is dual to $\mathcal{I}_5$.

6.6.1 Degree

First we show that *degree of $\mathcal{I}_5^\vee = 10m+4k$*. Quite generally, one can say the following. Suppose we are given a variety $X \subset \mathbb{P}^n$ which has singular locus consisting of a set of *lines*, meeting each other in a set of *points*, and let us

Table 6.5: Combinatorics of $\mathcal{I}_5$ and the 27 lines

Locus on a cubic surface (see Table 4.1)	Locus on $\mathcal{I}_5$
27 lines a_i, b_i, c_{ij}	27 hyperplane sections $\{a_i = 0\} \cap \mathcal{I}_5$, etc.
2 lines are skew	the hyperplanes intersect in one of 216 n-solids, which intersect $\mathcal{I}_5$ in the union of five planes.
two lines are in a tritangent	the hyperplanes intersect in one of the 45 $\mathbb{P}^3$'s of 6.1.5.8
45 tritangents	the 45 $\mathbb{P}^3$'s of 6.1.5.8
Two tritangents meet in a line of the cubic surface	two of the 45 $\mathbb{P}^3$'s meet in a $\mathbb{P}^2$; this is one of the planes in the $\mathbb{P}^3$ defining the arrangement $\mathcal{A}(W(\mathbf{D_4}))$ as discussed in 6.1.5.8
Two tritangents meet in a line outside of the cubic surface	two of the 45 $\mathbb{P}^3$'s are *skew*, i.e., meet only in a line; this line is part of the singular locus of the arrangement $\mathcal{A}(W(\mathbf{D_4}))$ just mentioned
36 double sixes	36 triple points of $\mathcal{I}_5$ AND 36 copies of the Nieto quintic $\mathcal{N}_5$
Two double sixes are azygetic	two of the triple points are connected by a κ-line
Two double sixes are syzygetic	two of the triple points are connected by an ε-line
A line is *not* contained in a double six	the hyperplane dual to the line contains the triple point which corresponds to the double six

further assume that the situation is symmetric, i.e., each line contains the same number of points, each point being hit by the same number of lines; let us denote these numbers by $N = \#$ lines, $M = \#$ points, $\nu = \#$ points in each line and $\mu = \#$ lines meeting at each point. Consider the dual variety $X^\vee$. We claim:

- There are N $\mathbb{P}^{n-2}$'s $\subset X^\vee$.

- Each $\mathbb{P}^{n-2}$ is cut out by ν hyperplanes.

- There are M such special hyperplane sections of $X^\vee$.

- Each of the M hyperplanes meets $X^\vee$ in μ of the N $\mathbb{P}^{n-2}$'s.

- Hence, $\deg(X^\vee) = m\mu + rest$,

where the *rest* is given in terms of the local geometry around the given point. The proofs of these are immediate: each of the points corresponds to a hyperplane (=set of all hyperplanes through the point), each line defines dually a

$\mathbb{P}^{n-2}$, and since X is singular along the line, each hyperplane through the line is *tangent* to X there $\Rightarrow$ the dual $\mathbb{P}^{n-2} \subset X^\vee$. The other statements are then clear. To determine *rest*, consider the following. The set theoretic image of the given point in the dual variety is the *total* transform (*not* the proper transform) of the given point. This is set theoretically easy to compute, but there may be a multiplicity coming in.

We apply these considerations to $\mathcal{I}_5$ and $\mathcal{I}_5^\vee$: on $\mathcal{I}_5$ we have singular lines, $N{=}120$, $M{=}36$, $\nu{=}3$, $\mu{=}10$, and hence $deg(\mathcal{I}_5^\vee) = m10 + rest$. In our case *rest* is easy to figure out: recall that we resolved the singularities of $\mathcal{I}_5$ by blowing up the 36 points, then the 120 lines; the resolving divisors over the points were copies of the Segre cubic. The variety dual to the Segre cubic is the Igusa quartic, and the image of the ten nodes on the Segre cubic are ten quadric surfaces (3.45) which are *tangent hyperplane sections*, i.e., the hyperplanes which meet the Igusa quartic in one such quadric and are tangent to it there. These ten hyperplanes are of course just the 10 $\mathbb{P}^3$'s on the dual variety being cut out by the chosen hyperplane section (see Proposition 6.6.1 below). This hyperplane section of $\mathcal{I}_5^\vee$ may be *tangent* to $\mathcal{I}_5^\vee$ along the Igusa quartic, hence

$$deg(\mathcal{I}_5^\vee) = 10m + k \cdot 4.$$

6.6.2   Singular locus

Consider the 45 $\mathbb{P}^3$'s on $\mathcal{I}_5$; since there is a pencil of hyperplanes through each, the dual variety $\mathcal{I}_5^\vee$ will have 45 singular lines, which meet in 27 points (which are dual to the 27 hyperplanes cutting out the 45 $\mathbb{P}^3$'s). These 27 points are of course A_i, B_i, C_{ij}. Applying our reasoning from above to this we see that $deg(\mathcal{I}_5^{\vee\vee}){=}5{+}$ rest. We conclude rest=0, or in other words, *a resolution of singularities of $\mathcal{I}_5^\vee$ is affected by blowing up the 45 lines simultaneously; there is no exceptional divisor over the 27 points.*

However, since we are dealing with fourfolds, $\mathcal{I}_5^\vee$ could even be normal and still have a singular locus of dimension two. For example, it is reasonable to believe that the ten quadrics on each copy of the Igusa quartic $\mathcal{I}_4$ on the reducible hyperplane sections discussed below might be *singular* on $\mathcal{I}_5^\vee$, but that is of course just a guess. Furthermore, there is no reason whatsoever why the dual variety should be normal. In fact, it is a case of great exception when the dual variety is normal, the general case being that there is a singular parabolic divisor (coming from the intersection $\mathrm{Hess}(X) \cap X$), as well as a double point locus, also (in general) a divisor, coming from the set of bitangents. In our case, however, since $\mathrm{Hess}(\mathcal{I}_5) \cap \mathcal{I}_5$ consists of the union of the 45 $\mathbb{P}^3$'s, all of which get blown down, there is no parabolic *divisor*. But there is no easy way to exclude a double point divisor.

6.6.3   Reducible hyperplane sections

As already mentioned, $\mathcal{I}_5^\vee$ contains 120 $\mathbb{P}^3$'s, each being cut out by three of the h's, (in fact by a triple of azygetic double sixes), and each such intersection

$h \cap \mathcal{I}_5^\vee$ consists of ten such $\mathbb{P}^3$'s, plus a copy of the Igusa quartic. There are 36 such hyperplane sections which decompose into ten $\mathbb{P}^3$'s and a copy of the Igusa quartic:

Proposition 6.6.1 *The 36 hyperplane sections* $\mathbf{h} \cap \mathcal{I}_5$, $\mathbf{h} = h$, h_{ij}, h_{ijk}, *are reducible, consisting of ten* $\mathbb{P}^3$ *'s and a copy of the Igusa quartic* $\mathcal{I}_4$. *The ten* $\mathbb{P}^3$ *'s are just the* K_{ijk} *of (3.46), each a bitangent plane to* $\mathcal{I}_4$.

Proof: These are the 36 hyperplanes dual to the 36 triple points of $\mathcal{I}_5$; at each such p ten of the 120 lines meet, and the triple point itself yields the copy of $\mathcal{I}_4$ (it is blown up with exceptional divisor $(\mathcal{S}_3)_p$, which is dual to $(\mathcal{I}_4)_p$, a copy of $\mathcal{I}_4$). □

So restricted to the triple point, the duality $\mathcal{I}_5 - - \to \mathcal{I}_5^\vee$ yields precisely the duality map φ mapping $\mathcal{S}_3$ birationally onto $\mathcal{I}_4$!

The 120 $\mathbb{P}^3$'s meet two at a time in 270 $\mathbb{P}^2$'s, each of which is cut out by six of the h's (2 triples of azygetic double sixes, two rows in a triple). Note that these 270 $\mathbb{P}^2$'s are the $t_6(2)$ of Table 6.1. Each $\mathbb{P}^2$ contains two nodes and five of the 27 points, as well as two of the 45 lines. Through each such line two of these $\mathbb{P}^2$'s pass (as each line is cut out by 12 of the h's). Therefore in each h we have ten $\mathbb{P}^3$'s meeting in $\binom{10}{2} = 45$ $\mathbb{P}^2$'s which meet in 15 of the 45 $\mathbb{P}^1$'s, and contain 15 of the 27 points. The 15 lines and 15 points are just the singular locus of the Igusa quartic, and the ten $\mathbb{P}^3$'s are tangent to the Igusa quartic along quadrics, as mentioned earlier.

6.6.4 Special hyperplane sections

Inspection of the 27 forms and 27 points in $\mathbb{P}^5$ shows that each of the 27 hyperplanes contains *none* of the 27 points and *none* of the 45 lines; it follows that hyperplane sections such as $\mathcal{K} := \mathcal{I}_5^\vee \cap \{a_1 = 0\}$ are irreducible hypersurfaces in $\mathbb{P}^4$ with 45 isolated singularities, coming from the intersections with the singular lines of $\mathcal{I}_5^\vee$. As mentioned above, there may also be a singular locus coming from other singularities on $\mathcal{I}_5^\vee$. Furthermore, there are 40 $\mathbb{P}^2$'s lying on this threefold, and 16 hyperplanes in a_i which cut out ten of these on $\mathcal{K}$. The 16 hyperplanes are those 16 of the 216 $\mathbb{P}^3$'s which lie in a_i, corresponding to the 16 lines which a_i is skew to. The symmetry group of this threefold is $W(D_5)$. This is a *degeneration* of a *generic* hyperplane section, which will contain 120 $\mathbb{P}^2$'s.

Appendix A

Rational groups of hermitian type

In this appendix we sketch the classification of $\mathbb{Q}$-groups of hermitian type in the classical cases. In section A.2 we recall the basic mechanism of the classification, given by a result of Weil. We sketch briefly the (standard) theory of central simple algebras, to introduce notations and conventions, in section A.1. In section A.3 we then sketch the classification. In section A.4 we recall a few other results on $\mathbb{Q}$-roots and $\mathbb{Q}$-parabolics, and in section A.5 we briefly indicate which additional data are necessary in defining an arithmetic group in a given rational group. For the reader interested in more details we suggest [T1] and §23 of [B3] (note that §23 is only contained in the second revised addition). The first source gives a classification of all k-simple groups, and in the book [B3] the k–root systems of most of the groups occuring here are computed.

A.1 Algebras

This section is preliminary, presenting standard facts from the theory of central simple algebras. For ease of reference we will list many results here without proof. Standard references are [A], [Bo2], and for a modern, brief presentation [Sch], which we will mostly refer to.

A.1.1 Hermitian forms over skew fields

A general reference for this section is [Sch], Ch. 7, §6, 7, 9 & 10.

Definition A.1.1 Let R be a ring. An *involution* on R is an involutive antiautomorphism $* : R \longrightarrow R$, i.e., $*$ fulfills:

$$(x + y)^* = x^* + y^*, \quad (xy)^* = y^* x^*, \quad x^{**} = x, \ \forall_{x,y \in R}.$$

Let $\lambda \in C(R)$ (the center of R) satisfy $\lambda\lambda^* = 1$. For an R-module M, a sesquilinear form

$$h : M \times M \longrightarrow R$$

is called λ-*hermitian*, if $h(x,y) = \lambda h(y,x)^*$, or equivalently, $h(x,y)^* = \lambda^* h(y,x)$.

Now consider the special case that $R = D$ is a skew field and $M = V$ is a (finite-dimensional) right D-vector space. The usual notions of orthogonality, orthogonal base, etc., hold also in this situation:

Facts 1.2

(i) (V,h) is regular $\iff V^\perp = 0$.

(ii) If (V,h) is regular, $W \subset V$, then $\dim(V) = \dim(W) + \dim(W^\perp)$.

(iii) If h is not skew-symmetric, and $h \neq 0$, then there exists $x \in V$ with $h(x,x) \neq 0$.

(iv) If h is not skew-symmetric, then (V,h) has a basis consisting of pairwise orthogonal vectors.

(v) If h is isotropic, then $V = E \perp V_0$, with E a hyperbolic plane, i.e., $\dim_D(E) = 2$ and h restricted to E is isotropic. The form h restricted to V_0 will be anisotropic if and only if the Witt index of h is $= 1$.

Definition A.1.3 Let $C = C(R)$ denote the center of R. The involution $*$ is said to be

a) of the *first kind*, if $*_{|C} = 1$.

b) of the *second kind* otherwise. In this case, C/C_0 is a quadratic extension, $C_0 := \{z \in C \,|\, z^* = z\}$.

It is well known that, if $*$ is of the first kind, we may assume $\lambda = \pm 1$, while if $*$ is of the second kind, we may even assume $\lambda = 1$. Thus it suffices to consider hermitian and skew-hermitian forms (with possibly trivial involution).

Remark A.1.4 The ± 1-hermitian forms are related to one another as follows. If $\delta \in R$ with $\delta = \varepsilon\delta^*$, then $*_\delta : R \longrightarrow R$, $x^{*_\delta} := \delta x^* \delta^{-1}$ is also an involution. If h is λ-hermitian, then δh is $\varepsilon\lambda$-hermitian. Therefore the $\pm\lambda$-hermitian forms over $(R,*)$ correspond bijectively to $(\varepsilon\lambda)$-hermitian forms over $(R, *_\delta)$.

We need the following basic result (Witt's theorem).

Theorem A.1.5 *Assume (V,h) is regular, $W \subset V$ a subspace, and suppose we are given an isometry $\sigma : W \longrightarrow V$. Then there exists an extension $\Sigma : V \longrightarrow V$ of σ, such that $\Sigma_{|W} = \sigma$.*

A.1.2 Central simple algebras

Reference for this section is [Sch], Ch. 8, §1-6. Let K be any field. Recall that a simple (associative) K-algebra is always isomorphic to a matrix algebra $A \cong M_n(D)$, where D is a skew field over K, and that if $A \cong M_{n_1}(D_1)$ and $A \cong M_{n_2}(D_2)$, then $n_1 = n_2$ and $D_1 \cong D_2$. A is central simple if K is the center, and in that case the skew field D is also central over K. In the sequel we deal only with finite dimensional A. If A and B are central simple algebras, so are $A \otimes B$ and if A^0 denotes the opposite algebra, then $A \otimes A^0 \cong M_n(K)$, where $n = \dim_K A$. A and B are said to be *similar*, if they come from the same skew field D, i.e., $A = M_n(D)$, $B = M_m(D)$.

Definition A.1.6 The *Brauer group* of K, denoted $Br(K)$, is the set of similarity classes of finite dimensional central simple algebras over K, with the group structure induced by $\otimes$ and inverse given by the opposite algebra.

Hence this is also the set of isomorphism classes of skew fields, or associative division algebras, over K.

Definition A.1.7 Let A be a central simple algebra over K.

a) The *index* of A, denoted $i(A)$, is the dimension of the central simple division algebra D over K for which $A \cong M_n(D)$.

b) The *order* of A, denoted $o(A)$, is the order of the class of A in $Br(K)$.

A *splitting field* of A is an extension $L|K$ such that $A \otimes_K L$ is trivial in the Brauer group, i.e., $A \otimes_K L \cong M_n(L)$, and if A is a division algebra of dimension d^2, the maximal commutative subfields are of dimension d. More precisely:

Proposition A.1.8 *If D is a d^2-dimensional skew field over K, then the maximal commutative subfields of degree d are all* splitting fields *of D.*

Definition A.1.9 Let L be a splitting field of A over K. For $a \in A$ (viewed as an element of $\phi(A \otimes_K L)$, where $\phi : A \otimes_K L \xrightarrow{\sim} M_n(L)$ is a fixed isomorphism), let $\chi(X, a) = X^n + \alpha_{n-1} X^{n-1} + \cdots + \alpha_0$ be the characteristic polynomial of the matrix $\phi(a)$. The coefficient $(-1)^n \alpha_0 =: n_{A|K}(a)$ is the *reduced norm* of the element a (and is equal to $det(\phi(a))$), and $-\alpha_{n-1} =: tr_{A|K}(a)$ is the *reduced trace* of the element a (and is equal to $tr(\phi(a))$).

See also (A.23) for the definition of the not reduced characteristic polynomial. The terms "norm" and "trace" without adjectives mean the norm and trace in the regular representation, i.e., viewing each element as an endomorphism of A. These will seldom be used, and would be denoted by capital letters, $N_{A|K}$ and $Tr_{A|K}$.

A.1.3 Algebras with involution

Reference: [Sch], Ch. 8, §7-9. Let A be central simple over K, and let $* : A \longrightarrow A$ be an involution. If $*$ is of the second kind, then $k := \{x \in K | x^* = x\}$ is a subfield of K and $K|k$ is a quadratic extension. If $*$ is of the first kind, one sets $K := k$ and speaks in both cases of a $K|k$-involution. The following result states essentially that given a central simple algebra with involution $(A, *)$, classifying involutions I on A is the same thing as classifying λ-hermitian forms on $(A, *)$.

Theorem A.1.10 *Let A be central simple over K, $* : A \longrightarrow A$ a $K|k$-involution. Then*

   (i) If $\lambda \in K, \lambda\lambda^* = 1$, $a \in A$, $a = \lambda a^*$, then

$$I_a : A \longrightarrow A, \ x^{I_a} := ax^*a^{-1}$$

   is a $K|k$-involution on A.

   (ii) If $I : A \longrightarrow A$ is an arbitrary $K|k$-involution, then there is a unit $a = \pm a^*$ in A such that $I = I_a$.

   (iii) If $*$ is of the first kind, a is unique up to scalars in $K^\times$. If $*$ is of the second kind, then $a = a^*$ $(\lambda = 1)$, and is unique up to scalars in $k^\times$.

Let A^+ (respectively A^-) be the $+1$ (respectively -1) eigenspaces for $*$, so $A = A^+ \oplus A^-$. It turns out that the dimensions of A^+ and A^- are: $(n^2 = \dim_K A)$

  • for $*$ of the second kind, $\dim_k(A^+) = \dim_k(A^-) = n^2$.

  • for $*$ of the first kind $\dim_k(A^+) = \frac{1}{2}n(n+1)$ or $\frac{1}{2}n(n-1)$.

Definition A.1.11 Let $(A, *)$ be a central simple algebra with involution $*$ of the first kind.

   (i) If $\dim_k(A^+) = \frac{1}{2}n(n+1)$, $*$ is of *orthogonal type*;

   (ii) If $\dim_k(A^+) = \frac{1}{2}n(n-1)$, $*$ is of *symplectic type*.

Under the correspondence of A.1.10 the former case corresponds to skew hermitian, the latter to hermitian forms. The next step is the question as to the existence of involutions, given an algebra A. The first basic result is the following.

Theorem A.1.12 *A central simple algebra over K admits an involution of the first kind $\iff A \cong A^o$.*

Phrased differently A admits an involution of the first kind exactly when A represents an element of order 2 in $Br(K)$. Over number fields one has in fact:

Lemma A.1.13 *A central simple algebra A over a number field K admits an involution of the first kind $\iff$ it represents an element of order 2 in $Br(K)$ $\iff$ A is a matrix algebra over a quaternion skew field.*

As to A admitting involutions of the second kind, (which by the above Lemma is a set disjoint from the previous), let $d = i(A)$ be the index of A (Definition A.1.7), which is $\neq 2$ by the above. Below we explain that D can be described as a degree d cyclic algebra (L, σ, a), where $a \in K^{\times}$, $L|K$ is a cyclic Galois extension of degree d, and σ is a generator of $Gal(L|K)$. There is then a degree d cyclic extension $\ell|k$ such that $K \cdot \ell = L$ (see (A.7)).

Theorem A.1.14 ([A], Theorem 10.18) *Let $A = (L, \sigma, a)$, a cyclic algebra. Then A admits an involution of the second kind $\iff$ there is an $\alpha \in (K')^{\times}$, such that*

$$N_{K|k}(a) = N_{\ell|k}(\alpha).$$

A.1.4 Cyclic algebras

These are defined as follows. $L|K$ is a cyclic Galois extension of degree d, and $\sigma \in Gal(L|K)$ is a generator of the Galois group. Given $a \in K^{\times}$, one defines the algebra

$$(L|K, \sigma, a) = A = L \oplus Le \oplus Le^2 \oplus \cdots \oplus Le^{d-1}, \qquad (A.1)$$

for $e \in A$ fulfilling the relations

$$e^d = a, \quad e \cdot z = z^{\sigma} \cdot e, \forall_{z \in L}. \qquad (A.2)$$

This can be constructed as a matrix algebra by setting

$$
e = \begin{pmatrix} 0 & 1 & 0 & \cdots & 0 \\ 0 & 0 & 1 & \cdots & 0 \\ \vdots & \vdots & & \ddots & 0 \\ \vdots & \vdots & & & 1 \\ a & 0 & \cdots & 0 & 0 \end{pmatrix}, z = \begin{pmatrix} z^{\sigma^{d-1}} & & 0 \\ & \ddots & \\ 0 & & z \end{pmatrix}. \qquad (A.3)
$$

Fact A.1.15 Every central simple division algebra over a number field is a cyclic algebra.

Furthermore one has the following results, which imply that the set of cyclic algebras $(L|K, \sigma, a)$ for fixed L and σ form a subgroup $\mathcal{C}(L|K) \subset Br(K)$, isomorphic to $K^{\times}/N_{L|K}(L^{\times})$.

Theorem A.1.16 (i) $(L, \sigma, a) \cong (L, \sigma, b) \iff b^{-1}a \in N_{L|K}(L^{\times})$;

(ii) $(L, \sigma, 1)$ splits, hence (L, σ, a) splits $\iff a \in N_{L|K}(L^{\times})$;

(iii) $(L, \sigma, a) \otimes (L, \sigma, b) \sim (L, \sigma, ab)$ (here $\sim$ means similar (A.1.6)).

Moreover, $A_{|K}$ splits $\iff A_{\mathfrak{p}}$ splits for all primes of K, and if the degree of A is prime, A is a division algebra $\iff$ it is not split. In this case we use the notation D for the division algebra A.

A.1.5   Quaternion algebras

Taking the subgroup of elements of order 2 in $Br(K)$ one gets all classes of quaternion algebras. We recall these briefly. By Lemma A.1.13, the involution on A is necessarily of the first kind, hence we have $K = k$, and in the sequel we will write k. In the case that A is a division algebra, we use the notation D for the quaternion algebra. Since $d = 2$, there exists a quadratic field $k(\sqrt{a}) \subset A$ and an element $e \in A$ with $e^2 = b$, $b \in k$. In other words, $A = \{x_0 e_0 + x_1 e_1 + x_2 e_2 + x_3 e_3, \ x_i \in k\}$, with e_0 an element in the center and multiplication given by

$$e_1 \cdot e_2 = e_3 = -e_2 \cdot e_1, \ e_1^2 = a, \ e_2^2 = b. \tag{A.4}$$

The canonical involution is $x = x_0 e_0 + x_1 e_1 + x_2 e_2 + x_3 e_3 \mapsto \overline{x} = x_0 e_0 - x_1 e_1 - x_2 e_2 - x_3 e_3$, and this is an involution of the first kind of symplectic type (Definition A.1.11). The norm of an element $x \in A$ is given by $n_{D|k}(x) = x\overline{x} = \overline{x}x$, which is a regular quadratic form, and as such the space A is isometric to the four-dimensional space $< 1, -a, -b, ab >$ over k. The trace of an element $x \in A$ is defined by $tr_{D|k}(x) = x + \overline{x}$. Two cases can occur:

- split case, $A \cong M_2(k)$;

- non-split case, $A = D$ is a skew field.

Lemma A.1.17   *(i) A is split $\iff$ the form $< 1, -a, -b, ab >$ is equivalent to $< 1, -1, -1, 1 >$.*

(ii) $A = D$ is a skew field $\iff$ the form $< 1, -a, -b, ab >$ is anisotropic over k.

The definite case over $\mathbb{Q}$ occurs if $a < 0$, $b < 0$, and one speaks then of a *definite* quaternion algebra. For a number field k, recall that D is said to be *totally definite* (respectively *totally indefinite*), if at all real primes ν, the local algebra D_ν is the skew field of the Hamiltonian quaternions $\mathbb{H}$ over $\mathbb{R}$ (respectively, if for all real primes ν, the local algebra D_ν is split). Recall also that a finite prime $\mathfrak{p}$ *ramifies* in D if $D_{\mathfrak{p}}$ is a division algebra; the isomorphism class of D is determined by the (finite) set of primes $\mathfrak{p}$ at which it ramifies and its isomorphism class at those primes ([A], Thm. 9.34). As a special case of cyclic algebras, quaternion algebras can be displayed as algebras in $M_2(\ell)$, where ℓ/k is a real quadratic extension, $\ell = k(\sqrt{a})$. Then there is some $b \in k^*$ such that

$$e = \begin{pmatrix} 0 & 1 \\ b & 0 \end{pmatrix}, \qquad z = \begin{pmatrix} z & 0 \\ 0 & z^\sigma \end{pmatrix}, \ z \in \ell,$$

where $z^\sigma = (z_1 + \sqrt{a} z_2)^\sigma = z_1 - \sqrt{a} z_2$. In other words, we can write for any $\alpha \in D$,

$$\alpha = \begin{pmatrix} a_0 + a_1\sqrt{a} & a_2 + a_3\sqrt{a} \\ b(a_2 - a_3\sqrt{a}) & a_0 - a_1\sqrt{a} \end{pmatrix}. \tag{A.5}$$

Then the canonical involution is given by the involution

$$\begin{pmatrix} a & b \\ c & d \end{pmatrix} \mapsto \begin{pmatrix} d & -b \\ -c & a \end{pmatrix} \tag{A.6}$$

on $M_2(\ell)$, and the norm and trace are just the determinant and trace of the matrix (A.5). We will use the notation (a,b) to denote this algebra $(\ell/k,\sigma,b)$, $\ell = k(\sqrt{a})$, if no confusion can arise from this.

A.1.5.1   Involutions of the second kind

In the case of involutions of the second kind, note first that the $K|k$-conjugation extends to the splitting field L; its invariant subfield ℓ is then a totally real extension of k, also cyclic with Galois group generated by σ. We have the following diagram:

$$\tag{A.7}$$

and the conjugations on L and K give the action of the Galois group on the extensions L/ℓ and K/k; these are ordinary imaginary quadratic extensions. There are precise relations known under which D admits a $K|k$-involution of the second kind.

Theorem A.1.18 ([A], Thm. 10.18) *A cyclic algebra $D = (L/K,\sigma,\gamma)$ has an involution of the second kind $\Longleftrightarrow$ there is an element $\omega \in \ell$ such that*

$$\gamma\bar{\gamma} = N_{K|k}(\gamma) = N_{\ell|k}(\omega) = \omega \cdot \omega^\sigma \cdots \omega^{\sigma^{d-1}}.$$

If this condition holds, then an involution is given explicitly by setting:

$$(e^k)^J = \omega \cdots \omega^{\sigma^{k-1}}(e^k)^{-1}, \left(\sum e^i z_i\right)^J = \sum \bar{z}_i(e^i)^J, \tag{A.8}$$

where $x \mapsto \bar{x}$ denotes the L/ℓ-involution. In particular for $x \in L$ we have

$$x^J = \bar{x}, \text{ and } x = x^J \Longleftrightarrow x \in \ell.$$

Later it will be convenient to have a description for when $x + x^J = 0$. This results from the following.

Theorem A.1.19 ([A], Thm. 10.10) *Given an involution J of the second kind on an algebra A, central simple of degree d over K, there are elements $u_1,\ldots,u_d$, with $u_i = u_i^J$, such that A is generated over K by $u_1,\ldots,u_d$. Furthermore, there is an element $q \in A$, $q^J = -q$, $q^2 \in k$, such that, as a k-vector space,*

$$A = A^+ + qA^+,$$

where $A^+ = \{x \in A | x = x^J\}$.

If $x \in A$ is arbitrary, then $\frac{1}{2}(x + x^J) \in A^+$, while $\frac{1}{2}(x - x^J) \in A^-$. For example, we have $e^i + (e^i)^J =: E^i \in A^+$, and then we have an isomorphism

$$A^+ \cong \ell \oplus E\ell \oplus \cdots \oplus E^{d-1}\ell. \qquad (A.9)$$

If, as above, $K = k(\sqrt{-\eta})$, $L = \ell(\sqrt{-\eta})$, then we may take $\sqrt{-\eta} = q$ in the theorem above, and for elements in

$$qA^+ \cong \sqrt{-\eta}\ell \oplus E\sqrt{-\eta}\ell \oplus \cdots \oplus E^{d-1}\sqrt{-\eta}\ell, \qquad (A.10)$$

we have $y = -y^J$. In particular, the dimension of qA^+ is d^2 as a k-vector space, and the dimension of A is $2d^2$ (see the remarks following Theorem A.1.10).

A.2   Weil's correspondence

For many purposes, the correspondence given in [Wl] between semisimple groups and semisimple algebras with involution is of fundamental importance in understanding rational groups of hermitian type. In this section we briefly describe this, and it will be utilized in the next section to give a classification of hermitian groups.

A.2.1   Over $\mathbb{C}$

In this section we consider algebras and groups over the universal domain, in our case usually just $\mathbb{C}$. All algebras are associative with unit element. An *involution* $* : A \longrightarrow A$ is an involutive antiautomorphism of A, i.e., fulfilling $*(x \cdot y) = *(y) \cdot *(x)$ (see section A.1.3). Because of this, one writes x^* instead of $*(x)$. Semisimple group will mean semisimple algebraic group (over $\mathbb{C}$). The basic result is the following:

Theorem A.2.1 ([Wl], p. 596-597) *There is a 1-1 correspondence between the following sets of objects:*

(I) semisimple groups with center reduced to unity, which have no simple factor isomorphic to an exceptional group or type $\mathbf{D_4}$;

(II) semisimple central algebras with involution whose simple factors are one of the following:

(a) $M_n \oplus M_n$, $n \geq 3$, with an involution which exchanges the two summands;

(b) M_{2n}, $n \geq 1$, with the involution $X^* = J^{-1} \cdot {}^t X \cdot J$, J an invertible alternating matrix;

(c) $M_n, n = 7$ or $n \geq 9$, with the involution $X^* = {}^t X$.

The proof of this is by construction: each of the groups listed is isomorphic to the connected component of the group of automorphisms of one of the algebras with involution (i.e., automorphisms of the algebra which commute with the given involution). Moreover, if A, A' are two such algebras, G, G' the corresponding groups, then any isomorphism between G and G' is induced by a uniquely determined isomorphism between A and A'.

A.2.2 Over k

Now let k be a subfield of the universal domain, say for us $k \subset \mathbb{C}$. The following result of the same paper now shows that the classification of semisimple groups over k is (for the groups of the above theorem) again equivalent to the classification of algebras over the field k.

Theorem A.2.2 ([Wl], Thm. 1, p. 592) *Let A be an algebra with involution defined over k, and $\xi \in H^1(Gal(K|k), \mathrm{Aut}(A))$ a cocycle (for some overfield K of k). Then there is an algebra with involution A^ξ, defined over k, and an isomorphism $\Phi : A \longrightarrow A^\xi$ defined over K, such that $\xi_\sigma = \Phi^\sigma \circ \Phi^{-1}$, $\forall_{\sigma \in Gal(K|k)}$.*

This result says essentially that the k-forms of a given algebra with involution correspond in a 1-1 manner to the k-forms of a semisimple group associated with the algebra by the Theorem A.2.1. Taking now $K = \mathbb{C}$, we have for any semisimple group G of the type listed in Theorem A.2.1 an algebra A over $\mathbb{C}$ and an isomorphism $G \cong (\mathrm{Aut}(A))^0$. We also get a sequence

$$1 \longrightarrow \mathrm{Aut}(A)^0 \longrightarrow \mathrm{Aut}(A) \longrightarrow \mathcal{B} \longrightarrow 1, \qquad (A.11)$$

and it is not difficult to see that in fact $\mathcal{B}$ is isomorphic to the group of outer automorphisms. It then follows that

$$H^1(Gal_k, \mathrm{Aut}(G)) \cong H^1(Gal_k, \mathrm{Aut}(A)). \qquad (A.12)$$

Hence the study of k-forms of a given G is equivalent to the study of k-forms of A. Recall also that, A being central simple, any automorphism is *inner*. In this way the classification problem is somewhat simplified.

Remark A.2.3 The results above can, in a sensible manner, be extended to the excluded cases, i.e., exceptional or type D_4, by means of consideration of certain non-associative algebras.

A.2.3 Lie algebras

A simple associative algebra A is said to be *non-degenerate* if it is not commutative and, if A is a quaternion algebra, then the dimension of anti-invariant elements under the canonical involution is *not* of dimension 1 (see Definition A.1.11 below). Then one has more precisely than the above the following:

Theorem A.2.4 ([Wl], Thm. 2, p. 598) *Let A be a non-degenerate semisimple algebra with involution. Set*

$$G_0 := (\mathrm{Aut}(A))^0, \quad U_0 := \left(\{ z \in A \big| z^* z = 1 \} \right)^0,$$

the connected components in the group of automorphisms of $(A, *)$ and of the multiplicative group of units. Then G_0 is semisimple, adjoint, and if $C \subset U_0$ denotes the center of U_0, we have

$$1 \longrightarrow C \longrightarrow U_0 \longrightarrow G_0 \longrightarrow 1.$$

In this way one has an explicit way to get the semisimple group from the corresponding algebra. Furthermore, to get the Lie algebra of G_0, it suffices to take the traceless elements:

Lemma A.2.5 *The Lie algebra $\mathfrak{g}_0$ of G_0 can be described as follows:*

$$\begin{aligned}
\mathfrak{g}_0 &= \{ \text{inner derivations of } A, \, D_u \text{ for } u^* = -u \} \\
&= \{ u \in A \big| tr(u) = 0, \, u^* = -u \},
\end{aligned}$$

where $tr(u)$ denotes the reduced trace of an element (see Definition A.1.9).

A.2.4 Over $\mathbb{R}$

Applying these ideas to $\mathbb{R}$-groups, Weil gets a description of the symmetric spaces in terms of the algebras. An involution α of $A_{\mathbb{R}}$ is said to be *positive*, if $tr(x^\alpha \cdot x) > 0, \forall_{x \neq 0}$ (see section 1.1.1).

Theorem A.2.6 ([Wl], Thm. 5, p. 619) *Let $(A, *)$ be a semisimple algebra with involution defined over $\mathbb{R}$, and let $G = \mathrm{Aut}(A)$, $G^0 = G(\mathbb{R})^0$, with G^0 denoting the topological connected component of unity. Set $\mathcal{R} := \{ \text{all positive involutions commuting with } * \text{ on } A \}$. For $\alpha \in \mathcal{R}$ set $K(\alpha) := \{ g \in G(\mathbb{R}) \big| g \cdot x^\alpha = (gx)^\alpha, \forall_{x \in G(\mathbb{R})} \}$. Then $\forall_{\alpha \in \mathcal{R}}$ the set $K(\alpha)$ is a maximal compact subgroup of $G(\mathbb{R})$, and $K(\alpha)^0 = G^0 \cap K(\alpha)$. Furthermore there are isomorphisms*

$$\mathcal{R} \cong G(\mathbb{R})/K(\alpha) \cong G^0/K(\alpha)^0$$

which is the symmetric space associated to $G(\mathbb{R})$; $G(\mathbb{R})$ acts on $\mathcal{R}$ by the prescription $(g, \alpha) \mapsto g^{-1} \alpha g$.

It then follows from [Wl], Thm. 4, that the Cartan involution of $\mathfrak{g}$ is induced by the involution $*$ of A, and the Lie algebra of $K(\alpha)$ is

$$\mathfrak{k}(\alpha) = \{ u \in A \big| tr(u) = 0, \, u\alpha = \alpha u \}. \tag{A.13}$$

In other words we have $*_{|\mathfrak{k}(\alpha)} = 1$, $*_{|\mathfrak{g}/\mathfrak{k}(\alpha)} = -1$.

A.3 Classification of simple Q-groups of hermitian type

Let G be a connected semisimple group (of the kind considered in A.2.1, i.e., without factors of type $\mathbf{D_4}$ or exceptional type), defined over $\mathbb{Q}$. By Weil's correspondence A.2.1, A.2.2 and A.2.4 there exists a semisimple algebra with involution A, such that the adjoint group G^{ad} fulfills $G^{ad} \cong (\text{Aut}(A))^0$, and there is an exact sequence

$$1 \longrightarrow C \longrightarrow U_0 \longrightarrow G^{ad} \longrightarrow 1$$

with the connected component of the group of units U_0 of A, and $C \subset U_0$ is the center. Hence classifying these $\mathbb{Q}$-groups up to isogeny amounts to classifying equivalence classes of pairs $(A, *)$ of semisimple $\mathbb{Q}$-algebras with involution $*$. By Theorem A.1.10, classifying pairs $(A, *)$ with A simple amounts to fixing the involution $*$ and classifying ± 1-hermitian (with respect to $*$) forms on A, i.e., equivalence classes of pairs (A, h), h a ± 1-hermitian form with respect to $*$. This in turn is related to hermitian forms on right vector spaces V over D, where D is the skew field for which $A \cong M_n(D) = \text{Hom}_D(V, V)$. The classification of pairs (A, h) can be divided into the following cases:

I. Orthogonal case

  I a) split case: quadratic forms over fields

  I b) non-split case: skew-hermitian forms over quaternion division algebras

II. Symplectic case

  II a) split case: alternating forms over fields

  II b) non-split case: hermitian forms over quaternion division algebras

III. Unitary case

  III a) split case: hermitian forms over fields

  III b) non-split case: hermitian forms over division algebras of degree $d \geq 2$ with involutions of the second kind

It is important that the notion of isotropy carries over from the $\mathbb{Q}$-groups (the notion of $\mathbb{Q}$-split torus) to the algebras (the notion of isotropy of a ± 1-hermitian form on V, $A = \text{Hom}_D(V, V)$).

Lemma A.3.1 *Let G be $\mathbb{Q}$-simple, corresponding to (V, h) as above. Then G is isotropic $\iff h$ is isotropic.*

Proof: (see [B3], p. 256) Suppose first that h is isotropic. Then V is the direct sum of a hyperbolic plane E defined over $\mathbb{Q}$ and of the orthogonal complement (see A.1.2 (v)). Let e_1, e_2 be a $\mathbb{Q}$-basis of E, with

$$h(e_1, e_1) = h(e_2, e_2) = 0, \quad h(e_1, e_2) = 1.$$

Then the transformations $s(x)$ $(x \in \mathbb{Q}^{\times})$ defined by

$$s(x)(e_1) = xe_1, \ s(x)(e_2) = x^{-1}e_2, \ s(x)(f) = f, \ f \in (E)^{\perp},$$

belong to G and form a $\mathbb{Q}$-split torus, hence G is isotropic over $\mathbb{Q}$.

Assume now G isotropic and let $S \subset G$ be a $\mathbb{Q}$-split torus. There exists a non-trivial character λ of S and $v \in V(\mathbb{Q})$ such that $s \cdot v = s^{\lambda}v$, $\forall_{s \in S}$. Then $h(v, v) = 0$, since

(a) $h(s \cdot v, s \cdot v) = h(v, v)$,

(b) $h(s \cdot v, s \cdot v) = s^{\lambda}h(v, v)\overline{s^{\lambda}}$, $\forall_{s \in S}$,

so this v is an isotropic vector of h.   □

In other words, isotropic vectors of h are weight spaces for the $\mathbb{Q}$-split torus S. Hence, to determine the isotropic (anisotropic) groups it is sufficient to determine the isotropic (anisotropic) ± 1-hermitian forms. One has the following criterion for anisotropy:

Theorem A.3.2 ([BHC], 12.1) *Let G' be absolutely simple over a number field k, and let G denote the $\mathbb{Q}$-group $Res_{k|\mathbb{Q}}G'$. Let $G(\mathbb{R}) \cong G_1 \times \ldots \times G_d$ be the decomposition of the real group into simple factors. If for some i the Lie group G_i is compact, then G, and hence G' also, is anisotropic.*

Definition A.3.3 Let G' be an absolutely almost simple k-group, $G = Res_{k|\mathbb{Q}}G'$ the corresponding $\mathbb{Q}$-group. We call G' or G a group *of hermitian type*, if $G(\mathbb{R})$ is a real group of hermitian type.

We now go through the cases listed above and determine for each which groups are of hermitian type.

A.3.1   Split orthogonal case

Let k be a totally real number field and V an n-dimensional k-vector space with a symmetric bilinear form h of Witt index r, i.e., for which there exists an r-dimensional totally isotropic (over k) subspace. Let G' be the orthogonal group of h, and G the $\mathbb{Q}$-group which is the restriction of scalars of G'. G' is isogenous to $SO(V, h)$, and if $\nu_1, \ldots, \nu_d$ denote the infinite places of k and h_{ν_i} the corresponding real quadratic form at the ith place, let (p_i, q_i), $p_i + q_i = n$, denote the signature of h_{ν_i}. Then we have for the group of real points of G:

$$G(\mathbb{R}) \sim SO(p_1, q_1) \times \cdots \times SO(p_d, q_d), \tag{A.14}$$

with $\sim$ denoting "isogenous".

Lemma A.3.4 *If there exists some i with $q_i = 0$, then G is anisotropic.*

Proof: This is a special case of Theorem A.3.2.   □

Lemma A.3.5 *G is of hermitian type* $\iff \forall_i q_i = 2$ *or* 0.

Proof: By Definition A.3.3 we must show that if $q_i > 0$ then the corresponding non-compact factor of the symmetric space determined by G is hermitian symmetric. But $SO(p,q)$ is of hermitian type if and only if $q = 2$. If $q_i = 0$ for some i, then $G(\mathbb{R})$ has a compact factor and G is anisotropic by A.3.2. $\qquad\square$

It follows from this that G may have $\mathbb{Q}$-rank $r=0$, 1 or 2, and all cases can occur. This r is just the Witt index of the symmetric bilinear form h.

A.3.2 Non-split orthogonal case

Here we are dealing with skew-hermitian forms on vector spaces over quaternion division algebras D. We take hermitian to mean hermitian with respect to the canonical involution on D (see (A.4)) $d \mapsto \bar{d}$, which induces an involution $*$ on $M_n(D)$ by

$$M^* = (\overline{m}_{ji}) \text{ for } M = (m_{ij}). \tag{A.15}$$

First, one has the following description of local skew-hermitian forms.

Theorem A.3.6 ([Ts], see also [Sch], 10.3.6) *Let K be a $\mathfrak{p}$-adic field and D the unique non-split quaternion algebra over K. For skew hermitian forms over $(D, -)$ the following statements hold:*

(i) *Two regular forms are isometric iff they have the same dimension and determinant.*

(ii) *Every form of dimension > 3 is isotropic.*

(iii) *In dimension 1 all regular forms are anisotropic; there are forms of any determinant $\neq 1$.*

(iv) *For any dimension > 1 there are forms of any determinant. In dimension 2 exactly the forms of determinant 1 are isotropic. In dimension 3 exactly the forms of determinant 1 are anisotropic.*

Unfortunately, the local-global principle does not hold for skew hermitian forms; there are finitely many global classes in each class of forms locally equivalent everywhere. Globally one has the following result, due to Kneser.

Theorem A.3.7 ([Sch], 10.4.1) *Let K be a global field of characteristic $\neq 2$ and let D be a quaternion division algebra central over K, with canonical involution $-$. Let (V, h) be a regular skew hermitian space over D.*

(i) *If $\dim(V) \geq 3$, and h is locally isotropic, then h is isotropic.*

(ii) *If $\dim(V) \geq 2$ and if λ is a skew-symmetric element $\neq 0$ of D which is represented locally by h, then λ is represented by h (i.e., $\exists_{v \in V} h(v,v) = \lambda$).*

Furthermore, the non-validity of the local-global principle results in the following:

Proposition A.3.8 ([Sch], p. 371) *For every dimension* $2n > 0$ *there exist exactly* 2^{s-2} *isometry classes of locally hyperbolic forms, where* s *is the even number of primes at which* D *does not split. Every class of locally isometric forms consists of* 2^{s-2} *isometry classes.*

Applying this to our case at hand we get

Lemma A.3.9 *Let* G *be of non-split orthogonal type. Then*

$$G(\mathbb{R}) \sim SO^*(2n) \times \cdots \times SO^*(2n) \times SO(p_1, q_1) \times \cdots \times SO(p_m, q_m).$$

Proof: Consider first the case in which G is absolutely almost simple over $\mathbb{Q}$. If D is definite, then $D_\infty \cong \mathbb{H}$, otherwise $D_\infty \cong M_2(\mathbb{R})$. G is isogenous to $SU(D^n, h)$, where h has index $r \leq [\frac{n}{2}]$. In the first case, as is well-known, $SU(\mathbb{H}^n, h)$, h skew hermitian, is isomorphic to $SO^*(2n)^1$. As for $SU(M_2(\mathbb{R})^n, h)$, this group gives rise to the orthogonal group of a quadratic form over the center (see [Sch], p. 361). Indeed, let $e_1 = \begin{pmatrix} 1 & 0 \\ 0 & 0 \end{pmatrix}$, so that $W := Ve_1$ is a $2n$-dimensional k-vector space. On W one defines the symmetric bilinear form b_h by the formula

$$h(xe_1, ye_1) = \begin{pmatrix} 0 & 0 \\ b_h(xe_1, ye_1) & 0 \end{pmatrix}. \tag{A.16}$$

(Recall that $h : V \times V \longrightarrow D$). Then h and b_h completely determine each other; they are *Morita equivalent*. At any rate, the statement of the Lemma follows now for G absolutely simple. In the general case we apply this to each real prime, yielding

Corollary A.3.10 *If* D *is totally definite, then the symmetric space associated with* G *is hermitian symmetric, all factors of which are of type* II. *If* D *is totally indefinite, then* $G(\mathbb{R})$ *is a product of orthogonal groups, and is hermitian iff the conditions of Lemma A.3.5 are satisfied.*

The general case now follows, completing the proof of the Lemma A.3.9.   □

Corollary A.3.11 *If* $n \geq 4$, *then* G *is isotropic.*

Proof: This follows from Theorems A.3.6 and A.3.7, as the form h is isotropic, hence so is also G, by Lemma A.3.1.   □

Finally we have the following description for the hermitian cases.

Corollary A.3.12 *Let* G *be of non-split orthogonal type, and let* $\nu_1, \dots, \nu_e$ *be the real primes for which* D *is definite,* $\nu_{e+1}, \dots \nu_m$ *the real primes for which* D *is indefinite. Then*

$$G(\mathbb{R}) \sim \prod_{i=1}^{e} SO^*(2n) \times \prod_{i=e+1}^{m} SO(p_i, q_i).$$

---

[1]the notation $SO^*(2n)$ for this group is used by Helgason; Satake in [S1] uses instead the notation $SU^-(\mathbb{H}^n)$, the "$-$" to indicate the form is *skew* hermitian

G is of hermitian type iff for all $i = e + 1, \ldots, m$, we have $q_i = 0$ or 2. In this case, the domain is a product of e factors of type II_n and $m - e$ factors of type IV_{2n-2}. For $n \geq 4$, G is anisotropic iff $m > e$ and for some $i \in \{e + 1, \ldots, m\}$ we have $q_i = 0$.

A.3.3 Split symplectic case

Let G' be absolutely simple over k of split symplectic type; then $G'(k) \sim Sp(2n, k)$, and letting G denote the restriction of scalars of G', we have $G(k) \sim (Sp(2n, k))^d$, $d = [k : \mathbb{Q}]$ (where k is totally real). Hence the corresponding symmetric space is a product of domains of type III_n. In particular, in this case we see that G', hence G, is always of hermitian type.

A.3.4 Non-split symplectic case

Let D be a division quaternion algebra, $A = M_n(D)$ the simple algebra of matrices, given the involution deriving from the canonical involution on D by (A.15). If G' is our group of non-split symplectic type, then $(G'^{ad})^0 \cong (\text{Aut}(A))^0$ as in Theorem A.2.4. There is a hermitian form on D^n, call it h, which is preserved by G'. As far as equivalence of hermitian forms is concerned, we have the following result of Ramanathan (see [Sh1], 4.4). Let $\nu_1, \ldots, \nu_u$, $u \leq d$ the real primes of k for which the localization D_{ν_i} is definite, i.e., isomorphic to the Hamiltonian quaternions $\mathbb{H}$. Let V be an n-dimensional right D-vector space; for each $i = 1, \ldots, u$ we have the localizations of V, $V_\nu \cong (\mathbb{H})^n$, $\nu \in \{\nu_1, \ldots, \nu_u\}$, and h_ν can be diagonalized, i.e., there is a base $\{v_1, \ldots, v_n\}$ of V_ν with $h_\nu(v_i, v_j) = \varepsilon_i \delta_{ij}$, $\varepsilon_i = 1$, $i = 1, \ldots, \mu_\nu$, $\varepsilon_i = -1$, $i = \mu_\nu + 1, \ldots, n$. The integer $\mu(h_\nu) = \mu_\nu$ is the signature at the real prime ν of h, and one sets $\mu_\nu(h) := \mu(h_\nu)$. Then one has

Theorem A.3.13 *Let f and g be non-degenerate hermitian forms on the D-space V. There exists $\sigma \in GL(V)$ with $f \circ \sigma = g \iff \mu_\nu(f) = \mu_\nu(g)$ for all real primes ν for which D is definite.*

Corollary A.3.14 *Let D be a $\mathbb{Q}$-quaternion division algebra. If D_∞ is indefinite, then any two hermitian forms on V are equivalent. If D_∞ is definite, then any two hermitian forms are equivalent $\iff$ they have the same signature.*

In the split (respectively non-split) case an absolutely simple $\mathbb{Q}$-group has the Tits index $C_{n,n}^{(1)}$ (respectively $C_{n,r}^{(2)}$).

Lemma A.3.15 *Suppose D (defined over $\mathbb{Q}$) is definite, and that the hermitian form h_∞ on V_∞ has signature (p, q) (i.e., p times +1 and q times -1 eigenvalue). Then*

$$G(\mathbb{R}) \sim Sp(p, q).$$

Proof: This is clear, as $Sp(p, q)$ is the group of symmetries of a hermitian form of index (p, q) on $\mathbb{H}^{p+q}$. □

On the other hand,

Lemma A.3.16 *Suppose D is indefinite. Then $G'(\mathbb{R}) \sim Sp(2n, \mathbb{R})$.*

Proof: This is well-known, since $D(\mathbb{R}) \cong M_2(\mathbb{R})$ and $U((M_2(\mathbb{R}))^n, \kappa) \cong Sp(2n, \mathbb{R})$, where κ is the bilinear form on $(M_2(\mathbb{R}))^n$ induced by the hermitian form on V.
$\square$

Corollary A.3.17 *Let D be a division quaternion algebra over k, and let $\nu_1, \ldots, \nu_u$ denote the real primes for which D is ramified (i.e., for which D_ν is definite), and let $\nu_{u+1}, \ldots, \nu_d$ be the remaining real primes of k. Then*

$$G(\mathbb{R}) \sim Sp(p_1, q_1) \times \cdots \times Sp(p_u, q_u) \times \underbrace{Sp(4n, \mathbb{R}) \times \cdots \times Sp(4n, \mathbb{R})}_{d-u \; factors}.$$

Proof: This follows from the results A.3.15 and A.3.16 and the general theory of the restriction of scalars.
$\square$

Corollary A.3.18 *Let G be simple of non-split symplectic type. Then G is of hermitian type $\iff$ for each ν_i where D is ramified, we have $p_i = 0$ or $q_i = 0$. If such a real prime exists, then G is anisotropic by Theorem A.3.2.*

A.3.5   Split unitary case

First we have the following result for $\mathbb{Q}$-groups of type $\mathbf{A_n}$.

Lemma A.3.19 *Let G be a $\mathbb{Q}$-group, all factors of $G(\mathbb{R})$ of which are of type $\mathbf{A_n}$. Then G is of hermitian type $\iff$ each factor of $G(\mathbb{R})$ is of outer type or compact.*

Proof: The $\mathbb{R}$-forms giving rise to symmetric spaces of type $\mathbf{AI}$ or $\mathbf{AII}$ in the notations of [H] are $SL(n, \mathbb{R})$ and $SU^*(2n)$, respectively. They are both of inner type. Hence each factor of $G_{\mathbb{R}}$ is of outer type or compact $\iff$ each factor of $G_{\mathbb{R}}$ is of type $SU(p, q)$ or compact $\iff$ each non-compact factor has associated symmetric space of type $\mathbf{AIII}$ in notations of [H] $\iff$ $G_{\mathbb{R}}$ is of hermitian type $\iff$ G is of hermitian type (see Defintion A.3.3).
$\square$

Remark A.3.20 The excluded $\mathbb{Q}$ groups are those denoted $^1A_{n,r}^{(d)}$ in [T1], and are realized as $SL(r+1, D)$, with $d(r+1) = n+1$, the split special linear case. By A.3.19 we need not concern ourselves with them, the case of $SL(1, D)$ being an exception.

If now G is of split type, then D is an imaginary quadratic field $K|k$, k totally real, and A is a matrix algebra over D and G is the set of automorphisms preserving an involution on A. Equivalently, G is the set of linear transformations preserving a hermitian form h on a K-vector space V. If G has $\mathbb{Q}$-rank r, then h has Witt index r (see Lemma A.3.1). Note that for *every* non-compact factor G_ν of $G(\mathbb{R})$, which by A.3.19 is isogenous to $SU(p_\nu, q_\nu)$, we must have $r \leq q_\nu$. This is because a $\mathbb{Q}$-split torus is in particular $\mathbb{R}$-split.

Lemma A.3.21 *Let G be of split unitary type, of $\mathbb{Q}$-rank r, and for each real prime ν_i of k let (p_i, q_i) be the signature of the local group G_{ν_i}. Then*

$$G(\mathbb{R}) \sim SU(p_1, q_1) \times \ldots \times SU(p_d, q_d),$$

and for all i it holds that $r \leq q_i$.

Proof: This follows from the above remarks. The notation does not exclude $q_i = 0$ for some values of i, in which case G is anisotropic by A.3.2 and in this case $r = 0$. ☐

Corollary A.3.22 *Let G be a simple $\mathbb{Q}$-group of split unitary type, and of $\mathbb{Q}$-rank r. Then G has the Tits index $^2A_{n,r}^{(1)}$, i.e., there is an isogeny*

$$G' \sim SU(V, h), \quad G \sim Res_{k|\mathbb{Q}}SU(V, h),$$

with (V, h) as above, and r is the Witt index of h.

A.3.6 Non-split unitary case

Here we are dealing with division algebras D with involutions of the second kind, of degree $d \geq 2$. Indeed, $d = 1$ together with an involution of the second kind is nothing but the split unitary case just discussed. We have an n-dimensional right D-vector space V with a hermitian fórm h. By A.1.15 we may take D to be a cyclic algebra of degree d. Let r denote the Witt index of h, i.e. the maximum dimension of a totally isotropic subspace.

Lemma A.3.23 *If D is a $\mathbb{Q}$-algebra, then $G \sim SU(V, h)$ is absolutely $\mathbb{Q}$-simple and $G(\mathbb{R}) \sim SU(p, q)$ with $rd^2 \leq q$.*

Proof: The first statement is obvious, the second follows from Lemma A.3.19, and $rd^2 \leq q$ follows as above from the fact that an r-dimensional isotropic D-subspace is an rd^2-dimensional isotropic $\mathbb{R}$-subspace (note that $p + q = nd^2$, where $n = \dim_D(V)$), or from the fact that a $\mathbb{Q}$-split torus is all the more $\mathbb{R}$-split. ☐

Clearly, if (V_∞, h_∞) denotes the unique $\mathbb{R}$-localization in the situation above, then $q =$ index of h_∞.

Corollary A.3.24 *Let G be almost simple of non-split unitary type and of $\mathbb{Q}$-rank r. Then*
$$G(\mathbb{R}) \sim SU(p_1, q_1) \times \cdots \times SU(p_d, q_d),$$
where (p_i, q_i) is the index of h_{ν_i}, the localization of h at the real prime ν_i, and for all i it holds that $r \leq q_i$.

Proof: The point here is that if $W \subset V$ is isotropic, then for each embedding $k \hookrightarrow \mathbb{R}$, $W_\nu \subset V_\nu$ is still an $\mathbb{R}$-isotropic subspace, since the zero form localizes to the zero form. Therefore this follows from Lemma A.3.23. ☐

As for isotopy in this case we have

Lemma A.3.25 *If G is absolutely simple and $n \geq 2$, then G is isotropic.*

Proof: Assuming G to be of hermitian type implies that there are non-compact factors of $G(\mathbb{R})$; assuming G to be absolutely simple implies there is only one factor of $G(\mathbb{R})$, so it must be non-compact, i.e., G is isotropic at ∞, that is G_∞ is isotropic. Hermitian forms with $n \geq 2$ are locally isotropic at all finite primes ($D_{\mathfrak{p}}$ is hyperbolic if $\mathfrak{p}$ is decomposed in $K|k$, and split if $\mathfrak{p}$ is ramified in $K|k$, see [Sch], 10.6.3), and the strong local global principle holds for this case ([Sch], 10.6.2), so in fact h is isotropic over $\mathbb{Q}$, as claimed.   □

Lemma A.3.26 *With D as above, the group $SL(1, D)$ is an anisotropic group of outer $\mathbf{A_n}$ type.*

Proof: This is well-known, see for example [B3], p. 254 for a proof of anisotropy, and the statement "of outer type" results from the existence of the involution of the second kind.   □

Corollary A.3.27 *In the case of absolutely simple G of non-split unitary type, we have that $G \sim SU(V, h)$ is anisotropic $\iff n = dim_D(V) = 1$.*

Corollary A.3.28 *Let G be of non-split unitary type, of $\mathbb{Q}$-rank r. Then*

$$G(\mathbb{R}) \sim SU(p_1, q_1) \times \cdots \times SU(p_d, q_d),$$

and $r \leq q_i, \forall_i$. G is anisotropic if either (i) $n = 1$, or (ii) $\exists_i$, $q_i = 0$.

Proof: The form of $G(\mathbb{R})$ is clear by the above. The first possibility is just A.3.25 and A.3.26, the second (ii) follows from Theorem A.3.2.   □

A.4   Rational parabolics and boundary components

In this section we present a few well-known facts on the $\mathbb{Q}$-roots and rational parabolics. G denotes a k-group, which we assume to be semisimple and connected.

Proposition A.4.1 **([BB], 2.6)** *Let k be a subfield of $\mathbb{R}$. Then G has a maximal torus defined over k which contains a maximal $\mathbb{R}$-split torus.*

Let $T_k \subset T_\mathbb{R}$ be a maximal split torus contained in a maximal $\mathbb{R}$-split torus $T_\mathbb{R}$, and let $\Phi_k = \Phi(T_k, G)$, $\Phi_\mathbb{R} = \Phi(T_\mathbb{R}, G)$ be the root systems. Let $r : X(T_\mathbb{R}) \longrightarrow X(T_k)$ be the restriction map on the character groups; give $X(T_\mathbb{R})$ and $X(T_k)$ compatible orderings and let $\Delta_\mathbb{R}$ and Δ_k denote the corresponding sets of simple roots. Then one has

Proposition A.4.2 **([BB], 2.9)** *Assume $dim_k T_k > 0$. Then*

  a) Φ_k is of type $\mathbf{BC_s}$ if: either

(i) $\Phi_{\mathbb{R}}$ *is of type* $\mathbf{BC_r}$, *or*

(ii) $\Phi_{\mathbb{R}}$ *is of type* $\mathbf{C_r}$ *and* $r(\eta_r) = 0$;

Φ_k *is of type* $\mathbf{C_s}$ *otherwise.*

b) *Each* $\beta \in \Delta_k$ *is the restriction of one and only one of the* $\eta_i \in \Delta_{\mathbb{R}}$.

Remark A.4.3 The highest roots of the two types of restricted root systems are:

- $\delta = 2(\beta_1 + \cdots + \beta_s)$, type $\mathbf{BC_s}$;
- $\delta = 2(\beta_1 + \cdots + \beta_{s-1}) + \beta_s$, type $\mathbf{C_s}$.

Corollary A.4.4 ([BB], 2.10) *(i) The proper maximal parabolic k-subgroups of G are also proper maximal among $\mathbb{R}$-parabolics.*

(ii) Let $\eta \in \Delta_{\mathbb{R}}$ map under r to a k-root $\beta \in \Delta_k$. Let Ψ be the segment of $\Delta_{\mathbb{R}}$ of all roots which come before (respectively after) η, and set $\theta := r(\Psi) \cap \Delta_k$. Then $(L_\Psi)_{\mathbb{R}} = (L_\theta)_k$ (these are the Levi subgroups corresponding to subsets of $\Delta_{\mathbb{R}}$ and Δ_k) and is defined over k.

Definition A.4.5 Let $\mathcal{D}$ be a bounded symmetric domain, and let $F \subset \overline{\mathcal{D}}$ be a boundary component. F is a *rational boundary component*, if $\mathcal{N}_{G(\mathbb{R})}(F)_{\mathbb{C}}$ is defined over $\mathbb{Q}$.

In particular, a $\mathbb{Q}$-parabolic P of G satisfies $P_{\mathbb{R}} = \mathcal{N}_{G(\mathbb{R})}(F)$ for some rational boundary component F of the domain defined by $G(\mathbb{R})$.

Proposition A.4.6 *Suppose $G = Res_{k|\mathbb{Q}}G'$, with G' absolutely simple, G of hermitian type. Then if we let $G(\mathbb{R}) = G_1 \times \cdots \times G_d$ be the decomposition of the real group into simple factors, and $\mathcal{D} = \mathcal{D}_1 \times \cdots \times \mathcal{D}_d$ the corresponding decomposition of the domain, then a rational boundary component $F = F_1 \times \cdots \times F_d$ has the property that $F_i \subsetneqq \mathcal{D}_i$, hence every maximal $\mathbb{Q}$-parabolic decomposes as $P = P_1 \times \cdots \times P_d$, with $P_i \subsetneqq G_i$ a maximal $\mathbb{R}$-parabolic.*

This follows from the fact that G' is absolutely simple, see [SC], p. 220.

Remark A.4.7 For the groups of classical type discussed in the last sections, one can find explicit calculations of the k-roots in [B3], §23.

A.4.1 Boundary components

We briefly discuss the rational boundary components occuring in each of the cases. Again we tabulate this, giving the Tits index in each case and describing the boundary components. We also describe, in the classical cases, the corresponding isotropic subspaces of the vector space V. Throughout, $\mathcal{D}^*$ denotes the union of $\mathcal{D}$ and the rational boundary components.

O.1 The Tits index is $D_{n,s}$ (for $n \equiv 2(4)$), $^2D_{n,s}$ (for $n \equiv 0(4)$) or $B_{n,s}$ (for n odd), where s is the Witt index of h. The corresponding diagrams are (the top diagrams are for the case $s = 2$, the lower ones giving the left ends for $s = 1$):

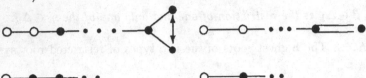

where the Galois action in the left-hand diagram is present only for $n \equiv 0(4)$. The boundary components of $\mathcal{D}' = G'(\mathbb{R})/K'$ are:

- $\{pt\} \subset \{1\text{-disc}\}^*$, $(s = 2)$
- $\{pt\}$, $(s = 1)$.

O.2 The index in this case is $D_{\frac{n}{2},s}^{(2)}$ (n even) or $^2D_{\frac{n-1}{2},s}^{(2)}$ (n odd), where s is the Witt index of h. The corresponding diagrams are (with non-trivial Galois action identifying the two right most vertices for n odd):

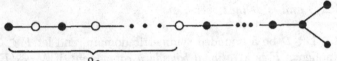

The corresponding boundary components are $\text{II}_{n-2}^* \supset \text{II}_{n-4}^* \supset \cdots \supset \text{II}_{n-2s}$.

S.1 The index is $C_{n,n}$, with the usual diagram and the following boundary components: $\{pt\} \subset \text{III}_1^* \subset \cdots \subset \text{III}_{n-1}^*$.

S.2 The index is $C_{n,s}^{(2)}$, with diagram

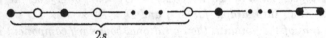

The boundary components are then the following: $\text{III}_{n-2}^* \supset \cdots \supset \text{III}_{n-2s}$.

U.1 The index is $^2A_{n-1,s}$, with the diagram

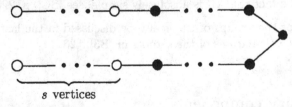

As above, let (p_ν, q_ν) denote the signature of h_ν, then in the factor $\mathcal{D}_\nu$ of $\mathcal{D}$ we have boundary components of the type $\text{I}_{p_\nu-b,q_\nu-b}$ for $1 \leq b \leq s$. Hence a flag of boundary components will be

$$\prod \text{I}_{p_\nu-1,q_\nu-1}^* \supset \prod \text{I}_{p_\nu-2,q_\nu-2}^* \supset \cdots \supset \prod \text{I}_{p_\nu-s,q_\nu-s}.$$

U.2 The index is in this case $^2A^{(d)}_{nd-1,s}$, with diagram

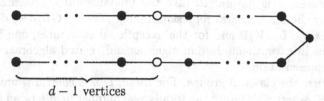

$$d - 1 \text{ vertices}$$

where there are $2s$ white vertices altogether. Letting the notations be as for the case **U.1**, we have the following boundary components:

$$\prod I^*_{p_\nu\text{-}d,q_\nu\text{-}d} \supset \prod I^*_{p_\nu\text{-}2d,q_\nu\text{-}2d} \supset \cdots \supset \prod I_{p_\nu\text{-}sd,q_\nu\text{-}sd}.$$

We now describe briefly the parabolics in terms of the geometry of (V, h) for all the cases above. Fixing a maximal k-split torus and an order on it amounts to fixing a maximal totally isotropic (s-dimensional) subspace $H_1 \subset V$ and a basis $v_1, \ldots, v_s$ of H_1. There are then k-vectors $v'_1, \ldots, v'_s$ spanning a complementary totally isotropic subspace H_2 such that $h(v_i, v'_j) = \delta_{ij}$. Then each pair (v_i, v'_i) spans a hyperbolic plane V_i (over D), and V decomposes:

$$V = V_1 \oplus \cdots \oplus V_s \oplus V', \quad V' \text{ anisotropic for } h. \tag{A.17}$$

Furthermore, $V_1 \oplus \cdots \oplus V_s = H_1 \oplus H_2$. With these notations, for $1 \leq b \leq s$ the standard k-parabolic $P'_b \subset G'$ is given as follows:

$$P'_b = \mathcal{N}_{G'}(< v_1, \ldots, v_b >), \tag{A.18}$$

where $< v_1, \ldots, v_b >$ denotes the span, a b-dimensional totally isotropic subspace. The hermitian Levi factor of P'_b is

$$L'_b = \mathcal{N}_{G'}(V_{b+1} \oplus \cdots \oplus V_s \oplus V')/\mathcal{Z}_{G'}(V_{b+1} \oplus \cdots \oplus V_s \oplus V'). \tag{A.19}$$

It reduces to the k-anisotropic kernel for $b = s$.

For the exceptional cases we have the following possibilities:

- E_6: Index: $^2E^{16'}_{6,2}$, boundary components: $\{pt\} \subset \mathbb{B}^*_5$.

  Index: $^2E^{35}_{6,1}$, boundary components: $\mathbb{B}_5$.

- E_7: Index: $E^{28}_{7,3}$, boundary components $\{pt\} \subset \mathbf{IV}^*_1 \subset \mathbf{IV}^*_{10}$.

  Index: $E^{31}_{7,2}$, boundary components $\mathbf{IV}_1 \subset \mathbf{IV}^*_{10}$.

A.5 Orders and arithmetic groups

By definition, an arithmetic subgroup $\Gamma \subset G(\mathbb{Q})$ is one which is commensurable with $\varrho^{-1}(GL(V_{\mathbf{Z}})) \cap G(\mathbb{Q})$, for some (equivalently, for any) faithful rational representation $\varrho : G \longrightarrow GL(V)$, where V is a finite-dimensional $\mathbb{Q}$-vector space, and $V_{\mathbb{Z}}$ is a $\mathbb{Z}$-structure, i.e., a $\mathbb{Z}$-lattice such that $V_{\mathbb{Z}} \otimes_{\mathbb{Z}} \mathbb{Q} = V_{\mathbb{Q}}$. In the classical cases, it is natural to take the fundamental representation as ϱ (more precisely the fundamental representation $\varrho' : G' \longrightarrow GL_D(V)$ determines $\varrho : G \longrightarrow Res_{k|\mathbb{Q}} GL_D(V)$), and for the exceptional structures, one has either representations in exceptional Jordan algebras and related algebras, or simply the adjoint representation.

Consider first the classical groups. For these, D is a central simple division algebra over K, where K is either the totally real number field k or an imaginary quadratic extension of k, and D has a $K|k$-involution. The rational vector space V is an n-dimensional right D-vector space, $A = M_n(D)$ is a central simple algebra over K with a $K|k$-involution extending the involution on D by (A.15). We have a $\pm$symmetric/hermitian form $h : V \times V \longrightarrow D$ such that

$$G' = \{g \in GL_D(V) | \forall_{x,y \in V}, \ h(x,y) = h(gx, gy)\} \tag{A.20}$$

is the unitary group of the situation. We take the natural inclusion given by (A.20), $\varrho' : G' \longrightarrow GL_D(V)$ and let the representation $\varrho : G \longrightarrow Res_{k|\mathbb{Q}} GL_D(V)$ determined by ϱ' be our rational representation. We now consider $\mathbb{Z}$-structures on V, for which we require an *order* $\Delta \subset D$, i.e., a lattice that is a subring of D, and consider Δ-lattices $\mathcal{L} \subset V$. The analog of (A.20), after fixing the $\mathbb{Z}$-structure on V, is

$$\Gamma_{\mathcal{L}} = \{g \in G | g\mathcal{L} \subseteq \mathcal{L}\}. \tag{A.21}$$

Then $\Gamma_{\mathcal{L}} \subset G(\mathbb{Q})$ is an arithmetic subgroup, as it is the set of elements which preserve the Δ-structure on V defined by $\mathcal{L}$, which itself is a $\mathbb{Z}$-lattice in the rational vector space V (viewing V as a $\mathbb{Q}$-vector space). If, for example, $\Gamma'_{\Delta} \subset \Gamma_{\Delta}$ is a normal subgroup of finite index, we get an induced representation of $\Gamma_{\Delta}/\Gamma'_{\Delta}$ in $\mathcal{L}/\mathcal{L}'$, where $\mathcal{L}'$ is the sublattice of $\mathcal{L}$ preserved by Γ'_{Δ}. This is the general formulation of an occurance which is well-known in specific cases. For example, if $\Gamma'_{\Delta} = \Gamma(N) \subset Sp(2n, \mathbb{Z})$ is the principal congruence subgroup of level N, there is a representation of $\Gamma/\Gamma(N) \cong Sp(2n, \mathbb{Z}/N\mathbb{Z})$ in $(\mathbb{Z}/N\mathbb{Z})^{2n}$.

Now consider the exceptional groups. In the case of E_6 we have the 27-dimensional representation in the exceptional Jordan algebra $\mathfrak{J}$, while in the case of E_7 we have the 56-dimensional representation in the exceptional algebra of 2×2 matrices over $\mathfrak{J}$. This is what W. Baily utilized in his beautiful paper [Bai]. In both cases we can also use the adjoint representation, so we require a $\mathbb{Z}$-structure on the Lie algebra itself. Such can be readily constructed, utilizing the Tits algebras, from lattices on the constituents, composition algebras and (exceptional) Jordan algebras.

After these introductory remarks we proceed to give a few details, which in particular allow us to give some relevant references in each case. We start by discussing orders, then describe the arithmetic groups these give rise to.

A.5.1   Orders in associative algebras

A general reference for this section is [R]. We first fix some notations. k is a totally real Galois extension of degree f over $\mathbb{Q}$, and $\mathcal{O}_k$ will denote the ring of integers in k. D will denote a division algebra (skew field), central simple of degree d over K, with a $K|k$ involution ($K = k$ for involutions of the first kind, and K is an imaginary quadratic extension of k for involutions of the second kind, see Definition A.1.3). V denotes an n-dimensional right D-vector space, so that $Hom_D(V, V) \cong M_n(D)$. $A = M_n(D)$ is a central simple algebra over K with involution extending the involution on D (see (A.15)), and (V, h) is a λ-hermitian space with λ-hermitian form h (with respect to the involution on D). Hence $[D : K] = d^2$, $[A : K] = (nd)^2 = t^2, t = nd$.

Let F be a number field, for example $F = K, k$ as above. Let W be an F-vector space. A full $\mathcal{O}_F$-lattice $\mathcal{L}$ in W is an $\mathcal{O}_F$-module, finitely generated, such that $F \cdot \mathcal{L} = W$. Usually we work with full lattices and delete the word full. If W is an F-algebra, then an $\mathcal{O}_F$-lattice $\mathcal{L}$ is an $\mathcal{O}_F$-order, if $\mathcal{L}$ is a subring of W. In particular in $W = D$, an $\mathcal{O}_F$-order is a (full) lattice which is a subring. Let $\Delta \subset D$ denote an order in D, and let V be an n-dimensional vector space over D. Then a (full) Δ-lattice in V is a Δ-module $\mathcal{M}$ with $\mathcal{M} \cdot D = V$; if again A is the algebra $M_n(D)$, then a Δ-lattice in A is a Δ-order, if it is a subring of A.

Let an $\mathcal{O}_F$-lattice $\mathcal{L} \subset A$ be given. $\mathcal{L}$ determines a right (respectively left) $\mathcal{O}_F$-order:

$$\mathcal{O}_r(\mathcal{L}) = \{x \in A \,\big|\, \mathcal{L} \cdot x \subset \mathcal{L}\}, \text{(respectively } \mathcal{O}_l(\mathcal{L}) = \{x \in A \,\big|\, x \cdot \mathcal{L} \subset \mathcal{L}\}). \quad (A.22)$$

If $\mathcal{L}$ is a Δ-lattice, then $\mathcal{O}_r(\mathcal{L})$ and $\mathcal{O}_l(\mathcal{L})$ are Δ-orders. If an $\mathcal{O}_F$-order $\mathcal{O} \subset A$ is given, and $\mathcal{L} \subset A$ is a lattice with $\mathcal{O} = \mathcal{O}_r(\mathcal{L})$ (respectively $\mathcal{O}_l(\mathcal{L})$), then one also calls $\mathcal{L}$ an $\mathcal{O}$-lattice, and says that $\mathcal{L}$ and $\mathcal{O}$ are associated. An element $a \in A$ is called integral, if its characteristic polynomial has integer coefficients[2], $\chi_a \in \mathcal{O}_F[X]$. It is a basic result that every element $a \in \mathcal{O}$ is integral for any $\mathcal{O}_F$-order $\mathcal{O}$ in A. An order $\mathcal{O}$ is maximal, if it is not properly contained in any other order. It is a basic fact that maximal orders exist in D and in A, and that any order is contained in a maximal one ([R], 10.4).

The characteristic poynomial of $a \in A$ (viewing a as a left multiplication operator on A) is

$$\chi_a(X) = \chi_{A|F,a}(X) = X^{t^2} - (Tr_{A|F}a)X^{t^2-1} + \cdots + (-1)^{t^2} N_{A|F}a, \quad (A.23)$$

where $Tr_{A|F}$, $N_{A|F}$ are the trace and norm (in the regular representation) of A over F (cf. Definition A.1.9). One has

$$Tr_{A|F}a = t(tr_{A|F}(a)), \quad N_{A|F}a = (n_{A|F}(a))^t. \quad (A.24)$$

The reduced trace gives rise to a non-degenerate symmetric F-bilinear form $\tau : A \times A \longrightarrow F$, defined by $\tau(a, b) = tr_{A|F}(ab)$, and using this one gets the following characterisation of orders:

Lemma A.5.1 ([R], 10.3) *Let* $\Gamma \subset A$ *be a subring,* $\mathcal{O}_F \subset \Gamma \subset A$, *such that* $F \cdot \Gamma = A$. *Then* Γ *is an order in* A $\Longleftrightarrow$ $\forall_{a \in \Gamma}$, a *is integral over* $\mathcal{O}_F$.

---

[2]for this definition it is irrelevant whether one takes the reduced characteristic polynomial (Definition A.1.9) or the non-reduced one (A.23)

Let $\mathcal{O} \subset A$ be an order. The *discriminant* $d(\mathcal{O})$ is the ideal in $\mathcal{O}_F$ generated by $\{det(tr_{A|F}(x_i \cdot x_j)), 1 \le i, j \le t^2 \big| x_i \in \mathcal{O}\}$, which for linearly independent x_i is non-vanishing. Moreover, if $\mathcal{O}$ has a free $\mathcal{O}_F$-basis $u_1, \ldots, u_{t^2}$, then $d(\mathcal{O}) = \mathcal{O}_F(det(tr_{A|F}(u_i u_j)))$ is a principal ideal. One has the following description of maximal orders in A:

Theorem A.5.2 ([R], 21.6) *Notations as obove, let $\Delta \subset D$ be a fixed maximal $\mathcal{O}_F$-order in D, and let M be any (full) right Δ-lattice in V. Then $Hom_\Delta(M, M)$ is a maximal $\mathcal{O}_F$-order in A, and for any maximal $\mathcal{O}_F$-order $\mathcal{O}$ in A, there exists a (full) right Δ-lattice $N \subset V$ with $\mathcal{O} = Hom_\Delta(N, N)$.*

This result gives a one to one correspondence between the set of maximal (right) orders $\mathcal{O} \subset A$ and the set of full right Δ-lattices $N \subset V$,

$$\{\text{maximal orders } \mathcal{O}\} \overset{1-1}{\longrightarrow} \{\text{right } \Delta\text{-lattices } N\}$$
$$\mathcal{O} \mapsto N: \quad \mathcal{O} = Hom_\Delta(N, N)$$
$$\mathcal{O} = Hom_\Delta(N, N) \longleftarrow N.$$

The following result of Chevally, describes maximal orders in associative algebras.

Theorem A.5.3 ([R], 27.6) *Let $\Delta \subset D$ be a maximal $\mathcal{O}_K$-order in D, and for each right ideal $J \subset \Delta$, set $\Delta' = \mathcal{O}_l(J)$. Then every maximal order of $A = M_n(D)$ is isomorphic to one of the form*

$$\mathcal{O}_J = \begin{pmatrix} \Delta & \cdots & \Delta & J^{-1} \\ \vdots & \ddots & \vdots & \vdots \\ \Delta & \cdots & \Delta & J^{-1} \\ J & \cdots & J & \Delta' \end{pmatrix},$$

for some right ideal J, and for each right ideal J, the lattice $\mathcal{O}_J$ above is a maximal order.

In other words, to give a maximal order in A is the same as giving a maximal order $\Delta \subset D$, together with a right ideal $J \subset \Delta$, i.e., the same as giving a pair (Δ, J). In particular if the class number $h(\Delta) = 1$ (note that $h(\Delta)$ is also equal to the number of right Δ-ideal classes, see [R], Ex. 7 iii), p. 232), then there is a one to one correspondence between Δ-isomorphism classes of maximal orders in D and A.

A.5.2   Arithmetic groups – classical cases

In this subsection G' will denote an absolutely (almost) simple k-group (k a totally real number field) which we assume is classical, $G = Res_{k|\mathbb{Q}}G'$ the $\mathbb{Q}$-simple group it defines, which we assume is of hermitian type. We let $\varrho' : G' \longrightarrow GL_D(V)$ be the natural inclusion and $\varrho : G \longrightarrow Res_{k|\mathbb{Q}}GL_D(V)$ be the natural representation of G defined by ϱ'. Fix a maximal order $\Delta \subset D$, and let $\mathcal{L} \subset V$ be a Δ-lattice (which is in particular a $\mathbb{Z}$-lattice of the underlying $\mathbb{Q}$-vector space). As above, $\mathcal{O}_r(\mathcal{L})$ (respectively $\mathcal{O}_l(\mathcal{L})$) will denote the right (respectively left) order of $\mathcal{L}$, given by the equation (A.22). First of all, we have the arithmetic subgroup $GL_\Delta(\mathcal{L}) \subset GL_D(V)$, and we define the subgroup

$$\Gamma'_{\mathcal{L}} := \{g \in G'(k) \big| \varrho(g)(\mathcal{L}) \subseteq \mathcal{L}\} = \varrho^{-1}(GL_\Delta(\mathcal{L})) \subset G'(k),$$

and similarly $\Gamma_{\mathcal{L}} \subset G(\mathbb{Q})$. By definition these are arithmetic subgroups of $G'(k)$ and $G(\mathbb{Q})$, respectively. Our central simple algebra is in this case $A = M_n(D)$, and $\mathcal{O}_r(\mathcal{L})$ is a maximal order in A. Recall how the group G' and the algebra are related (Theorem A.2.4). Let $U = \{z \in A | zz^* = 1\}$, U_0 the connected component of U, $((G')^0 =)G_0 := (\mathrm{Aut}(A))^0$, and let $C \subset U_0$ be the center of U_0. Then we have an exact sequence

$$1 \longrightarrow C \longrightarrow U_0 \longrightarrow G_0 \longrightarrow 1.$$

As a lattice in A we consider $\mathcal{O} := \mathcal{O}_r(\mathcal{L})$ and its intersection with U_0,

$$\mathcal{O}_0 = \mathcal{O} \cap U_0.$$

Similarly, $\mathcal{C} := C \cap \mathcal{O}_0$ is the center of $\mathcal{O}_0$, and we have the sequence

$$1 \longrightarrow \mathcal{C} \longrightarrow \mathcal{O}_0 \longrightarrow \Gamma' \longrightarrow 1,$$

where $\Gamma' \cong \mathcal{O}_0/\mathcal{C}$ is the arithmetic subgroup $\Gamma' \subset G'(k)$, showing how the maximal orders are related to the arithmetic groups. In our situation here, $(G')^0$ plays the role of G_0, while $(U')^0 = (\{z \in A \big| z^*z = 1\})^0$ plays the role of U_0. Let further $C' \subset (U')^0$ be the center. We have $\mathcal{O} \cong \mathcal{O}_J$ for some right ideal J (Theorem A.5.3), and $(\mathcal{O}_J)_0 = \mathcal{O}_J \cap U_0$ plays the role of $\mathcal{O}_0$. Then $C' \cap (\mathcal{O}_J)_0 = \mathcal{C}$ is the center of $(\mathcal{O}_J)_0$, and we have sequences:

$$
\begin{array}{ccccccccc}
1 & \longrightarrow & C' & \longrightarrow & (U')^0 & \longrightarrow & (G')^0 & \longrightarrow & 1 \\
 & & \cup & & \cup & & \cup & & \\
1 & \longrightarrow & \mathcal{C} & \longrightarrow & (\mathcal{O}_J)_0 & \longrightarrow & \Gamma'_{\mathcal{L}} & \longrightarrow & 1.
\end{array}
$$

In this sense, maximal orders give rise to arithmetic subgroups. Viewing the $\mathcal{O}_k$-lattice $\mathcal{L}$ as a $\mathbb{Z}$-lattice gives the corresponding diagram for the $\mathbb{Q}$-groups (with hopefully obvious notations)

$$
\begin{array}{ccccccccc}
1 & \longrightarrow & C & \longrightarrow & U^0 & \longrightarrow & G^0 & \longrightarrow & 1 \\
 & & \cup & & \cup & & \cup & & \\
1 & \longrightarrow & Z(\mathcal{O}_0) & \longrightarrow & \mathcal{O}_0 & \longrightarrow & \Gamma_{\mathcal{L}} & \longrightarrow & 1.
\end{array}
$$

We now describe this more precisely for the following special cases:

a) Siegel modular groups.

b) Picard modular groups.

c) Hyperbolic plane modular groups.

These are examples of $\mathbb{Q}$-groups which are of both inner type (for a)) and outer type (for b) and c)), of split over $\mathbb{R}$-type, meaning the $\mathbb{Q}$-rank is equal to the $\mathbb{R}$-rank (for a) and b)) and more or less the *opposite* of split over $\mathbb{R}$-type ($\mathbb{Q}$-rank equal to one, $\mathbb{R}$-rank unbounded) (for c)). Case a) is well-known, b) is also to a certain extent, while c) was introduced in [H2]. In cases b) or c) we are describing the unitary groups which are not simple but reductive. Taking the subgroup of matrices with determinant 1 gives the simple, special unitary, group.

a) Siegel case:

- $A = M_{2n}(\mathbb{Q})$ with the involution $* : X \mapsto JX^t J$, $J = \begin{pmatrix} 0 & 1 \\ -1 & 0 \end{pmatrix}$.

- $\mathrm{Aut}(A, *) \cong PSp(2n, \mathbb{Q})$, $V = \mathbb{Q}^{2n}$.

- $D = \mathbb{Q}$, a maximal order is $\Delta = \mathbb{Z}$, $V_{\mathbb{Z}} = \mathbb{Z}^{2n}$.

- $\Gamma = PSp(2n, \mathbb{Z})$.

The sequence above becomes:

$$
\begin{array}{ccccccccc}
1 & \longrightarrow & \mathbb{Z}/(2) & \longrightarrow & Sp(2n, \mathbb{Q}) & \longrightarrow & PSp(2n, \mathbb{Q}) & \longrightarrow & 1 \\
 & & \| & & \cup & & \cup & & \\
1 & \longrightarrow & \mathbb{Z}/(2) & \longrightarrow & Sp(2n, \mathbb{Z}) & \longrightarrow & \Gamma & \longrightarrow & 1.
\end{array}
$$

b) Picard case:

- $A = M_n(K)$ with involution $* : X \mapsto HX^t H$, H hermitian, where $K|\mathbb{Q}$ is imaginary quadratic.

- $\mathrm{Aut}(A, *) \cong PU(K^n, h)$, $V = K^n$, h is a hermitian form represented by H.

- $D = K$, a maximal order is $\Delta = \mathcal{O}_K$, $V_{\mathbb{Z}} = \mathcal{O}_K^n$.

- $\Gamma = PU(\mathcal{O}_K^n, h)$.

The sequence above becomes:

$$
\begin{array}{ccccccccc}
1 & \longrightarrow & \mathcal{C} & \longrightarrow & U(K^n, h) & \longrightarrow & PU(K^n, h) & \longrightarrow & 1 \\
 & & \cup & & \cup & & \cup & & \\
1 & \longrightarrow & \mathcal{C} & \longrightarrow & U(\mathcal{O}_K^n, h) & \longrightarrow & PU(\mathcal{O}_K^n, h) & \longrightarrow & 1.
\end{array}
$$

Note that $\mathcal{C}$ is given essentially by $\mathcal{O}_K \cap U(1)$, which is ± 1 except for the two fields $K = \mathbb{Q}(\sqrt{-1})$, $K = \mathbb{Q}(\sqrt{-3})$ which contain fourth (respectively sixth) roots of unity.

c) Hyperbolic plane case:

- D is a division algebra, central simple of degree $d \geq 2$ over K, with a $K|\mathbb{Q}$-involution (K imaginary quadratic), $\Delta \subset D$ is a maximal order.

- $A = M_2(D)$ with involution $* : X \mapsto {}^t\overline{X}$, where ${}^t\overline{X} = (\overline{x}_{ji})$, if $X = (x_{ij})$, and $\overline{x}$ denotes the involution in D.

- $\text{Aut}(A, *)$ is a $\mathbb{Q}$-form of $PU(d, d)$, and $V = D^2$, with a hermitian form $h : V \times V \longrightarrow D$ which is isotropic, $V_{\mathbb{Z}} = \Delta^2$.

- $\Gamma = PU(\Delta^2, h)$.

The above sequence becomes in this case

$$
\begin{array}{ccccccccc}
1 & \longrightarrow & \mathcal{C} & \longrightarrow & U(D^2, h) & \longrightarrow & PU(D^2, h) & \longrightarrow & 1 \\
& & \cup & & \cup & & \cup & & \\
1 & \longrightarrow & \mathcal{C} \cap \Delta & \longrightarrow & U(\Delta^2, h) & \longrightarrow & PU(\Delta^2, h) & \longrightarrow & 1.
\end{array}
$$

As D is central simple over K, the center is as in the last case, $\mathcal{C} \cap \Delta \cong \mathcal{O}_K \cap U(1)$, hence it is ± 1 except for the case $K = \mathbb{Q}(\sqrt{-1})$ and $K = \mathbb{Q}(\sqrt{-3})$ as above.

Appendix B

Some classical algebraic geometry

B.1 Invariants

B.1.1 Invariants and covariants

B.1.1.1 Invariants

Let $\rho : G \longrightarrow GL_n(K)$ be a linear representation of a group G.

Definition B.1.1 A polynomial $f \in K[x_1, \ldots, x_n]$ is an *invariant* of the group action of G, if

$$f(\rho(g)x) = f(x), \; \forall_{g \in G}.$$

f is called a *relative* invariant, if $f(\rho(g)x) = \chi(g) \cdot f(x)$, for some character χ of G.

If, for example, G is finite we get the notion of invariant hypersurfaces (a hypersurface $f = 0$ for an invariant f), while if ρ is the representation in $\mathrm{Sym}^m(K^n)$, we get the notion of invariants of homogenous polynomials of degree m of $\mathbb{P}^{n-1}(K)$. In this case, one wants to consider the action on the coefficients of the form, so one needs additional sets of variables. Accordingly one makes the following definition, generalizing B.1.1.

Definition B.1.2 Given a representation

$$\begin{aligned} \rho : G &\longrightarrow GL_{n_1}(K) \times \cdots \times GL_{n_h}(K) \\ g &\mapsto (\rho_1(g), \ldots, \rho_h(g)), \end{aligned}$$

a polynomial $f \in K[x_i^j]$, $i = 1, \ldots, n$, $j = 1, \ldots, h$ is called a *simultaneous (relative) invariant* of the ρ_j, if

$$f(\rho_1(x^1), \ldots, \rho_h(x^h)) = \chi_1(g) \cdots \chi_h(g) \cdot f(x^1, \ldots, x^h),$$

where we have written $x^j = (x_1^j, \ldots, x_n^j)$. If $\chi_i(g) \equiv 1$ for all i, f is called an *absolute* invariant.

It is a basic fact that for $G = GL_n(K)$ and each ρ_i the standard representation in $GL_n(K)$, that there are no invariants for ρ (other than the constants) if $h < n$, and the only invariants are $f \equiv c \cdot (det)^w$, $w \in \mathbb{Z}$ if $h = n$. In this case one calls the w the *weight* of the invariant. An absolute invariant is thus one of weight 0. For $n > h$ there are such invariants, for example the 2×2 minors of the determinant for $h = 4, n = 2$ (these remarks under the assumption that all ρ_i coincide). One uses the more general definition of B.1.2 for example by letting $G = GL_n(K)$ act on some sets of variables covariantly and on others contravariantly. This will give the notion of covariants.

In this generality one can define *polarization*: let $f \in K[x_1^1, \ldots, x_n^h]$ be given, and let $x_1^{h+1}, \ldots, x_n^{h+1}$ be a $(h+1)^{st}$ set of coefficients. Then

$$\Delta_{x^\sigma, x^{h+1}} f := \sum_{i=1}^n \frac{\partial f}{\partial x_i^\sigma} x_i^{h+1} \in K[x_1^1, \ldots, x_n^{h+1}], \qquad \text{(B.1)}$$

is called the polarization of f with respect to $\{x_i^\sigma\}$ by $\{x_i^{h+1}\}$. Usually one takes $\sigma = 1$ (by changing coordinates if necessary). Let us call an invariant f as in Definition B.1.2 for the choice: $\rho_j =$ the identical representation for $j = 1, \ldots, h$, a simultaneous invariant of $\{x_i^1\}_{i=1,\ldots,n}, \ldots, \{x_i^h\}_{i=1,\ldots,n}$. Then one has the basic result:

Theorem B.1.3 ([SIV], Satz 1.6, p.12) *Let $f \in K[x_1^1, \ldots, x_n^h]$ be a simultaneous invariant of*
$\{x_i^1\}_{i=1,\ldots,n}, \ldots, \{x_i^h\}_{i=1,\ldots,n}$. *Then*

$$\overline{f} := \Delta_{x^1, x^{h+1}} f$$

is a simultaneous invariant of $\{x_i^1\}_{i=1,\ldots,n}, \ldots, \{x_i^{h+1}\}_{i=1,\ldots,n}$.

B.1.1.2 Invariants of homogenous forms

We now specialize to the case of homogenous forms $p \in \text{Sym}^m(K^n)$ and their invariants. Here we are considering $G = GL_n(K)$, and the representation

$$\rho := (\text{Sym}^m)^* \times (\text{Sym}^m) : G \longrightarrow GL_N(K) \times GL_N(K),$$

where Sym^m denotes the usual representation on the polynomials of degree m, $(\text{Sym}^m)^*$ is the contragredient representation which is the representation induced on the coefficients of polynomials of degree m, and $N = \binom{n+m}{n}$ is the dimension of the space of homogenous polynomials of degree m. If we write a polynomial p of degree m as

$$p(a|x) = p(x_1, \ldots, x_n) = \sum a_{k_1 \cdots k_n} x_1^{k_1} \cdots x_n^{k_n}, \quad k_1 + \cdots + k_n = m,$$

then we consider p as a linear form in the coeffients $a_{k_1 \cdots k_n}$ and as a linear form in the monomials $x_1^{k_1} \cdots x_n^{k_n}$, on which G is acting by the symmetric representation (on the monomials) and the contragredient representation (on the coefficients).

Definition B.1.4 A polynomial $F \in K[a_i^j | x_1, \ldots, x_n]$, $i = 1, \ldots, \binom{m_j + n}{n}$, $j = 1, \ldots, h$ is a *projective simultaneous covariant* of h polynomials $p_1, \ldots, p_h$, with $p_j \in \mathrm{Sym}^{m_j}(K^n)$ homogenous, if

$$F\left(\rho_{m_1}(g)(a_i^1), \ldots \rho_{m_h}(g)(a_i^h) | g(x_1, \ldots, x_n)\right) = \det(g)^w \cdot F(a_i^j | x_1, \ldots, x_n),$$

where $g \in G = GL_n(K)$ and $\rho_{m_i} = (\mathrm{Sym}^{m_i})^*$. The degree, order and weight are defined as:

$$
\begin{aligned}
degree\ F &= (d_1, \ldots, d_h), \text{ the degrees of } F \text{ in the variables } \{a_i^1\}, \ldots, \{a_i^h\} \\
order\ F &= \text{degree of } F \text{ in the } x_i \\
weight\ F &= w.
\end{aligned}
$$

In case the order of F is zero, one speaks of *relative invariants*, and if in addition the weight is zero, of (absolute) *invariants*.

So for example a linear covariant of a polynomial p is a covariant of degree 1 in the coefficients of p (regardless of its degree in the variables x_i).

The following relation holds between the degrees, order and weight of a covariant:

Lemma B.1.5 ([SIV], Satz 1.11, p. 25) *Let* $p_1, \ldots, p_h$ *be as above, of degrees* $m_1, \ldots, m_h$. *Let* $F(\{a\}^1, \ldots, \{a\}^h | x)$ *be a projective simultaneous covariant of degree* $d = (d_1, \ldots, d_h)$, *and set*

$$
\begin{aligned}
d_x &= order\ F \\
w &= weight\ F.
\end{aligned}
$$

Then

$$m_1 d_1 + \cdots + m_h d_h = nw + d_x.$$

In particular, for an invariant *we have* $\sum m_i d_i = nw$.

An important fact about covariants is that *the covariants of a covariant C of f are again covariants of f.* For example, the *discriminant D of the Hessian* (see next section) is a covariant of the Hessian, hence of f. Since its degree is zero, this is in fact an *invariant* of f.

B.1.1.3 Covariants

We give some important examples of covariants. First of all, the polars $\Delta_y f(x)$ will be covariants by Aronhold's theorem B.1.3 (here $\Delta_y f(x)$ is what was denoted $\Delta_{x,y} f$ above, i.e., $\Delta_y f(x) = \sum \frac{\partial f}{\partial x_i} y_i$).

Definition B.1.6 Let $f(x_1, ..., x_n)$ be homogenous of order d in the variables x_i and $(y_1, ..., y_n) = y \in \mathbb{P}^{n-1}$ be any point. The *first polar of y with respect to* f is defined by:

$$\Delta_y f = \frac{1}{d}(\frac{\partial f}{\partial x_1} y_1 + ... + \frac{\partial f}{\partial x_n} y_n),$$

and iteratively the rth *polar* is defined by $\Delta_y^r f = \Delta_y(\Delta_y^{r-1} f)$. The rth *polar variety* is $\{\Delta_y^r f = 0\}$. The degree of $\Delta_y^r f$ is 1 (recall f is linear in its coefficients) its order is r in y and $d - r$ in x.

Note in particular, that

$$\Delta_y^r f(x) = \Delta_x^{d-r} f(y), \tag{B.2}$$

i.e., if x lies on the rth polar of y, then y lies on the $(d-r)^{th}$ polar of x. In the particular case where $r = d - 2$ one speaks of the polar quadric, for $r = d - 1$ of the polar plane of y. The geometric meaning of the polar is as follows. The intersection $\{p = 0\} \cap \{\Delta_y p = 0\}$ consists of the set of points $x \in \{p = 0\}$ such that the tangent plane at x passes through y. In other words, the *condition* on $x \in \{p = 0\}$ that the tangent plane at x pass through $y \in \mathbb{P}^{n-1}$ is

$$\sum \frac{\partial f}{\partial x_i}(x) y_i = 0. \tag{B.3}$$

A natural simultaneous covariant of n forms $f_1, \ldots, f_n$ in n variables is the *Jacobian determinant*.

Definition B.1.7 Given $f_1(x_1, \ldots, x_n), \ldots, f_n(x_1, \ldots, x_n)$, let

$$\mathcal{J}(f_1, \ldots, f_n) = \det\left(\frac{\partial f_i}{\partial x_j}\right).$$

Then $\mathcal{J}(f_1, \ldots, f_n)$ is a simultaneous projective covariant of $f_1, \ldots, f_n$ of degree one in each set of coefficients, order $d_x = \sum m_\nu - n$, $m_\nu =$ degree(f_ν), and of weight one.

A particular case of this is given by taking $f_i = \frac{\partial p}{\partial x_i}$ for a homogenous polynomial p.

Definition B.1.8 The *Hessian* of a polynomial p of degree d in n variables is the determinant

$$\text{Hess}(p) = \det\left(\frac{\partial^2 p}{\partial x_i \partial x_j}\right) = \mathcal{J}\left(\frac{\partial p}{\partial x_1}, \ldots, \frac{\partial p}{\partial x_n}\right).$$

It is of degree n, order $n(d - 2)$ and weight two. The *Hessian variety* $\text{Hess}(p)$ is the hypersurface $\{\text{Hess}(p) = 0\}$, a variety of degree $n(d - 2)$.

By (B.2) we have

$$\Delta_y^{d-2} p(x) = \Delta_x^2 p(y); \tag{B.4}$$

the former is the polar quadric of y with respect to p. From this we get the following, the proof of which is left to the reader as an exercise.

Lemma B.1.9 *The Hessian variety* Hess(p) *is the locus of* $y \in \mathbb{P}^{n-1}$ *for which the polar quadric of* y *with respect to* p *is singular.*

Another covariant is the *Steinerian* of f.

Definition B.1.10 The *Steinerian variety* of f is the locus of the vertices of the singular quadric cones C_x, $x \in$ Hess(f). Equivalently it can be defined as follows. For each $x \in$ Hess(f), the Hessian determinant vanishes at x, i.e., generically there is a one-dimensional kernel of the symmetric matrix Hess(f), spanned by, say $a := (a_1, ..., a_n)$. a then determines a point $[a_1 : ... : a_n]$ in $\mathbb{P}^{n-1}$, and the closure of the locus of these generic points is the Steinerian of f, which will be denoted by $\mathcal{ST}(f)$.

Furthermore, the following more general statement is true: Let $x \in$ Hess(f), $\alpha :$ Hess$(f) \longrightarrow \mathcal{ST}(f)$ be the Steinerian map, which sends x to the vertex of the cone C_x. Then the polar of $\alpha(x)$ with respect to f is *singular* in the point x. In other words, the polar variety (of $\alpha \in \mathcal{ST}(f)$) meets the Hessian in the locus of singular points of that polar.

Let us now describe the equation defining the Steinerian. Let H$= (\frac{\partial^2 f}{\partial x_i \partial x_j})_{i,j}$ be the Hessian matrix. The matrix $\bigwedge^{n-1}$H has n rows and columns, each entry consisting of an $(n-1) \times (n-1)$-minor of H. This matrix fulfills the relation:

$$\overset{n-1}{\bigwedge} \text{H} \cdot \text{H} = \overset{n}{\bigwedge} \text{H} = det\text{H},$$

so one gets the implication: $x \in$ **Hess**$(V) \Rightarrow (\bigwedge^{n-1} \text{H})(x) \in \mathcal{ST}(V)$. From this one deduces that the form defining the Steinerian, F, is characterised by:

$$F \left((\overset{n-1}{\bigwedge} \text{H})_{i,1}(x), ..., (\overset{n-1}{\bigwedge} \text{H})_{i,n}(x) \right) = 0, \quad i = 1, ..., n \qquad (\text{B.5})$$

where $(\bigwedge^{n-1} \text{H})_{i,j}(x)$ denotes the i,j^{th} entry of the matrix $(\bigwedge^{n-1} \text{H})$. Hence $\mathcal{ST}(V)$ is defined by the equation which the rows of the $(n-1) \times (n-1)$ minors of the Hessian fulfill.

The form F can also be calculated as follows. From the n equations

$$\frac{\partial^2 f}{\partial x_i \partial x_1} y_1 + \cdots + \frac{\partial^2 f}{\partial x_i \partial x_n} y_n = 0, \ i = 1, ..., n,$$

eliminate the x_i. The Hessian form can also be derived by eliminating the y_i from these equations.

B.1.1.4   The order

Let f be a form of degree d in n variables, and let $V = \{f = 0\} \subset \mathbb{P}^{n-1}$ be the hypersurface (of degree d) defined by f. Suppose V is smooth. Then for a fixed line L, the number of hyperplanes tangent to V through the line L can be calculated as follows. The hyperplanes will be spanned by $n-3$ points and the line, and for each such point, the set of lines through the point and tangent to V at a point is determined by the intersection $V \cap \{\Delta_y f = 0\}$; if the hyperplane is tangent to V, say at a point $z \in V$, then the line from z to each of the $n - 3$ points $y_1, \ldots, y_{n-3}$ as well as the line from z to any point on the line L must be tangent to V, hence the number of planes is given by the number of points in the intersection:

$$V \cap \{\Delta_{y_1} f = 0\} \cap \cdots \cap \{\Delta_{y_{n-3}} f = 0\},$$

which is $d(d-1)^{n-3}$. Each (isolated) multiple point of V (i.e., singular point) reduces the order by the multiplicity of the point.

B.1.1.5 Mixed concomitants

Let us view $x_1, \ldots, x_n$ as homogenous coordinates on $\mathbb{P}^{n-1}$. Then the coordinates on the dual projective space are linear forms $u_1, \ldots, u_n$ with the property that

$$x_1 u_1 + \cdots + x_n u_n = 0. \tag{B.6}$$

These variables are, just like the coefficients of a polynomial, contravariant with respect to the variables x_i, that is, if the variables x_i are changed by a projective transformation g, then the variables u_i change by the adjoint of g. With these variables one can define, more generally than simultaneous covariants, so-called mixed concomitants.

Definition B.1.11 Let p_j be polynomials of the x_i and u_i, homogenous in each set of variables, say of degrees d_j, e_j. Then a polynomial $F \in K[a_i^j; x_i; u_i]$ is a simultaneous *mixed concomitant* of the p_j, if

$$F\left(\rho^*(g)(\{a_i^j\}); \rho(g)(x_i); \rho^*(g)(u_i)\right) = (\det(g))^w \cdot F(\{a_i^j\}; x_i; u_i).$$

Once again w is the weight, while the mixed concomitant has orders in the x_i variables as well as in the u_i variables, and degrees in the coefficients $\{a_i^j\}$ of the polynomials p_j. One denotes such a mixed concomitant by $C_{d, \nu_x; \nu_u}$ if C has degree d, order ν_x in the x variables and order ν_u in the u variables.

B.1.1.6 The dual variety

An important mixed concomitant of the given form f is the dual variety.

Definition B.1.12 Given a hypersurface $V \subset \mathbb{P}^{n-1}$, the *dual variety* $\check{V} \subset \check{\mathbb{P}}^{n-1}$ is defined to be the union of all hyperplanes of $\mathbb{P}^{n-1}$ which are *tangent* to V. As a mixed concomitant of f it is a $C_{d, 0; \nu}$, where d is the degree in the coefficients of f and ν is the order in the u_i (see below).

The coordinates of a tangent plane are given by:

$$\rho u_1 = \frac{\partial f}{\partial x_1}, \ \rho u_2 = \frac{\partial f}{\partial x_2}, \ \ldots, \rho u_n = \frac{\partial f}{\partial x_n}.$$

Assuming V is not ruled, the dual variety $\check{V}$ is itself a hypersurface whose equation in $(u_1, ..., u_n)$ is gotten by eliminating the x and ρ out of the above equations and of (B.6). The *order* of $\check{V}$ is the number of hyperplanes passing through a fixed generic line and tangent to V, see section B.1.1.4. This is also called the *class* of V. An s-fold point of V corresponds to an s-fold tangent plane of $\check{V}$. If V is smooth, the order of $\check{V}$ is $d(d-1)^{n-2}$, and each isolated multiple point of V reduces the order of $\check{V}$ by the multiplicity of that point; in particular if V has only κ nodes, then the order of $\check{V}$ is $d(d-1)^{n-3} - 2\kappa$.

The singularities of the dual variety arise from (at least) two phenomena on V.

- If V contains *linear* $\mathbb{P}^{n-2}$'s, then $\check{V}$ has a singular line, corresponding to the $\mathbb{P}^1$ of hyperplanes through the given $\mathbb{P}^{n-2}$ which are tangent to V there.

- The parabolic divisor on V, i.e., the locus of points for which the tangent plane touches V with higher multiplicity, yields singularities on the dual variety.

B.1.2   Invariants of finite group actions

B.1.2.1   Absolute and relative invariants

Let G be a finite group, acting on K^n, i.e., with a faithful representation $\rho : G \longrightarrow GL_n(K)$. An invariant in this case is a polynomial unchanged by the action of $\rho(G)$ on the coordinates. We will be especially interested in *projective* actions, i.e., $\mathbb{P}\rho : G \longrightarrow PGL_n(K)$. Then $\mathbb{P}\rho(G) \cong G/Z$, where Z is the set of elements for which $\rho(g)$ is a scalar matrix. An absolute invariant for G yields an absolute invariant for G/Z, but also a relative invariant of G can yield an absolute invariant for G/Z. The degrees of the absolute invariants of G can be found as follows. Let $G_1, ..., G_k$ be the distinct conjugacy classes ($\sum |G_i| = |G|$); the characteristic polynomial

$$\phi(\lambda) = \det|1 - \lambda \cdot \rho(g)|$$

is constant on each conjugacy class G_i, so we denote this restriction by $\phi_i(\lambda)$. Then consider

$$\Phi(\lambda) = \frac{1}{|G|} \sum_i \frac{|G_i|}{\phi_i(\lambda)}.$$

The number of linearly independent invariants of G of degree ν is the coefficient of λ^ν in $\Phi(\lambda)$. With some care the same method carries over to projective groups: if f is an invariant of G/Z, then it gives an absolute invariant for some subgroup $G' \subset G$, and G/G' is cyclic. Hence to determine the invariants of G/Z, one must determine the invariants of G and of its subgroups G' for which G/G' is cyclic.

B.1.2.2   Unitary reflection groups

We now specialize to the case of unitary reflection groups G and associated collineation (projective) groups.

Definition B.1.13 A group $\rho(G) \subset GL_n(\mathbb{C})$ is called a *unitary reflection group*, if it is generated by elements $g_1, \ldots, g_k$, such that each g_j has $(n-1)$ eigenvalues $=1$ and one eigenvalue $= \exp(2\pi i/n_j)$, $j = 1, \ldots, k$, $n_j \in \mathbb{Z}$ (each g_j is a unitary reflection).

Each such unitary reflection fixes a hyperplane; a usual reflection is a unitary reflection of order 2 ($n_j = 2$). Finite unitary reflection groups have the following properties, which are proved in [ST].

Theorem B.1.14 *Let G be a finite unitary reflection group in n variables. Then:*

(i) *There are algebraically independent invariants $I_{m_1}, \ldots, I_{m_n}$, such that*

$$\prod_{i=1}^{n} m_i = |G|.$$

(ii) *The number of reflections in G is $s := \sum_{i=1}^{n} (m_i - 1)$.*

(iii) *Considering the Jacobian (Definition B.1.7) of these, one has*

$$\mathfrak{J}(I_{m_1}, \ldots, I_{m_n}) = \prod_{j=1}^{s} H_j^{n_j - 1},$$

where H_j is the reflection hyperplane of a reflection of order n_j.

(iv) *The number of elements fixing some linear subspace of dimension $n - k$ is the coefficient of t^k in the product $\prod_{i=1}^{n}((m_i - 1)t + 1)$.*

The numbers $m_i - 1$ are called the exponents of the group. Aside from certain infinite series, the finite groups in dimensions $n \geq 4$ (so projective in dimensions ≥ 3) are listed in Table B.1.

   We note that the *unique* invariants for $n \geq 5$ are (neglecting the quadrics I_2):

$$\begin{aligned} I_4, \ I_6 \quad &\text{in } \mathbb{P}^4(\mathbb{C}), \quad G = G_{25,920} \\ I_5 \quad &\text{in } \mathbb{P}^5(\mathbb{C}), \quad G = W(E_6) \\ I_6' \quad &\text{in } \mathbb{P}^5(\mathbb{C}), \quad G = \#34. \end{aligned}$$

The invariants I_4 and I_5 have concerned us in Chapters 5 and 6; the latter invariant I_6' has been considered by Todd in [Td2] and by Hartly in [Ht]. While the projective varieties given by the vanishing of I_4 and I_5 have interesting

Table B.1: Finite unitary reflection groups in dimensions ≥ 4

n	group	$|G|$	$|G/Z|$	degrees of invariants	$\mathbb{R}$	$\mathbb{C}$
$\mathbb{C}^4$	$W(F_4)$	1152	576	$2, 6, 8, 12$	×	
	#29	7680	1920	$4, 8, 12, 20$		×
	#30	14400	7200	$2, 12, 20, 30$	×	
	#31	$64 \cdot 6!$	11520	$8, 12, 20, 24$		×
	#32	$216 \cdot 6!$	25920	$12, 18, 24, 30$		×, 3
$\mathbb{C}^5$	$G_{25,920}$	$72 \cdot 6!$	25920	$4, 6, 10, 12, 18$		×, 2
$\mathbb{C}^6$	#34	$108 \cdot 9!$	$18 \cdot 9!$	$6, 12, 18, 24, 30, 42$		×, 3
	$W(E_6)$	$72 \cdot 6!$	$72 \cdot 6!$	$2, 5, 6, 8, 9, 12$	×	
$\mathbb{C}^7$	$W(E_7)$	$8 \cdot 9!$	$4 \cdot 9!$	$2, 6, 8, 10, 12, 14, 18$	×	
$\mathbb{C}^8$	$W(E_8)$	$192 \cdot 10!$	$96 \cdot 10!$	$2, 8, 12, 14, 18, 20, 24, 30$	×	

The numbers #xy refer to the list given in [ST]. The groups denoted $W(X_y)$ are the Weyl groups of the root system of type X_y. The last two colums indicate whether the group has a real representation (in which case it is generated by genuine reflections) or not. If not, the order of the generating reflections is also listed.

singularities, both invariants of degree 6 are smooth, and $\{I_6 = 0\} \subset \mathbb{P}^4$ is a hyperplane section of $\{I_6' = 0\} \subset \mathbb{P}^5$.

Given any unitary reflection group G acting on $\mathbb{C}^{n+1}$, we can consider the following problems:

Definition B.1.15 The *equation problem* for G is: given values $(v_1, \ldots, v_{n+1})$ of the invariants of G, find a point $p = (p_1, \ldots, p_{n+1}) \in \mathbb{C}^{n+1}$ such that

$$(I_{m_1}(p), \ldots, I_{m_{n+1}}(p)) = (v_1 \ldots, v_{n+1}).$$

The *form problem* is: given the ratios $(w_1 : \cdots : w_{n+1})$ of the values of the invariants, find a point $p \in \mathbb{P}^n$ such that

$$(I_{m_1}(p) : \cdots : I_{m_{n+1}}(p)) = (w_1 : \cdots : w_{n+1}).$$

The first problem is more general and a solution gives also a solution of the second problem.

B.2   Symbolic notation

The symbolic notation was introduced in the last century by Clebsch and Gordan, and turns out to be ideally suited to the study of covariants. It is based on Sylvester's theory of hyperdeterminants. A general reference for more details than we can provide is [GY]. For a more mordern treatment based on representation theory see §4 of [How]. The sybolic method is quite unpopular with modern mathematicians, and putting it on a sound foundation is quite difficult. In the paper [KR], a method of making it rigorous for *binary* forms is presented and this turns out to get quite involved. However, we feel that, applied in an appropriate manner, the symbolic method is simply *convenient*. It may well be compared with computer algebra, which is often applied these days to perform manipulations which would be too difficult to perform by hand, getting "computational proofs" of certain facts. The symbolic method does essentially the same thing – its manipulations amount to short-cutting a computational proof of certain facts or just the computation of something which is too difficult to do by hand. In what follows we shall understand identities of symbolic expressions as expressions between polynomials in the corresponding variables, which all can be "checked" by computer algebra computations. So while we shall not give a rigorous foundation for "the symbolic method", we shall use it to give elegant derivations of certain polynomial identities and no more, bypassing the problem of rigor.

B.2.1 Notations

The idea is quite simple. Instead of writing, for example

$$f(x_1, x_2) = a_0 x_1^n + n a_1 x_1^{n-1} x_2 + \cdots + \binom{n}{r} a_r x_1^{n-r} x_2^r + \cdots + a_n x_2^n,$$

one writes f "symbolically" as

$$f(x_1, x_2) = (\alpha_1 x_x + \alpha_2 x_2)^n,$$

which means for the coefficients α_i: $\alpha_1^n = a_0, \ldots, \alpha_1^{n-r} \alpha_2^r = a_r, \ldots, \alpha_2^n = a_n$. Of course the latter has only two coefficients while the first has n, i.e., there are relations like

$$a_0 a_2 = \alpha_1^{2n-2} \alpha_2^2 = a_1^2,$$

which do not necessarily hold among the a_i. This difficulty arises as in the expression $\alpha_1^{2n-2} \alpha_2^2$, α_1 occurs to some power higher than n. This problem is alleviated symbolically by setting

$$f(x_1, x_2) = (\alpha_1 x_x + \alpha_2 x_2)^n = (\beta_1 x_x + \beta_2 x_2)^n = \cdots,$$

and stipulating that in the symbolic equivalent of an expression for the a_i, no higher that n^{th} power of α's, β's, etc. are allowed to occur. So, for example, the symbolic equivalent of $a_0 a_2$ is *not* $\alpha_1^{2n-2} \alpha_2^2$, but rather $\alpha_1^n \beta_1^{n-2} \beta_2^2$ (or $\beta_1^n \alpha_1^{n-2} \alpha_2^2$), while a_1^2 is given by $\alpha_1^{n-1} \alpha_2 \beta_1^{n-1} \beta_2$. So, in general, to represent an expression of the a_i of degree m one utilizes m symbolic expressions $\alpha, \beta, \cdots$.

Although there is no unique way to symbolically represent a given monomial of coefficients, there is a canonical choice. This is based on Euler's theorem

$$f(x) = \frac{1}{d} \sum_i \frac{\partial f}{\partial x_i} x_i,$$

for a homogenous form of degree d. Namely, if P is a homogenous function of degree m in $a_0, \ldots, a_n$, then

$$P_1 = \sum_i \lambda_i^1 \frac{\partial}{\partial a_i} P \text{ is of degree } (m-1) \text{ in } a_0, \ldots, a_n, \text{ linear in } \lambda_i^1$$

$$P_2 = \sum_i \lambda_i^2 \frac{\partial}{\partial a_i} P_1 \text{ is of degree } (m-2) \text{ in the } a_i \text{ and linear in the } \lambda_i^1, \lambda_i^2$$

$$\vdots$$

$$P_{m-1} = \sum_i \lambda_i^{m-1} \frac{\partial}{\partial a_i} P_{m-2} \text{ is linear in } a_i \text{ and } \lambda_i^j.$$

Furthermore, by Euler's relation we have

$$P_{j \mid \lambda_i^1 = \cdots = \lambda_i^j = a_i} = (j+1)! P.$$

Now in P_{m-1} replace λ_i^j by its symbolic expression in $(\alpha)^j$ (which is unique since P_{m-1} is linear in the a_i), giving the canonical choice of symbolic expression of α_i's. Precisely the same can be done for any number of variables, i.e., $(\alpha_1 x_1 + \cdots + \alpha_n x_n)^d$ represents symbolically a homogenous form of degree d.

Definition B.2.1 If $f \in k[x_1, \ldots, x_n]$ is homogenous of degree d, and

$$f(x_1, \ldots, x_n) = (\alpha_1 x_1 + \cdots + \alpha_n x_n)^d = (\beta_1 x_1 + \cdots + \beta_n x_n)^d = \cdots$$

is the symbolic notation as above, we write

$$f \equiv (\alpha x)^d = (\beta x)^d = \cdots.$$

An easy calculation then gives

Lemma B.2.2 If $f \equiv (\alpha x)^d = (\beta x)^d = \cdots$, then the rth polar of $y = (y_1, \ldots, y_n)$ with respect to f is given by

$$\Delta_y^r f = (\alpha x)^{n-r} (\alpha y)^r.$$

Now suppose we let $g \in GL_n(K)$ act on the variables x. Then $(\alpha x)^d$ transforms to $(\alpha^g X)^d$, where α^g is the transform of α and X is the transform of x. Write $x_i = \sum_j g_{ij} X_j$ and $\alpha_j^g = \sum_i \alpha_i g_{ij}$. To write this again symbolically, note that

$$\alpha_1 x_1 + \cdots + \alpha_n x_n = \alpha_1 \sum_j g_{1j} X_j + \cdots + \alpha_n \sum_j g_{nj} X_j$$

$$= \left(\sum \alpha_i g_{i1} \right) X_1 + \cdots + \left(\sum \alpha_i g_{in} \right) X_n$$

$$= (\alpha_1)^g X_1 + \cdots + (\alpha_n)^g X_n.$$

From this it follows that
$$(\alpha x)^d = ((\alpha)^g X)^d,$$
which just expresses the fact that if x is transformed covariantly, the coefficients α transform contravariantly. From this it is clear that one must consider certain determinants formed as follows. Let us now denote by $\alpha^1, \alpha^2, \ldots$ a set of symbolic variables (which were denoted $\alpha, \beta, \ldots$ above). We are working with forms of n variables, so each α^λ is a n-vector, $\alpha^1 = (\alpha_1^1, \ldots, \alpha_n^1)$, etc. For each α^λ, let $(\alpha^\lambda)^g$ denote the transformed symbolic coefficient, i.e., in the notations above, $\alpha_j^\lambda = \sum_i g_{ij}(\alpha_i^\lambda)^g, \lambda = 1 \ldots$. Then one has the following equality:

$$\det \begin{pmatrix} \alpha^1 \\ \vdots \\ \alpha^n \end{pmatrix} \cdot \det(g) = \det \begin{pmatrix} (\alpha^1)^g \\ \vdots \\ (\alpha^n)^g \end{pmatrix}. \tag{B.7}$$

Definition B.2.3 The determinant

$$\det \begin{pmatrix} \alpha^1 \\ \vdots \\ \alpha^n \end{pmatrix}$$

is called a *symbolical determinant factor*, and is denoted $(\alpha^1 \cdots \alpha^n)$.

So the equation (B.7) becomes

$$(\alpha^1 \cdots \alpha^n) \cdot \det(g) = ((\alpha^1)^g \cdots (\alpha^n)^g). \tag{B.8}$$

The symbolic determinant $(\alpha^1 \cdots \alpha^n)$ is skew symmetric upon permutation of two sets α^i and α^j of symbolic coefficients. Equation (B.8) shows in particular that $(\alpha^1 \cdots \alpha^n)$, if non vanishing, is a relative invariant of f of weight one.

B.2.2 The fundamental theorem

Every simultaneous covariant $F(\{a\}^1, \ldots, \{a\}^h; x)$ of homogenous forms $f_1, \ldots, f_h$ can be expressed as a product, and, written in symbolic form, each factor is either a symbolic determinant factor or a form $(\alpha x)^d$. The proof is easy, using the operator

$$\Omega = \begin{vmatrix} \frac{\partial}{\partial x_1^{(1)}} & \cdots & \frac{\partial}{\partial x_n^{(1)}} \\ \vdots & \ddots & \vdots \\ \frac{\partial}{\partial x_1^{(n)}} & \cdots & \frac{\partial}{\partial x_n^{(n)}} \end{vmatrix}, \tag{B.9}$$

where $x_i^{(j)}$, $j = 1, \ldots, n$ are n sets of variables $i = 1, \ldots, n$. First one notes the action

$$\Omega \cdot \begin{vmatrix} x_1^{(1)} & \cdots & x_n^{(1)} \\ \vdots & \ddots & \vdots \\ x_1^{(1)} & \cdots & x_n^{(n)} \end{vmatrix}^d = d(d+1) \cdots (d+(n-1)) \begin{vmatrix} x_1^{(1)} & \cdots & x_n^{(1)} \\ \vdots & \ddots & \vdots \\ x_1^{(1)} & \cdots & x_n^{(n)} \end{vmatrix}^{d-1} \tag{B.10}$$

and similarly, denoting the matrix occuring in (B.10) by $\mathbf{X}$, the relation

$$\Omega^r |\mathbf{X}|^d = const. |\mathbf{X}|^{d-r},$$

so in particular,

$$\Omega^d |\mathbf{X}|^d = \text{a numerical constant.} \tag{B.11}$$

Secondly one considers expressions $\Omega^r S$, where S is a product of m_i factors of type $(\alpha^{(i)} x^{(i)})$. Then $\Omega^r S$ is a product of:

- r factors of symbolic determinants $(\alpha^1 \cdots \alpha^n)$;

- $m_i - r$ factors of type $(\alpha^{(i)} x^{(i)})$.

$$\tag{B.12}$$

Now suppose that $F(\{a\})$ is an invariant of a form $f(x) \equiv (\alpha x)^d = (\beta x)^d = \cdots$. Then

$$F(\rho^*(g)(\{a\})) = (\det(g))^w \cdot F(\{a\}), \tag{B.13}$$

where w is the weight. Apply Ω^w to this expression, where now we are thinking of the matrix $\mathbf{X}$ as the matrix of g, i.e., insert g into $\mathbf{X}$ in equation (B.10) and apply Ω^w. Then Ω^w(right hand side of (B.13)) $= const.F(\{a\})$, while the left hand side of (B.13) is written symbolically as a sum of terms, each of which contains w factors of each type $\rho^*(g)(\alpha^{(i)} x^{(i)})$. Hence by (B.12) Ω^w(left hand side of (B.13)) is a product of terms, each of which is a determinant $(\alpha^1 \cdots \alpha^n)$. The same reasoning can be applied to the covariants (cf. [GY], §213, p. 261), and the result is the fundamental theorem of invariants.

Theorem B.2.4 *Let $f(x) = f(a|x)$ be a homogenous form of degree d in n variables. Then any projective covariant of f can be written as a product of terms, each of which is either*

a) a symbolic determinant $(\alpha^1 \cdots \alpha^n)$, or

b) a form $(\alpha^{(i)} x^{(i)})$.

Conversely, any such product is a covariant of f. The same holds for h polynomials and simultaneous covariants.

This result, which was proved by Gordan for binary forms (a very complicated, tricky proof) and given a general proof by Hilbert (no longer tricky), shows the usefulness of the symbolic notation: one can write down explicitly all possible covariants, and must only show non-vanishing of the invariants so determined as well as checking for existing relations.

The symbolic notation may also be viewed as a differential calculus. Just replace, in the symbolic notation $(\alpha x) = \alpha_1 x_1 + \cdots + \alpha_n x_n$ the α_i by the differential operators

$$\alpha_i \mapsto \frac{\partial}{\partial y_i}, \tag{B.14}$$

and a symbolic factor $\alpha_1^{k_1} \cdots \alpha_n^{k_m} = \left(\frac{\partial}{\partial y_1}\right)^{k_1} \cdots \left(\frac{\partial}{\partial y_n}\right)^{k_n}$ acts on f as a function of y, i.e.,

$$(\alpha x)^d f(y) = (\alpha x)^d (\alpha y)^d$$

is gotten by differentiating $f(y)$ according to the differential operators (B.14), and then setting $y = x$. Given, for example, n forms

$$f_1(x) = (\alpha^1 x)^{d_1}, \ldots, f_n(x) = (\alpha^n x)^{d_n},$$

the *transvectant* (Überschiebung) is the symbolic expression

$$(\alpha^1 \cdots \alpha^n)^r (\alpha^1 x)^{d_1 - r} \cdots (\alpha^n x)^{d_n - r}, \quad (r \le \min_i(d_i)), \qquad (B.15)$$

and it is calculated by

$$const. \begin{vmatrix} \frac{\partial}{\partial x_1^{(1)}} & \cdots & \frac{\partial}{\partial x_n^{(1)}} \\ \vdots & \ddots & \vdots \\ \frac{\partial}{\partial x_1^{(n)}} & \cdots & \frac{\partial}{\partial x_n^{(n)}} \end{vmatrix}^r \cdot \left(\sum_i x_i \frac{\partial}{\partial x_i^{(1)}}\right)^{d_1 - r} \cdots \left(\sum_i x_i \frac{\partial}{\partial x_i^{(n)}}\right)^{d_n - r} \cdot$$

$$f_1(x^{(1)}) \cdots f_n(x^{(n)}),$$

and then setting after differentiation $x_j^{(i)} = x_j$ for all $i = 1, \ldots, n$. This covariant is usually denoted $(f_1, \ldots, f_n)^r$. Let us note the following two special cases.

Lemma B.2.5 *Let $f \equiv (\alpha x)^{d_1}$ be a binary form, and let $C = C_{a,b}$ and $D = D_{c,d}$ be covariants, a, c the degrees and b, d the orders in x. Then the tranvectant $(C, D) = E_{e,f}$ is a covariant of degree $e = a + c$ and order $f = b - d$.*

Lemma B.2.6 *Let $f \equiv (\alpha^1 x)^d = (\alpha^2 x)^d = \ldots$ be homogenous of degree d in n variables. Then the transvectant $\underbrace{(f, \ldots, f)}_{n \text{ times}}{}^d$ is given by $(\alpha^1 \cdots \alpha^n)^d$. If the degree d is even, then $(\alpha^1 \cdots \alpha^n)^d$ is nonvanishing and gives an invariant of degree n.*

B.2.3 Mixed concomitants

For applications to geometry it is important to consider, in addition to the homogenous coordinates the dual coordinates u_i, related to the x_i by the relation

$$x_1 u_1 + \cdots + x_n u_n = 0.$$

Then one can consider, in addition to the determinants $(\alpha^1 \cdots \alpha^n)$ (see Definition B.2.3), determinants of the form $(\alpha^1 \cdots \alpha^{n-1} u)$, which, the u_i's being contragredient to the x_i, hence cogredient to the α_i, are themselves relatively invariant. Hence one has the following possibilities for factors of weight one:

$$(\alpha^1 \cdots \alpha^n), \quad (\alpha^1 \cdots \alpha^{n-r} u^1 \cdots u^r), \quad (\alpha x), \quad (\alpha u^1, \cdots u^{n-1}) \sim (\alpha x). \quad (B.16)$$

This leads to the following vocabulary:

Definition B.2.7 Let $F(\{a\}; x; u)$ be a product of factors of the types expressed in (B.16). Then F

a) is an *invariant* $\iff$ F has no x or u dependence;

b) is a *covariant* $\iff$ F contains no factor in the u's;

c) is a *contravariant* $\iff$ F contains no factor in the x's;

d) F is a *mixed concomitant* otherwise (see Definition B.1.11).

The *class* of F is its order in the u variables.

Example B.2.8 For a quadric in n variables, $f(x) = (\alpha^1 x)^2 = \cdots = (\alpha^n x)^2 = 0$, we have:

1) $(\alpha^1 \cdots \alpha^n)^2 = 0$ $\iff$ the quadric is singular, i.e., $(\alpha^1 \cdots \alpha^n)^2$ is the discriminant.

2) $(\alpha^1 \cdots \alpha^{n-1} u)^2 = 0$ $\iff$ the hyperplane $u = 0$ touches the quadric.

$$\vdots$$

r) $(\alpha^1 \cdots \alpha^{n-r+1} u^1 \cdots u^{r-1})^2 = 0$ $\iff$ $\mathbb{P}^{n-r+1} = \{u^1 = 0\} \cap \cdots \cap \{u^{r-1} = 0\}$ touches the quadric.

$$\vdots$$

n) $(\alpha^1 u^1 \cdots u^{n-1})^2 = 0$ $\iff$ the point $\{u^1 = 0\} \cap \cdots \cap \{u^{n-1} = 0\}$ lies on the quadric.

In particular, the dual variety introduced above is an example of a contravariant of a given form.

B.2.4   The transference principle of Clebsch

We can now formulate a principle which allows one to relate covariants of a given form in n variables with the covariants of the form in $(n-1)$ variables which is obtained by a hyperplane section $u = (u_1, \ldots, u_n)$ (i.e., by the *variable hyperplane* $u_1 x_1 + \cdots + u_n x_n = 0$; for a particular hyperplane, say $\xi_1 x_1 + \cdots \xi_n x_n = 0$, the u_i's are replaced by the ξ_i's after the symbolic manipulation). Let $F(\alpha^1, \ldots, \alpha^n; u^1, \ldots, u^{n-1}; x)$ be a mixed concomitant of the form $f(x)$; writing it as a sum of products, each factor is of type (B.16). The factors of F are modified as follows:

$$(\alpha^1 \cdots \alpha^{n-1} u) \;\mapsto\; (\alpha^1 \cdots \alpha^{n-1})$$

$$\vdots \qquad \vdots \tag{B.17}$$

$$(\alpha^1 \cdots \alpha^{n-r} u^1 \cdots u^r) \;\mapsto\; (\alpha^1 \cdots \alpha^{n-r} u^2 \cdots u^r),$$

which transforms mixed concomitants of f to mixed concomitants of a variable hyperplane section; if $(\xi x) = \xi_1 x_1 + \cdots + \xi_n x_n = 0$ is a particular hyperplane, then to get the particular mixed concomitants one must use the relation $(\xi x) = 0$ to reduce the number of coefficients α^i to $n - 1$, and inserting in the modified F gives the desired concomitants. Conversely, given a form $g(x)$ of $n - 1$ variables, if f is a form of n variables such that

$$g(x) = 0 \iff \begin{cases} f(x) = 0 \\ (\xi x) = 0 \end{cases}$$

for a hyperplane (ξx), then the modification

$$(\alpha^1 \cdots \alpha^{n-1}) \mapsto (\alpha^1 \cdots \alpha^{n-1}\xi)$$

$$\vdots \qquad \vdots \qquad\qquad\qquad\qquad (\text{B.18})$$

$$(\alpha^1 \cdots \alpha^{n-r} u^1 \cdots u^r) \mapsto (\alpha^1 \cdots \alpha^{n-r}\xi u^1 \cdots u^r),$$

transforms concomitants of g into such of f. Note that these are symbolic manipulations, so that the coefficients of f and (ξx) are inserted in the expression after the transformation (B.17) or (B.18). We will see examples of this below in section B.4.2.

B.3   Binary forms

B.3.1   Invariants as functions of the roots

In the case of a binary form of degree d, say

$$f(x) = a_0 x_1^d + \binom{d}{1} a_1 x_1^{d-1} x_2 + \cdots + a_d x_2^d, \qquad (\text{B.19})$$

f can be written in terms of the *roots* of the equation, i.e.,

$$f(x) = \prod_{i=1}^{d} (\xi_2^i x_1 - \xi_1^i x_2), \qquad (\text{B.20})$$

where (ξ_1^i, ξ_2^i), $i = 1, \ldots, d$ are the roots (in homogenous coordinates). Since, setting $\eta_i = \xi_1^i / \xi_2^i$,

$$(-1)^k \binom{d}{k} a_k a_0^{-1} = \sigma_k(\eta_1, \ldots, \eta_d) \qquad (\text{B.21})$$

can be expressed in terms of the elementary symmetric functions of the roots, it follows that an invariant of f, say $F(a)$, can be expressed in terms of the σ_k:

$$F(a) = a_0^r F\left(1, \frac{a_1}{a_0}, \ldots, \frac{a_d}{a_0}\right) = a_0^r \Phi(\eta_1, \ldots, \eta_d),$$

where r is the degree of F and Φ is symmetric in the roots, of degree r. Conversely, one can give necessary and sufficient conditions on Φ in order for $a_0^r \Phi(\eta_1, \ldots, \eta_d)$ to be an invariant of the form f in (B.19) (cf. [SIV], II, §4). These are given as follows:

Theorem B.3.1 ([SIV], 2.10) *If $a_0^r \Phi(\eta_1, \ldots, \eta_d)$ is an invariant of $f(x)$ in (B.19), then:*

1) Φ *is of degree r in each η_i;*

2) Φ *is homogenous of degree $p = \frac{dr}{2}$;*

3) Φ *depends only on the differences $\eta_i - \eta_j$;*

4) Φ *fulfills the following reciprocity property:*

$$(\eta_1 \cdots \eta_d) \Phi(\frac{1}{\eta_1}, \ldots, \frac{1}{\eta_d}) = (-1)^p \Phi(\eta_1, \ldots, \eta_d),$$

where $(\eta_1 \cdots \eta_d)^r$ is the rth power of the product $\eta_1 \cdots \eta_d$ (not a determinant).

Sufficient conditions are given by either:

(i) 1), 3) and 4);

(ii) 2), 3) and 4).

For example, the *discriminant*

$$\Phi = \prod_{i<j} (\eta_i - \eta_j)^2$$

fulfills the above for $r = 2(d-1)$, $p = d(d-1)$, so $a_0^r \Phi$ is an invariant.

B.3.2   Quintics

Given a binary quintic $f(x) = (\alpha x)^5 = (\beta x)^5 = \cdots$, we have the following symbolic expressions for the invariants (see [E], Chapter 13). First one has a quadratic covariant $C_{2,2}$, which is the fourth transvection of f with itself (see (B.15)), i.e.,

$$C_{2,2} = (f, f)^4 = (\alpha\beta)^4 (\alpha x)(\beta x). \qquad (B.22)$$

As an exercise, let us compute this in terms of the coefficients a_i. The first factor $(\alpha\beta)^4$ is the fourth power of the symbolic determinant $(\alpha_1\beta_2 - \beta_1\alpha_2)$, so

$$(\alpha\beta)^4 = (\alpha_1\beta_2 - \beta_1\alpha_2)^4 = \alpha_1^4\beta_2^4 - 4\alpha_1^3\beta_2^3\beta_1\alpha_2 + \ldots + \beta_1^4\alpha_2^4,$$

while the product $(\alpha x)(\beta x)$ is

$$\alpha_1\beta_1 x_1^2 + (\alpha_2\beta_1 + \alpha_1\beta_2)x_1 x_2 + \alpha_2\beta_2 x_2^2,$$

and multiplying these together, *then* replacing the symbolic expressions by their equivalents in the coefficients a_i, for example $\alpha_1^5\beta_2^4\beta_1 = a_0 a_4$, etc., we get the expression

$$(2a_0 a_4 - 8a_1 a_3 + 10a_2^2)x_1^2 + (-6a_1 a_4 + 12a_2 a_3)x_1 x_2 + (2a_1 a_5 - 8a_2 a_4 + 10a_3^2)x_2^2.$$

The first invariant, of degree 4, is the discriminant of $C_{2,2}$, (viewing $C_{2,2}$ as a quadric in x)[1]

$$I_4 = (C_{2,2}C_{2,2})^2. \tag{B.23}$$

The vanishing of I_4 implies that f can be transformed into one of the quintics

$$x_1^5 + 5x_1x_2(bx_1^3 + \tfrac{1}{b}x_2^3) + x_2^5$$
$$x_1^5 + 5x_1x_2(bx_1^3 + \tfrac{1}{9b}x_2^3) + x_2^5. \tag{B.24}$$

Next, letting

$$C_{2,6} = (f, f)^2 = (\alpha\beta)^2(\alpha x)^3(\beta x)^3$$

be the second transvection of f with itself, the invariant of degree 8 is given by

$$I_8 = (C_{2,6}, C_{2,2}^3)^6 = ((C_{2,6}, C_{2,2}^2)^4, C_{2,2})^2.$$

Its vanishing implies that the quintic can be brought into the form

$$\frac{x_1^5}{\lambda} + \frac{x_2^5}{1-\lambda} + (x_1 + x_2)^5.$$

The invariant of degree 12 is given by

$$I_{12} = (f^2, C_{2,2}^5)^{10};$$

its vanishing implies that the quintic can be reduced to Bring's form

$$ax_1^5 + ex_1x_2^4 + fx_2^5. \tag{B.25}$$

Finally, there are two linear covariants

$$\alpha := (f, C_{2,2}^2)^4, \quad \beta := -(f, C_{2,2}^3)^5$$

whose transvectant is the invariant of degree 18:

$$I_{18} = (\alpha, \beta).$$

Quintics for which I_{18} vanish are soluable, i.e., can be written

$$(x_1 + x_2)(x_1^2 + x_2^2 + px_1x_2)(x_1^2 + x_2^2 + qx_1x_2),$$

and the first (linear) factor is one of the common harmonic conjugates of the pairs of zeros of the quadratic factors. Conversely, if one of the five roots of a quintic is one of the common harmonic conjugates of two pairs of zeros which together form the other four roots, then $I_{18} = 0$.

If both I_{12} and I_{18} vanish, then the quintic has one of the forms:

$$x_1(x_1^4 + x_2^4) \ (\Rightarrow I_4 = 0), \quad x_1^5 + x_2^5 \ (\Rightarrow I_8 = 0), \quad x_1^4(x_1 + x_2).$$

---

[1] here and in the following we a abusing notations somewhat. Recall the discriminant of a quadric $f = (\alpha x)^2 = (\beta x)^2$ is $(\alpha\beta)^2$; here we view $C_{2,2}$ as a quadric and write $(C_{2,2}C_{2,2})$

B.3.3   Sextics

First we mention the degree 2 invariant of any binary n-tic, n even (see Lemma B.2.6):

$$(\alpha\beta)^n = I_2.$$

Written in terms of the coefficients, this is

$$(\alpha\beta)^n = (\alpha_1\beta_2 - \beta_1\alpha_2)^n = \sum_{j=0}^{n} \binom{n}{j}(-1)^j a_j a_{n-j}.$$

So, in particular, for the sextic, $n = 6$, we have

$$\frac{1}{2}(\alpha\beta)^6 = a_0 a_6 - 6a_1 a_5 + 15a_2 a_4 - 10a_3^2, \tag{B.26}$$

an invariant of degree 2. Next we have the fourth transvection of f with itself,

$$k := (f, f)^4 = (\alpha\beta)^4 (\alpha x)^2 (\beta x)^2 = C_{2,4}, \tag{B.27}$$

of degree 2 and order 4. There are 3 quadratic covariants

$$
\begin{aligned}
\ell : \; &= \; (\alpha k)^4 (\alpha x)^2 = C_{3,2} \\
m : \; &= \; (k\ell)^2 (kx)^2 = C_{5,2} \\
n : \; &= \; (km)^2 (kx)^2 = C_{7,2}.
\end{aligned}
\tag{B.28}
$$

Here we are abusing notations by writing (αk), meaning the symbolic determinant (Definition B.2.3) of α=coefficients of f and k=coefficients of k. Note that these may be described as follows. In $f(x_1, x_2)$ replace x_1 by $-\frac{\partial}{\partial x_2}$, replace x_2 by $\frac{\partial}{\partial x_1}$ and let f act on the Hessian $C_{2,8} = (\alpha\beta)^2 (\alpha x)^4 (\beta x)^4$; let us denote this action by $f \vdash C_{2,8}$. Then the covariants of (B.28) can also be derived as:

$$f \vdash C_{2,8} = \ell; \quad \ell \vdash k = m; \quad m \vdash k = n.$$

Symbolic expressions in terms of $f(x) \equiv (\alpha^1 x) = (\alpha^2 x) = \ldots$ are:

$$
\begin{aligned}
k \; &= \; (\alpha^1\alpha^2)^4 (\alpha^1 x)^2 (\alpha^2 x)^2, \\
\ell \; &= \; (\alpha^1\alpha^2)^4 (\alpha^1\alpha^3)^2 (\alpha^2\alpha^3)^2 (\alpha^3 x)^2, \\
m \; &= \; (\alpha^1\alpha^2)^4 (\alpha^1\alpha^3)^2 (\alpha^2\alpha^3)^2 (\alpha^4\alpha^5)^4 (\alpha^3\alpha^4)^2 (\alpha^5 x)^2, \\
n \; &= \; (\alpha^1\alpha^2)^4 (\alpha^1\alpha^3)^2 (\alpha^2\alpha^3)^2 (\alpha^4\alpha^5)^4 (\alpha^3\alpha^4)^2 (\alpha^6\alpha^7)^4 (\alpha^5\alpha^6)^2 (\alpha^7 x)^2,
\end{aligned}
$$

and from this we see these covariants are quite systematic.

In terms of these we can express all invariants of the sextic:

$$
\begin{aligned}
I_2 &= (\alpha\beta)^6 \\
I_4 &= (k^1 k^2)^4 = (C_{2,4}, C_{2,4})^4 \\
I_6 &= (k^1 k^2)^2 (k^1 k^3)^2 (k^2 k^3)^2 = \text{discriminant of } C_{3,2}(=\ell) \\
I_{10} &= (\ell n)^2 = \text{discriminant of } C_{5,2}(=m) \\
I_{15} &= (\ell m)(\ell n)(mn) = \mathcal{J}(C_{3,2}, C_{5,2}, C_{7,2}) = \mathcal{J}(\ell, m, n) \\
I_5 &= \sqrt{\Delta} = \prod_{i<j}(\eta_i - \eta_j),
\end{aligned}
$$

$$\tag{B.29}$$

where $\eta_1, \ldots, \eta_6$ are the roots of the sextic. I_{15} is a skew invariant. There exists the following relation among the invariants:

$$
-\frac{1}{2 \cdot 3^4} I_5^2 = 3 \cdot 2^7 I_2^5 - 3 \cdot 2^4 \cdot 5^3 I_2^3 \cdot I_4 - 2^4 \cdot 5^4 I_2^2 \cdot I_6 + 3 \cdot 2 \cdot 5^2 (I_2 I_4^2 + I_4 I_6) + 3^2 \cdot 5^5 I_{10}.
$$

$$\tag{B.30}$$

Other interesting invariants are

$$
I_2' = \frac{5}{9} I_2 \tag{B.31}
$$

$$
I_4' = \frac{1}{2 \cdot 9^2}(8 I_2^2 - 75 I_4) \tag{B.32}
$$

$$
I_6' = \frac{5}{2 \cdot 9^3}(-2^3 \cdot 13 I_2^3 + 3^3 \cdot 5^2 I_2 I_4 + 2 \cdot 3^2 \cdot 5^3 I_6);
$$

the two invariants I_4' and I_6' vanish if f has a triple root.

Letting $\eta_1, \ldots, \eta_6$ be the roots of the sextic, consider the differences $(ij) = \eta_i - \eta_j$. As mentioned above, the invariants are expressible in terms of these differences. Consider the quadrics

$$
\begin{aligned}
2Q_1 &= (25)(13) + (51)(42) + (14)(35) + (43)(21) + (32)(54) \\
2Q_2 &= (12)(53) + (23)(14) + (34)(25) + (45)(31) + (51)(42) \\
2Q_3 &= (53)(41) + (34)(25) + (42)(13) + (21)(54) + (15)(32) \qquad \text{(B.33)} \\
2Q_4 &= (31)(45) + (24)(53) + (25)(41) + (15)(32) + (43)(21) \\
2Q_5 &= (31)(24) + (12)(53) + (25)(41) + (54)(32) + (43)(15) \\
2Q_6 &= (42)(35) + (23)(14) + (31)(52) + (15)(43) + (54)(21);
\end{aligned}
$$

these are quadrics in the differences above. From symmetry it follows that the linear relation which must hold between the six quadrics of (B.33) is

$$
Q_1 + \cdots + Q_6 = 0. \tag{B.34}
$$

The symmetric group acting on the first five roots η_i, $i = 1, \ldots, 5$ acts on the quadrics Q_i like the icosahedral group permuting the six diagonals of the icosahedron, while the Cremona transformation $T_{(16)}$ (which with Σ_5 generates Σ_6) permutes the quadrics Q_i by $(Q_1 Q_5)(Q_2 Q_3)(Q_4 Q_6)$. The stabilizer of Q_i is a Σ_5, and hence, for a cubic to be invariant it must allow a Σ_5 action after fixing Q_1, i.e.,

$$Q_1^3 + \cdots + Q_6^3 = 0. \tag{B.35}$$

Now one sees that these are the quadrics of (3.31) giving a map of $\mathbf{P}_6^1$ onto the Segre cubic.

The quadrics $Q_1, \ldots, Q_6$ fulfill a sextic equation ([C2] p. 317)

$$\Sigma = \prod (X - Q_i) = X^6 + 15q_2 X^4 + 15q_4 X^2 + 6q_5 X + q_6, \tag{B.36}$$

and the coefficients $q_2, \ldots, q_6$ are integral in I_2, I_4, I_6, I_5 and the square root of the discriminant of S (the original sextic) is a constant times I_{10}. Thus Σ is a resolvent of S (see section 4.2.2), whose roots are $\eta_1, \ldots, \eta_6$. Coble then finds the expressions

$$15q_2 = \frac{5}{3}I_2, \quad 15q_4 = \frac{5}{3}I_2^2 - \frac{5^3}{2 \cdot 3^2}I_4, \quad 6q_5 = \frac{1}{3^5}I_5,$$

$$q_6 = -\frac{5}{3^3}I_2^3 + \frac{5^3}{2 \cdot 3^3}I_2 I_4 + \frac{5^4}{3^4}I_6.$$

In terms of the invariants Σ can be expressed

$$\Sigma = (X^2 - I_2')^3 + 15I_4' X^2 + \frac{1}{3^5}I_5 X + I_6'.$$

B.4   Cubic surfaces

In this section we sketch briefly the known invariant system for cubic surfaces. This gives a method to "abstract" from a cubic surface to a finite set of *numbers*, the invariants of the surface. With these, two cubic surfaces which are not isomorphic can be distinguished. From a different point of view this amounts to distinguishing between the equations of degree 27 for the corresponding sets of 27 lines of the surfaces.

B.4.1   Invariants and linear covariants

The invariants and linear covariants of a quatenary cubic form are given explicitly in [Sal], II, Chapter 5. For this he uses the pentahedral form (Theorem 4.1.1), with coefficients $a_1, \ldots, a_5$. The proof given by Clebsch that the cubic can indeed be put in the above mentioned form used the fact that from the Sylvester pentahedron, one can construct the cubic surface. Also these five hyperplanes (viewed in $\mathbb{P}^3$) have a special meaning in terms of the Hessian of the cubic, a quartic surface with ten double points and containing ten lines (see B.5.3.2).

These ten lines are just the edges of the pentahedron and the ten points are the vertices. The five planes meet the Hessian each in the union of four lines. Clebsch shows that one can determine this pentahedron from the equation of the cubic, *after* solving an equation of degree five which has the alternating group A_5 as Galois group (see section 4.2). The solution of this quintic equation gives values for either the ten nodes of the Hessian variety or the five planes which form the Sylvester pentahedron of the cubic surface. We will assume the cubic to be given in the pentahedral form of Theorem 4.1.1 and sketch the construction of the invariants in these terms.

Let us fix some notations for the different coordinate systems which are to be used. For a quatenary form F we use homogenous coordinates $[u_0 : \ldots : u_3]$; for the pentahdral form as in Theorem 4.1.1 we use coordinates $[y_0 : \ldots : y_4]$, and for the hexahedral form $[x_1 : \ldots : x_6]$. To determine the covariants of the quatenary form F one can use different methods. The first, used by Salmon, is to take a cubic surface in pentahedral form and to determine its covariants, by using the linear relation $y_0 + \cdots + y_4 = 0$ to eliminate one of the variables. A second method, developed by Coble, is to use the hexahedral form, which displays the cubic surface as a hyperplane section of the Segre cubic threefold, and then to apply the Clebsch transference principle. In the following we present the algebraic determination of the covariants for both methods.

The equation for the dual surface of the cubic can be written

$$64S^3 - T^2 = 0, \tag{B.37}$$

where S, T are contravariants of class 4 and 6, respectively. The tangent planes to S meet the cubic surface C in elliptic curves which are *equiharmonic*, i.e., for which the modulus $\tau = \exp(2\pi i/3)$; this is the elliptic curve with automorphism group isomorphic to $\mathbb{Z}/6\mathbb{Z}$. Similarly, the tangent planes to T meet C in *harmonic* elliptic curves, i.e., for which $\tau = i$; this is the elliptic curve with automorphism group isomorphic to $\mathbb{Z}/4\mathbb{Z}$. The equations for the contravariants S and T were derived by Salmon. They are:

$$S = \sum a_{i_1} a_{i_2} a_{i_3} a_{i_4} (v_{i_1} - v_{i_5})(v_{i_2} - v_{i_5})(v_{i_3} - v_{i_5})(v_{i_4} - v_{i_5})$$

$$T = \sum a_{i_1}^2 a_{i_2}^2 a_{i_3}^2 (v_{i_4} - v_{i_5})^6 - 2\sum a_{i_1} a_{i_2} a_{i_3}^2 a_{i_4}^2 (v_{i_5} - v_{i_1})^3 (v_{i_5} - v_{i_2})^3$$

$$+ 2a_1 a_2 a_3 a_4 a_5 \sum a_{i_5} (\varepsilon - \varepsilon')(\varepsilon' - \varepsilon'')(\varepsilon'' - \varepsilon)$$

$$\tag{B.38}$$

where v_i, $i = 0, \ldots, 4$ are variables which are contravariant to the $y_i, i = 0, \ldots, 4$, and the abbreviations $\varepsilon = (v_{i_1} - v_{i_3})(v_{i_2} - v_{i_4})$, $\varepsilon' = (v_{i_1} - v_{i_4})(v_{i_3} - v_{i_2})$, $\varepsilon'' = (v_{i_1} - v_{i_2})(v_{i_4} - v_{i_3})$ are used. The first contravariant S can be written in symbolic notation as

$$S = \frac{1}{6}(\alpha^1 \alpha^2 \alpha^3 v)(\alpha^1 \alpha^2 \alpha^4 v)(\alpha^1 \alpha^3 \alpha^4 v)(\alpha^2 \alpha^3 \alpha^4 v). \tag{B.39}$$

As to the geometry of the dual we have the following. Since all planes passing through a given line of the cubic surface are (at least) bitangents and form a pencil of planes, it follows that the dual variety has 27 singular lines. Furthermore, the *parabolic curve* of the cubic surface, the intersection with the Hessian variety (and hence of degree 12) corresponds dually to a singular curve, of degree 24. Hence the entire singular locus of the dual of the cubic has degree 51, consisting of 27 lines and an (in general) irreducible curve of degree 24. By operating with these contravariants on covariants one can construct new covariants. We again denote this by, for example

$$S \vdash f = p,$$

the result of replacing v_i in S with $\frac{\partial}{\partial y_i}$ and operating with this on f.

As a first covariant we have the Hessian determinant (see Definition B.1.8),

$$H = \sum a_{i_1} a_{i_2} a_{i_3} a_{i_4} y_{i_1} y_{i_2} y_{i_3} y_{i_4}, \tag{B.40}$$

and the symbolic expression is

$$(\alpha^1 \alpha^2 \alpha^3 \alpha^4)^2 (\alpha^1 y)(\alpha^2 y)(\alpha^3 y)(\alpha^4 y).$$

The Hessian can also be found as follows. Consider the following mixed concomitant of degree 3, order 3 in y and order 2 in v, written in symbolic notation:

$$(\alpha^1 \alpha^2 \alpha^3 v)^2 (\alpha^1 y)(\alpha^2 y)(\alpha^3 y).$$

The first factor is the condition that the plane $v = 0$ touches the quadric $(\alpha^1 y)^2 = 0$ (see Example B.2.8), and this quadric is, from the last three factors of the expression, the polar quadric of y with respect to the cubic. Expressed in terms of the coefficients a_i, this is

$$C_{3,3;2} = \sum a_{i_1} a_{i_2} a_{i_3} y_{i_1} y_{i_2} y_{i_3} (v_{i_4} - v_{i_5})^2. \tag{B.41}$$

Letting this act on the original cubic, the result will have degree 3+1=4 ($C_{3,3;2}$ has degree 3, the cubic degree 1), order in y=3+(3-2)=4 (since $C_{3,3;2}$ has order 3 in y, and order 2 in v, which reduces the order of C in y by 2), and order 0 in v. It is in fact just the Hessian:

$$C_{3,3;2} \vdash C = H.$$

Acting in turn on H with $C_{3,3;2}$ one gets a covariant of degree 7 and order 5 as follows. Taking $H = (\alpha^1 \alpha^2 \alpha^3 \alpha^4)^2 (\alpha^1 y) \cdots (\alpha^4 y)$, and

$$C_{3,2;2} = (\alpha^5 \alpha^6 \alpha^7 v)^2 (\alpha^5 y)(\alpha^6 y)(\alpha^7 y),$$

the substitution $v_i \mapsto \frac{\partial}{\partial y_i}$ does away with $(\alpha^1 y)$ and $(\alpha^2 y)$ in H, v is replaced by α^1, α^2, respectively, and the other factors remain. Hence

$$C_{7,5} = (\alpha^1 \alpha^2 \alpha^3 \alpha^4)^2 (\alpha^5 \alpha^6 \alpha^7 \alpha^1)(\alpha^5 \alpha^6 \alpha^7 \alpha^2)(\alpha^3 y)(\alpha^4 y)(\alpha^5 y)(\alpha^6 y)(\alpha^7 y).$$

In terms of the coefficients a_i this becomes

$$C_{7,5} = C_{3,3;2} \vdash H = \Phi = \left(\sum_{i_1 \neq i_2 \neq i_3} a_{i_1} a_{i_2} y_{i_1}^2 y_{i_2}^2 y_{i_3} \right) \cdot a_1 a_2 a_3 a_4 a_5. \tag{B.42}$$

Acting with S on H we get an invariant (note that $P_1 \vdash P_2$ has degree = degree P_1 + degree P_2, and order = order P_1 − order P_2), which can be symbolically expressed by applying

$$(\alpha^1 \alpha^2 \alpha^3 v)(\alpha^1 \alpha^2 \alpha^4 v)(\alpha^1 \alpha^3 \alpha^4 v)(\alpha^2 \alpha^3 \alpha^4 v)$$

to $(\alpha^5 \alpha^6 \alpha^7 \alpha^8)^2 (\alpha^5 y)(\alpha^6 y)(\alpha^7 y)(\alpha^8 y)$; replacing v_i by $\frac{\partial}{\partial y_i}$ eliminates the linear factors, and v in the sybolic expression gets replaced by $\alpha^5, \ldots, \alpha^8$; a factor $(\alpha^5 \alpha^6 \alpha^7 \alpha^8)^2$ remains:

$$I_8 = (\alpha^1 \alpha^2 \alpha^3 \alpha^5)(\alpha^1 \alpha^2 \alpha^4 \alpha^6)(\alpha^1 \alpha^3 \alpha^4 \alpha^7)(\alpha^2 \alpha^3 \alpha^4 \alpha^8)(\alpha^5 \alpha^6 \alpha^7 \alpha^8)^2.$$

In terms of the coefficients this becomes

$$S \vdash H = I_8 = \sum_{i_1 \neq i_2 \neq i_3 \neq i_4} a_{i_1}^2 a_{i_2}^2 a_{i_3}^2 a_{i_4}^2 - 2a_1 a_2 a_3 a_4 a_5 \sum_{i_1 \neq i_2 \neq i_3} a_{i_1} a_{i_2} a_{i_3}. \tag{B.43}$$

Operating with S on the square of the cubic C, replacing v_i by $\frac{\partial}{\partial y_i}$ reduces the degree of y in $C^2 = (\alpha y)^6$ by 4, so the result is

$$C_{6,2} = (\alpha^1 \alpha^2 \alpha^3 \alpha)(\alpha^1 \alpha^2 \alpha^4 \alpha)(\alpha^1 \alpha^3 \alpha^4 \alpha)(\alpha^2 \alpha^3 \alpha^4 \alpha)(\alpha y)^2. \tag{B.44}$$

The expression in terms of the coordinates is then

$$S \vdash C^2 = C_{6,2} = \left(\sum a_i y_i^2 \right) a_1 a_2 a_3 a_4 a_5. \tag{B.45}$$

Operating with $C_{6,2}$ on S we get a quadratic contravariant; the symbolic expression is derived by applying (B.44) to S in the form

$$(\alpha^5 \alpha^6 \alpha^7 v)(\alpha^5 \alpha^6 \alpha^8 v)(\alpha^5 \alpha^7 \alpha^8 v)(\alpha^6 \alpha^7 \alpha^8 v),$$

by replacing y_i by $\frac{\partial}{\partial v_i}$; this replaces the first two v's by α, while the last two are unaffected. As a result, we get

$$C_{10,0;2} = (\alpha^1 \alpha^2 \alpha^3 \alpha)(\alpha^1 \alpha^2 \alpha^4 \alpha)(\alpha^1 \alpha^3 \alpha^4 \alpha)(\alpha^2 \alpha^3 \alpha^4 \alpha)(\alpha^5 \alpha^6 \alpha^7 \alpha) \cdot$$

$$(\alpha^5 \alpha^6 \alpha^8 \alpha)(\alpha^5 \alpha^7 \alpha^8 v)(\alpha^6 \alpha^7 \alpha^8 v).$$

Expressed in terms of the coefficients this is

$$C_{6,2} \vdash S = C_{10,0;2} = (a_1 a_2 a_3 a_4 a_5)^2 \sum (v_i - v_j)^2. \tag{B.46}$$

Operating with $\mathcal{C}_{10,0;2}$ on the cubic C we get a linear covariant; here we apply $\mathcal{C}_{10,0;2}$ to $(\beta y)^3$; the two factors of v reduce the degree of $(\beta y)^3$ by two and v is replaced by β:

$$\mathcal{C}_{11,1} = (\alpha^1\alpha^2\alpha^3\alpha)(\alpha^1\alpha^2\alpha^4\alpha)(\alpha^1\alpha^3\alpha^4\alpha)(\alpha^2\alpha^3\alpha^4\alpha)(\alpha^5\alpha^6\alpha^7\alpha)(\alpha^5\alpha^6\alpha^8\alpha)\cdot$$

$$(\alpha^5\alpha^7\alpha^8\beta)(\alpha^6\alpha^7\alpha^8\beta)(\beta y).$$

In terms of the coefficients this becomes

$$\mathcal{C}_{10,0;2} \vdash C = \mathcal{C}_{11,1} = (a_1 a_2 a_3 a_4 a_5)^2 \sum a_i y_i. \tag{B.47}$$

Operating with $\mathcal{C}_{6,2}$ on $\mathcal{C}_{10,0;2}$ we get an invariant of degree 16, which symbolically is a mess to write down, so we give directly the expression in terms of the coefficients,

$$\mathcal{C}_{6,2} \vdash \mathcal{C}_{10,0;2} = I_{16} = (a_1 a_2 a_3 a_4 a_5)^3 \left(\sum a_i\right). \tag{B.48}$$

Continuing in this manner one gets five fundamental invariants, $I_8, I_{16}, I_{24}, I_{32}, I_{40}$ and a skew invariant I_{100}. All invariants can be written in terms of the elementary symmetric functions of the five coefficients a_i. Let σ_i be the elementary symmetric function of the coefficients a_i. Then the invariants are

$$I_8 = \sigma_4^2 - 4\sigma_3\sigma_5, \quad I_{16} = \sigma_5^3\sigma_1, \quad I_{24} = \sigma_5^4\sigma_4, \quad I_{32} = \sigma_5^6\sigma_2, \quad I_{40} = \sigma_5^8, \quad \text{(B.49)}$$

with degrees as indicated by the subscripts.

A further invariant, the discriminant Δ of the cubic, takes the form

$$\Delta = (I_8^2 - 64 I_{16})^2 - 2^{14}(I_{32} + 2 I_8 I_{24}).$$

Finally the *skew* invariant of degree 100, I_{100}, has the property that its square can be rationally expressed in terms of the five fundamental invariants, but itself cannot be so expressed. The weights of the invariants $I_8, ..., I_{100}$ are: 6, 12, 18, 24, 30 and 75.

There are four *linear* (in y) covariants of degrees 11, 19, 27 and 43, respectively, the first of which is $\mathcal{C}_{11,1}$ above. We now denote these by L_j:

$$L_{11} = \sigma_5^2 \Sigma a_i y_i, \quad L_{19} = \sigma_5^3 \sum_{i \neq j \neq k \neq l \neq m} a_i a_j a_k a_l y_m \tag{B.50}$$

$$L_{27} = \sigma_5^5 \Sigma a_i^2 y_i, \quad L_{43} = \sigma_5^8 \Sigma a_i^3 y_i.$$

These linear covariants have the property that the four planes (in the y variables) they define meet in one point iff $I_{100}(a) = 0$. There is a covariant of degree 15 and order 5,

$$\mathcal{C}_{15,5} = \sigma_5^3 y_1 y_2 y_3 y_4 y_5,$$

which when set equal to zero is just the Sylvester pentahedron. Finally one has a covariant Θ, which is the locus of points whose polar plane with respect to H (B.40) is tangent to the polar quadric with respect to the cubic C. This condition is expressed by the following determinant, which is of order 9 and degree 11:

$$\Theta = \begin{vmatrix} & & \frac{\partial H}{\partial u_1} \\ & Hess(C) & \vdots \\ & & \frac{\partial H}{\partial u_4} \\ \frac{\partial H}{\partial u_1} & \cdots & \frac{\partial H}{\partial u_4} & 0 \end{vmatrix},$$

where $u_1, \ldots, u_4$ are coordinates on $\mathbb{P}^3$ derived from the y_i by elimination.

Using the covariants Φ, Θ and H one finds a covariant surface of 9th order

$$\Theta - 4H\Phi = 0 \tag{B.51}$$

whose intersection with the cubic surface is the union of the 27 lines, where $\Phi = \mathcal{C}_{7,5}$ as in (B.42).

B.4.2 Hexahedral form

It was mentioned in the text that the equations for the 27 lines on a cubic surface given in the hexahedral form are known. If one can find a representation of a given quatenary cubic g in the hexahedral form, then one can determine the equations for the 27 lines on $\{g = 0\}$ in terms of those of the hexahedral form. This requires the calculation of the invariants and linear covariants of the hexahedral form and their *identification* with the corresponding invariants and covariants of g. For details on the matters presented here we refer the reader to [Sou].

The basis for doing this is the Clebsch transference principle (section B.2.4). Let $[x_0 : \ldots : x_3]$ denote variables contravariant to the $[u_0 : \ldots : u_3]$ on $\mathbb{P}^3$, and let

$$F(\alpha^1, \ldots, \alpha^k, u^1, \ldots, u^s; x)$$

be any concomitant of g. Then F is a sum of products, all factors of which are of types (see Theorem B.2.4)

$$(\alpha^1 \cdots \alpha^4), \quad (\alpha^1 \alpha^2 \alpha^3 u), (\alpha x). \tag{B.52}$$

The Clebsch transference principle states that for the embedding of the surface $C \subset \mathbb{P}^5$, determined by a cubic fourfold C^4 and two linear forms (ξx) and (ηx), the factors are transformed into

$$\begin{aligned} (\alpha^1 \cdots \alpha^4) &\mapsto (\alpha^1 \cdots \alpha^4 \xi \eta) \\ (\alpha^1 \alpha^2 \alpha^3 u) &\mapsto (\alpha^1 \alpha^2 \alpha^3 \xi \eta u) \\ (\alpha x) &\mapsto (\alpha x). \end{aligned} \tag{B.53}$$

Now let $a, \ldots, f$ be cubics on six points, and

$$\left. \begin{aligned} a^3 + \cdots + f^3 &= 0 \\ a + \cdots + f &= 0 \\ a\bar{a} + \cdots f\bar{f} &= 0 \end{aligned} \right\} C \subset \mathbb{P}^5. \tag{B.54}$$

be the hexahedral form. An *invariant* here is required to give the same result for all 36 choices of double six; it will be a polynomial expression in $\bar{a}, \ldots, \bar{f}$, which are *dual* coordinates. Furthermore the expression must be symmetric in $\bar{a}, \ldots, \bar{f}$.

Consider the degree, order and weight of a concomitant obtained from a given one, say $C^w_{k,l;c}$ of the quatenary cubic, by means of the transference above. Since the various factors of (B.52) contribute $1, 1, 0$, respectively, to the weight, and since a covariant of g of degree k (in a) contains $3k$ symbols α, which are distributed among the factors of the type of (B.52), of which there will be $w - c, c$ and l, respectively, we have the relation

$$3k = 4(w - c) + 3c + l = 4w - c + l.$$

For the transformed invariant (B.53), as it is an invariant of (B.54), satisfies:

- the a are numerical;

- the $\xi_i = 1$;

- coefficients of η are $\bar{a}, \ldots, \bar{f}$.

Let $\widetilde{C^w_{k,l;c}}$ be the transformed concomitant and let d denote its degree in $\bar{a}, \ldots, \bar{f}$. Then these occur linearly in the factors of type $(\alpha^1 \cdots \alpha^4 \xi \eta)$ and $(\alpha^1 \alpha^2 \alpha^3 \xi \eta u)$, hence

$$d = w = \frac{1}{4}(3k + c - l).$$

Let $i_j, j = 8, 16, 24, 32, 40, 100$ denote the invariants (B.49) of g, and let l_{11}, l_{19}, l_{27}, and l_{43} denote the linear covariants of g. We then have

$$\widetilde{i_8} = I_6, \quad \widetilde{i_{16}} = I_{12}, \quad \widetilde{i_{24}} = I_{18}, \quad \widetilde{i_{32}} = I_{24}, \quad \widetilde{i_{40}} = I_{30}, \quad \widetilde{i_{100}} = I_{75},$$
$$\widetilde{l_{11}} = L_8, \quad \widetilde{l_{19}} = L_{14}, \quad \widetilde{l_{27}} = L_{20}, \quad \widetilde{l_{43}} = L_{32}.$$

$$(\text{B.55})$$

In *loc. cit.* these invariants and covariants are described. For example, the first invariant is derived precisely as above: by acting with S on H. In the expression (B.39) for S the transformation (B.53) is applied; we have

$$\widetilde{S} = \frac{1}{6}(\alpha^1 \alpha^2 \alpha^3 \xi \eta u)(\alpha^1 \alpha^2 \alpha^4 \xi \eta u)(\alpha^1 \alpha^3 \alpha^4 \xi \eta u)(\alpha^2 \alpha^3 \alpha^4 \xi \eta u)$$

$$= 4 \sum_1^{15} (imn)(jmn)(kmn)(lmn)$$

where the determinant (ijk) is

$$(ijk) = \begin{vmatrix} u_i & u_j & u_k \\ \xi_i & \xi_j & \xi_k \\ \eta_i & \eta_j & \eta_k \end{vmatrix},$$

and the sum is over all pairs $(mn) \in \{1, \ldots, 6\}$. Note that the coefficients $\bar{a}, \ldots, \bar{f}$ occur in the second expression for $\widetilde{S}$ through their occurance as η_i in (ijk). The tranformation for H yields

$$\widetilde{H} = (\alpha^1 \alpha^2 \alpha^3 \alpha^4 \xi \eta)^2 (\alpha^1 x)(\alpha^2 x)(\alpha^3 x)(\alpha^4 x),$$

where the equation of the quatenary cubic is given symbolically by $C \equiv (\alpha^1 x)^3 = (\alpha^2 x)^3 = (\alpha^3 x)^3 = (\alpha^4 x)^3$, and this expression can also be written in terms of the $\bar{a}, \ldots, \bar{f}$ of the hexahedral form:

$$\widetilde{H} = \sum x_i x_j x_k x_l (\eta_m - \eta_n)^2,$$

where $(\eta_1, \ldots, \eta_6) = (\bar{a}, \ldots, \bar{f})$ and we have used coordinates $(x_1, \ldots, x_6) = (a, \ldots, f)$. Hence this expression is quadratic in the coefficients $\bar{a}, \ldots, \bar{f}$ of the hexahedral form. By operating with $\widetilde{S}$ on $\widetilde{H}$ we get I_6,

$$\widetilde{S} \vdash \widetilde{H} = I_6.$$

Symbolically, one takes the expression of I_8 of (B.43) and transforms it according to (B.53), yielding

$$I_6 = (\alpha^1 \alpha^2 \alpha^3 \alpha^5 \xi \eta)(\alpha^1 \alpha^2 \alpha^4 \alpha^6 \xi \eta)(\alpha^1 \alpha^3 \alpha^4 \alpha^7 \xi \eta)(\alpha^2 \alpha^3 \alpha^4 \alpha^8 \xi \eta)(\alpha^5 \alpha^6 \alpha^7 \alpha^8 \xi \eta)^2,$$

and each factor is linear in the coefficients $\bar{a}, \ldots, \bar{f}$, as just mentioned, showing it is of degree 6. This can be calculated and expressed in terms of the elementary symmetric functions $\sigma_i(\bar{a}, \ldots, \bar{f})$. Of course by definition $\sigma_1 = 0$. Sously finds:

$$I_6 = 24 \left[4\sigma_2^3 - 3\sigma_3^2 - 16\sigma_2\sigma_4 + 12\sigma_6 \right]. \tag{B.56}$$

As above, the next invariant is the result of operating with $\mathcal{C}_{6,2}$ on $\mathcal{C}_{10,0;2}$; we have

$$\widetilde{\mathcal{C}_{6,2}} = \widetilde{S} \vdash (\alpha x)^3 = \mathcal{C}_{4,2}$$

$$\widetilde{\mathcal{C}_{10,0;2}} = \widetilde{\mathcal{C}_{6,2}} \vdash \widetilde{S} = \mathcal{C}_{8,0;2}$$

and

$$I_{12} = \widetilde{\mathcal{C}_{6,2}} \vdash \widetilde{\mathcal{C}_{10,0;2}}. \tag{B.57}$$

Likewise the first linear covariant is

$$L_8 = \widetilde{\mathcal{C}_{10,0;2}} \vdash \mathcal{C}_{1,3} = \mathcal{C}_{8,1;0}.$$

Next set $M = \frac{1}{6}(\alpha^1 \alpha^2 \alpha^3 \xi \eta u)(\alpha^1 x)(\alpha^2 x)(\alpha^3 x)$, (this is $\widetilde{\mathcal{C}}_{3,3;2}$, see (B.41)) $\widetilde{\mathcal{C}_{6,2}} \vdash M = N$, and $N \vdash \widetilde{S} = \widetilde{\mathcal{C}_{13,0;1}} = \mathcal{C}_{10,0;1}$. Then the linear covariant of order 14 is given by

$$\mathcal{C}_{10,0;1} \vdash \mathcal{C}_{4,2} = \widetilde{\mathcal{C}_{19,0;1}} = L_{14}.$$

Finally, to get the other covariants one utilizes a *collineation*, i.e., a covariant of order one and class one. This is obtained by taking the polar of $C_{4,2}$, $\Delta_y C_{4,2}$. Operating with this on $C_{8,0;2}$, one gets the mixed concomitant just mentioned,

$$\Delta_y C_{4,2} \vdash C_{8,0;2} = C_{12,1;1}.$$

Then the remaining linear covariants are

$$L_{20} = C_{12,1;1} \vdash L_8,$$

$$L_{32} = C_{12,1;1} \vdash L_{20}.$$

The remaining invariants are then

$$I_{18} = C_{10,0;1} \vdash L_8, \quad I_{24} = C_{10,0;1} \vdash L_{14}, \quad I_{30} = C_{10,0;1} \vdash L_{20}. \tag{B.58}$$

All of the above computations are sufficiently explicit to allow a calculation by means of computer algebra systems.

B.5   Some quartic surfaces

B.5.1   Kummer surfaces

The theory of Kummer surfaces is sufficiently well-known that we just recall a few fundamental facts used in the text. First recall that the space of polarized K3-surfaces is 19-dimensional for any polarization $2e$. For $e = 1$ one has the family of double covers of $\mathbb{P}^2$ which are branched along a sextic plane curve. For $e = 2$ one has the family of quartic surfaces in $\mathbb{P}^3$, for $e = 3$ the complete intersections of type $(2,3)$ in $\mathbb{P}^4$, and for $e = 4$ the family of complete intersections of type $(2,2,2)$ in $\mathbb{P}^5$. The higher polarizations of K3's yield surfaces which are not complete intersections.

Consider a quartic K3-surface S with an ordinary double point p; projecting from this node displays S_p (this is S blown up at p) as a double cover of $\mathbb{P}^2$, with branch locus a sextic plane curve. Hence these quartics lie in the intersection of the $e = 1$ and $e = 2$ polarizations. The sextic is necessarily everywhere tangent to a conic, namely the image of the blown up point. Adding additional nodes amounts to introducing singularities in the branch sextic. A maximally singular sextic with ordinary double points is a configuration of six lines in $\mathbb{P}^2$. There is a four-dimensional family of such configurations, and a three-dimensional subfamily with the additional property that all lines are tangent to a conic. This three-dimensional family of singular sextics coincides with the three-dimensional family of Kummer surfaces, namely quartics with the maximal number (16) of nodes. The image of 15 of the nodes are the intersections of the six lines, the image of the node p is the conic tangent to the six lines. Dual to the six lines are six points, and by taking $\mathbb{P}^2$ blown up in these six points one gets a cubic surface, and hence a correspondence between the family of Kummer surfaces and the three-dimensional family of cubic surfaces which arise when the six points are dual to six lines, all tangent to a conic. One sees easily that this means precisely that the six points lie on a conic, and the corresponding cubic surface has an ordinary double point.

An important fact about Kummer surfaces is the geometric form that the corresponding moduli take. The 16_6-configuration consists of 16 planes and 16 nodes, each plane containing six points, each point being contained in six nodes. This configuration was studied by Kummer, hence the name Kummer surfaces. Each of these 16 planes is a *trope* of the Kummer, i.e., a bitangent. In other words, the intersection of the plane and the Kummer surface consists of a conic counted with multiplicity two. Since the six nodes in the plane are on the Kummer surface, they also lie on the conic. The relation is then given by

Lemma B.5.1 *The Kummer surface is* $S = A_\tau / < i >$, *where the abelian surface* A_τ *with the period* τ *is determined as the Jacobian of the following genus two curve: the double cover of the conic (in the trope of the Kummer) branched at the six nodes on that conic.*

B.5.2   Desmic surfaces

There is a one-dimensional family of quartic surfaces with 12 nodes with the property that these 12 nodes form three sets of four vertices of four tetrahedra, which are in perspective to one another, so-called *desmic tetrahedra*.

B.5.2.1   Desmic tetrahedra

Let $\Delta_1, \Delta_2, \Delta_3$ be 3 tetrahedra in $\mathbb{P}^3$.

Definition B.5.2 *The three tetrahedra* $\Delta_1, \Delta_2, \Delta_3$ *are said to be desmically related, or a set of desmic tetrahedra, if there are constants* α, β, γ *such that* $\alpha\Delta_1 + \beta\Delta_2 + \gamma\Delta_3 \equiv 0$.

The following geometric facts about desmically related tetrahedra follow easily (see [Je], §13).

1) There are 16 lines through which a face of *each* of $\Delta_1, \Delta_2, \Delta_3$ pass.

2) There are 16 lines each containing a vertex of *each* of $\Delta_1, \Delta_2, \Delta_3$; these 16 lines are dual to those of 1).

3) Any two of $\Delta_1, \Delta_2, \Delta_3$ have four centers of perspective, the vertices of the third tetrahedron.

4) Conversely, any three tetrahedra with property 3) are desmically related.

(B.59)

Examples of equations for desmic tetrahedra are:

$$\Delta_1 = xyzt,$$

$$\Delta_2 = (x+y+z+t)(x+y-z-t)(x-y+z-t)(x-y-z+t),$$

$$\Delta_3 = (x+y+z-t)(x+y-z+t)(x-y+z+t)(-x+y+z+t), \qquad \text{(B.60)}$$

$$16\Delta_1 - \Delta_2 - \Delta_3 = 0.$$

$$\Delta_1 = (x-y)(x+y)(z-t)(z+t), \quad \Delta_2 = (x-z)(x+z)(y-t)(y+t), \qquad \text{(B.61)}$$

$$\Delta_3 = (x-t)(x+t)(y-z)(y+z), \quad \Delta_1 + \Delta_2 + \Delta_3 = 0.$$

The vertices of the desmic tetrahedra (B.60) are:

$$\Delta_1 \;:\; (0:0:0:1),\; (0:0:1:0),\; (0:1:0:0),\; (1:0:0:0)$$

$$\Delta_2 \;:\; (1:1:1:1),\; (1:1:-1:-1),\; (1:-1:1:-1),\; (1:-1:-1:1)$$

$$\Delta_3 \;:\; (1:1:1:-1),\; (1:1:-1:1),\; (1:-1:1:1),\; (-1:1:1:1)$$

while those of (B.61) are

$$\Delta_1 \;:\; (1:1:0:0),\; (1:-1:0:0),\; (0:0:1:1),\; (0:0:1:-1)$$

$$\Delta_2 \;:\; (1:0:1:0),\; (1:0:-1:0),\; (0:1:0:1),\; (0:1:0:-1)$$

$$\Delta_3 \;:\; (1:0:0:1),\; (1:0:0:-1),\; (0:1:1:0),\; (0:1:-1:0).$$

Remark B.5.3 Note that of the 16 lines in 2) of (B.59), four pass through any vertex, accounting for the eight vertices of the other two tetrahedra, while there are three edges of the tetrahedron through the vertex. Dually, four of the 16 lines of 1) are contained in each face, while three of the 16 lines of type 2) are contained in each face.

B.5.2.2 Desmic surfaces

A desmic surface is, by definition, one which can be written

$$a\Delta_1 + b\Delta_2 + c\Delta_3 = 0, \qquad \text{(B.62)}$$

for desmic tetrahedra $\Delta_1, \Delta_2, \Delta_3$. Utilizing the relation $\alpha\Delta_1 + \beta\Delta_2 + \gamma\Delta_3 \equiv 0$, Δ_3 can be eliminated and this equation becomes

$$(a - \alpha)\Delta_1 + (b - \beta)\Delta_2 = 0$$

or

$$\Delta_1 + \frac{b-\beta}{a-\alpha}\Delta_2 = 0.$$

This shows that desmic surfaces form a one-dimensional family with parameter $k = \frac{b-\beta}{a-\alpha}$. In particular, for the desmic tetrahedra (B.61) we get the equations

$$S_k = \{(x^2 - y^2)(z^2 - t^2) + k(x^2 - z^2)(y^2 - t^2) = 0\}. \qquad \text{(B.63)}$$

The 12 vertices of Δ_1, Δ_2 and Δ_3 are nodes for each of the surfaces S_k, and the surface S_k contains the 16 lines of (B.59) 2) joining the vertices in threes. Note that (B.63) can be written

$$k(x^2 y^2 + z^2 t^2) + (x^2 z^2 + y^2 t^2) - (1+k)(y^2 z^2 + x^2 t^2),$$

and after multiplying by λ/k and setting $\mu = \lambda/k$, $\nu = \frac{-\lambda(1+k)}{k}$, this takes the form

$$\lambda(x^2 y^2 + z^2 t^2) + \mu(x^2 z^2 + y^2 t^2) + \nu(y^2 z^2 + x^2 t^2) = 0, \quad \lambda + \mu + \nu = 0. \quad \text{(B.64)}$$

There are important curves on these surfaces, which are cut out by quadrics on eight of the points. More precisely,

Lemma B.5.4 ([Je], §14) *There is a pencil of quadrics through each of eight vertices of a pair of $(\Delta_1, \Delta_2, \Delta_3)$, which intersect S_k in the union of two quartic curves on these quadrics (i.e., of class (2,2) in the Picard group of each quadric; these are called quadri-quartics); through each point $x \in S_k$ there pass one curve of each system.*

Remark B.5.5 The lines in $\mathbb{P}^3$ which form the fibers of the rulings of these quadrics form a *cubic line complex*, i.e., an intersection of the quadric $\mathbf{G}(2,4) \subset \mathbb{P}^5$ with a cubic hypersurface. From this it follows that any line $l \in$ {cubic line complex} is a generator of a ruling of each of three quadrics, and hence that this line is a chord to each of three pairs of quadri-quartics mentioned in Lemma B.5.4.

B.5.2.3 Desmic surfaces as Kummer surfaces

Theorem B.5.6 *Each desmic surface (B.62), excepting the tetrahedra themselves, is a Kummer surface of an abelian surface of the form $A_\tau = E_\tau \times E_\tau$, $\tau \in \mathbb{S}_1$. The images of the 16 two-torsion points are the 16 lines (B.59), 2), and the 12 nodes are the images of three sets of four copies of $E_\tau \subset A_\tau$, which are given as follows:*

(i) two-torsion points on the second factor: $(p, E_\tau), p \in E_\tau, 2p \equiv 0$.

(ii) two-torsion points on the diagonal (p, p) times the antidiagonal.

(iii) two-torsion points on the antidiagnal $(p, -p)$ times the diagonal.

The apparent asymmetry of statement (i) can be explained by noting the existence of the exchange involution $\mathcal{I} : A_\tau \longrightarrow A_\tau; (x, y) \mapsto (y, x)$, under which the fibres of (i) are mapped to (E_τ, p), while (ii) and (iii) remain fixed.

Proof: This is explained in [Je], §15. One makes use of the four sigma functions with characteristics $\sigma_{00}, \ldots, \sigma_{11}$, which are certain multiples of the corresponding theta functions with characteristics. Number these $\sigma_0 := \sigma_{00}, \ldots, \sigma_3 := \sigma_{11}$. Then these functions fulfill

$$\left(\frac{\sigma_i(u)}{\sigma_0(u)}\right)^2 = \wp(\tau; u) + e_i, \ i = 1, 2, 3,$$

where $e_1 = 1/2, e_2 = \tau/2, e_3 = (\tau + 1)/2$ are the two-division points of E_τ. Hence setting

$$x = \frac{\sigma_1(u)}{\sigma_1(v)}, \quad y = \frac{\sigma_2(u)}{\sigma_2(v)}, \quad z = \frac{\sigma_3(u)}{\sigma_3(v)}, \quad t = \frac{\sigma_0(u)}{\sigma_0(v)}, \tag{B.65}$$

one has

$$\frac{x^2}{x^2 - t^2} = \frac{\wp(\tau; u) + e_1}{\wp(\tau; u) - \wp(\tau; v)}, \quad \frac{y^2}{y^2 - t^2} = \frac{\wp(\tau; u) + e_2}{\wp(\tau; u) - \wp(\tau; v)},$$

$$\frac{z^2}{z^2 - y^2} = \frac{\wp(\tau; u) + e_3}{\wp(\tau; u) - \wp(\tau; v)},$$

which after elimination of $\wp(\tau; u)$ and $\wp(\tau; v)$ takes the form

$$\begin{vmatrix} 1 & x^2 & e_1(x^2 - t^2) \\ 1 & y^2 & e_2(y^2 - t^2) \\ 1 & z^2 & e_3(z^2 - t^2) \end{vmatrix} = 0,$$

or:

$$(e_1 - e_2)(x^2 y^2 + z^2 t^2) + (e_3 - e_1)(x^2 z^2 + y^2 t^2) + (e_2 - e_3)(x^2 t^2 + y^2 z^2) = 0.$$

Comparing with (B.64) we see this is the desmic surface S_k as in (B.64) with $(\lambda, \mu, \nu) = (e_1 - e_2, e_3 - e_1, e_2 - e_3) = (\frac{1-\tau}{2}, \frac{\tau}{2}, -\frac{1}{2})$. This shows that the Kummer of $E_\tau \times E_\tau$ is a desmic surface. From the definition (B.65) we see that the x, y, z, t become indeterminant at the two-division points. Consequently they are in the base locus of the σ's and get blown up to the 16 lines. The statement about the 12 nodes is then a straightforward calculation. $\quad\square$

Corollary B.5.7 *The desmic pencil (B.63) may be viewed as a universal symmetric square of the elliptic curve E_τ, $\tau \in \mathbb{S}_1$, modulo $\Gamma(2) \subset SL(2, \mathbb{Z})$, i.e. there is a smooth threefold $X \longrightarrow X(2) = \Gamma(2) \backslash \mathbb{S}_1 = \mathbb{P}^1 - \{0, 1, \infty\}$, each fibre of which is a desmic surface, which can be compactified to*

$$X^* \longrightarrow X(2)^* = \mathbb{P}^1.$$

The singular fibres over $\{0, 1, \infty\}$ are the three desmic tetrahedra Δ_1, Δ_2 and Δ_3.

Remark B.5.8 The special curves of Lemma B.5.4 are given in terms of (u, v) by the equations

$$v = \text{constant}, \quad u + v = \text{constant}, \quad u - v = \text{constant}.$$

B.5.2.4 Projection from a node

Theorem B.5.9 *Projecting a desmic surface from a node presents* $\widetilde{S}_k = \{S_k$ *blown up in the node of projection} as a double cover of* $\mathbb{P}^2$ *branched along the union of a three-nodal quartic with a conic, which are tangent at four points (so tangent everywhere).*

Proof: This is an easy calculation; the tangent cone at the node is given by inserting $s(x - x_0), \ldots, s(t - t_0)$ into the equation and expanding in powers of s. This can also be seen geometrically: the seven lines through each node mentioned in remark B.5.3 all correspond to singularities of the branch locus. The three edges of the tetrahedron of which the node is a vertex give ordinary double points, while four of the 16 lines through the node give tangent singularities (as each maps two nodes to a point). It is then easy to see that the mentioned configuration is the only possible sextic having these properties. $\square$

Corollary B.5.10 *Let* S_k *be a desmic surface,* $C \subset \mathbb{P}^2$ *the conic of Theorem B.5.9, and let* $p_1 \ldots, p_4$ *be the four points of tangency. Then* S_k *is the Kummer of* $E_\tau \times E_\tau$, *where* $E_\tau \longrightarrow C$ *is the double cover branched at the four points* p_i.

Proof: Indeed, the inverse image $\widetilde{C}$ of C in $E_\tau \times E_\tau$ under the composition $i \circ p \circ \pi^{-1}$, where $i : E_\tau \times E_\tau \longrightarrow K$ is the quotient by the involution, $\pi : \widetilde{S}_k \longrightarrow S_k$ is the blow up at the 12 nodes of S_k, and $p : \widetilde{S}_k \longrightarrow K$ is the blow down of the proper transform in $\widetilde{S}_k$ of the 16 lines on S_{k_2} is intersected by each of the four lines of Remark B.5.3 through the node, i.e., C passes through the corresponding two-torsion points on $E_\tau \times E_\tau$. Since branched at four points, $\widetilde{C}$ must be elliptic, hence $\widetilde{C} \cong E_\tau$. $\square$

B.5.3 Symmetroids and Weddle surfaces

Given a Kummer quartic S, there are two birational maps of S onto another quartic (see [Hu], p. 165):

  i) The rational map onto the dual, also a Kummer surface (isomorphic to S);

  ii) The rational map onto a quartic with six nodes, a so-called Weddle surface.

This Weddle surface is a Jacobian (see Definition B.1.7) of four quadrics, which are in turn derived from a determinental quartic called a symmetroid; this symmetroid is the Kummer surface itself.

B.5.3.1   General symmetroids

Definition B.5.11 A quartic surface S is called *determinental*, if the homogenous form defining it is a determinant of a 4×4 matrix of linear forms. S is called a *symmetroid*, if, in addition, the 4×4 matrix is symmetric.

It is easily seen that the determinental surfaces form an 18-dimensional subfamily of the family of all quartics. Determinental surfaces have special properties arising from linear algebra. Let $M(x)$ be a matrix of linear forms; then

$$S = \{x \in \mathbb{P}^3 \big| \det(M(x)) = 0\} \tag{B.66}$$

is a quartic surface. But since M is a 4×4 matrix, we know by linear algebra that $x \in S \iff \exists_{v \neq 0},\ M(x) \cdot v = (0,\ldots,0) \in \mathbb{C}^4$. If $M_i(x)$ denotes the ith row of $M(x)$, then

$$M(x) \cdot v = 0 \iff M_1(x) \cdot v = \cdots = M_4(x) \cdot v = 0, \tag{B.67}$$

a set of four linear equations. Since the $M_i(x)$ are also linear in the x_i it follows that we may also write

$$M(x) \cdot v = 0 \iff M_1'(v) \cdot x = \cdots = M_4'(v) \cdot x = 0. \tag{B.68}$$

Then

$$\Sigma := \{v \in \mathbb{P}^3 \big| \det(M'(v)) = 0,\ \ M'(v) := (M_1'(v),\ldots,M_4'(v))\} \tag{B.69}$$

is a *different* quartic surface, and $x \mapsto v$ gives a birational map of S onto Σ.

Let ${}^t M(x)$ denote the transposed matrix. Since $\det(M(x)) = \det({}^t M(x))$, the equation (B.66) for S can also be written

$${}^t M(y) \cdot w = 0 \iff {}^t M_1(y) \cdot w = \cdots {}^t M_4(y) \cdot w = 0, \tag{B.70}$$

and one gets a second surface,

$$\Sigma' := \{w \in \mathbb{P}^3 | \det({}^t M'(w)) = 0,\ \ {}^t M'(w) = ({}^t M_1'(w),\ldots,{}^t M_4'(w))\}, \tag{B.71}$$

where as in (B.68),

$${}^t M(y) \cdot w = 0 \iff {}^t M_1'(w) \cdot y = \cdots = {}^t M_4(w) \cdot y = 0. \tag{B.72}$$

Hence we have associated to the determinental surface S two other quartic surfaces Σ and Σ'. Supoose $x \in S$, and let $v_x \in \mathbb{P}^3$ be the point in the kernel, i.e., the solution of (B.67), and let $w_y \in \mathbb{P}^3$ be the solution of (B.70), and $y \in S$ the solution of (B.72), viewed as an equation in y. Then the correspondence $x \mapsto y$ (i.e., $x \mapsto v_x \mapsto w_y \mapsto y$) is an involution of S.

Now suppose that S is a symmetroid, i.e., that $M(x) = {}^t M(x)$. Then clearly Σ and Σ' coincide, and this surface is in fact a Jacobian determinant of four quadrics,

$$\Sigma = \mathcal{J}(Q_1, Q_2, Q_3, Q_4).$$

This Jacobian posseses an involution which derives from $v_x \mapsto w_y$, which defines an involution on the symmetroid. The involution on Σ is that given on any Jacobian of four quadrics $Q_1, \ldots, Q_4$: for $x \in \Sigma$, the four polar planes $\Delta_x^1 Q_i$ meet at a point $y \in \Sigma$, hence $x \mapsto y = \cap \Delta_x^1 Q_i$ gives an involution. This in turn gives an involution of S:

Lemma B.5.12 *Using the notations above, let $x \in S$ and $v \in \Sigma$ (see (B.68)), and let $y \in S$, $w \in \Sigma$ as in (B.71). Finally, let $v \mapsto i(v)$ denote the involution on Σ just defined. Then*

$$S \ni x \mapsto v \mapsto i(v) \mapsto y \in S$$

gives an involution on S. If $\alpha : S - -- \to \Sigma$ denotes the birational map $x \mapsto v$, then we have the diagram

$$
\begin{array}{ccc}
S & \longrightarrow & S \\
\downarrow & & \uparrow \\
\Sigma & \longrightarrow & \Sigma'.
\end{array}
$$

For a proof, see [Je], p. 166.

Remark B.5.13 Taking the quotient of the symmetroid by the involution just defined, the result is an *Enriques* surface. So all symmetroids (the family of which is nine-dimensional as we will see below) are K3-surfaces which are covers of Enriques surfaces.

Lemma B.5.14 ([Je], p. 166) *The surface Σ contains ten lines, the surface S has ten nodes.*

If the quadrics Q_i have a point in common, then the surface Σ aquires a node, and the symmetroid aquires an additional node. If the quadrics have more that one point in common, the symmetroid aquires an additional node for each, and the line joining any two of these additional nodes lies on the Jacobian Σ. Hence,

Lemma B.5.15 ([Je], p. 173) *If the quadrics Q_i have k points in common $(k \leq 6)$, then the Jacobian Σ has*

- *k nodes, and*

- *$\binom{k}{2}$ additional lines (so $10 + \binom{k}{2}$ lines altogether).*

The symmetroid has

- *k additional nodes (so $10 + k$ nodes altogether), and*

- *$\binom{k}{2}$ tropes.*

In particular, if the quadrics have six points in common, then the symmetroid has 16 nodes and is a Kummer surface. In this case the Jacobian has six nodes and contains 25 lines; this surface is known as a Weddle surface, and is the birational image of the Kummer alluded to at the begining of this section.

Projecting the symmetroid from a node, one has

Lemma B.5.16 ([Je], §101, [Co], 2.4.3) *Let S be a symmetroid, $p \in S$ one of the nodes, and consider the projection from that node. Then the branch locus in $\mathbb{P}^2$ is a sextic curve which is the union of two cubic curves. Conversely, this branch locus characterizes the symmetroid, i.e., if a nodal quartic is projected from a node and the branch locus consists of two cubic curves, then the quartic is a symmetroid.*

Since the symmetroids have ten nodes, it follows that the dimension of the family of all such is $9(= 19 - \#$ nodes)$-$ dimensional (see Proposition 2.3.2).

B.5.3.2   The Hessian surface of a cubic surface

A special case of a symmetroid is given by the Hessian of a cubic; but while the family of symmetroids is nine-dimensional, the space of cubics, hence the space of Hessians, is only four-dimensional. We would like a nice characterization of those symmetroids which are Hessians of cubic surfaces. To describe these recall the Sylvester pentahedron $\mathcal{P}$ of a cubic surface C (Theorem 4.1.1). An easy calculation shows the following.

- **Hess**(C) has nodes at the ten vertices of $\mathcal{P}$.

- **Hess**(C) contains the ten edges of $\mathcal{P}$.

A characterization is given by the following.

Theorem B.5.17 *Let a symmetroid S be given, and let Σ denote the corresponding Jacobian. Then the following are equivalent:*

(i) $S = \Sigma$.

(ii) S is the Hessian of a cubic surface.

Proof: This amounts to the statement that, quadrics $Q_1, \ldots, Q_4$ fulfill $\frac{\partial Q_i}{\partial x_j} = \frac{\partial Q_j}{\partial x_i}$ if and only if $Q_i = \frac{\partial C}{\partial x_j}$ for a cubic form C. This follows from Euler's theorem $C = \frac{1}{4} \sum \frac{\partial C}{\partial x_j} x_j$: if $\frac{\partial Q_i}{\partial x_j} = \frac{\partial Q_j}{\partial x_i}$ for all i, j, then $C := \sum Q_i x_i$ is a well-defined cubic with $\frac{\partial C}{\partial x_i} = Q_i$. The converse is obvious.   □

Remark B.5.18 Clearly the same holds in any dimension, i.e., any Jacobian Σ of N quadrics in N variables is symmetric if and only if Σ is the Hessian of a cubic. In particular, for threefolds these are quintics, and we find the following special sets of quintics:

$\mathcal{S}$ Symmetroids, i.e., quintic threefolds given as the zero set of a 5×5 determinant $|M(x)|$.

$\mathcal{J}$ Jacobians, $T = \mathcal{J}(Q_1, \ldots, Q_5)$.

$\mathcal{H}$ Hessians of cubic threefolds.

By the above we have $\mathcal{H} = \mathcal{J} \cap \mathcal{S}$. It would be interesting to determine the structure of the symmetroids and Hessians in this case.

Pictures of the Hessians of the Cayley and Clebsch cubics are found in Figures 4.3 and 4.6. Looking at these and comparing with the cubics themselves persuades one that they "mirror" one another, in particular that they are covariant.

B.5.3.3 Kummer symmetroids and Weddle surfaces

As mentioned above, if the four quadrics Q_i have six points in common, then the symmetroid (corresponding to the Jacobian Σ of those four quadrics) is a Kummer surface. The Jacobian Σ is a so-called Weddle surface, with six nodes and containing 25 lines. 15 of these lines join the six "additional" nodes of the symmetroid in pairs, while the other ten lines are the intersection of the ten pairs of planes which contain each three of the six nodes of the Jacobian surface. One has the following, a proof of which can be found in Hudson's book.

Lemma B.5.19 *The Weddle surface Σ is the image of S under the rational map $\mathbb{P}^3 - - \to \mathbb{P}^3$ given by the cubic surfaces on the ten "even" nodes. The images of these nodes are the ten lines which are the intersections of the planes containing each three of the six nodes on Σ.*

However, the birational map $S - - \to \Sigma$ may also be viewed as coming from a geometric *projection*. Consider the map

$$\beta : \mathbb{P}^3 \longrightarrow \mathbb{P}^4$$
$$(z_0 : \ldots : z_3) \mapsto (\xi, \eta, \zeta, \xi', \eta', \zeta'),$$

where the ξ etc. are given by (3.30). Under this map $\mathbb{P}^3$ is mapped onto the Segre cubic, and we can combine this map with the map from $\mathbb{P}^4$ to itself given by the Jacobian ideal of the Segre cubic (i.e., by the quadrics in $\mathbb{P}^4$ on the ten nodes of the Segre). The Segre cubic is mapped to the Igusa quartic, and

Lemma B.5.20 ([Je], p. 184) *The image $\beta(\Sigma)$ in $\mathbb{P}^4$ is mapped by the rational map given by the quadrics on the ten nodes of the Segre cubic onto the tangent hyperplane section of the Igusa quartic, giving an explicit birational map of the Weddle surface onto the corresponding Kummer.*

Bibliography

[A]   A. A. Albert, "Structure of algebras", AMS Colloqium Pub. XXIV, AMS: Providence 1961.

[Ara]   T. Arakawa, *The dimension of the space of cusp forms on the Siegel upper half plane of degree two related to a quaternion unitary group*, J. Math. Soc. Japan **33** (1981) 125-145.

[Art]   E. Artin, "Geometric algebra", Wiley Interscience: New York 1957.

[Bai]   W. Baily, *An exceptional arithmetic group and its Eisenstein series*, ANN. OF MATH. **91** (1970), 512-549.

[BB]   W. Baily & A. Borel, *Compactification of arithmetic quotients of bounded symmetric domains*, ANN. OF MATH. **84** (1966), 442-528.

[Ba1]   H. Baker, "Principles of geometry", Vol IV, Cambridge, at the University Press 1940.

[Ba2]   H. Baker, "A locus with 25,920 linear self-transformations", Cambridge, at the University Press 1946.

[BN]   W. Barth & I. Nieto, *Abelian surfaces of type (1,3) and quartic surfaces with 16 skew lines*, J. Alg. Geo. **3** (1994), 173-222.

[BL]   W. Beynon & G. Lusztig: *Some numerical results on the characters of exceptional Weyl groups*, Math. Proc. Camb. Phil. Soc. **84** (1978), 417-426.

[BS]   F. v. d. Blij, T. Springer, *Arithmetics of octaves and of groups of type* G_2, Indag. Math. **21** (1959), 406-418.

[B1]   A. Borel, *Compact Clifford-Klein forms of symmetric spaces*, Topology **2** (1963), 111-122.

[B2]   A. Borel, *Ensembles fondamentaux pour les groupes arithmétiques*, Colloque sur la théorie des groupes algébriques, Bruxelles 1962, 23-40.

[B3]   A. Borel, "Linear algebraic groups", Second enlarged edition, Springer: New York 1991.

[BHC]   A. Borel & S. Harish-Chandra, *Arithmetic subgroups of algebraic groups*, ANN. OF MATH. **75** (1962), 485-535.

[BT]   A. Borel & J. Tits, *Groupes réductives*, Publ. IHES **27** (1965), 55-150.

[Bo1]   N. Bourbaki, "Éléments de mathématique, Groupes et algèbres de Lie", Hermann: Paris 1968 & 1975.

[Bo2]   N. Bourbaki, "Éléments de mathématique, Algebré", Chaptier 8, Hermann: Paris 1972.

[Br]   K. Brown, "Buildings", Springer: Berlin 1990.

[Bu]   H. Burkhardt, *Untersuchungen aus dem Gebiet der hyperelliptischen Modulfunctionen, II*, Math. Ann. **38** (1890), 161-224. III, Math. Ann. **40** (1892), 313-343.

[C1]   A. Coble, *Point sets and allied Cremona transformations I, II and III*, Trans. AMS **16** (1915), 155-198, **17** (1916), 345-385 and **18** (1917), 331-372.

[C2]   A. Coble, *An application of Moore's cross-ratio group to the solution of the sextic equation*, Tran. AMS, **12** (1911), 311-325.

[C]   J. Conway et al., "Atlas of finite simple groups", Clarendon Press: Oxford 1985.

[Co]   F. Cossec, *Reye congruences*, Transactions A.M.S., **280** (1993), 737-751.

[D]   P. Deligne, *Variétés de Shimura: Interprétation modulaire, et techniques de construction de modèles canonique*, Proc. Symp. Pure Math **33**, AMS: Providence 1979.

[DM]   P. Deligne & G. Mostow, *Hypergeometric equation and non-arithmetic monodromy*, Publ. IHES **63** (1986), 5-89.

[DO]   I. Dolgachev & D. Ortland, "Point sets in projective spaces and theta functions", Asterisque **165** (1988).

[E]   E. Elliot, "An introduction to the algebra of quantics", Second edition, Chelsea: New York 1964

[Fe]   M. Feustel, "Picardsche Modulflächen", Habilitationsschrift, Humboldt Universität, 1988.

[Fi]   H. Finkelnberg, "On the geometry of the Burkhardt quartic", PhD. Thesis, Leiden, 1989.

[Ge]   B. v. Geemen, *Projective models of Picard modular varieties*, in: Classification of Irregular varieties, Proceedings Trento 1990, Springer Lecture Notes in Mathematics **1515**, pp. 68-99.

[GM]   M. Goresky & R. MacPherson, "Stratified Morse theory," Ergebnisse der Mathematik und ihre Grenzgebiete, 3. Folge 14, Springer: Berlin 1988.

[GY]   J. Grace & A. Young, "The algebra of invariants", Chelsea: New York 1903.

[Go]   L. Godeaux, *Sur la surface du quatrième ordre contenant trete-deux droites*, Bull. Acad. Royale de Belgique 25 (1939), 539-553.

[GS]   B. Grünbaum & G. Shepard, *Simplicial Arrangements in 3-space*, Mitt. Math. Sem. Univ. Giessen 166 (1984), 49-101.

[Ht]   E. Hartly, *A sextic primal in five dimensions*, Proc. Cambridge Phil. Soc. 46 (1950), 91-105.

[H]   S. Helgason, "Differential geometry, Lie groups and symmetric spaces," Academic Press: New York 1978.

[Hem]   J. C. Hemperly, *The parabolic contribution to the number of linearly independent automorphic forms on a certain bounded doamin*, Amer. J. Math. 94 (1978), 1078-1100.

[Hen]   A. Henderson, "The 27 lines upon a cubic surface", Cambridge, at the University Press 1911.

[Hi]   F. Hirzebruch, *Automorphe Formen und der Satz von Riemann Roch*, Symp. Inter. de Topologia Algebraica (Mexico 1958), 129-144.

[Hö]   T. Höfer, "Ballquotienten als verzweigte Überlagerungen der projektiven Ebene", PhD. Thesis, Bonn 1986.

[How]   R. Howe, *"The classical groups" and Invariants of Binary Forms*, Proc. Symp. Pure Math 48, AMS: Providence, Rhode Island 1988.

[Hu]   R. Hudson, "Kummers quartic surface", Reissue in Cambridge Mathematical Library, Cambridge, at the University Press 1990.

[HKW]   K. Hulek & C. Kahn & S. Weintraub, "Moduli spaces of abelian surfaces: compactifications, degenerations and theta functions", de Gruyter: Berlin 1993.

[H0]   B. Hunt, *Coverings and Ball quotients*, Bonner Math. Schriften 176 (1986)

[H1]   B. Hunt, *Complex manifold geography in dimensions 2 and 3*, J. Diff. Geom. 30 (1989), 51-153.

[H2]   B. Hunt, *Hyperbolic planes*, preprint, alggeom/9504001.

[I]   J. Igusa, *On Siegel modular forms of genus two*, I, Am. J. Math. 86 (1964), 219-246, II, 88 (1964), 392-412.

[Ig]   J. Iguas, *A desingularization problem in the theory of Siegel modular functions*, Math. Ann. **168** (1967), 228-260.

[J]   B. Hunt & S. Weintraub, *Janus-like algebraic varieties*, J. Diff. Geo. **39** (1994), 509-557.

[Je]   C. Jessop, "Quartic surfaces with singular points", Cambridge, at the University Press 1916.

[JVS]   J. de Jong & N. Shepherd-Baron & A. ven de Ven, *On the Burkhardt quartic*, Math. Ann. **286** (1990), 309-328.

[Kl1]   F. Klein, *Sur la résolution, par fonctions hyperelliptique, de l'équation du vingt-septième degré, de laquelle dépend la détermination des vingt-sept droites d'une surface cubique*, Jour. de math. pure et appl., ser. 4, **4** (1888).

[Kl2]   F. Klein, *Über die Auflösung der allgemeinen Gleichungen fünften und sechsten Grades*, Math. Ann. **61** (1905), 50-71.

[Kl3]   F. Klein, "Vorlesungen über das Ikosaeder", Teubner: Leipzig 1884.

[Kl4]   F. Klein, *Weitere Untersuchungen über das Ikosaeder*, Math. Ann. **12** (1877), 503-560.

[KN]   S. Kobayashi & K. Nomizu, "Foundations of differential geometry", John Wiley & sons: New York 1969.

[KR]   J. Kung & G.-C. Rota, *The invariant theory of binary forms*, Bull. AMS **10** (1984), 27-85.

[L]   S. Lang, "Elliptic functions", Second edition, Springer: New York 1987.

[Man]   Y. Manin, "Cubic forms", North-Holland: Amsterdam 1974.

[Mar]   G. Margulis, "Discrete subgroups of semisimple Lie groups", Ergebnisse der Mathematik und ihre Grenzgebiete, 3. Folge **17**, Springer: Berlin 1991.

[Ma1]   H. Maschke, *Über die lineare Gruppe der Borchardt'schen Moduln*, Math. Ann. **31** (1887), 496-515.

[Ma2]   H. Maschke, *Aufstellung des vollen Formensystems einer quaternären Gruppe von 51840 lineare Substitutionen*, Math. Ann. **33** (1889), 317-344.

[Mi]   J. S. Milne, *Abelian varieties*, Chapter V in "Arithmetic Geometry", G. Cornell, J. SIlverman, eds. Springer: New York 1986.

[Mat]   K. Matsumoto, *Theta functions on the bounded symmetric domain of type $I_{2,2}$ and the period map of a 4-parameter family of K3 surfaces*, Math. Ann. **295** (1993), 383-409.

[MSY]   K. Matsumoto & T. Sasaki & M. Yoshida, *The monodromy of the period map of a 4-parameter family of K3 surfaces and the hypergeometric function of type (3,6)*, Inter. J. Math. **3** (1992), 1-164.

[M1]   D. Mumford, G. Fogarty, "Geometric invariant theory", Ergebnisse der Mathematik und ihrer Grenzgebiete, 2. Folge **34**, Springer: Berlin 1982.

[M2]   D. Mumford, "Abelian varicties", Tata Institute of Fundamendtal Research, Studies in Mathematics **5**, Oxford University Press: Oxford, 1970.

[M3]   D. Mumford, *Hirzebruch's proportionality in the non-compact case*, Inv. Math. **42** (1977), 239-272.

[Na]   I. Naruki, *On smooth quartic embedding of Kummer surfaces*, Proc. J. Acad. **67 A** (1991), 223-225.

[N]   I. Nieto, "Invariante Quadriken unter der Heisenberggruppe", Thesis, Erlangen (1989)

[Ni]   V. Nikulin, *On Kummer surfaces*, Transl. Math USSR Izv. **9** (1975), 261-275.

[OS1]   P. Orlik & L. Solomon, *Arrangements defined by unitary reflection groups*, Math. Ann. **261** (1982), 339-357.

[OS2]   P. Orlik & L. Solomon, *Coxeter Arrangements*, Proc. Symp. Pure Math. **40**, (Part 2), (1983), 269-292.

[Pas]   Pascal, "Repertorium der höheren Mathematik", Bd. II, Geometrie, Teubner: Leipzig 1922.

[R]   I. Reiner, "Maximal orders", Acedemic Press: New York 1969.

[Ra]   M. Racine, "Arithmetics of quadratic Jordan algebras", Memoirs. AMS, **136** (1973).

[Ri]   H. Richmond, *On canonical forms*, Quart. J. math. **33** (1902), 331-340.

[Sal]   G. Salmon, "Analytische Geometrie des Raumes", II, german translation by W. Fiedler, Teubner: Leipzig 1898.

[S1]   I. Satake, "Arithmetic structures of symmetric domains," Publ. of Math. Soc. Japan **14**, Iwanami Shoten Publ.: Tokyo 1980.

[Sc]   F. Scattone, "On the compactification of moduli spaces for algebraic K3-surfaces", Memoirs of AMS, **374** (1987).

[Sch]   W. Scharlau, "Quadratic and hermitian forms", Grundlehren der Math. Wiss. **270**, Springer: Berlin 1988.

[SIV] I. Schur, "Vorlesungen über Invariantentheorie", Grundlehren der math. Wissenschaften **143**, Springer, Berlin 1968.

[Seg] B. Segre, "The non-singular cubic surfaces", Clarendon Press: Cambridge 1941.

[Se1] J.-P. Serre," Trees", Springer: Berlin 1980.

[Se2] J.-P. Serre, "Local fields", Springer: Berlin 1978

[Sh] N. Shephard-Barron, *The rationality of certain spaces associated to trigonal curves*, Proc. Symp. Pure Math. **46** (1987), 165-171.

[ST] G. Shephard & J. Todd, *Unitary reflection groups*, Can. J. Math. **6** (1954), 274-304.

[Sh1] G. Shimura, *Arithmetic of alternating forms and quaternion hermitian forms*, J. Math. Soc. Japan **15** (1963), 33-65.

[Sh2] G. Shimura, *On analytic families of polarized abelian varieties and automorphic functions*, ANN. OF MATH. **78** (1963),149-192.

[Sh3] G. Shimura, *Moduli and fibre systems of abelian varieties*, ANN. OF MATH. **83** (1966), 294-338.

[Sh4] G. Shimura, *Moduli of abelian varieties and number theory*, in "Algebraic groups and discontinuous subgroups", Proc. Symp. in Pure Math. **9**, AMS, Providence, Rhode Island 1966.

[Sh5] G. Shimura, *The arithmetic of automorphic forms with respect to a unitary group*, ANN. OF MATH. **107** (1978), 569-605.

[SC] A. Ash, D. Mumford, M. Rapoport, Y. Tai, "Smooth compactification of locally symmetric spaces," Math. Sci. Press: Brookline 1975.

[Sou] C. Sously, *Invariants and covariants of the Cremona cubic surfaces*, Amer. J. Math. **41** (1919), 135-146.

[Te] T. Terada, *Fonctions hypergéométriques F_1 et fonctions automorphes I*, J. Math. Soc. Japan **35** (1983), 451-475.

[TY] G. Tian & S.-T. Yau, *Existence of Kähler-Einstein metrics on complete Kähler manifolds and their applications to algebraic geometry*, in "Mathematical aspects of string theory", edited by S.-T. Yau, World Scientific: Singapur 1986.

[Td1] J. Todd, *On the simple group of order 25,920*, Proc. Royal Soc.(A) **189** (1947), 326-358.

[Td2] J. Todd, *The invariants of a finite collineation group in five dimensions*, Proc. Camb. Phil. Soc. **46** (1950), 73-90.

[Tr]   M. Traynard, *Sur les fonctions theta de deux variables et les surfaces hyperelliptiques*, Annales de l'Ecole Normale supérieure **24** (1907), 77-177.

[Ts]   T. Tsukamoto, *On the local theory of quaternionic anti-hermitian forms*, J. Math. Soc. Japan, **13**, (1961), 387-400.

[T1]   J. Tits, *Classification of simple algebraic groups*, Proc. Symp. Pure Math. **9**, AMS: Providence 1966.

[V]   E. Viehweg, "Quasi-projective moduli for polarized manifolds", Ergebnisse der Mathematik und ihrer Grenzgebiete, 3. Folge **30**, Springer: Berlin 1995.

[Wl]   A. Weil, *Algebras with involution and semisimple groups*, J. Indian Math. Soc. **24** (1960), 589-623.

[We]   J. Werner, "Kleine Auflösungen spezieller dreidimensionaler Varietäten", Dissertation, Bonn 1987.

[W]   J. Wolf, *Fine structure of hermitian symmetric spaces*, In: "Symmetric Spaces", pp. 271-357, Dekker: New York 1972.

[Ya]   T. Yamazaki, *On Siegel modular forms of degree 2*, Amer. J. Math. **98** (1976), 39-53.

[Y1]   S.-T. Yau, *Uniformisation of geometric structures*, Proc. Symp. Pure Math. **48**, AMS: Providence 1988.

[Y2]   S.-T. Yau, *Calabi's conjecture and some new results in algebraic geometry*, Proc. Nat. Acad. Sci. USA **74** (1977), 1798-1799.

[Yo]   M. Yoshida, *Discrete reflection groups in the parabolic subgroup of $SU(n,1)$ and generalized Cartan matrices of Euclidean type*, Journal of the Fac. of Sci, Univ. Tokyo, **30** (1983), 25-52.

[Ze]   H. Zeltinger, "Spitzenanzahlen und Volumina Picardscher Modulvarietäten", Bonner Math. Schriften **136**, 1981.

Index

Springer
and the
environment

At Springer we firmly believe that an
international science publisher has a
special obligation to the environment,
and our corporate policies consistently
reflect this conviction.
We also expect our business partners –
paper mills, printers, packaging
manufacturers, etc. – to commit
themselves to using materials and
production processes that do not harm
the environment. The paper in this
book is made from low- or no-chlorine
pulp and is acid free, in conformance
with international standards for paper
permanency.

 Springer

Lecture Notes in Mathematics

For information about Vols. 1–1449
please contact your bookseller or Springer-Verlag

Vol. 1491: E. Lluis-Puebla, J.-L. Loday, H. Gillet, C. Soulé, V. Snaith, Higher Algebraic K-Theory: an overview. IX, 164 pages. 1992.

Vol. 1492: K. R. Wicks, Fractals and Hyperspaces. VIII, 168 pages. 1991.

Vol. 1493: E. Benoît (Ed.), Dynamic Bifurcations. Proceedings, Luminy 1990. VII, 219 pages. 1991.

Vol. 1494: M.-T. Cheng, X.-W. Zhou, D.-G. Deng (Eds.), Harmonic Analysis. Proceedings, 1988. IX, 226 pages. 1991.

Vol. 1495: J. M. Bony, G. Grubb, L. Hörmander, H. Komatsu, J. Sjöstrand, Microlocal Analysis and Applications. Montecatini Terme, 1989. Editors: L. Cattabriga, L. Rodino. VII, 349 pages. 1991.

Vol. 1496: C. Foias, B. Francis, J. W. Helton, H. Kwakernaak, J. B. Pearson, H_∞-Control Theory. Como, 1990. Editors: E. Mosca, L. Pandolfi. VII, 336 pages. 1991.

Vol. 1497: G. T. Herman, A. K. Louis, F. Natterer (Eds.), Mathematical Methods in Tomography. Proceedings 1990. X, 268 pages. 1991.

Vol. 1498: R. Lang, Spectral Theory of Random Schrödinger Operators. X, 125 pages. 1991.

Vol. 1499: K. Taira, Boundary Value Problems and Markov Processes. IX, 132 pages. 1991.

Vol. 1500: J.-P. Serre, Lie Algebras and Lie Groups. VII, 168 pages. 1992.

Vol. 1501: A. De Masi, E. Presutti, Mathematical Methods for Hydrodynamic Limits. IX, 196 pages. 1991.

Vol. 1502: C. Simpson, Asymptotic Behavior of Monodromy. V, 139 pages. 1991.

Vol. 1503: S. Shokranian, The Selberg-Arthur Trace Formula (Lectures by J. Arthur). VII, 97 pages. 1991.

Vol. 1504: J. Cheeger, M. Gromov, C. Okonek, P. Pansu, Geometric Topology: Recent Developments. Editors: P. de Bartolomeis, F. Tricerri. VII, 197 pages. 1991.

Vol. 1505: K. Kajitani, T. Nishitani, The Hyperbolic Cauchy Problem. VII, 168 pages. 1991.

Vol. 1506: A. Buium, Differential Algebraic Groups of Finite Dimension. XV, 145 pages. 1992.

Vol. 1507: K. Hulek, T. Peternell, M. Schneider, F.-O. Schreyer (Eds.), Complex Algebraic Varieties. Proceedings, 1990. VII, 179 pages. 1992.

Vol. 1508: M. Vuorinen (Ed.), Quasiconformal Space Mappings. A Collection of Surveys 1960-1990. IX, 148 pages. 1992.

Vol. 1509: J. Aguadé, M. Castellet, F. R. Cohen (Eds.), Algebraic Topology - Homotopy and Group Cohomology. Proceedings, 1990. X, 330 pages. 1992.

Vol. 1510: P. P. Kulish (Ed.), Quantum Groups. Proceedings, 1990. XII, 398 pages. 1992.

Vol. 1511: B. S. Yadav, D. Singh (Eds.), Functional Analysis and Operator Theory. Proceedings, 1990. VIII, 223 pages. 1992.

Vol. 1512: L. M. Adleman, M.-D. A. Huang, Primality Testing and Abelian Varieties Over Finite Fields. VII, 142 pages. 1992.

Vol. 1513: L. S. Block, W. A. Coppel, Dynamics in One Dimension. VIII, 249 pages. 1992.

Vol. 1514: U. Krengel, K. Richter, V. Warstat (Eds.), Ergodic Theory and Related Topics III, Proceedings, 1990. VIII, 236 pages. 1992.

Vol. 1515: E. Ballico, F. Catanese, C. Ciliberto (Eds.), Classification of Irregular Varieties. Proceedings, 1990. VII, 149 pages. 1992.

Vol. 1516: R. A. Lorentz, Multivariate Birkhoff Interpolation. IX, 192 pages. 1992.

Vol. 1517: K. Keimel, W. Roth, Ordered Cones and Approximation. VI, 134 pages. 1992.

Vol. 1518: H. Stichtenoth, M. A. Tsfasman (Eds.), Coding Theory and Algebraic Geometry. Proceedings, 1991. VIII, 223 pages. 1992.

Vol. 1519: M. W. Short, The Primitive Soluble Permutation Groups of Degree less than 256. IX, 145 pages. 1992.

Vol. 1520: Yu. G. Borisovich, Yu. E. Gliklikh (Eds.), Global Analysis – Studies and Applications V. VII, 284 pages. 1992.

Vol. 1521: S. Busenberg, B. Forte, H. K. Kuiken, Mathematical Modelling of Industrial Process. Bari, 1990. Editors: V. Capasso, A. Fasano. VII, 162 pages. 1992.

Vol. 1522: J.-M. Delort, F. B. I. Transformation. VII, 101 pages. 1992.

Vol. 1523: W. Xue, Rings with Morita Duality. X, 168 pages. 1992.

Vol. 1524: M. Coste, L. Mahé, M.-F. Roy (Eds.), Real Algebraic Geometry. Proceedings, 1991. VIII, 418 pages. 1992.

Vol. 1525: C. Casacuberta, M. Castellet (Eds.), Mathematical Research Today and Tomorrow. VII, 112 pages. 1992.

Vol. 1526: J. Azéma, P. A. Meyer, M. Yor (Eds.), Séminaire de Probabilités XXVI. X, 633 pages. 1992.

Vol. 1527: M. I. Freidlin, J.-F. Le Gall, Ecole d'Eté de Probabilités de Saint-Flour XX – 1990. Editor: P. L. Hennequin. VIII, 244 pages. 1992.

Vol. 1528: G. Isac, Complementarity Problems. VI, 297 pages. 1992.

Vol. 1529: J. van Neerven, The Adjoint of a Semigroup of Linear Operators. X, 195 pages. 1992.

Vol. 1530: J. G. Heywood, K. Masuda, R. Rautmann, S. A. Solonnikov (Eds.), The Navier-Stokes Equations II – Theory and Numerical Methods. IX, 322 pages. 1992.

Vol. 1531: M. Stoer, Design of Survivable Networks. IV, 206 pages. 1992.

Vol. 1532: J. F. Colombeau, Multiplication of Distributions. X, 184 pages. 1992.

Vol. 1533: P. Jipsen, H. Rose, Varieties of Lattices. X, 162 pages. 1992.

Vol. 1534: C. Greither, Cyclic Galois Extensions of Commutative Rings. X, 145 pages. 1992.

Vol. 1535: A. B. Evans, Orthomorphism Graphs of Groups. VIII, 114 pages. 1992.

Vol. 1536: M. K. Kwong, A. Zettl, Norm Inequalities for Derivatives and Differences. VII, 150 pages. 1992.

Vol. 1537: P. Fitzpatrick, M. Martelli, J. Mawhin, R. Nussbaum, Topological Methods for Ordinary Differential Equations. Montecatini Terme, 1991. Editors: M. Furi, P. Zecca. VII, 218 pages. 1993.

Vol. 1538: P.-A. Meyer, Quantum Probability for Probabilists. X, 287 pages. 1993.

Vol. 1539: M. Coornaert, A. Papadopoulos, Symbolic Dynamics and Hyperbolic Groups. VIII, 138 pages. 1993.

Vol. 1540: H. Komatsu (Ed.), Functional Analysis and Related Topics, 1991. Proceedings. XXI, 413 pages. 1993.

Vol. 1541: D. A. Dawson, B. Maisonneuve, J. Spencer, Ecole d´ Eté de Probabilités de Saint-Flour XXI - 1991. Editor: P. L. Hennequin. VIII, 356 pages. 1993.

Vol. 1542: J.Fröhlich, Th.Kerler, Quantum Groups, Quantum Categories and Quantum Field Theory. VII, 431 pages. 1993.

Vol. 1543: A. L. Dontchev, T. Zolezzi, Well-Posed Optimization Problems. XII, 421 pages. 1993.

Vol. 1544: M.Schürmann, White Noise on Bialgebras. VII, 146 pages. 1993.

Vol. 1545: J. Morgan, K. O'Grady, Differential Topology of Complex Surfaces. VIII, 224 pages. 1993.

Vol. 1546: V. V. Kalashnikov, V. M. Zolotarev (Eds.), Stability Problems for Stochastic Models. Proceedings, 1991. VIII, 229 pages. 1993.

Vol. 1547: P. Harmand, D. Werner, W. Werner, M-ideals in Banach Spaces and Banach Algebras. VIII, 387 pages. 1993.

Vol. 1548: T. Urabe, Dynkin Graphs and Quadrilateral Singularities. VI, 233 pages. 1993.

Vol. 1549: G. Vainikko, Multidimensional Weakly Singular Integral Equations. XI, 159 pages. 1993.

Vol. 1550: A. A. Gonchar, E. B. Saff (Eds.), Methods of Approximation Theory in Complex Analysis and Mathematical Physics IV, 222 pages, 1993.

Vol. 1551: L. Arkeryd, P. L. Lions, P.A. Markowich, S.R. S. Varadhan. Nonequilibrium Problems in Many-Particle Systems. Montecatini, 1992. Editors: C. Cercignani, M. Pulvirenti. VII, 158 pages 1993.

Vol. 1552: J. Hilgert, K.-H. Neeb, Lie Semigroups and their Applications. XII, 315 pages. 1993.

Vol. 1553: J.-L- Colliot-Thélène, J. Kato, P. Vojta. Arithmetic Algebraic Geometry. Trento, 1991. Editor: E. Ballico. VII, 223 pages. 1993.

Vol. 1554: A. K. Lenstra, H. W. Lenstra, Jr. (Eds.), The Development of the Number Field Sieve. VIII, 131 pages. 1993.

Vol. 1555: O. Liess, Conical Refraction and Higher Microlocalization. X, 389 pages. 1993.

Vol. 1556: S. B. Kuksin, Nearly Integrable Infinite-Dimensional Hamiltonian Systems. XXVII, 101 pages. 1993.

Vol. 1557: J. Azéma, P. A. Meyer, M. Yor (Eds.), Séminaire de Probabilités XXVII. VI, 327 pages. 1993.

Vol.´ 1558: T. J. Bridges, J. E. Furter, Singularity Theory and Equivariant Symplectic Maps. VI, 226 pages. 1993.

Vol. 1559: V. G. Sprindžuk, Classical Diophantine Equations. XII, 228 pages. 1993.

Vol. 1560: T. Bartsch, Topological Methods for Variational Problems with Symmetries. X, 152 pages. 1993.

Vol. 1561: I. S. Molchanov, Limit Theorems for Unions of Random Closed Sets. X, 157 pages. 1993.

Vol. 1562: G. Harder, Eisensteinkohomologie und die Konstruktion gemischter Motive. XX, 184 pages. 1993.

Vol. 1563: E. Fabes, M. Fukushima, L. Gross, C. Kenig, M. Röckner, D. W. Stroock, Dirichlet Forms. Varenna, 1992. Editors: G. Dell'Antonio, U. Mosco. VII, 245 pages. 1993.

Vol. 1564: J. Jorgenson, S. Lang, Basic Analysis of Regularized Series and Products. IX, 122 pages. 1993.

Vol. 1565: L. Boutet de Monvel, C. De Concini, C. Procesi, P. Schapira, M. Vergne. D-modules, Representation Theory, and Quantum Groups. Venezia, 1992. Editors: G. Zampieri, A. D'Agnolo. VII, 217 pages. 1993.

Vol. 1566: B. Edixhoven, J.-H. Evertse (Eds.), Diophantine Approximation and Abelian Varieties. XIII, 127 pages. 1993.

Vol. 1567: R. L. Dobrushin, S. Kusuoka, Statistical Mechanics and Fractals. VII, 98 pages. 1993.

Vol. 1568: F. Weisz. Martingale Hardy Spaces and their Application in Fourier Analysis. VIII, 217 pages. 1994.

Vol. 1569: V. Totik, Weighted Approximation with Varying Weight. VI, 117 pages. 1994.

Vol. 1570: R. deLaubenfels, Existence Families, Functional Calculi and Evolution Equations. XV, 234 pages. 1994.

Vol. 1571: S. Yu. Pilyugin, The Space of Dynamical Systems with the C^0-Topology. X, 188 pages. 1994.

Vol. 1572: L. Göttsche, Hilbert Schemes of Zero-Dimensional Subschemes of Smooth Varieties. IX, 196 pages. 1994.

Vol. 1573: V. P. Havin, N. K. Nikolski (Eds.), Linear and Complex Analysis – Problem Book 3 – Part I. XXII, 489 pages. 1994.

Vol. 1574: V. P. Havin, N. K. Nikolski (Eds.), Linear and Complex Analysis – Problem Book 3 – Part II. XXII, 507 pages. 1994.

Vol. 1575: M. Mitrea, Clifford Wavelets, Singular Integrals, and Hardy Spaces. XI, 116 pages. 1994.

Vol. 1576: K. Kitahara, Spaces of Approximating Functions with Haar-Like Conditions. X, 110 pages. 1994.

Vol. 1577: N. Obata. White Noise Calculus and Fock Space. X, 183 pages. 1994.

Vol. 1578: J. Bernstein, V. Lunts, Equivariant Sheaves and Functors. V, 139 pages. 1994.

Vol. 1579: N. Kazamaki, Continuous Exponential Martingales and *BMO*. VII, 91 pages. 1994.

Vol. 1580: M. Milman, Extrapolation and Optimal Decompositions with Applications to Analysis. XI, 161 pages. 1994.

Vol. 1581: D. Bakry, R. D. Gill, S. A. Molchanov, Lectures on Probability Theory. Editor: P. Bernard. VIII, 420 pages. 1994.

Vol. 1582: W. Balser, From Divergent Power Series to Analytic Functions. X, 108 pages. 1994.

Vol. 1583: J. Azéma, P. A. Meyer, M. Yor (Eds.), Séminaire de Probabilités XXVIII. VI, 334 pages. 1994.

Vol. 1584: M. Brokate, N. Kenmochi, I. Müller, J. F. Rodriguez, C. Verdi, Phase Transitions and Hysteresis. Montecatini Terme, 1993. Editor: A. Visintin. VII, 291 pages. 1994.

Vol. 1585: G. Frey (Ed.), On Artin's Conjecture for Odd 2-dimensional Representations. VIII, 148 pages. 1994.

Vol. 1586: R. Nillsen, Difference Spaces and Invariant Linear Forms. XII, 186 pages. 1994.

Vol. 1587: N. Xi, Representations of Affine Hecke Algebras. VIII, 137 pages. 1994.

Vol. 1588: C. Scheiderer, Real and Étale Cohomology. XXIV, 273 pages. 1994.

Vol. 1589: J. Bellissard, M. Degli Esposti, G. Forni, S. Graffi, S. Isola, J. N. Mather, Transition to Chaos in Classical and Quantum Mechanics. Montecatini Terme, 1991. Editor: S. Graffi. VII, 192 pages. 1994.

Vol. 1590: P. M. Soardi, Potential Theory on Infinite Networks. VIII, 187 pages. 1994.

Vol. 1591: M. Abate, G. Patrizio, Finsler Metrics – A Global Approach. IX, 180 pages. 1994.

Vol. 1592: K. W. Breitung, Asymptotic Approximations for Probability Integrals. IX, 146 pages. 1994.

Vol. 1593: J. Jorgenson & S. Lang, D. Goldfeld, Explicit Formulas for Regularized Products and Series. VIII, 154 pages. 1994.

Vol. 1594: M. Green, J. Murre, C. Voisin, Algebraic Cycles and Hodge Theory. Torino, 1993. Editors: A. Albano, F. Bardelli. VII, 275 pages. 1994.

Vol. 1595: R.D.M. Accola, Topics in the Theory of Riemann Surfaces. IX, 105 pages. 1994.

Vol. 1596: L. Heindorf, L. B. Shapiro, Nearly Projective Boolean Algebras. X, 202 pages. 1994.

Vol. 1597: B. Herzog, Kodaira-Spencer Maps in Local Algebra. XVII, 176 pages. 1994.

Vol. 1598: J. Berndt, F. Tricerri, L. Vanhecke, Generalized Heisenberg Groups and Damek-Ricci Harmonic Spaces. VIII, 125 pages. 1995.

Vol. 1599: K. Johannson, Topology and Combinatorics of 3-Manifolds. XVIII, 446 pages. 1995.

Vol. 1600: W. Narkiewicz, Polynomial Mappings. VII, 130 pages. 1995.

Vol. 1601: A. Pott, Finite Geometry and Character Theory. VII, 181 pages. 1995.

Vol. 1602: J. Winkelmann, The Classification of Three-dimensional Homogeneous Complex Manifolds. XI, 230 pages. 1995.

Vol. 1603: V. Ene, Real Functions – Current Topics. XIII, 310 pages. 1995.

Vol. 1604: A. Huber, Mixed Motives and their Realization in Derived Categories. XV, 207 pages. 1995.

Vol. 1605: L. B. Wahlbin, Superconvergence in Galerkin Finite Element Methods. XI, 166 pages. 1995.

Vol. 1606: P.-D. Liu, M. Qian, Smooth Ergodic Theory of Random Dynamical Systems. XI, 221 pages. 1995.

Vol. 1607: G. Schwarz, Hodge Decomposition – A Method for Solving Boundary Value Problems. VII, 155 pages. 1995.

Vol. 1608: P. Biane, R. Durrett, Lectures on Probability Theory. VII, 210 pages. 1995.

Vol. 1609: L. Arnold, C. Jones, K. Mischaikow, G. Raugel, Dynamical Systems. Montecatini Terme, 1994. Editor: R. Johnson. VIII, 329 pages. 1995.

Vol. 1610: A. S. Üstünel, An Introduction to Analysis on Wiener Space. X, 95 pages. 1995.

Vol. 1611: N. Knarr, Translation Planes. VI, 112 pages. 1995.

Vol. 1612: W. Kühnel, Tight Polyhedral Submanifolds and Tight Triangulations. VII, 122 pages. 1995.

Vol. 1613: J. Azéma, M. Emery, P. A. Meyer, M. Yor (Eds.), Séminaire de Probabilités XXIX. VI, 326 pages. 1995.

Vol. 1614: A. Koshelev, Regularity Problem for Quasilinear Elliptic and Parabolic Systems. XXI, 255 pages. 1995.

Vol. 1615: D. B. Massey, Lê Cycles and Hypersurface Singularities. XI, 131 pages. 1995.

Vol. 1616: I. Moerdijk, Classifying Spaces and Classifying Topoi. VII, 94 pages. 1995.

Vol. 1617: V. Yurinsky, Sums and Gaussian Vectors. XI, 305 pages. 1995.

Vol. 1618: G. Pisier, Similarity Problems and Completely Bounded Maps. VII, 156 pages. 1996.

Vol. 1619: E. Landvogt, A Compactification of the Bruhat-Tits Building. VII, 152 pages. 1996.

Vol. 1620: R. Donagi, B. Dubrovin, E. Frenkel, E. Previato, Integrable Systems and Quantum Groups. VIII, 488 pages. 1996.

Vol. 1621: H. Bass, M. V. Otero-Espinar, D. N. Rockmore, C. P. L. Tresser, Cyclic Renormalization and Auto-morphism Groups of Rooted Trees. XXI, 136 pages. 1996.

Vol. 1622: E. D. Farjoun, Cellular Spaces, Null Spaces and Homotopy Localization. XIV, 199 pages. 1996.

Vol. 1623: H.P. Yap, Total Colourings of Graphs. VIII, 131 pages. 1996.

Vol. 1624: V. Brînzănescu, Holomorphic Vector Bundles over Compact Complex Surfaces. X, 170 pages. 1996.

Vol.1625: S. Lang, Topics in Cohomology of Groups. VII, 226 pages. 1996.

Vol. 1626: J. Azéma, M. Emery, M. Yor (Eds.), Séminaire de Probabilités XXX. VIII, 382 pages. 1996.

Vol. 1627: C. Graham, Th. G. Kurtz, S. Méléard, Ph. E. Protter, M. Pulvirenti, D. Talay, Probabilistic Models for Nonlinear Partial Differential Equations. X, 301 pages. 1996.

Vol. 1628: P.-H. Zieschang, An Algebraic Approach to Association Schemes. XII, 189 pages. 1996.

Vol. 1629: J. D. Moore, Lectures on Seiberg-Witten Invariants. VII, 105 pages. 1996.

Vol. 1630: D. Neuenschwander, Probabilities on the Heisenberg Group: Limit Theorems and Brownian Motion. VIII, 139 pages. 1996.

Vol. 1631: K. Nishioka, Mahler Functions and Transcendence. VIII, 185 pages. 1996.

Vol. 1632: A. Kushkuley, Z. Balanov, Geometric Methods in Degree Theory for Equivariant Maps. VII, 136 pages. 1996.

Vol.1633: H. Aikawa, M. Essén, Potential Theory – Selected Topics. IX, 200 pages. 1996.

Vol. 1634: J. Xu, Flat Covers of Modules. IX, 161 pages. 1996.

Vol. 1635: E. Hebey, Sobolev Spaces on Riemannian Manifolds. X, 116 pages. 1996.

Vol. 1636: M. A. Marshall, Spaces of Orderings and Abstract Real Spectra. VI, 190 pages. 1996.

Vol. 1637: B. Hunt, The Geometry of some special Arithmetic Quotients. XIII, 332 pages. 1996.

Druck: STRAUSS OFFSETDRUCK, MÖRLENBACH
Verarbeitung: GANSERT, WEINHEIM/SULZBACH